# 4th International Conference on Mechanics and Mechatronics (ICMMR 2017)



IOP Conference Series: Materials Science and Engineering
Volume 224

Xi'an, China
20 – 24 June 2017

ISBN: 978-1-5108-4646-3
ISSN: 1757-8981

**Printed from e-media with permission by:**

Curran Associates, Inc.
57 Morehouse Lane
Red Hook, NY 12571

**Some format issues inherent in the e-media version may also appear in this print version.**



Printed by Curran Associates, Inc. (2017)

For permission requests, please contact the Institute of Physics
at the address below.

Institute of Physics
Dirac House, Temple Back
Bristol BS1 6BE UK

Phone: 44 1 17 929 7481
Fax: 44 1 17 920 0979

techtracking@iop.org

**Additional copies of this publication are available from:**

Curran Associates, Inc.
57 Morehouse Lane
Red Hook, NY 12571 USA
Phone: 845-758-0400
Fax: 845-758-2633
Email: curran@proceedings.com
Web: www.proceedings.com

# Table of contents

## Volume 224

**4th International Conference on Mechanics and Mechatronics Research (ICMMR 2017)**

**20–24 June 2017, Xi'an, China**

**Accepted papers received: 12 July 2017 Published online: 7 August 2017**

## Modeling and Simulation of mechanical system

## Mechanical design-manufacture and automation

## Fluid mechanics and computation



## Control theory and robot development technology

## Power and electrical engineering

## Mechatronics



## Microwave engineering and electronic information technology

## Medical electronics and devices



## Computer science and Information Engineering

## Civil and Structural Engineering



## Environmental Science and chemistry

IOP Conf. Series: Materials Science and Engineering **224** (2017) 011001 doi:10.1088/1757-899X/224/1/011001

# PREFACE

It is our great pleasure to introduce you the proceedings of 2017 the 4th International Conference on Mechanics and Mechatronics Research (ICMMR 2017) held in Xi'an, China during June 20-24, 2017.

The conference program included keynote, oral and poster presentations from scholars working in the areas to establish platforms for collaboration. This conference covered recent trends and progresses made in the field of mechanics and mechatronics research.

The objective of the conference is to bring academia, industry together to exchange and share their experiences and research results, and discuss the practical challenges encountered and the potential solutions. Professors from Canada and China were invited to deliver keynote speeches regarding latest development of the research in their areas.

International scientific activities, like conference, symposium and workshop are big scientific platforms for knowledge exchange among the scientists from all over the world. International scientific activities are the best opportunities for the students, researchers and engineers to interact with the experts and specialists to get their advice or consultation on technical matters. The theme of this conference was the key issues associated with science and technology in this rapidly evolving field of research and to promote contact between basic researches and technological needs for real advanced mechanics and mechatronics.

We want to express our gratitude to all the members of conference committee, and reviewers who spent their valuable time, for their advice which have certainly helped to improve the quality, accuracy, relevance and the sincere efforts to maintain the quality of each paper selected for conference program. Also, we would like to thank all the authors who have contributed to this conference and also to the organizing committee, reviewers, speakers, chairpersons, sponsors and all the conference participants for their support to ICMMR 2017.

Prof. Dan Zhang, York University, Canada
Prof. Bin CHEN, Xi'an Jiaotong University, China
June, 2017

# Committee

**Conference Committee Chairs**
Prof. Dan Zhang, York University, Canada
Prof. Bin CHEN, Vice director, Xi'an Jiaotong University, China

**Program Committee Chairs**
Prof. Songyi Dian, Sichuan University, China
Prof. Shih-Chieh Lin, National Tsing-Hua University, Taiwan
Prof. Jian Yang, Chongqing University, China

**Technical Committee Chairs**
Prof. Kamran Behdinan, University of Toronto, Canada
Prof. ZHANG Wei-she, Chang' an University, China
Prof. Dr. Md Mustafizur Rahman, Universiti Malaysia Pahang, Malaysia
Prof. V.Rambabu, GMR Institute of Technology, India
Assoc. Prof. Xiuling Wang, Purdue University Calumet, Hammond, IN, USA
Assoc. Prof. Shengyong Zhang, College of Engineering and Sciences, Purdue University Northwest, USA
Assoc. Prof. Dr. J. Amudhavel, KL University, India
Asst. Prof. Ye Sun, Michigan Technological University, USA
Asst. Prof. Balhassn S. M. Ali, Bursa Orhangazi University, Turkey
Asst. Prof. Atilla Altintas, Chalmers University, Sweden
Dr. Gai-Ge Wang, Jiangsu Normal University, China
Dr. HUANG Lei, Tianjin University of Technology and Education, China
Dr. Muhammad Naufal Bin Mansor, Universiti Malaysia Perlis (UniMAP), Malaysia
Dr. Lei Cui, Department of Mechanical Engineering, Curtin University, Australia
Dr. Weiji Wang, University of Sussex, United Kingdom
Dr. Yu Bi, University of Central Florida, USA
Dr. Ma Le, Northeast Electric Power University, China
Dr. Bin Yang, Zaozhuang University, China
Dr. Lei Zhang, University of Regina, Canada
Dr. Behrouz Takabi, Texas A&M University, College Station, Texas, USA
Dr. S Krishnakumar, Ministry of Defence, Defence Research and Development Organisation, India
Hassan Elahi, Sapienza University of Rome, Italy
Uğur Arabacı, Gazi University, Department of Metallurgical and Materials Engineering, Ankara, Turkey
Zhou Jiehua, Nanchang Hangkong University, China
Alexander Popov, Lobachevsky State University of Nizhny Novgorod, Russia
Hossein Bakhshi Khaniki, Iran University of Science and Technology (IUST), Iran

IOP Conf. Series: Materials Science and Engineering **224** (2017) 011003 doi:10.1088/1757-899X/224/1/011003

# Peer review statement

All papers published in this volume of *IOP Conference Series: Materials Science and Engineering* have been peer reviewed through processes administered by the proceedings Editors. Reviews were conducted by expert referees to the professional and scientific standards expected of a proceedings journal published by IOP Publishing.

# Analytical Kinematics and Coupled Vibrations Analysis of Mechanical System Operated by Solar Array Drive Assembly

**M Sattar[1,5], C Wei[2], A Jalali[3] and R Sattar[4]**

[1,2]School of Aeronautics Science & Engineering, Beihang University, Beijing, China
[3]University of Engineering and Technology, Taxila, Pakistan
[4]Department of Mathematics, Quaid-e-Azam University, Islamabad, Pakistan

[5]msattar@buaa.edu.cn

**Abstract.** To address the impact of solar array (SA) anomalies and vibrations on performance of precision space-based operations, it is important to complete its accurate jitter analysis. This work provides mathematical modelling scheme to approximate kinematics and coupled micro disturbance dynamics of rigid load supported and operated by solar array drive assembly (SADA). SADA employed in analysis provides a step wave excitation torque to activate the system. Analytical investigations into kinematics is accomplished by using generalized linear and Euler angle coordinates, applying multi-body dynamics concepts and transformations principles. Theoretical model is extended, to develop equations of motion (EoM), through energy method (Lagrange equation). The main emphasis is to research coupled frequency response by determining energies dissipated and observing dynamic behaviour of internal vibratory systems of SADA. The disturbance model captures discrete active harmonics of SADA, natural modes and vibration amplifications caused by interactions between active harmonics and structural modes of mechanical assembly. The proposed methodology can help to predict true micro disturbance nature of SADA operating rigid load. Moreover, performance outputs may be compared against actual mission requirements to assess precise spacecraft controller design to meet next space generation stringent accuracy goals.

## 1. Introduction

SADA is a particular spacecraft component which rotates SA to maintain a perpendicular position to incidence direction of sunlight in order to obtain maximum solar power. Over the course of sun-tracking drive process vibration disturbances caused by low-and-dense natural frequencies of SA, attached to SADA, simultaneously react on spacecraft body thus influencing its dynamic environment [1]. These must be mitigated effectively to achieve high pointing accuracy [1]. Current expedition is to research & apply some cost effective and reliable methods of analysis that may help to understand disturbance behaviour of SADA. Most of research in SADA analysis domain is focused on its electric and electronic systems [2]. Xia and Bodson worked on mechanism of electromagnetic vibration [3, 4] of SADA and methods to reduce its disturbance amplitude using appropriate drive circuit [3]. Other contributions include simulation procedures to study slewing phenomenon [5] and design of controllers to enhance sensitivity [6], improve resolution [7] and dynamics of drive assembly [8]. However, SA is composed of mechanical systems [9, 10] which get influenced by external force / torque provided by drive assemblies. It is verified experimentally [11] that emitted vibration forces can degrade performance of precision instruments. In [12] and [13] coupled disturbance produced by SADA driving a rigid load through a transmission shaft of variable stiffness is modelled and simulated. Mariyam proposed theoretical formulations [13] and experimental analysis procedures [14] to investigate disturbance behaviour of different subdivisions (SD) SADA operated without and with rigid load. However, there are few limitations associated with experimental tests and analytical techniques. Testing of interruptions generated by SADA in atmospheric environment / ground is hard to carry out, costly and time consuming task. The constraint of formerly proposed theoretical disturbance models is the assumption that systems operated by it possess single degree of freedom (DoF) and moves only along the spin axis which restricts research to obtain its realistic behaviour during space mission. In actual scenario, SADA controlled systems generate multi-DoF motion in which residual rotations and lateral translations also exist besides an ideal movement of body about spin axis. To the author's knowledge, this gap has not been covered yet in any of its proposed dynamic analysis methods. This research attempts to bridge short comings in earlier projected evaluation schemes by providing comprehensive analytical investigation strategies to determine SADA coupled vibration behaviour. The proceeding sections provide SADA specifications and problem description, then analytical kinematics and coupled dynamics for SADA disruption is conversed, simulation method to validate analytical model is discussed next, followed by conclusion and future scope.

## 2. SADA details & problem description

SADA, an equipment of Electric Power Subsystem, [15] consists of Solar Array Drive Mechanism (SADM) and Solar Array Drive Electronics that drive SA. In this investigation SADA consists of 50 rotor teeth ($r_t$), 4 beats, 32 subdivisions, current amplitude ($C_{amp}$) of 0.1 A and electromagnetic torque coefficient ($E_{tc}$) of 10. When an electric pulse signal is contributed to SADA stepper it induces electronic and electromagnetic coupling vibrations [13, 14] in system as stated in (1);

$$f_{32} = \sum_{m=1}^{\infty} 7.41m \qquad ; \qquad k_{em} = E_{tc} C_{amp} r_t \tag{1}$$

where; $m = 1,2,..,\infty$ , $f_{32}$ is active frequency and $k_{em}$ is the electromagnetic stiffness. The problem geometry for analytical model consists of SADA operating rigid rectangular plate of 600*500*90 $mm^3$ attached to transmission shaft, of length 300 mm and diameter of 100 mm, fixed to top of motor rotor figure (1). SADA ideally provides a step wave excitation torque to mounted assembly along spin axis. $k_{em}$ in motor is approximated as a torsional spring between rotor and stator. After input of excitation force, mechanical system is under influence of lateral disturbances and angular motions about roll (x), pitch (y) and yaw (z) axis because of shaft flexibility. Analytical modelling for disturbance analysis of SADA provides an estimate of vibrations that are possibly aroused by drive mechanism in actual space mission when it is made to operate flexible SA.

## 3. SADA disturbance (analytical model): Transformations and coordinate systems

For kinematic analysis, geometry is defined by 3 cartesian and 3 angular coordinates. From figure (1), *XYZ* is inertial coordinate system (CS), $X^*Y^*Z^*$ is fixed to point *O* of body, $x_b y_b z_b$ is body fixed CS with origin at *O*. $x_b y_b z_b$ is related to $X^*Y^*Z^*$ system through a set of Euler angles i.e. $\alpha$, $\beta$ and $\gamma$ which help to prescribe body axis with respect to (wrt) $X^*Y^*Z^*$ and hence wrt *XYZ* system. Intermediate systems i.e., $l_1^* l_2^* l_3^*$, and $s_1^* s_2^* s_3^*$ are used to facilitate computation of composite axis transformations. $\alpha$, $\beta$ and $\gamma$ define an orthogonal CS i.e. $\underline{i}\,\underline{j}\,\underline{k}$ that result from three successive rotations from a fixed CS i.e. $\underline{I}\,\underline{J}\,\underline{K}$. Three successive positive rotations i.e., about $Z^*$ axis, new y-axis i.e., $l_2^*$ and about final x-axis i.e., $s_1^*$ axis result in final body fixed system. $x_b y_b z_b$ is related to *XYZ* by orthogonal transformation provided below. *C* & *S* in (2) correspond to *Cos* & *Sine* functions respectively.

$$\begin{bmatrix} \underline{i} \\ \underline{j} \\ \underline{k} \end{bmatrix} = \begin{bmatrix} C\beta C\gamma & C\beta S\gamma & -S\beta \\ S\alpha S\beta C\gamma - S\gamma C\alpha & S\alpha S\beta S\gamma + C\alpha C\gamma & S\alpha C\beta \\ C\alpha S\beta C\gamma + S\alpha S\gamma & C\alpha S\beta S\gamma - S\alpha C\gamma & C\alpha C\gamma \end{bmatrix} \begin{bmatrix} \underline{I} \\ \underline{J} \\ \underline{K} \end{bmatrix} \tag{2}$$

## 4. Kinematic analysis of SADA operated system

Suppose the body undergoes small rotation $d\chi$ in time *dt*. The position of every point in rotated body is determined if position of a single point in it is specified and orientation of $x_b y_b z_b$ body axis wrt fixed *XYZ* axis. An infinitesimal rotation of body during time interval *dt* is;

$$\omega_b = \frac{d\chi}{dt} \quad \Rightarrow \quad d\chi = \omega_b dt \quad ; \quad \underline{\omega}_b = \dot{\alpha}\underline{e}_\alpha + \dot{\beta}\underline{e}_\beta + \dot{\gamma}\underline{e}_\gamma \tag{3}$$

Using figure (1) and (2) and putting (3.1) in (3) gives of $\omega_b$ in $x_b y_b z_b$ components wrt fixed frame.

$$\underline{e}_\alpha = \underline{i} \quad ; \quad \underline{e}_\beta = \cos\alpha \underline{j} - \sin\alpha \underline{k} \quad ; \quad \underline{e}_\gamma = -\sin\beta \underline{i} + \sin\alpha\cos\beta \underline{j} + \cos\alpha\cos\beta \underline{k} \tag{3.1}$$

$$\underline{\omega}_b = \omega_{xb}\underline{i} + \omega_{yb}\underline{j} + \omega_{zb}\underline{k} \quad \Rightarrow \quad \begin{bmatrix} p \\ q \\ r \end{bmatrix} \underline{\underline{\Delta}} \begin{bmatrix} \omega_{xb} \\ \omega_{yb} \\ \omega_{zb} \end{bmatrix} = \begin{bmatrix} 1 & 0 & -\sin\beta \\ 0 & \cos\alpha & \sin\alpha\cos\beta \\ 0 & -\sin\alpha & \cos\alpha\cos\beta \end{bmatrix} \begin{bmatrix} \dot{\alpha} \\ \dot{\beta} \\ \dot{\gamma} \end{bmatrix} = \{\omega_b\} \tag{4}$$

*p, q, r* are the components of angular velocity vector. $\alpha$, $\beta$ and $\gamma$ are the angles associated with roll, pitch and yaw. An assumption of constant rotational motion, ($Sx \approx x$ & $Cx \approx 1$), reduces (2) to;

$$\begin{bmatrix} C\beta C\gamma & C\beta S\gamma & -S\beta \\ S\alpha S\beta C\gamma - S\gamma C\alpha & S\alpha S\beta S\gamma + C\alpha C\gamma & S\alpha C\beta \\ C\alpha S\beta C\gamma + S\alpha S\gamma & C\alpha S\beta S\gamma - S\alpha C\gamma & C\alpha C\gamma \end{bmatrix} = [a] \quad ; \quad [a] = \begin{bmatrix} 1 & \gamma & -\beta \\ -\gamma & 1 & \alpha \\ \beta & -\alpha & 1 \end{bmatrix} \quad ; \quad \begin{bmatrix} \underline{i} \\ \underline{j} \\ \underline{k} \end{bmatrix} = \begin{bmatrix} 1 & \gamma & -\beta \\ -\gamma & 1 & \alpha \\ \beta & -\alpha & 1 \end{bmatrix} \begin{bmatrix} \underline{I} \\ \underline{J} \\ \underline{K} \end{bmatrix} \tag{5}$$

Considering figure (3) for position vectors i.e., $R_o$, position vector of transmission shaft attachment point to rigid load with respect to origin of inertial frame then;

$$R_o = X_o\underline{I} + Y_o\underline{J} + Z_o\underline{K} \quad ; \quad \dot{R}_o = \dot{X}_o\underline{I} + \dot{Y}_o\underline{J} + \dot{Z}_o\underline{K} \quad ; \quad \ddot{R}_o = \ddot{X}_o\underline{I} + \ddot{Y}_o\underline{J} + \ddot{Z}_o\underline{K} \tag{6}$$

$X_o Y_o Z_o$ denote components of lateral and linear motion of position *O* wrt inertial *XYZ* axis. These three linear independent coordinates in addition to $\alpha$, $\beta$ and $\gamma$ completely specify position and orientation of mechanical system. Let point *P* ($x_p y_p z_p$), in figure (3), be any fixed point to lateral / linear end of rigid load wrt position ($X_o Y_o Z_o$). $R_p$ is position vector of point *P* wrt inertial frame.

$$\underline{R}_p = \underline{R}_o + \underline{r}_p \quad ; \quad \underline{r}_p = x_p\underline{i} + y_p\underline{j} + z_p\underline{k} \quad ; \quad \dot{r}_p = \frac{\partial r_p}{\partial t} + \omega_b \times r_p \quad ; \quad \frac{\partial r_p}{\partial t} = \dot{x}_p\underline{i} + \dot{y}_p\underline{j} + \dot{z}_p\underline{k} = 0 \tag{7}$$

$$\ddot{r}_p = \dot{\omega}_b \times r_p = \begin{bmatrix} \ddot{\alpha} \\ \ddot{\beta} \\ \ddot{\gamma} \end{bmatrix} \times \begin{bmatrix} x_p\underline{i} \\ y_p\underline{j} \\ z_p\underline{k} \end{bmatrix} = \begin{vmatrix} \underline{i} & \underline{j} & \underline{k} \\ \ddot{\alpha} & \ddot{\beta} & \ddot{\gamma} \\ x_p & y_p & z_p \end{vmatrix} \quad \Rightarrow \quad \ddot{r}_p = \underline{i}(z_p\ddot{\beta} - y_p\ddot{\gamma}) + \underline{j}(x_p\ddot{\gamma} - z_p\ddot{\alpha}) + \underline{k}(y_p\ddot{\alpha} - x_p\ddot{\beta}) \tag{8}$$

Using $\underline{i}\,\underline{j}\,\underline{k}$ from (5) in (8), and formulation for $R_p$ from (7) gives acceleration of shaft at point *P*.

$$\ddot{R}_p = \underbrace{\underline{I}(\ddot{X}_o + z_p\ddot{\beta} - y_p\ddot{\gamma})}_{I} + \underbrace{\underline{J}(\ddot{Y}_o + x_p\ddot{\gamma} - z_p\ddot{\alpha})}_{II} + \underbrace{\underline{K}(\ddot{Z}_o + y_p\ddot{\alpha} - x_p\ddot{\beta})}_{III} \tag{9}$$

Part *I* and *II* show lateral acceleration, while *III* is along bounce axis. $x_p$, $y_p$ and $z_p$ are distances of point *P* from centre *O* along the $x_b$, $y_b$ and $z_b$ axis respectively. The acceleration equation of any point *P* reflects influence of pitch, roll and yaw motion in addition to translational accelerations. Since, point $R_p$ is defined wrt to centre *O* on rigid body and geometry is mechanically coupled so, the displacement equation of point $R_p$ can be used to show deflections generated in the whole system or the coupled oscillations because of input motion by rotor. So, deflection equation $\Delta_s$ at point *P* on rigid body is;

$$\Delta_s = (X_o + z_p\beta - y_p\gamma) + (Y_o + x_p\gamma - z_p\alpha) + (Z_o + y_p\alpha - x_p\beta) \tag{10}$$

$X_o Y_o$ are the lateral displacements and $Z_o$ is bounce motion of point *O*, $\gamma$ is angular displacement of body about $z_b$ axis, while $\beta$ and $\alpha$ are angular displacements of body about $y_b$ and $x_b$ axis respectively.

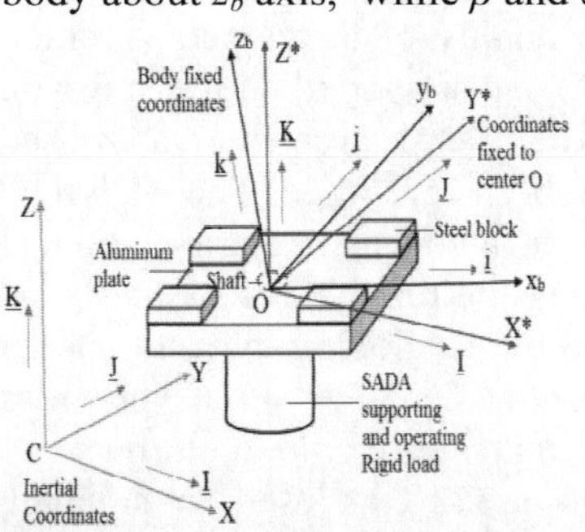


**Figure 1.** Problem geometry for SADA driving rigid load

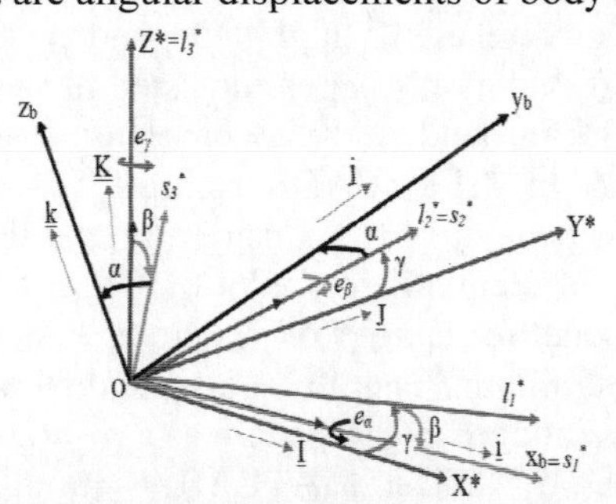

**Figure 2.** Coordinates and Euler angles

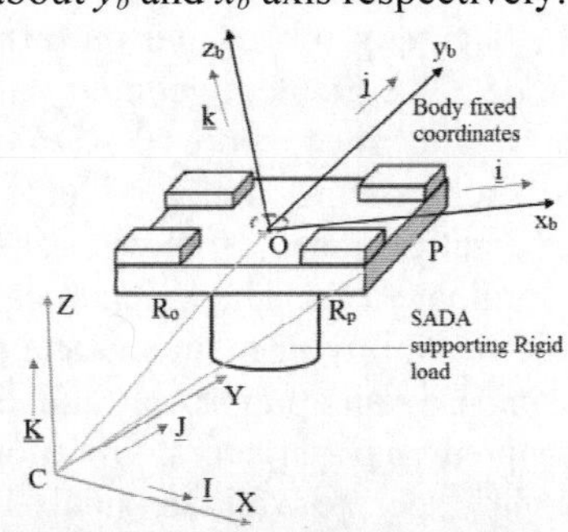


**Figure 3.** Analytical kinematics, position vectors

IOP Conf. Series: Materials Science and Engineering **224** (2017) 012001 doi:10.1088/1757-899X/224/1/012001

## 5. Dynamic analysis

Figure (4) provides approximate dynamic model of SADA supporting rigid load. Rotor strictly rotates along spin axis while mechanical system mounted atop shows coupled motion as explained in kinematic analysis. The electromagnetic effect between rotor and stator of stepper motor is approximated as torsional spring possessing electromagnetic stiffness. Moment of inertias of rigid load and SADA rotor are considered to determine coupled frequencies of mechanical system. The seven EoM of system are derived using Lagrange's method which in its generalized form is;

$$\frac{d}{dt}\left(\frac{\partial KE}{\partial \dot{q}_i}\right)-\frac{\partial KE}{\partial q_i}+\frac{\partial PE}{\partial q_i}+\frac{\partial DE}{\partial \dot{q}_i}=Q_i \tag{11}$$

$$KE=\frac{1}{2}m(\dot{X}_o^2+\dot{Y}_o^2+\dot{Z}_o^2)+\frac{1}{2}I_o\dot{\gamma}_o{}^2+\frac{1}{2}I_{1z}\dot{\gamma}_1^2+\frac{1}{2}I_{1y}\dot{\beta}^2+\frac{1}{2}I_{1x}\dot{\alpha}^2 \tag{12}$$

$$PE=\frac{1}{2}Kx^2=\frac{1}{2}K_o\gamma_o^2+\frac{1}{2}K_1\left[\{X_o+z_p\beta-y_p(\gamma_1-\gamma_o)\}+\{Y_o+x_p(\gamma_1-\gamma_o)-z_p\alpha\}+\{Z_o+y_p\alpha-x_p\beta\}\right]^2 \tag{13}$$

$$DE=\frac{1}{2}C\dot{x}^2=\frac{1}{2}C\left[\{\dot{X}_o+z_p\dot{\beta}-y_p(\dot{\gamma}_1-\dot{\gamma}_o)\}+\{\dot{Y}_o+x_p(\dot{\gamma}_1-\dot{\gamma}_o)-z_p\dot{\alpha}\}+\{\dot{Z}_o+y_p\dot{\alpha}-x_p\dot{\beta}\}\right]^2 \tag{14}$$

Where; *KE, PE* and *DE* are kinetic, potential and dissipation function / dissipation energies respectively, $Q_i$ is generalized external force acting on system, $K$ is stiffness and $C$ is damping coefficient. Equations (12~14) govern coupled torsional-lateral motion behaviour of SADA supporting rigid load with transmission shaft. For an initial approximation of system's coupled frequency response, Part-I inside square brackets of *PE*, *DE* and corresponding acceleration functions from KE are used to formulate EoM for yaw, pitch and lateral motion in x-direction using Lagrange method.

$$\begin{bmatrix} I_o & 0 & 0 & 0 \\ 0 & I_{1z} & 0 & 0 \\ 0 & 0 & I_{1y} & 0 \\ 0 & 0 & 0 & m_x \end{bmatrix}\begin{bmatrix}\ddot{\gamma}_o\\ \ddot{\gamma}_1\\ \ddot{\beta}\\ \ddot{X}_0\end{bmatrix}+\begin{bmatrix} K_o+K_1y_p^2 & -K_1y_p^2 & K_1y_pz_p & K_1y_p \\ -K_1y_p^2 & K_1y_p^2 & -K_1y_pz_p & -K_1y_p \\ K_1y_pz_p & -K_1y_pz_p & K_1z_p^2 & K_1z_p \\ K_1y_p & -K_1y_p & K_1z_p & K_1 \end{bmatrix}\begin{bmatrix}\gamma_o\\ \gamma_1\\ \beta\\ X_0\end{bmatrix}+\begin{bmatrix} C_o+C_1y_p^2 & -C_1y_p^2 & C_1y_pz_p & C_1y_p \\ -C_1y_p^2 & C_1y_p^2 & -C_1y_pz_p & -C_1y_p \\ C_1y_pz_p & -C_1y_pz_p & C_1z_p^2 & C_1z_p \\ C_1y_p & -C_1y_p & C_1z_p & C_1 \end{bmatrix}\begin{bmatrix}\dot{\gamma}_o\\ \dot{\gamma}_1\\ \dot{\beta}\\ \dot{X}_0\end{bmatrix}=\begin{bmatrix}T_e\\ 0\\ 0\\ 0\end{bmatrix} \tag{15}$$

Moment of inertia of SADA rotor, electromagnetic torque coefficient, stiffness of transmission shaft and damping coefficient are obtained by experimental testing. System is designed in solid works to get moment of inertia and mass of rigid load. The values of $x_p$, $y_p$ and $z_p$ depend upon dimensions of rigid load. Later, program will be written in MATLAB / MuPad to determine vibration behaviour of SADA driven assembly. Equation (16) provides probable disturbance response of SADA influenced by structural coupling. It reveals that both electronic harmonics of SADA and natural frequencies of mechanical system contribute to SADA disturbance.

$$f_{SADA}=\{f_{electronic}, f_{natural}\} \tag{16}$$

## 6. Conclusion & future scope

This work has proposed a mathematical SADA disturbance analysis scheme for incorporation into a performance assessment and enhancement framework of satellite motion. In that context, disturbance model is used to drive model of spacecraft and outputs are compared against requirements to assess controller design. The development of disturbance model is an important step in this process because the accuracy of original performance results is highly dependent on quality of developed vibration model. The problem geometry analysed in this paper consisted of rigid load supported on transmission shaft and functioned by SADA. Background study is completed to extract least targeted areas. Based on short comings a new disturbance model for SADA controlled assembly is proposed. The analytical kinematic model captures motion behaviour by considering translation along and rotation about 3 coordinate axes. The kinematics of system utilizes Euler transformations and relative motion analysis. The feasibility of mathematical model & stability of system is ensured by developing dynamic model using Lagrange approach and determining energies dissipated in system. Dynamic model provides coupled torsional lateral vibrations induced in the system as a result of input step excitation torque.

In order to validate analytical dynamics, MATLAB / Simulink toolbox is used to simulate disturbance model of SADA figure (5). This model will be based on dynamic equations of stepper motor dragging rigid load with transmission shaft. The Simulink model has advantage that results from it can be represented in either time or frequency domain and is applicable to any geometry of rigid loading. The input step excitation torque will be considered from [11, 13]. Earlier studies have considered SADA supporting rigid load as 2-DoF system. The disturbance response consisted of natural frequencies and discrete active harmonics of SADA [12, 13]. However, present investigation is an effort to contribute its more complete and realistic disruption behaviour by considering influence of coupled torsional and lateral / linear vibrations in analytical model. The disturbance feedback will consist of mode excited by electromagnetic frequency, three natural mechanical frequencies of coupled equivalent SADA driven system and harmonics of its electronic frequencies. The expected coupled frequency response for 2-DoF system (active harmonics and two natural frequencies) is depicted in figure 6. Analysis provides a theoretical basis for prediction of instabilities emitted by SADA during in orbit operation. The Simulink model will verify that fundamental active harmonic of SADA has the highest disturbance amplitude and it will decrease in higher frequency zone. In addition, at any point during analysis, interference of electronic and natural frequencies will tend to multiply disturbance amplitude observed by SADA supported system.

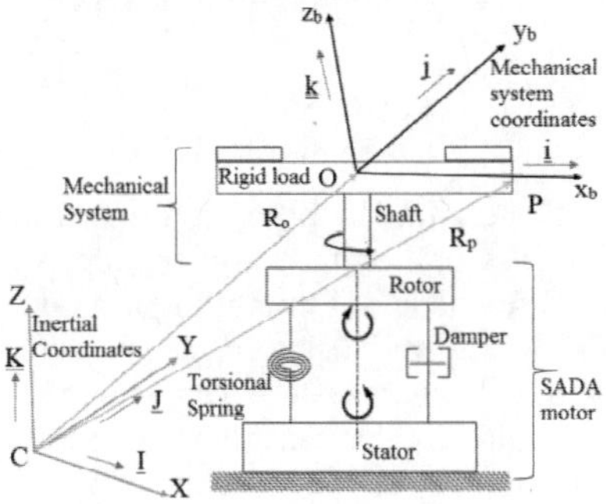


**Figure 4.** Approximate dynamic model of mechanical system

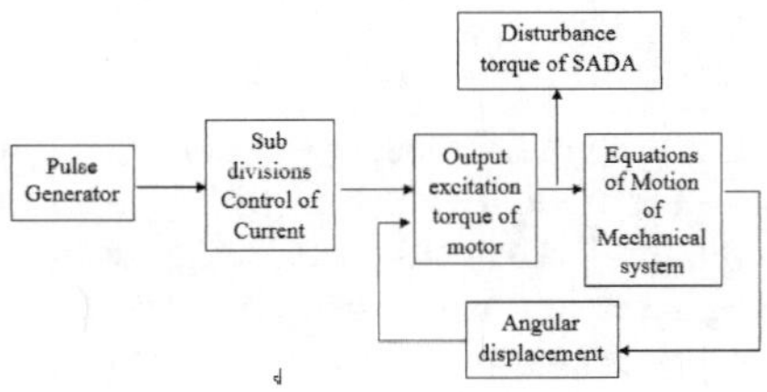


**Figure 5.** Simulation model for SADA driving rigid load

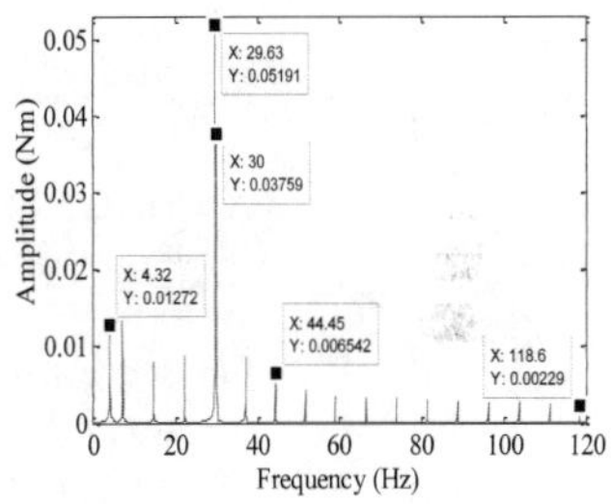


**Figure 6.** Disturbance of SADA 2-dof system [13]

Researchers have worked on the influence of electromagnetic stiffness and increase in subdivisions number on the disturbance amplitude of SADA operated systems [13]. The same factors can be applied in this work to evaluate influence of coupled torsional lateral dynamics of system supported, on, and driven by SADA. Models of SADA induced disturbances can be used in jitter analysis to predict effects of vibrations on spacecraft, allow development of drive circuits and design of suitable isolation and damping technologies.

## References

[1] Zhu S and Lei Y 2015 Proc. *IMechE, Part G: J. of Aerospace Engineering* 229 14: p. 2646-58.
[2] Breitenstein O, Bauer J, Wagner J, Zakharov N, Blumtritt H, Lotnyk A, Kasemann M, Kwapil W and Warta W 2010 *IEEE Transactions on Electron Devices* 57 9: p. 2227-34.
[3] Xia L 1994 *Journal of Shenyang University* 16 3: p. 73-8.
[4] Bodson M and Sato J S 2006 *IEEE Transactions on Control Systems Technology* Vol. 14 No. 2.
[5] Maharana P and Goel P 1991 *Acta Astronautica* 25 3: p. 157-64.
[6] Kumar R R, Paul A and Lim T W 1992 *J. of guidance, control and dynamics* 15 6: p. 1427-33.
[7] Rajagopal K, Krishnaswamy M, Singh B and Singh B P 1997 *IEEE Transactions on Industry Applications* 33 4: p. 906-13.
[8] Goel P, Maharana P K and Sathyanarayana M N 1994 *Acta Astronautica* 32 4: p. 267-73.
[9] Xiao N C, Li Y F and Huang Z H 2009 *J. of Astronautics* 4: p. 065.
[10] Yuan J 2004 Design and analysis of satellite structures, *China Astronautic Publishing House*.
[11] Sattar M, Wei C and Sattar R 2017 *14th IBCAST IEEE Conference Publications* p. 50-7.
[12] Chen J and Wei C 2014 *Applied Mechanics and Materials* Trans Tech Publ 256: p. 103-08.
[13] Sattar M, Wei C and Jalali A 2016 *Vibro engineering procedia* JVE Int.: Moscow. p. 6.
[14] Sattar M, Wei C, Chen J and Israr A 2016 *Advances in Engineering Research* Atlantis Press: Phuket, Thailand. p. 4.
[15] Larson W and Wertz J R 1992 *Space mission analysis and design* Microcosm Inc. Torrance CA.

IOP Conf. Series: Materials Science and Engineering **224** (2017) 012002 doi:10.1088/1757-899X/224/1/012002

# The Sensitivity Analysis of Inherent Frequency of Roll Mill's Main Drive System

**Hou Dongxiao**

School of Control Engineering, Northeastern University at Qinhuangdao, Qinhuangdao 066004, China

houdongxiao1982@163.com

**Abstract:** In order to avoid the beat vibration of rolling mill, taking F1 rolling mill of 1780 tandem mill as an example, a twelve DOF branched torsional vibration model of main drive system of F1 rolling mill was constructed. Then the inherent frequency of the rolling mill and the corresponding vibration modes of every inherent frequency were calculated. Meanwhile, the sensitivity of first and second inherent frequency with the change of equivalent rotational inertias and torsional stiffness of rolling mill were analyzed, and the most sensitive parameters which influence on first and second inherent frequency were found out. The result shows that the equivalent inertias of upper and lower work rolls and equivalent stiffness of upper and lower joint shaft of drive system were most sensitive parameters, which are easier to change first and second inherent frequency of drive system. The research will provide theory reference for adopting proper mechanical parameters in equipment design.

## 1. Introduction

In recent years, as the increase of rolling speed and rolling product quality, the vibration of rolling mill become more frequently. The vibration of rolling mill not only affects the surface quality of products and the precision of control, severe vibration may lead to serious accidents, such as strip break and equipment damage. Vibration has become the major bottleneck in productivity and strip yield. For avoiding vibration, many theory models and control methods are proposed [1-2]. Consider the influence of the nonlinear friction damping, a single DOF torsional vibration model is constructed, and the vibration characteristics were analyzed [3]. Based on finite element model, the spatial vibration which include vertical, horizontal, axial, reverse, cross and swinging vibration were analyzed [4].

In this paper, a twelve DOF branched torsional vibration model of F1 rolling mill of 1780 tandem mill was constructed, and the inherent frequency and vibration modes were obtained. Then the sensitivity of first and second frequency with the change of equivalent rotational inertias and stiffness were analyzed, it will provide theory reference for designing drive system of rolling mill.

## 2. The torsional vibration model of F1 rolling mill of 1780 tandem mill

According to the actual mechanical structure of F1 rolling mill of 1780 tandem mill, a twelve DOF branched torsional vibration mechanical model of the main drive system can be constructed as Figure 1.

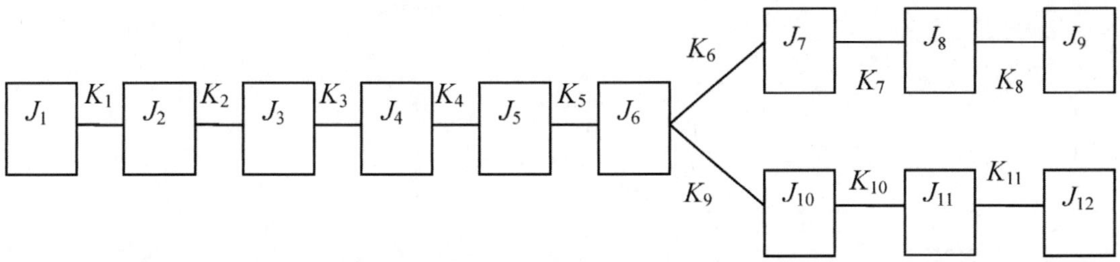


**Figure 1.** The torsional vibration mechanical model of main drive system of F1 rolling mill

The parameters of the mechanical model in Fig.1 are shown as Table 1.

**Table 1.** The mechanical parameters of main drive system of F1 rolling mill

| rotational inertia | kg.m$^2$ | torsional stiffness | N.m/rad |
|---|---|---|---|
| $J_1$(Motor) | 9895.6 | $K_1$ | 7.2884e8 |
| $J_2$(Motor coupling) | 1013.8 | $K_2$ | 5.0356e8 |
| $J_3$(gear reducer) | 1543.7 | $K_3$ | 1.27e8 |
| $J_4$ (The main coupling(left)) | 138.0556 | $K_4$ | 8.175e7 |
| $J_5$ (The main coupling(right)) | 136.6309 | $K_5$ | 6.037e7 |
| $J_6$(Gear base) | 87.0458 | $K_6$ | 4.5848e7 |
| $J_7$(upper joint shaft(left)) | 24.2059 | $K_7$ | 5.9155e6 |
| $J_8$(upper joint shaft (right)) | 18.2076 | $K_8$ | 2.0681e7 |
| $J_9$(upper work roll) | 229.6021 | $K_9$ | 4.5848e7 |
| $J_{10}$(lower joint shaft(left)) | 24.2059 | $K_{10}$ | 5.9155e6 |
| $J_{11}$(lower joint shaft(left)) | 18.2076 | $K_{11}$ | 2.0681e7 |
| $J_{12}$(lower work roll) | 229.6021 | | |

Based on the Lagrange equation, the dynamic equation in Fig.1 can be expressed as

$$[J]\{\ddot{\theta}\}+[K]\{\dot{\theta}\}=\{F\} \tag{1}$$

Where, $J$ and $K$ are the matrix of rotational inertia and torsional stiffness, respectively. $F$ is the matrix of external force.

## 3. The inherent frequency and vibration mode of drive system of F1 rolling mill

Based on MATLAB, the inherent frequency of drive system of F1 rolling mill can obtained from Eq. (1), and it is shown as Table 2.

**Table 2.** The inherent frequency of drive system of F1 rolling mill (Hz)

| $\omega_1$ | $\omega_2$ | $\omega_3$ | $\omega_4$ | $\omega_5$ | $\omega_6$ | $\omega_7$ | $\omega_8$ | $\omega_9$ | $\omega_{10}$ | $\omega_{11}$ |
|---|---|---|---|---|---|---|---|---|---|---|
| 18.3 | 20.9 | 60.3 | 82.6 | 149.8 | 188.8 | 190.1 | 197.7 | 223.4 | 238.5 | 295.7 |

The corresponding vibration mode of every inherent frequency are shown as Figure 2.

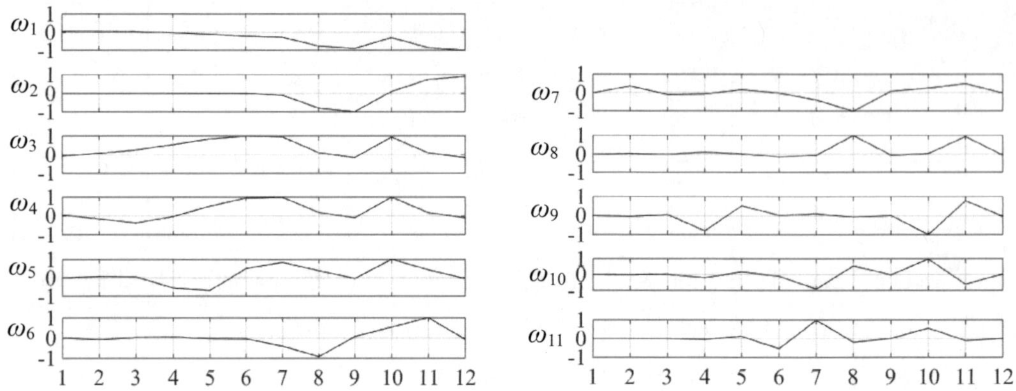


**Figure 2.** The vibration mode of the corresponding frequency of drive system of F1 rolling mill

If the first inherent frequency of drive system is close to the second inherent frequency, the beat vibration easier to arise. The first and second inherent frequencies may change with the change of rotational inertia and stiffness of drive system. When design a new rolling mill, it should select proper mechanical parameters and the beat vibration will be avoided in the design stage.

## 4. The sensitivity of inherent frequency of drive system of F1 rolling mill

### 4.1. *The sensitivity of inherent frequency with the change of rotational inertias*

The change of every rotational inertia of drive system may change the inherent frequency, it has different sensitivity chart for every rotational inertia. The sensitivity comparison chart of the first and second inherent frequency with the change of rotational inertia are shown as Figure 3(a) and Figure 3(b), respectively. Where the abscissa axis is rotational inertia of $J_1$ to $J_{12}$ which range from 90% to 110%.

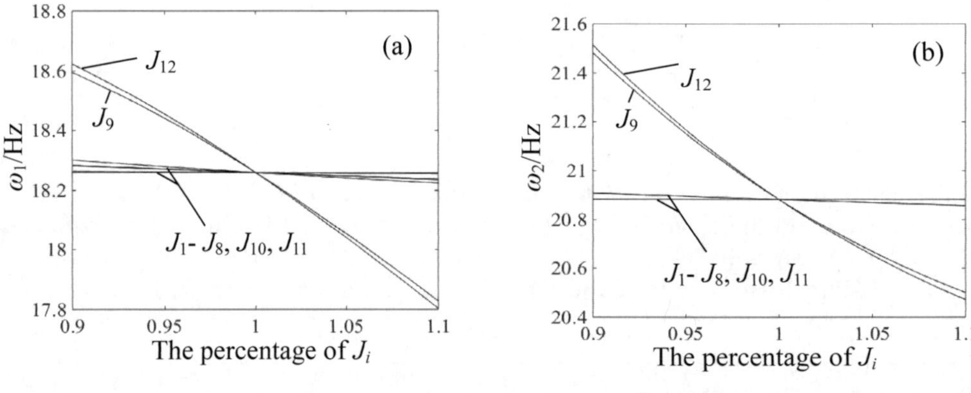


**Figure 3.** The sensitivity comparison chart of first and second inherent frequency in different rotational inertias from $J_1$ to $J_{12}$

In Figure 3, It is found that $J_9$ and $J_{12}$ are more sensitive to change first and second inherent frequency, so adjusting $J_9$ and $J_{12}$ are easier to change the first and second inherent frequency in mechanical design. Others rotational inertias are less impact on the change of the first and second inherent frequency.

Because of upper and lower structure of the drive system is symmetrical, the sensitivity charts of inherent frequency are same when varying $J_9$ and $J_{12}$. Figure 4 is sensitivity Comparison chart of first and second inherent frequency with the change of J9.

IOP Conf. Series: Materials Science and Engineering **224** (2017) 012002 doi:10.1088/1757-899X/224/1/012002

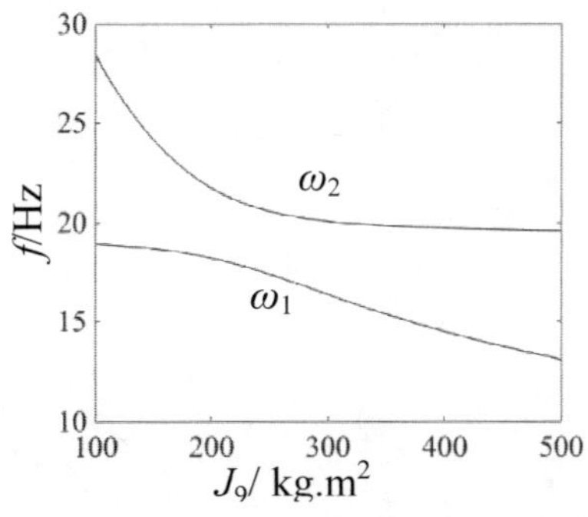


**Figure 4.** The sensitivity Comparison chart of the first and second frequency with the changes of $J_9$

In Figure 4, it could found that the first and second inherent frequency have different change tendency. Where $J_9$ is equivalent rotational inertia of upper working roll, it may be changed as rolling mill wears off. When $J_9$ is equal to 220 kg.m$^2$, the first inherent frequency is close to the second inherent frequency, and the beat vibration may arise in this situation.

### 4.2. *The sensitivity of inherent frequency with the change of torsional stiffness*

The inherent frequency of drive system also vary with the change of torsional stiffness. The sensitivity chart of the first and second inherent frequency with the change of torsional stiffness are shown as Figure 5(a) and Figure 5(b), respectively.

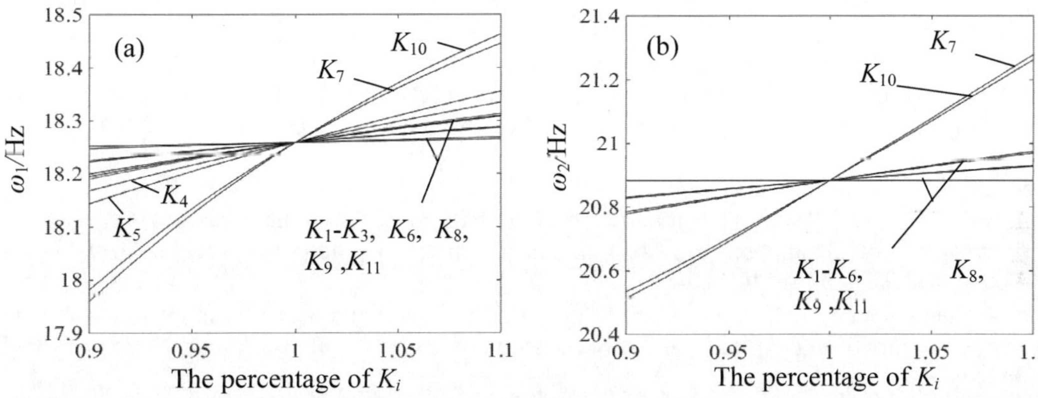


**Figure 5.** The sensitivity comparison chart of first and second inherent frequency with different stiffness from $K_1$ to $K_{11}$

In Figure 5, the abscissa axis is torsional stiffness from $K_1$ to $K_{11}$ which range from 90% to 110%, it is found that $K_7$ and $K_{10}$ are more sensitive for varying the first and second inherent frequency. Where $K_7$ and $K_{10}$ are equivalent torsional stiffness of upper and lower joint shaft, respectively.

Figure 6 is sensitivity comparison charts of first and second inherent frequency with the change of $K_7$. It is found that the first and second inherent frequency has different change tendency with the change of $K_7$. The equivalent stiffness $K_7$ may vary with the change of length, diameter or another mechanical structure size of the upper joint shaft. When $K_7$ is equal to 5.2e10 N.m/rad, the first inherent frequency is close to the second inherent frequency, then the beat vibration arises.

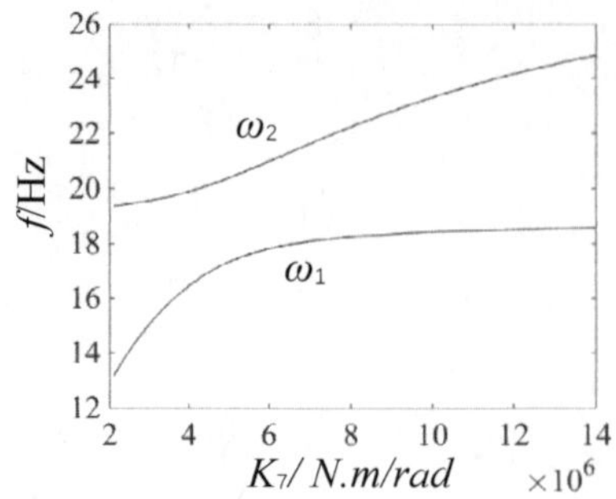


**Figure 6.** The sensitivity Comparison chart of the first and second frequency with the change of $K_7$

## 5. Conclusion

(1) According to the actual mechanical structure, a twelve DOF branched torsional vibration model of main drive system of F1 rolling mill is constructed.

(2) The inherent frequency and the corresponding vibration mode of every inherent frequency were obtained by means of MATLAB.

(3) The sensitivity of first and second inherent frequency in different rotational inertias and torsional stiffness were analyzed, and it is found that rotational inertias of the upper and lower working rolls, the torsional stiffness of upper and lower joint shaft are more sensitive to the first and second inherent frequency. Based on the results, it could provide theoretical way for avoiding beat vibration in design of drive system of rolling mill.

## Acknowledgments

This research is supported by National Natural Science Foundation of China (Grant No. 51405068), Natural Science Foundation of Hebei Province of China (Grant No.E2014501006), and Technology Research Project of Colleges and Universities of Hebei Province of China (Grant No. 2014202).

## References

[1] P.M. Shi, J.Z. Li, J.S. Jiang, B. Liu and D.Y. Han 2013 Nonlinear dynamics of torsional vibration for rolling mill's main drive system under parametric excitation *Jounal of Iron and Steel Research, International* **20**(1) 7-12

[2] Liu Shuang, Wang Zhaolong and Wang Jinjin 2016 Sliding bifurcation research of a horizontal-torsional coupled main drive system of rolling mill *Nonlinear Dynamics* **83**(1-2) 441-455

[3] Lu Xiaoyan, Ye Qianyuan and Qu Zihao 2004 Analysis and solution of the self-excited vibration on a rollingnmill *Journal of University of Shanghai for Science and Technology* **26**(2) 141-144

[4] Yongjiang Zheng, Zhaohui Xie, Yigeng Li, Guanxian Shen and Hongmin Liu 2013 Spatial vibration of rolling mills *Journal of Materials Processing Technology* **213** 581– 588

IOP Conf. Series: Materials Science and Engineering **224** (2017) 012003 doi:10.1088/1757-899X/224/1/012003

# Solid Propellant Burning Equilibrium Ingredients Calculation Based on Temperature Iteration

**HU Kuan[1], ZHANG Lin[2], Chang Xinlong[1], Wang Chao[1]**

[1]Xi'an High-tech Institute, Xi'an 710025, China

[2] Xi'an Aerospace Composites Research Institute, Xi'an 710025, China

huk1976@sina.com

**Abstract.** Aiming at shortcomings of depending on experiences and inferior computation accuracy brought by using linear interpolation to calculate fixed pressure burning temperature, a new method using iteration was put forward. Moreover, the optimization model to calculate equilibrium ingredients was formulated based on the new method, and sequential quadric programming method rather than Lagrange multiplier was used to compute the model. At last, a numerical example was given to validate the method in this paper. Results make show that more perfect outcomes can be achieved through the method in this paper than classical method.

## 1. Introduction

Equilibrium ingredients and burning temperature of propellant is necessary data for ignition engine design, inner ballistic computation and performance evaluation of solid rocket motor, at present there are three classical methods to calculate equilibrium ingredients and burning temperature of propellant, which are gradual approximation method, minimum G-free energy method and Brinkley method [1]. Burning temperature must be certain beforehand in all of above-mentioned method, however burning temperature is also unknown. Therefore, a simple linear interpolation method was frequently adopted to calculate burning temperature based two presupposed temperatures in all of above-mentioned method. In this way, analyst's experiences and the two preselected temperatures would have remarkable influences on computation accuracy. To solve this problem, a new method to calculate burning temperature was put forward in this paper, which linear interpolation was substituted by iteration. Furthermore, the equilibrium ingredients optimization model was set up based on minimum G-free energy and the iteration method. Finally, the method in this paper was used to calculate equilibrium ingredients and burning temperature of a given solid propellant components.

## 2. Fixed pressure burning temperature calculation

The principle of solving fixed pressure burning temperature of $T_c$ is that solid propellant's total enthalpy of $I_p$ must be equal to burning products total enthalpy of $I_m$, namely

$$\sum_{i=1}^{m}\left[H_{fi}^{T_s}+c_i(T_i-T_s)\right]\cdot q_i=\sum_{i=1}^{N}\left(H_{i-298}^{\theta}+\int_{T_s}^{T_c}C_{pi}dT\right)\cdot n_i \tag{1}$$

Where $T_i$ is initial temperature of propellant, K.

$T_s$ is benchmark temperature, which is commonly 298K.

$c_i$ is specific heat of the $i$-th component of propellant, which varies hardly with temperature, J/(kg.K).

$q_i$ is mass percentage of the $i$-th component of propellant.

$H_{fi}^{T_s}$ is standard creation enthalpy of the *i*-th component of propellant, which can be found in reference [2].

$n_i$ is molar number of the *i*-th burning equilibrium ingredient.

$H_{i-298}^{\theta}$ is standard creation enthalpy of the *i*-th burning equilibrium ingredient.

$C_{pi}$ is fixed pressure specific heat of the *i*-th burning outcome, which is function of temperature and can be calculated through reference [3].

*2.1. The method of classic linear interpolate*

In the method of calculating burning temperature by classic linear interpolate, two temperatures were presumed firstly. Then burning ingredients and their total enthalpy could be obtained when propellant burning at the two presumed temperatures. If total enthalpy of burning ingredients was linear function of temperature between two presumed temperatures, using interpolate could easily get the fixed pressure temperature based on conversation of energy. The process can be figured by Fig.1.

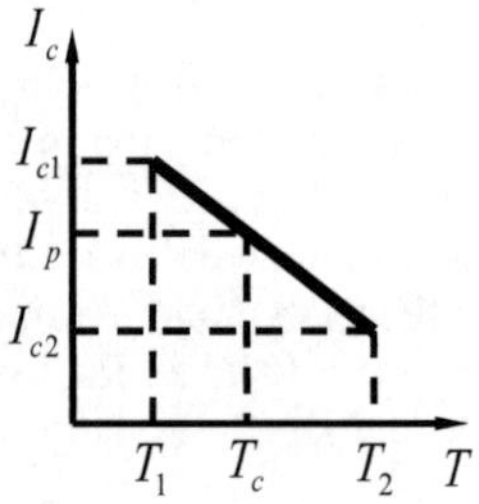


**Fig.1** method of linear interpolation to calculate fixed pressure burning temperature

Obviously, the two presumed temperatures are vital to computation accuracy in this method. If analysts had abundant experiences, the presumed two temperatures would approximate true fixed pressure burning temperature and computation error of linear interpolate would be little. Otherwise, linear interpolate would bring non-ignorable errors. Even though for proficient analysts, it was also difficult to presume appropriate temperatures when facing a new type of propellant. Therefore, it is necessary to find new method to solve burning temperature.

*2.2. Iteration method to solve burning temperature*

In practice, calculating burning temperature is to solve the following equation

$$f(T)=\sum_{i=1}^{m}\left[H_{fi}^{T_s}+c_i(T_i-T_s)\right]\cdot q_i-\sum_{i=1}^{N}\left(H_{i-298}^{\theta}+\int_{T_s}^{T_c}C_{pi}dT\right)\cdot n_i=0 \tag{2}$$

Here, the formula (2) is a higher nonlinear equation with regard to $T_c$, because $C_{pi}$ and $n_i$ are all function of $T_c$, and there are no visible function between $n_i$ and $T_c$. The equation can't be solved by analytical method; it can only be solved by iteration. However, iteration based on gradient is not advisable because of no visible function between $n_i$ and $T_c$, we can use secant method [4] to solve the equation. So, the iteration formation to calculate $T_c$ can be expressed as following

$$T_{k+1}=T_k-\frac{f(T_k)(T_k-T_{k-1})}{f(T_k)-f(T_{k-1})} \tag{3}$$

When absolute value of difference between $I_p$ and $I_m$ is less than $\boldsymbol{\varepsilon}_2$, where $\boldsymbol{\varepsilon}_2$ denotes iteration accuracy and is a given small decimal fraction.

## 3. Equilibrium ingredients calculation based on temperature iteration

Now, there are three classical methods to calculate equilibrium ingredients and burning temperature of propellant, which are gradual approximation method, minimum G-free energy method and Brinkley method. Gradual approximation method is simplest among three methods, but this method isn't suitable for thermodynamics calculation with complex ingredients. Minimum G-free energy method has explicit physics meanings, this method considers equilibrium ingredients are ingredients which minimize G-free energy of burning products, so the problem of calculating equilibrium ingredients is transferred to minimum optimization of G-free energy. The number of equations to be solved in Brinkley method is fewest, but Brinkley method is most complicated among three methods, the method demands to input a good many coefficient matrix, which easily creates mistake, and linearization in solving equations will bring calculation errors. So, the minimum G-free energy method would be used to calculate equilibrium ingredients in this paper. In this way, we can set up following optimization model to calculate equilibrium ingredients.

$$\min \quad \Phi = \sum_{i=1}^{L}\left(-Y_i^c n_i\right) + \sum_{i=L+1}^{N}\left(-Y_i + \ln n_i + \ln p_c - \ln n_g\right)\cdot n_i \tag{4a}$$

$s.t.$

$$\begin{cases} N_k = \sum_{i=1}^{N} A_{ki}\cdot n_i & (k = 1,2,\cdots,m) \\ n_i \geq 0 & (i = 1,2,\cdots,N) \end{cases} \tag{4b}$$

Where $\Phi = G/RT$, G is G-free energy.

$Y_i^c = -G_{Tic}^{\theta}/RT$, $G_{Tic}^{\theta}$ is standard molar G-free energy of agglomerate, which can be calculated through reference [3].

$Y_i = -G_{Ti}^{\theta}/RT$, $G_{Ti}^{\theta}$ is standard molar G-free energy of gas, which can be calculated through reference [3].

$A_{ki}$ is molar number of the *k*-th element contained in one mole of the *i*-th burning ingredient.

$m$ is the number of propellant element.

$N_k$ is molar number of element in assumed chemical formula.

It is noted that $\ln n_i$ was involved in formula (4a), but $n_i$ could be zero, it would bring calculation difficulty. To avoid the problem, we can make $n_i$ greater than $\varepsilon_1$ (a given small decimal fraction) in the process of optimization, namely

$$n_i \geq \varepsilon_1 \qquad i = 1,2,\cdots,N$$

If result of $n_i$ was less than $\xi$, where, $\xi$ is an another given small decimal fraction and greater than $\varepsilon_1$, $n_i$ should be zero, namely this burning ingredient didn't exist.

In engineering, Lagrange multiplier was usually used to solve above optimization model. In this way, $n_i$ could be less than zero in the process of optimization, which can make optimization calculation terminate. In this paper, Lagrange multiplier was substituted by SQP (Sequence Quadratic Programming) to solve the optimization model.

It is noted that burning temperature must be known when we use SQP to solve the optimization model, but burning temperature is also parameter to be solved. So, we must integrate formula (4) with the new method in section 2.2 to calculate equilibrium ingredients. The step of calculation can be expressed as following.

(1) Input initial datum of propellant;
(2) Calculate assumed chemical formula and total enthalpy of propellant;
(3) Assume two initial burning temperatures ($T_0, T_1$);
(4) Use SQP to solve formula (4), get burning ingredients at $T_0, T_1$;
(5) Calculate total enthalpy of burning ingredients at $T_0, T_1$;

(6) If $\left|I_p - I_m\right| \le \boldsymbol{\varepsilon}_2$, transfer to (9). Otherwise calculate a new burning temperature $T_{k+1}$ according to formula (3);

(7) Use SQP to solve formula (4), get burning ingredients at $T_{k+1}$;

(8) Calculate total enthalpy of burning ingredients at $T_{k+1}$, and transfer to (6);

(9) Terminate, and output results.

## 4. Example

Assume solid propellant's components were showed in Tab.1; please calculate its equilibrium ingredients and burning temperature when the propellant burns at pressure of 6.08Mpa. ($T_i$ is 298K.)

**Tab.1** Component of solid propellant

| Number | Component | Mass percent | Standard creation enthalpy (KJ/mol) |
|---|---|---|---|
| 1 | $NH_4ClO_4$ | 68% | -291.56 |
| 2 | Al | 15.5% | 0 |
| 3 | $CaCO_3$ | 3.5% | -1211.66 |
| 4 | $C_{220}H_{441}O_{61}$ | 13% | -14001.83 |

Via calculation, we can achieve that assumed chemical formula and total enthalpy of the propellant are respectively $Ca_{0.3497}Al_{5.745}C_{7.389}H_{37.262}O_{26.1523}N_{5.7878}Cl_{5.7878}$ and $-6.0935\times10^5$cal.

Assume burning products include fifteen types of ingredients, which are CaO（s）, $Al_2O_3$（s）, CO, $H_2$, $H_2O$, $N_2$, HCl, $CO_2$, H, AlOCl, Cl, HO, NO, CHO and O, their G-free energy can be calculated through reference 3.

Here, $\varepsilon_1$, $\varepsilon_2$ and $\xi$ are respectively $10^{-5}$, $10^{-4}$ and $10^{-4}$. We can achieve that fixed burning temperature is 3634.5 K and equilibrium ingredients are shown in Tab.2.

**Tab. 2** The result of equilibrium ingredients calculation

| burning ingredient | the number of moore | burning ingredient | the number of moore |
|---|---|---|---|
| CaO（s） | 0.3497 | H | 0.5486 |
| $Al_2O_3$（s） | 2.7392 | AlOCl | 0.2665 |
| CO | 6.4232 | Cl | 0.2735 |
| $H_2$ | 6.9687 | HO | 0.3413 |
| $H_2O$ | 8.5868 | NO | 0.0333 |
| $N_2$ | 2.8772 | CHO | 0.0134 |
| HCl | 5.2478 | O | 0.0155 |
| $CO_2$ | 0.9524 | | |
| total enthalpy of burning ingredients | | -6.09346×105 cal | |
| G-free energy of burning ingredients | | -1.13375×103 J | |

From tab.2 we can see that relative error between total enthalpy of burning ingredients achieved by the method in this paper and total enthalpy of propellant is only $6.56\times10^{-6}$. So, the method in this paper can be used to calculate equilibrium ingredients and fixed burning temperature of solid propellant.

## 5. Conclusions

We can obtain following conclusions via above analysis.

(1) Compare to classic linear interpolate, iteration has no special requirements on initial assumed burning temperature, which don't rely on analyst's experiences, and can get high calculation accuracy.
(2) When solving equilibrium ingredients optimization model, SQP avoids the problem brought by Lagrange multiplier, and is more effective method.

**References**

[1] Chen Ruxun 1991 *Design and research of solid rocket motor* (Beijing: China Astronautic Publishing House)
[2] QJ1814-89 *Standard Moore enthalpy of some composite propellant component* (Beijing: Aerospace Industry Corporation of China)
[3] QJ 2036-91 *Temperature index of propellant burning production* (Beijing: Aerospace Industry Corporation of China)
[4] Deng Jianzhong 2010 *Numerical method* (Xi'an: Press of Xi'an Jiaotong University)

IOP Conf. Series: Materials Science and Engineering **224** (2017) 012004 doi:10.1088/1757-899X/224/1/012004

# Study on Active Suppression Control of Drivetrain Oscillations in an Electric Vehicle

**HUANG Lei[1], CUI Ying[2]**

[1]Tianjin Key Laboratory of Information Sensing and Intelligent Control, School of Automation and Electrical Engineering, Tianjin University of Technology and Education, Tianjin, China
[2]Urban Communications Engeneering CO.LTD of CCCC First Harbor Engeneering CO.LTD, Tianjin, China

huanglei@tute.edu.cn

**Abstract.** Due to the low damping in a central driven electric vehicle and lack of passive damping mechanisms as compared with a conventional vehicle, the vehicle may endure torsional vibrations which may deteriorates the vehicle's drivability. Thus active damping control strategy is required to reduce the undesirable oscillations in an EV. In this paper, the origin of the vibration and the design of a damping control method to suppress such oscillations to improve the drivability of an EV are studied. The traction motor torque that is given by the vehicle controller is adjusted according to the acceleration rate of the motor speed to attenuate the resonant frequency. Simulations and experiments are performed to validate the system. The results show that the proposed control system can effectively suppress oscillations and hence improve drivability.

## 1. Introduction

The research of electric vehicles (EVs) to achieve green transportation is increasing in recent years [1]. In an EV, the high voltage battery storages energy, the integration of the motor and transmission system configuring the powertrain provides the ability of driving. In addition, a highly coordinated electronic control unit (ECU) is used for torque control, speed control and power distribution to meet the vehicle requirement.

Conventional none EV vehicle drivetrain system mainly consists of diesel engine, transmission system, driveshaft, clutch, half shafts, differential, and wheels, etc. The existence of the clutch and the driveline system's elastic nature performs a passive way of damping the oscillations in conventional vehicles. In a central driven pure EV, no clutches is present and the overall system configuration is less elastic inherently due to the fact that the motor and the transmission system are direct mechanical coupled. This configuration has little passive damping effect that can dampen disturbances and avoid oscillations, which are mostly noticeable during low speed range. Therefore, due to the low damping in an EV and lack of passive damping hardware as compared with a conventional vehicle, a damping control strategy is needed to minimize the drivetrain oscillations.

Active damping approach using a control strategy that can be implemented as software is more preferable than passive ones most of which need extra hardware to dampen oscillations. Several methods and configurations to minimize the oscillation have been studied [2-4]. Some of them introduce certain notch filter to attenuate the resonant poles by cancelling the resonant poles of the system transfer function to produce poles that are located in a more well-damped position in the s-plane left part. Such kind of method may be useful and effective, but the drawback of heavy computational burden makes it not easy to implement in real-time applications. The suppression of the oscillations in an EV is achieved by an easy-to-implement active damping torque control strategy in this paper. The origin of the vibration and the design of a damping control method to suppress such oscillations to improve the drivability of an EV are studied. The drive motor torque that is given by the ECU is adjusted according to the acceleration rate of the motor speed to attenuate the resonant frequency [5,6]. The proposed control strategy is validated by simulation and experimental tests. The results clearly show that the proposed approach is easy to implement and effective in suppressing the oscillations in an EV.

## 2. Oscillations in central driven EV

Several factors can contributed to the oscillation of EV, especially noticeable in low speed range. However, the deep reason is that the whole EV drivetrain system is an under-damping system. When the motor speed go across the inherent resonant frequency, oscillation will be inevitable if no damping is added to the system.

It is shown from the Figure 1 that the most sever vehicle oscillation happens when the speed of the drive motor is around 80-100 rpm, and the vehicle resonant frequency is about 5~6Hz. Figure 2 showed the drive motor current waveform during the oscillation

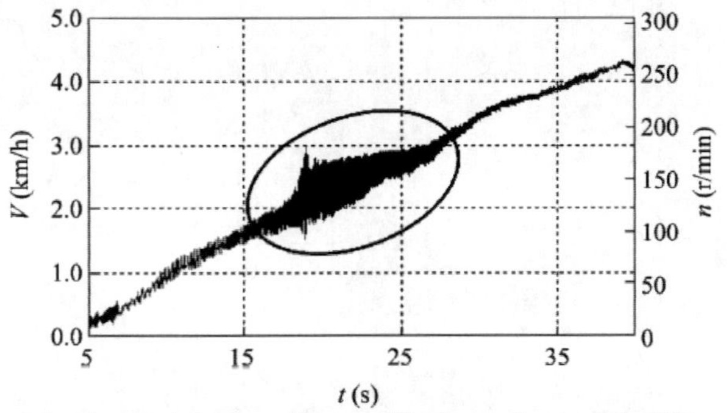


**Figure 1.** Acceleration oscillation of a certain EV.

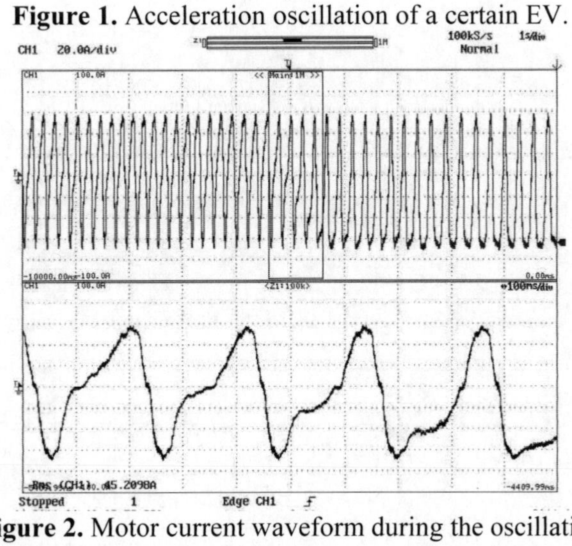


**Figure 2.** Motor current waveform during the oscillation

## 3. Active suppression control strategy and verification

The active-damping-capable control method is to command a traction motor with a provisional torque being adjusted by a proportion-differential control based on the difference between the measured motor speed and the calculated vehicle speed. This difference is proportion to the acceleration rate of the motor speed by further analysis. The calculated vehicle speed is based on an average of wheel speed signals through a low pass filter. Unlike other active damping controllers that use wheel speed input to perform the active damping functionality, only two parameters are needed in this method.

The target torque is generated based on the differential of measured motor speed. The target torque is filtered then to attenuate the drivetrain resonant frequency as discussed before. Actually this method is a proportion plus derivative (PD) control with the input to the controller the calculation of drivetrain oscillation. For the drivetrain system, the proportional control of speed difference will introduce additional damping effect to the whole system.

The feasibility of the above proposed method is verified both by simulation and experiments. Figure 3 showed the simulation model for the proposed damping control method. Figure 4 showed the simulation result of speed acceleration of motor speed without and with vibration damping with the upper plot without damping and lower plot with damping. Figure 5 showed the comparison of experimental result of torque and speed of the drive motor for the effectiveness of the proposed method in mitigating drivetrain oscillations at low vehicle speed with the upper plot without damping and lower plot with damping. Figure 5 showed the comparison of experimental result of vehicle response to throttle tip-in/tip-out without and with vibration damping.

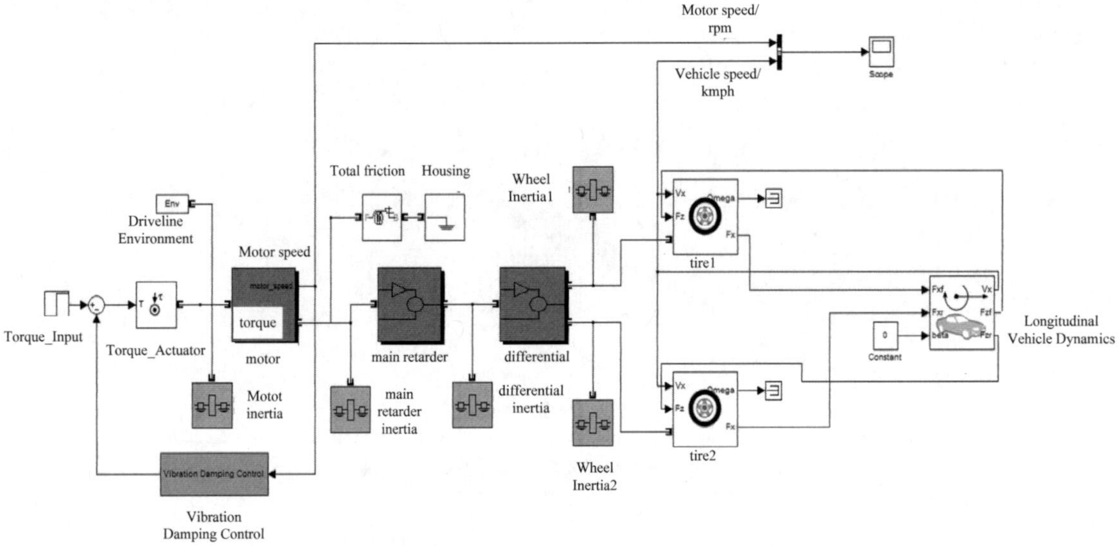


**Figure 3.** Simulation model for the proposed damping control method.

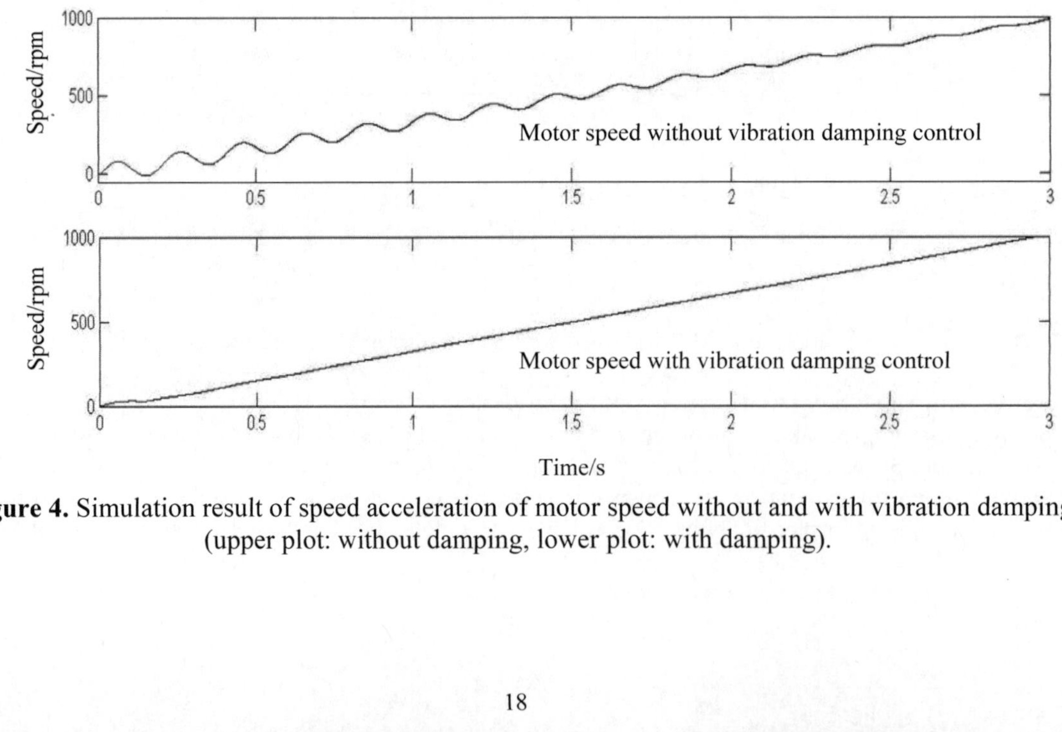


**Figure 4.** Simulation result of speed acceleration of motor speed without and with vibration damping (upper plot: without damping, lower plot: with damping).

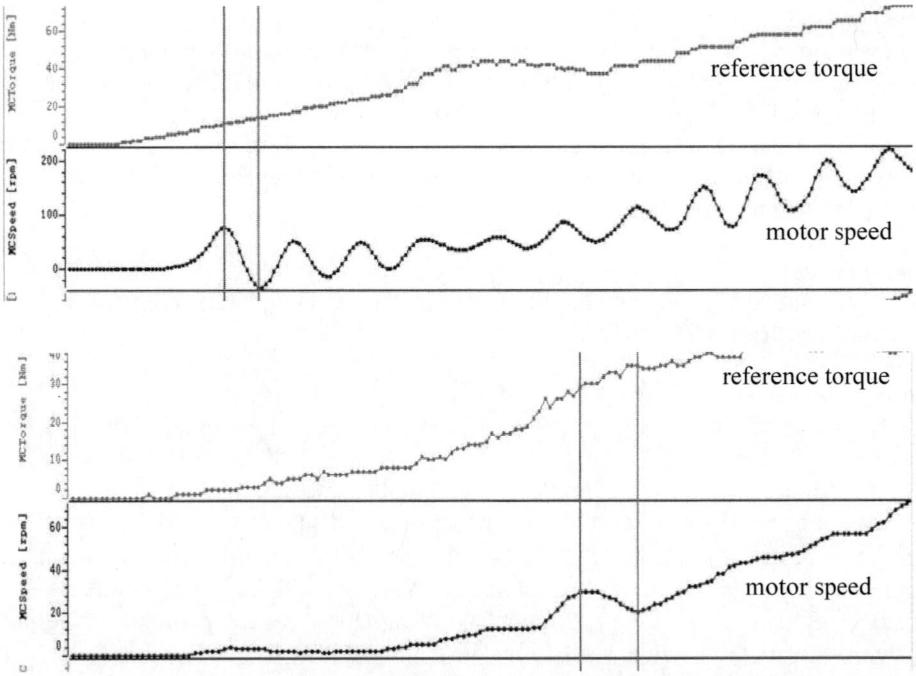


**Figure 5.** Comparison of experimental result of torque and speed of the drive motor for the effectiveness of the proposed method in mitigating drivetrain oscillations at low vehicle speed (upper plot: without damping, lower plot: with damping).

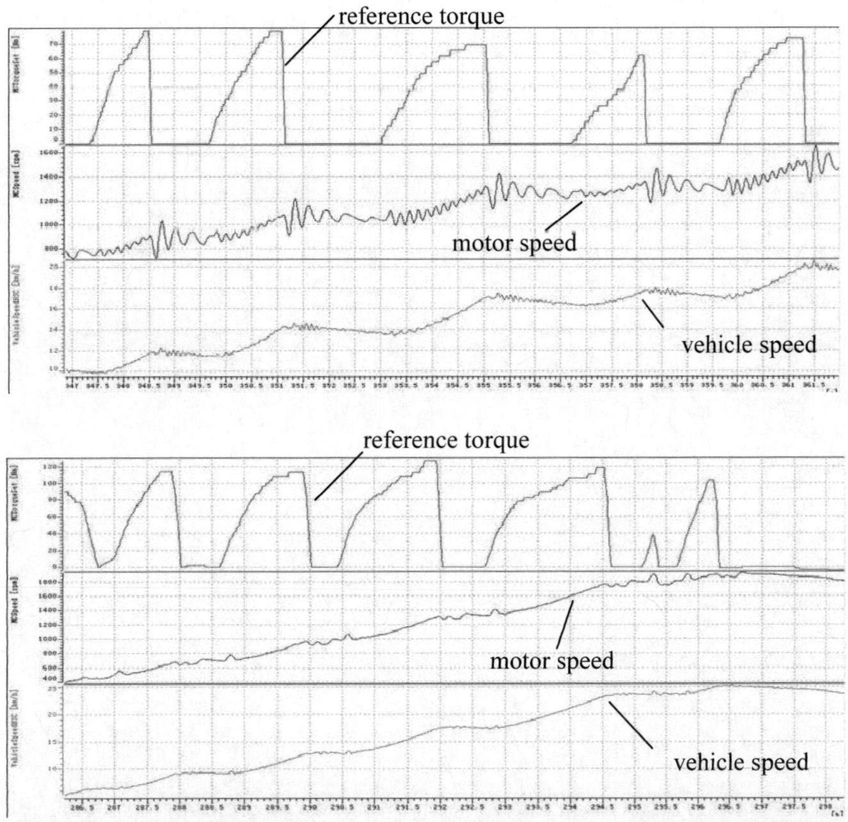


**Figure 6.** Comparison of experimental result of vehicle response to throttle tip-in/tip-out without and with vibration damping (upper: without damping, lower: with damping).

## 4. Conclusion

This paper has proposed a simple yet effective approach to actively damping the oscillation of EV powertrain, which can improve drivability and customer satisfaction.. The acceleration of speed of the drive motor is used to form a PD controller to compensate the damping. This active damping controller was verified by simulation and experimental results. The results showed that the proposed damping control method can improve the performance of the vehicle by reducing the resonance component that inducing driveline oscillations.

**Acknowledgments**
This work was supported in part by the Foundation of Tianjin University of Technology and Education under contract KYQD1624.

## References

[1] C. C. Chan and Y. S. Wong 2004 Electric vehicles charge forward *IEEE Power Energy Mag* **2** 24.
[2] M. A. Valenzuela, J. M. Bentley, A. Villablanca, and R. D. Lorenz 2005 Dynamic compensation of torsional oscillation in paper machine sections *IEEE Trans. Ind. Appl* **41** 1458.
[3] Karikomi T, Itou K，Okubo T 2006 Development of the shaking vibration control for electric vehicles *SICE-ICASE International Joint Conference* 2434.
[4] Fazal U. Syed, Ming L. Kuang, and Hao Ying 2009 Active damping wheel-torque control system to reduce driveline oscillations in a power-split hybrid electric vehicle *IEEE Trans. Vehicular Technology* **58** 4769.
[5] M. Morandin, S. Bolognani, and A. Faggion 2015 Active Torque Damping for an ICE-Based Domestic CHP System With an SPM Machine Drive *IEEE Trans. Ind. Appl* **51** 3137.
[6] Hong Fu, Chao Feng and Shan Xue 2014 Vibration Suppression Control of an Integrated Powertrain of Electric and Hybrid Vehicles Using LQR Controller and Reduced-Order Observer, *ITEC Asia-Pacific* 2014.

IOP Conf. Series: Materials Science and Engineering **224** (2017) 012005 doi:10.1088/1757-899X/224/1/012005

# Design and Analysis of Hydraulic Chassis with Obstacle Avoidance Function

**Yingjie Hong, Xiang Zhang**

Fujian Agriculture and Forestry University, China

1025587638@qq.com

**Abstract.** This article mainly expounds the design of hydraulic system for the hydraulic chassis with obstacle avoidance function. Including the selection of hydraulic motor wheels, hydraulic pump, digital hydraulic cylinder and the matching of engine power. And briefly introduces the principle of obstacle avoidance.

## 1. Introduction

Modern walking machinery mainly adopts four ways of transmission, Including the traditional pure mechanical transmission, electric drive, hydrodynamic drive and Hydraulic transmission. Chassis as the core part of the walking machinery, the selection and combination of the drive is particularly important. Most of the existing chassis can only achieve the vehicle walking function. In actual work, often have to face different road conditions. When some uneven surface road, may encounter some rocks and obstacles, such as nails, easy to cause damage to the tire, and even have significant impact to burst or rollover. So the design of a wheel track adjustable chassis can effectively reduce the incidence of accidents.

The traditional pure mechanical transmission is difficult to realize the driving, steering and expansion function of the wheel at the same time. Hydraulic transmission belongs to the field of fluid power transmission, drive arrangement the advantages of flexible and can be stepless speed regulation[1], Used to the hydraulic cylinder and hydraulic motor can well realize the chassis of the drive and obstacle avoidance, and other functions.

## 2. General structure design of hydraulic drive chassis

Hydraulic drive chassis includes engine, variable hydraulic pump, fuel tank, digital hydraulic cylinder, hydraulic motor, manual valve and other components. Hydraulic system drive as shown in figure 1. Engine connected to the variable hydraulic pump by belt transmission, drive hydraulic pump draws oil from the fuel tank, can convert mechanical energy into pressure. The hydraulic pump provides power for two digital cylinders and the two hydraulic motors by tubing. By controlling three position four-way directional control valve manually walking three working position to realize mechanical forward backward and parking. A sensor is arranged at the front end of the wheel, and the left and right expansion of the digital hydraulic cylinder is adjusted by the pulse signal which is fed back by the sensor, and the left and right movement of the driving wheel is realized. So as to avoid the direct contact between the walking machine and the obstacle, and can effectively reduce the failure rate of the vehicle. At the same time equipped with a relief valve with the function of overload protection and hydraulic auxiliary components such as filter, cooler, guarantee the normal work of the system[2]. Hydraulic system structure diagram as shown in figure 2.

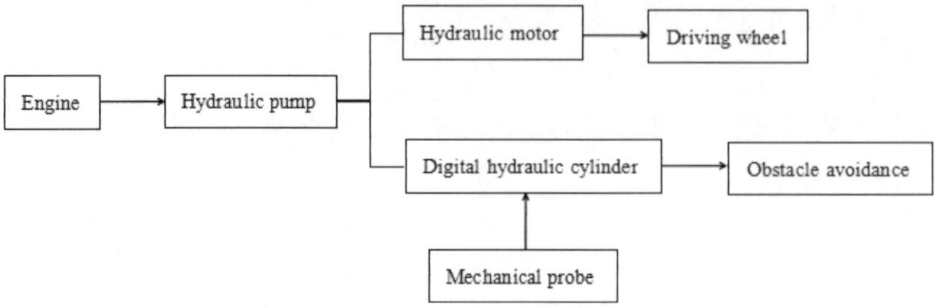


**Figure 1.** Schematic diagram of hydraulic system drive.

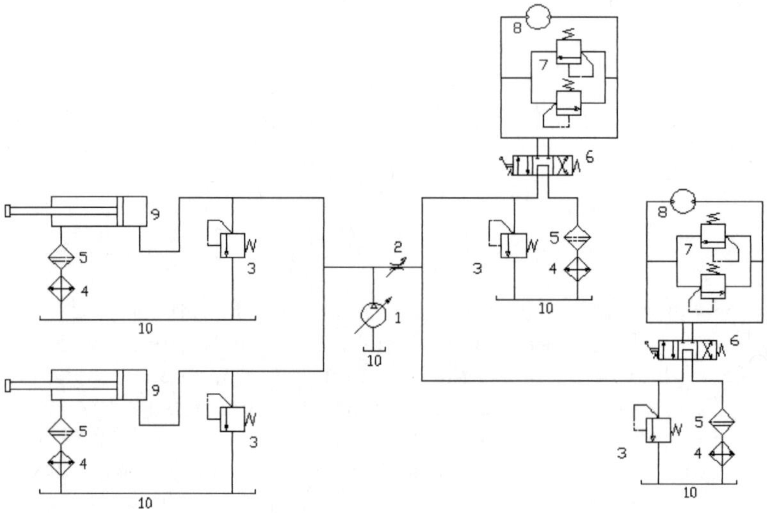

1--Hydraulic pump 2--Throttle valve 3--Safety valve 4--cooler 5--Filter 6--Three position four-way directional control valve 7--Overload valve 8--Wheel hydraulic motor 9--Digital hydraulic cylinder 10--Tank

**Figure 2.** Hydraulic system.

## 3. Main hydraulic components[3]

### *3.1 Hydraulic motor*

Hydraulic motor as the implementation of components, the direct drive wheel rotation, to drive the machinery. Hydraulic motor type and the choice of relevant parameters directly affect the efficiency of the walking machine, The design of hydraulic chassis is mainly guarantee walking machine can adapt to all kinds of road conditions, In areas where the road is more complex, a hydraulic motor is required to provide a greater torque to accommodate the ascent and mud, At the same time, the two driving wheels of the walking machine to achieve obstacle avoidance function, must ensure that its independence, so the two wheels driven by hydraulic motor respectively.

Common low speed high torque hydraulic motor mainly include vane hydraulic motor and radial piston hydraulic motor, the vane hydraulic motor can keep stable in the process of low speed rotation speed, suitable for low voltage system, but the leakage is more serious, is not conducive to the harsh environment conditions. Radial piston hydraulic motor has the advantages of high volumetric efficiency, large starting torque and stronger adaptability[4].

In order to ensure the normal running of the walking machine, it is necessary to meet the requirement of traction. If the maximum climbing angle is α, the maximum tangential traction:

$$F_{max} = (f+\sin\alpha+a)m_t g \quad (1)$$

IOP Conf. Series: Materials Science and Engineering **224** (2017) 012005 doi:10.1088/1757-899X/224/1/012005

In the formula: $F_{max}$ is the most effective traction(N); f is the coefficient of rolling resistance; α is the acceleration resistance conversion factor; $m_t$ is the full load mass of walking machine(kg); g is the gravitational acceleration.

According to the requirement of the maximum traction force of the walking machine, the torque that the motor needs to be supplied is:

$$M_{max}=F_{max}r_1/\eta_{mm} \tag{2}$$

In the formula: r1 is the driving wheel radius of walking machine(m); ηmm is the mechanical efficiency of hydraulic drive motor.

Hydraulic motor displacement:

$$q_m=M_{max}Z_q/(159\Delta p_m i_j Z_L \eta_{mm}\eta_{jt}) \tag{3}$$

In the formula:$Z_q$ is the number of drive wheels per hydraulic motor drive; $\Delta p_m$ is the hydraulic motor inlet and outlet pressure difference(MPa); $i_j$ is the mechanical transmission ratio between the hydraulic motor and the actual wheel driven by motor; $Z_L$ is the number of driving wheel; $\eta_{jt}$ is the total transmission efficiency between the output shaft of the hydraulic motor and the drive wheel.

According to the motor displacement, determine the minimum size of the hydraulic motor to meet the mechanical requirements, Considering the load characteristics of the walking machine, the internal curve hydraulic motor can be selected(Part of the motor parameters such as table 1), facilitate directly installed on the driven wheel. It has the advantages of small size, light weight, reliable operation, large speed range, etc.. It has been widely used in mines, ships, cranes and other large walking equipment, and it has strong adaptability to environment change.

**Table 1.** QJM type hydraulic motor.

| Type | Theoretical displacement | Rated pressure | Peak pressure | Speed range | Rated torque | maximum power |
|---|---|---|---|---|---|---|
| | ml/r | MPa | MPa | r/min | N.m | KW |
| 1QJM02-0.4Z | 0.406 | 10 | 16 | 5-320 | 483 | 13 |
| 1QJM12-1.25Z | 1.33 | 10 | 16 | 4-160 | 1968 | 25 |
| 1QJM21-1.6Z | 1.65 | 10 | 16 | 2-100 | 2442 | 25 |
| 1QJM32-3.2Z | 3.3 | 10 | 16 | 1-125 | 4884 | 62 |
| 1QJM52-6.3Z | 6.36 | 10 | 16 | 1-125 | 9413 | 120 |

### *3.2. Variable hydraulic pump*

Variable hydraulic pump as a power component plays an important role in the hydraulic system. Driven by the engine, drain the oil from the fuel tank, can convert mechanical energy into hydraulic pressure, provide power for hydraulic motors and digital hydraulic cylinders[5]. And variable hydraulic pump can control the output power of the hydraulic motor by adjusting the swashplate opening.

There are three common types of hydraulic pumps, gear, vane and piston type. Gear pump has the characteristics of small size and simple structure. However, due to the imbalance of the pump shaft force, easy to wear. The vane pump is running smoothly, but it is difficult to adapt to the high pressure working conditions. Therefore, it is possible to use the axial piston variable pump with high volumetric efficiency and small leakage to improve the working efficiency of the walking machine.

According to the different condition of speed and total displacement calculation, a set of corresponding maximum main circuit flow value can be obtained:

$$Q_{max}=1000V_{max}\sum q_{md}/(120\pi r_{10}\eta_{vm}\eta_{\delta}) \qquad (4)$$

In the formula: Vmax is the maximum travel speed of walking machine(km/h); ∑qmd is the Total displacement of hydraulic motor(L/r); ηvm is the Volumetric efficiency of hydraulic motor; ηδ is the driving efficiency of driving wheel at maximum speed in no-load condition.

The displacement of the hydraulic pump is estimated based on the maximum displacement of the main circuit:

$$q_{pmax}=1000Q_{max}/(n_{pmax}\eta_{vp}) \qquad (5)$$

In the formula: npmax is the actual maximum speed of hydraulic pump(r/min), The hydraulic pump is usually connected with the engine through a belt drive, and its maximum speed is determined by the rated speed of the engine and the transmission ratio between them; $\eta_{vp}$ is the volumetric efficiency of hydraulic pump.

According to the hydraulic pump displacement, select the appropriate axial piston variable pump(As shown in Table 2). Due to its high pressure, compact structure and convenient flow control, etc, has been widely used in engineering machinery such as need to high pressure, large flow, high power applications.

**Table 2.** Axial piston variable displacement pump.

| Type | Nominal pressure | Nominal displacement | Theoretical flow L/min | | Maximum power | Maximum torque |
|---|---|---|---|---|---|---|
| | MPa | ml/r | 1000r/min | 1500r/min | KW | N.m |
| 2.5YCY14-1B | 31.5 | 2.5 | 2.5 | 3.75 | 1.43 | 17.5 |
| 10YCY14-1B | 31.5 | 10 | 10 | 15 | 5.7 | 54.6 |
| 25YCY14-1B | 31.5 | 25 | 25 | 37.5 | 14.1 | 134.9 |
| 40YCY14-1B | 31.5 | 40 | 40 | 60 | 22.6 | 201.5 |
| 63YCY14-1B | 31.5 | 63 | 63 | 94.5 | 35.6 | 339.9 |

### 3.3 *Digital hydraulic cylinder*

Digital hydraulic cylinder is by controlling the stepper motor, the use of digital pulse signal to complete the length of the vector control. Digital hydraulic cylinder movement characteristics of the fully digital, can be directly by controlling the frequency and number of electrical pulses to achieve control the hydraulic cylinder piston rod movement speed and stroke. It has the advantages of high control precision, simple operation, convenient maintenance, etc.[6].

## 4. Obstacle avoidance principle

The walking machine is driven by hydraulic pressure, and the function of obstacle avoidance is achieved by controlling the expansion of the piston rod of the digital hydraulic cylinder. Respectively, in the front of the two drive wheel to install a mechanical probe, the end of the mechanical probe is equipped with an angle sensor, By measuring the deflection angle of the probe α, and using formula:

$$d=r\sin\alpha \qquad (6)$$

In the formula: r is the probe rod length. The offset distance of the probe rod d is the displacement of the driving wheel. First shift the distance into a digital pulse signal, by controlling the number of electrical pulses to control the number of digital hydraulic cylinder piston rod expansion and contraction, to achieve the hydraulic walking machine obstacle avoidance function. The working principle of the probe is shown in figure 3.

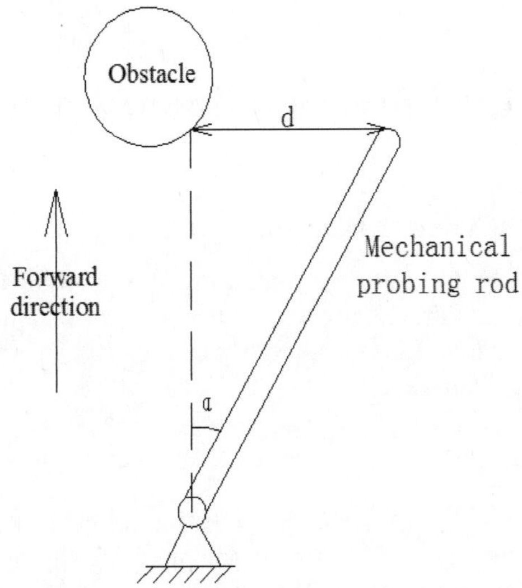


**Figure 3.** Mechanical probing rod.

## 5. Concluding remarks

The application of hydraulic drive to the obstacle avoidance function of walking machinery, you can use its low-speed high torque and high degree of digital, etc., to ensure that walking machinery can adapt to complex and changeable environment and better achieve obstacle avoidance function.

## Acknowledgments

Thanks to professor zhang xiang for his valuable Suggestions on the hydraulic system transmission scheme, and thank professor Lin jian for giving advice in the principle of obstacle avoidance control.

## References

[1] Yao Huaixin. Construction machinery chassis and hydraulic transmission theory of walking machine hydraulic transmission and control [M]. Beijing: China Communications Press, 2002.

[2] Wang Jiwei, Zhang Hongjia, Huang Yi. Hydraulic and pneumatic transmission [M]. Beijing: Mechanical Industry Press, 2005.

[3] Wang Yi. Vehicle and walking machine hydrostatic drive [M]. Beijing: Chemical Industry Press, 2014.5.

[4] Jiang Youshan, Zou Guangde. Study on hydraulic motor selection of full hydraulic bulldozer [J]. construction mechanization, 2009 (09): 45-48.

[5] Li Sheng, et al. Modeling and Simulation of hydraulic drive system for heavy duty truck hub motor [J]. proceedings of the Chinese society of agricultural machinery, 2013.43 (4): 10-14.

[6] Chen Jia, et al. Likun Peng penalty, the transfer function of the digital hydraulic cylinder modeling and analysis of [J]. China based on mechanical engineering, 2014.25 (1): 65-70.

IOP Conf. Series: Materials Science and Engineering **224** (2017) 012006 doi:10.1088/1757-899X/224/1/012006

# Substance-Field Model for Functional Pneumatic Design

**Z G Xu[1,a] , D Y Yang [2] , W D Shen[2] , T T Liu[1]**

[1]Shenzhen Research Institue of Shandong University, 518057, High-tech Zone A301, Virtual University Part, Shenzhen, China.

[2]School of Mechanical Eng, Shandong University, 250061, No 17923, Jingshilu, Lixiaqu, Jinan, Shandong, China

zhgxu@sdu.edu.cn

**Abstract**. The Substance-field analysis is put forward. The functional analysis method is studied to find out the problem, and an improved algorithm is put forward. The internal combustion engine is taken as an example, the harmful function is recognized, and is improved, i.e, high pressure gas is introduced to remove polluted air. The working principle of pneumatic engine is described, the thermodynamic engineering is analyzed, the energy release amounts are analyzed in the isothermal, polymorphism and adiabatic processes. It is concluded that, the isothermal process releases the most energy than the others. The expansion process of the pneumatic engine should be as close as possible to the isothermal process.

## 1. Introduction

With the increasingly serious environmental problems, the transformation and upgrading of China's auto industry is becoming more and more urgent. In recent years, due to the deterioration of the natural environment, especially the impact of fog and haze, new energy vehicles are becoming more and more concerned. Y.Shimomura,1998 [1] suggested a functional body, functional modification and target entity building method, a FBS/M graph is to describe a function-behavior-form model, and to describe the design intent. J.S.Gero,2 002, [2]put forward 8 extended original frame models to rebuild the functional model. S.B.Tor,2002, [3] put forward the B-FES model based on the FEBS, three levels of function, behavior and environment are put forward. University of Washington developed a liquid nitrogen powered vehicle prototype[4]. Their energy source is liquid nitrogen. University of North Texas did the study of liquid nitrogen powered vehicle. A team by Dr.Carfos Ordonez,[5]uses liquid nitrogen as power source, to drive vane type pneumatic motor. Tsu-Chin Tsa [6] from the University of California, Los Angeles, lead a R & D team, funded by Ford Motor Company. developed a hybrid power engine with no cam shaft. In China, Liu Hao of Zhejiang University studied the working principle of a pneumatic engine, The characteristics of energy release of high pressure air under different conditions are analyzed, A number of experiments have been carried out on the pneumatic engine, It provides the theoretical basis and experimental data for the design and manufacture of pneumatic engine

## 2. Heat exchanging for air powered vehicles

In the TRIZ based functional analysis, two kinds of material and a field are used to represent the function. In order to make the functional model more understandable, the field is replaced by "action". The standard function is to determine the designer based on the user's current or potential needs; harmful effects are the adverse effects on the whole system, which should be eliminated as far as possible. Any component in the system has its functional uses, to complete one or more functions.

Through functional analysis, functions of each component are recognized, problems with the elements are recognized, thereby the system is improved. The establishment of functional model is composed of the following two steps: Recognizing component, product, and super-system;

Recognizing the relationship among component, product, and super-system.

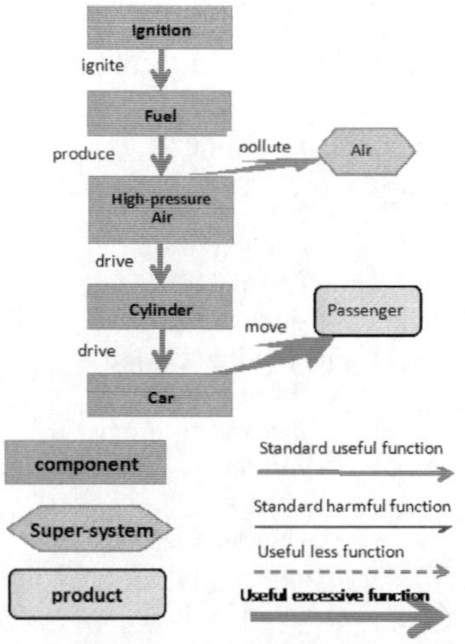


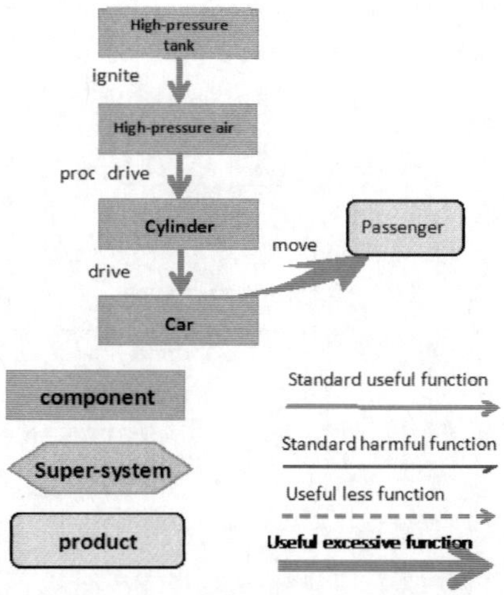


**Figure 1.** Combustion engine function model **Figure 2.** Function model of air powered vehicle

The use of internal combustion engines is the "explosion" effect, high pressure gas generated by explosion of air and fuel to drive cylinder; the use of "magnetic" effect and electromagnetic induction can be made of electric vehicles; the steam engine uses a "thermal expansion" effect. It can be seen that different effects can be used to design different products

The most harmful effect of an internal combustion engine is air pollution, its biggest function is to drive the cylinder an the car. The key word for searching is in a "Verb + noun" format, in the paper, "produce force"" is used. In order to simplify the engine, and do not pollute the air, first order phase transition effect is utilized. When the phase change occurs, there are volume changes at the same time, having the same amount of heat absorption or releasing. Combined with the actual situation in the project, the production of compressed air is utilized, the engine piston is driven by compressed air. Its functional model is shown in Figure2.

## 3. Working principle of air powered engine

The working principle of an air powered engine is similar to that of an internal combustion engine, however, there are differences, the structure is shown in Figure3. When compressed air enters the cylinder, the piston moves back and forth under the pushing force of the compressed air, which drives the crankshaft to rotate continuously, to move the car. High pressure gas in compressed air bottle 1 flows into the mechanical valve 2 through the air inlet of the mechanical reversing valve 2, then enter the first cylinder 3 though the first reversing interface. It pushes the piston 3 of the first cylinder 5 downward, drives the crankshaft 8 to rotate. Crankshaft 8 drives the second cylinder piston upward. Cam 9 reverses valve 2 through the ejector rod 10. The second piston in cylinder discharge the gas in into the atmosphere, through the second reversing interface of the mechanical reversing valve 2. When the first cylinder piston moves down to the lower stop point, the second cylinder piston moves up to the top stop point, cam 9 reverses the mechanical reversing valve 2 through the ejector rod 10. High pressure gas in compressed air bottle 1 flows into the second cylinder 4, through the mechanical valve 2, pushes the second cylinder piston downwards, to rotate the crankshaft 8. Crankshaft 8 drives the first cylinder piston upwards, Cam 9 Make mechanical reversing valve 2 change direction, through the ejector rod 10.

The first cylinder piston discharges the gas in the first cylinder 3 into the atmosphere, through the mechanical valve 2. A work cycle finished. In order to make the engine more stable, four or six cylinders could be set, which can be in a straight line, "V" or "asterisk" shaped arrangement. The air powered engine has the following two strokes. Intake stroke, in this process, high pressure gas enters the cylinder from the gas tank, expansion works, which is equivalent to the internal combustion engine intake stroke and power stroke; Exhaust stroke, in this process, low pressure gas is discharged from cylinder , which is equivalent to the compression stroke and exhaust stroke of an internal combustion engine.

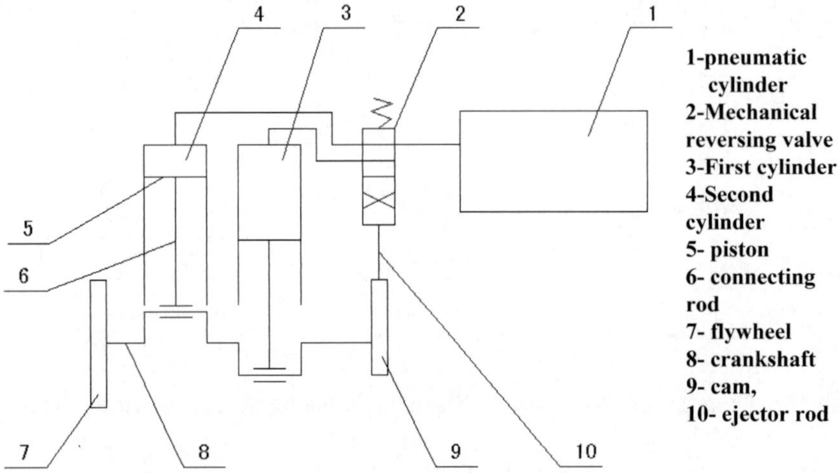


**Figure 3.** Schematic diagram of 4- air powered engine

The pressure/ volume change of the gas in the cylinder is shown in Figure 4. During the intake stroke, high pressure gas first enters the cylinder with pressure P1, push the piston down, performs works. When the piston moves to position 3, the mechanical reversing valve stops the inlet air. The cylinder becomes a closed system, the high-pressure gas inside the cylinder expands, pressure energy releases, and pushes the piston downward to continue to work outside.

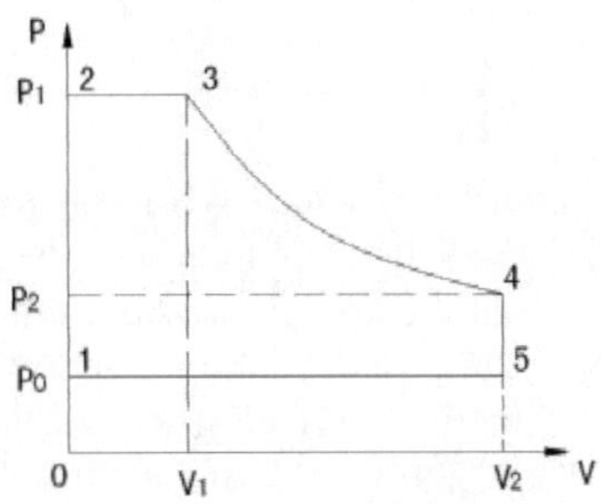


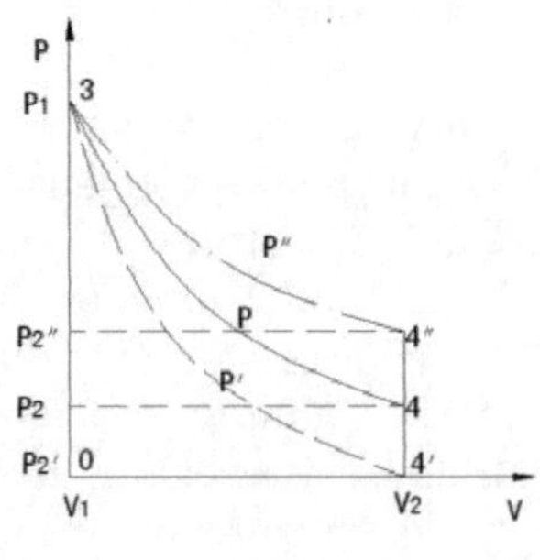


**Figure 4.** gas expansion process **Figure 5.** P- V curve of gas in cylinder

When the piston reaches the bottom dead point, i.e. point4 in the picture, the mechanical reversing valve controls the cylinder to exhaust, reduced the inner pressure to atmospheric pressure P0, in the mean while, the piston moves upward to discharge low-pressure gas in cylinder. The enclosed area by 1-2-3-4-5-1 in the chart is the theoretical work of high pressure gas W0 , during intake and exhaust, pneumatic engine exchange the refrigerant with the outside world. In an ideal state, gas pressure has no change, the output function by the cylinder equals to gas technical work Ws[2]; i.e.

$$W_s = P_1V_1 - P_0V_2 \tag{1}$$

When the cylinder is a closed system, i.e. the mechanical reversing valve closed completely. Gas expansion performs the external work. The Cylinder output function is the expansion function Wv:

$$W_v = \int_{V_1}^{V_2} pdv \tag{2}$$

Power generated by the compressed air W equals the sum of the technical power and the expansion power.

$$W = W_s + W_v \tag{3}$$

## Thermodynamic analysis of air powered engine

When the gas varies in different conditions, the thermodynamic processes are different. Under the same pressure and expansion ratio, different thermodynamic processes affect the energy efficiency of an air powered engine. In the process of high pressure, gas expanded from V1 to V2, the thermodynamic processes are different, the expansion work of the high-pressure gas is also different. The P-V curves of the high-pressure gas through different thermodynamic processes are shown in Figure 5. In this paper, three cases are discussed in the thermodynamic process of high pressure gas expansion.

$$\frac{P'}{P_1} = (\frac{V_1}{V})^k \tag{4}$$

1. Heat insulation process, as shown in the Figure5, curve 3-4 '. In this process, gas does not absorb ambient heat, gas pressure decreased from P1 to P2', the gas pressure P′ decreases rapidly with the change of volume, therefore, the external output expansion power of is less, equals to the area of 0-3-4 (-0) of the dotted line and the axis. In this process, the relationship between the gas pressure and volume is as the follows. Where, K is the molar heat capacity ratio, the formula (4) is added to the formula (2) to obtain the expansion work of the adiabatic process:

$$W_v = (\frac{P_1 V_1}{k-1})[1 - (\frac{V_1}{V_2})^{k-1}] \tag{5}$$

2. Isothermal process, as shown in figure 3-4. In this process, high pressure gas absorbs ambient heat, gas pressure decreased from P1 to P2″, the gas pressure P″ decreases gradually with the volume change, therefore, the expansion of external output power is larger, i.e. the area of 0-3-4 '-4' -0, the circumference of the dotted line and the axis. In this process, the equation of the gas pressure and volume are as follows

$$\frac{P''}{P_1} = \frac{V_1}{V} \tag{6}$$

The isothermal output power can be obtained by putting formula (4-6) into the formula (7).

$$W_v = P_1 V_1 \ln(\frac{V_1}{V_2}) \tag{7}$$

3. Polymorphic process, in the actual operation of the pneumatic engine, affected by various external conditions, the heat absorbed from the environment is between adiabatic and isothermal processes. Gas pressure decreased from P1 to P2 , as shown in Figure 3. At this time the output power of mechanical work is also between the above two conditions. i.e. the area of 0-3-4-4 '-0, the fine solid line and axis. In this process, the equation between the gas pressure and volume is as follows.

$$\frac{P}{P_1} = (\frac{V_1}{V})^n \tag{8}$$

Where, n is the polymorphism index, putting the formula (8) into the formula (8), the output expansion power for polymorphic process is obtained.

IOP Conf. Series: Materials Science and Engineering **224** (2017) 012006 doi:10.1088/1757-899X/224/1/012006

$$W_v = (\frac{P_1V_1}{n-1})[1-(\frac{V_1}{V_2})^{n-1}] \tag{9}$$

In order to compare the differences of energy output in different thermodynamic processes, the ratio of the output to the mechanical work of isothermal and adiabatic processes is set to Zw1. So,

$$Z_{w_1} = \frac{(k-1)\ln(\frac{V_1}{V_2})}{1-(\frac{V_1}{V_2})^{k-1}} \tag{10}$$

Zw2 is set as the ratio of the output power of the polymorphic process and adiabatic process. So,

$$Z_{w_2} = \frac{(k-1)(1-(\frac{V_1}{V_2})^{n-1})}{(n-1)(1-(\frac{V_1}{V_2})^{k-1})} \tag{11}$$

In the normal working of the engine, the total cylinder volume and the air intake are constant. In the process of gas expansion, the polymorphism index n, changes with the condition of gas and external environment. Because the volume expansion ratio is a fixed value, the formula is simplified as the follows,

$$e = \frac{V_2}{V_1} \tag{12}$$

So, the calculation of Zw1, Zw2 are as the follows.

$$Z_{w_1} = \frac{(k-1)\ln(e)}{1-e^{1(k-1)}} \tag{13}$$

In the above three thermodynamic processes, due to the absorption of energy from the outside world, therefore, the expansion process releases the most energy.

$$Z_{w_2} = \frac{(k-1)(1-e^{-(n-1)})}{(n-1)(1-e^{-(k-1)})} \tag{14}$$

The following table lists the energy released by gas at different expansion ratios. From table1, it could be found that, increase the heat exchange between the pneumatic engine and the outside air, make the expansion of the high pressure gas in cylinder approximately equals to the isothermal process.

**Table1** Energy releases in the isothermal, polymorphic and adiabatic processes

| Expansion ratio | 10 | 20 | 30 |
|---|---|---|---|
| When the polytropic index is 1.2, the surplus expansion power issued by the polymorphic process than that of the adiabatic processes | 22.63% | 29.09% | 32.76% |
| the surplus expansion power issued by the isothermal process than that of the adiabatic processes | 53.02% | 71.6% | 82.99% |

## 4. Conclusions

When solving the practical problems using sub-F model, first of all, it is necessary to analyze the improved technology system, find out the elements and their relationships, then the sub-F model is established. The transformed system consists of 3 basic elements and two material and field elements, according to the completeness of the three elements, four categories are classified, i.e. ideal sub-F model, incomplete sub-F model, sub-F model with insufficient effect and the harmful sub-F model. In this paper, the function model of the internal combustion engine is established. The harmful function is recognized, and the cutting method is used to improve, i.e. high pressure gas with pollution are removed. Use high pressure air as an energy source to drive a car, the working principle of pneumatic engine is described, the thermodynamic engineering properties are analyzed, energy release under different conditions are compared, i.e. isothermal process, polymorphic process, and adiabatic process. It could be concluded that, the isothermal process releases the most energy, the expansion process of the pneumatic engine should be as close as possible to the isothermal process

**Acknowledgment**
The above-mentioned research work is supported by the Science, Technology and Innovation Commission of Shenzhen Municipality, JCYJ20160510165328965 and the Chinese NSFC, 61272017, P.R. China.

**References**
[1] Wu Hongchao, Hu h L. research on the model of product configuration unit based on, Computer engineering and design, 2009, (CBR): 5482-5484+5488.
[2] Xie Qing, Tan Jianrong, Feng Yixiong. Research on configurable product structure mapping based on automata. computer integrated manufacturing system, 2007, (): 1722-1731.
[3] Liu Xijuan, Jiang Shouwei. Design of component library based on component reuse. machine design and manufacture, 2007, (): 57-59.
[4] Jiang Xianfeng, Yang Rong, Zhang Gong, he Yue. CAD in the journal, mapping study of computer aided design and computer graphics 1997, (05): 59-66.
[5] Tai, Chun, Tsao, Tsu-Chin, M. B. Levin, G. Barta,and M.M.Schechter,"Using Camless Valvetrain for Air Hybrid Optimization,"SAE Paper 2003-01-0038 presented in SAE World Congress, March 2003, Detroit, MI. Also in SAE World Congress 2003 Special Publications (SP-1752), ISBN: 0-7680-1175-2.
[6] Liu Hao. Air power engine research, doctoral dissertation of Zhejiang University, 2004:23-24.

IOP Conf. Series: Materials Science and Engineering **224** (2017) 012007 doi:10.1088/1757-899X/224/1/012007

# An Improved Model for Air Damping of Perforated Structures

**Cunhao Lu[1] and Pu Li[2 a]**

[12]School of Mechanical Engineering, Southeast University, China

[a]seulp@seu.edu.cn

**Abstract:** the prediction of air damping of micromachined mechanical resonant structures is significant in the design of high quality factor devices. In rarefied air, based on Bao's molecule model, Li gives an analytical model for air damping of perforated structures. By studying the action of molecules going through holes and reflected by the fixed plate, this paper gives a probability of molecules through holes going into the gap between the moving plate and the fixed one. Comparison with Li's model, the new model can play a better performance of air damping for perforated structures, at a wide range of size of holes.

## 1. Introduction

Micromachined mechanical resonant structures have been applied extensively due to their high sensitivity and low fabrication cost[1]. A common approach to evaluate resonant structures is to measure the quality factor of the structure defined as the ratio of total input energy to dissipated energy. However, even in rarefied air, the squeeze film damping cannot be neglected[2]. The character of evaluation of air rarefaction effect is *Kn* number which is the ratio of the molecule mean free path and the characteristic length of the structures. According to the value of *Kn*, the different air regimes can be briefly separated into 4 kinds[3]: continuum regime $Kn \leq 10^{-3}$; Slip regime $10^{-3} \leq Kn \leq 10^{-1}$; transition regime $10^{-1} \leq Kn \leq 10$; free molecule regime $Kn > 10$.

For the prediction of the quality factor of air damping in free molecule regime, owing to several reasonable assumptions, Bao *et al*[4] calculate the change of kinetic energy of molecules in the gap and give a simple but widely used expression of quality factor:

$$Q_{Bao} = (2\pi)^{\frac{3}{2}} \rho H \omega (\frac{d}{L})(\frac{RT}{M_m})^{\frac{1}{2}} \frac{1}{P} \tag{1}$$

where $\rho$, $H$, $L$ is the specific mass ,thickness and peripheral length of the moving plate, $\omega$ is the vibrating frequency, $d$ is the initial gap between the moving plate and the fixed one, $R$ the universal gas constant, $T$ is the absolute temperature, $M_m$ and $P$ are the molar weight and pressure. Based on Bao's model, Li *et al*[5]devised a modification for perforated structures, the total quality factor expression for perforated structures is presented:

$$Q_{Li} = r[f(c_1) + f(c_2) N_x N_y L_2 (2L)^{-1}]^{-1} Q_{Bao} \tag{2}$$

where $L_2$ is the peripheral length of the square hole, $N_{x,}$ $N_y$ is the number of holes along $x$ direction and $y$ direction. However, a limitation exists in Li's model that the model is valid only for $g>1$ or more, where $g$ is ratio of $l_{hole}$ and $d$, $l_{hole}$ is the size of the square hole.

This paper is an extension of Li's model[5] for air damping of perforated structures. In section 2, the brief description of Li's model is given. In section 3, a detailed description of molecules in gap between plates, and the probability of molecules through holes into the gap are found. In section 4, the improved model is found and the validation is proved compared with Li's model.

## 2. A description of Li's model

According to Li's model, there are two parts of molecules going into the gap. The first part is the molecules going through four side edges of the moving plate into the gap, $N_1$ is the molecule number; the second part is the molecules entering the gap through the holes, $N_2$ is the molecule number, so the total number of molecules entering the gap is $N=N_1+N_2$.

To the first part of molecules going into the gap, $N_1 = 1/4nvLd$, where $n$ is the concentration of the molecules and $v=[8kT/(\pi M_m)]^{1/2}$ is the average velocity of the molecules, $k$ is the Boltzmann constant. The quality factor of the first part of molecules is given:

$$Q_1 = 2\pi \frac{E_m}{\Delta E_1^{cycle}} = \frac{r}{f(c_1)} Q_{Bao} \tag{3}$$

$$f(c_1) = \frac{1}{c_1^2}\left[\frac{2[(1-a_1)-(1-a_1)^{c_1}]}{a_1^2} - \frac{2(c_1-1)(1-a_1)^{c_1}}{a_1} - \frac{[(1-a_1)-(1-a_1)^{c_1}]}{a_1} + (2c_1-1)(1-a_1)^{c_1}\right]$$

where $E_m = 1/2M_{plate}(Aw)^2$ is the total mechanical energy of the moving plate , $M_{plate} = r\rho L_x L_y H$, $L_x$, $L_y$ and $H$ are the length, width and thickness of the plate, $r = (L_x L_y - N_x N_y A_{hole})/(L_x L_y)$, $a_1=1-r$, $A_{hole} = l_{hole}^2$, $l_{hole}$ is the size of the hole, $c_1 = [L_x L_y/(4\pi d^2)]^{1/2}$ is the average number of collisions with the moving plate for the first part molecules leaving from the four edges of the plate.

To the second part of molecules entering the gap from the holes, Li [5] treated $N_2=(1/4nvL_2d)N_xN_y$, $L_2=4l_h$, The quality factor of the second part of molecules going through the holes is found:

$$Q_2 = 2\pi \frac{E_m}{\Delta E_2^{cycle}} = \frac{r}{f(c_2)} \frac{2L}{N_x N_y L_2} Q_{Bao} \tag{4}$$

$$f(c_2) = c_2^{-2}[2[(1-a_1)-(1-a_1)^{c_2}]a_1^{-2} - 2(c_2-1)(1-a_1)^{c_2}a_1^{-1} - [(1-a_1)-(1-a_1)^{c_2}]a_1^{-1} + (2c_2-1)(1-a_1)^{c_2}]$$

where $c_2 = \sqrt{L_x L_y/(8\pi d^2)}$ is the average number of collisions with the moving plate for the second part molecules and leaving through four edges.

After above calculations, the total quality factor of Li's model[5] is found:

$$Q_{Li} = (Q_1^{-1} + Q_2^{-1})^{-1} = r[f(c_1) + f(c_2)N_x N_y L_2 (2L)^{-1}]^{-1} Q_{Bao} \tag{5}$$

However, there is a limitation in Li's model, that the model is reasonable only for a high value of $g$. For a low value of $g$, equation (4) overestimates the number of molecules that enter the gap from the holes. A discussion in detail will be given in section 5.

## 3. An improved model

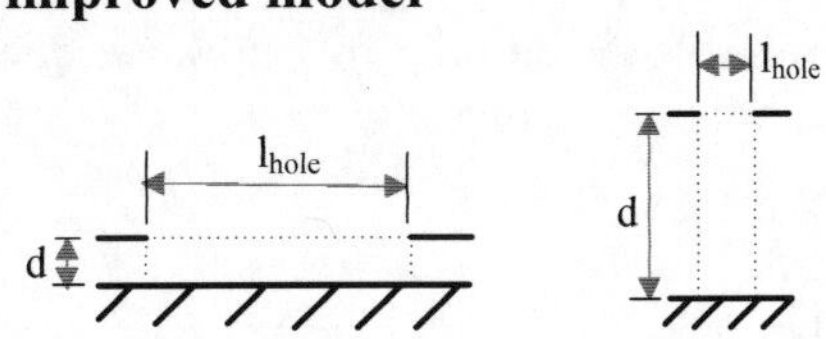


(a) hole at large $g$; (b) hole at small $g$

**Figure 1**.Two holes when $g$ is large or small

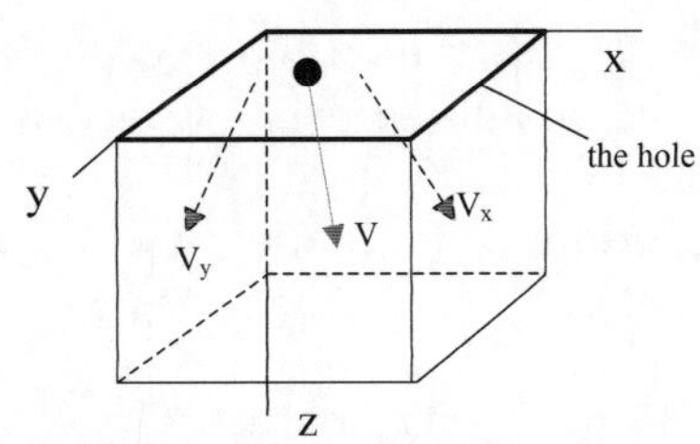


**Figure 2.** the motion of a molecule in a hole

Based on the above analysis, it is critical to find the exact number of molecules entering the gap through holes. When molecules enter the hole of structures, some of them can go straight into the gap; others collide with the fixed plate and go back to the space through the hole, only the molecules enter the gap that could count for the energy change with the moving plate. This paper gives two holes that $g$ is extremely large and small in figure 1.

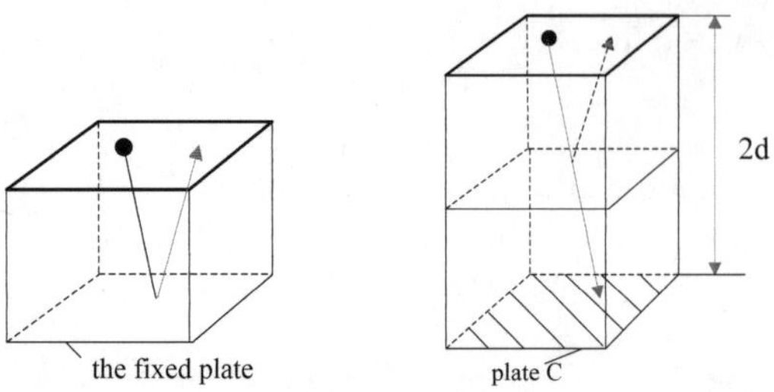


(a) the reflection motion (b) the movement of reaching plate C

**Figure 3.**Two equal models

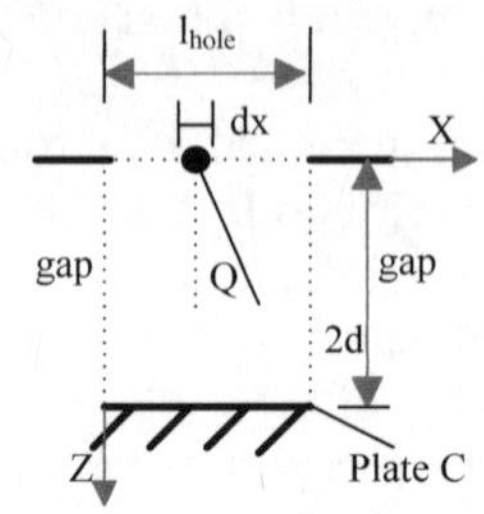


**Figure 4.** x-z plane of a single hole

The total number of molecules entering a hole is $N_{hole}=1/4nvA_{hole}$. When $g$ is extremely large in figure 1(a), molecules entering the hole can easily be reflected through the hole by the fixed plate. So the exact number of molecules that can go into the gap is $1/4nvL_2d$. Use the probability $P$ to estimate molecules go into the gap from the hole, $P_{L-g}\approx(1/4nvL_2d)/N_{hole}=4/g$, the subscript $L$-$g$ means $P$ at large $g$. When $g$ is extremely small in figure 1(b), molecules entering the hole are hardly reflected by the fixed plate and back into the space. As a result, most of molecules from the hole can easily reach the gap and $P_{s-g}\approx(1/4nvA_{hole})/N_{hole}=1$. According to the analysis above, using $P_{L-g}$ or $P_{S-g}$ to calculate the number of molecules that enters the gap from holes isn't reasonable theoretically. To get an exact value of $P$, a particular motion of a molecule in a hole is shown in figure 2. When a molecule goes through the hole, its velocity $V$ can be divided into two parts: $V_x$ in x-z plane and $V_y$ in y-z plane. The location of a molecule and velocity $V_xV_y$ decide whether the molecule goes into the gap or not. The depth of the hole in figure 3(b) is $2d$. When a molecule goes back to the space through the hole after the collision with the fixed plate in figure 3(a), it is equal that a molecule can reach plate C from the hole. The probability of molecules go into the gap from the hole is equal to that of molecules which cannot reach plate C from the hole.

Considering $x$-$z$ plane in figure 4, the velocity direction and locations of molecules are both uniformly distributed. To a special location $x$, the injection angle of a molecule is $Q$, and the small interval of length along x direction is $dx$. the probability of molecules from $dx$ reaching C plate is:

$$P_x=\frac{1}{\pi}[\arctan(\frac{x}{2d})+\arctan(\frac{l_{hole}-x}{2d})] \tag{6}$$

$$P_{xzplane}=\frac{1}{l_{hole}}\int_0^{l_{hole}}P_xdx=\frac{8d}{\pi a}[\frac{l_{hole}}{4d}\arctan(\frac{l_{hole}}{2d})+\frac{1}{2}\ln(\cos(\arctan(\frac{l_{hole}}{2d})))] \tag{7}$$

$$P=1-(P_{xzplane})^2=1-[\frac{4}{\pi g}[\frac{g}{2}\arctan(\frac{g}{2})+\ln(\cos(\arctan(\frac{g}{2})))]]^2 \tag{8}$$

where $P_{\text{xzplane}}$ is the probability of molecules reaching C plate from the hole in $x$-$z$ plane, due to the square hole, $P$ is the exact probability of molecules from the hole going to the gap.

## 4. Validation of the model of P

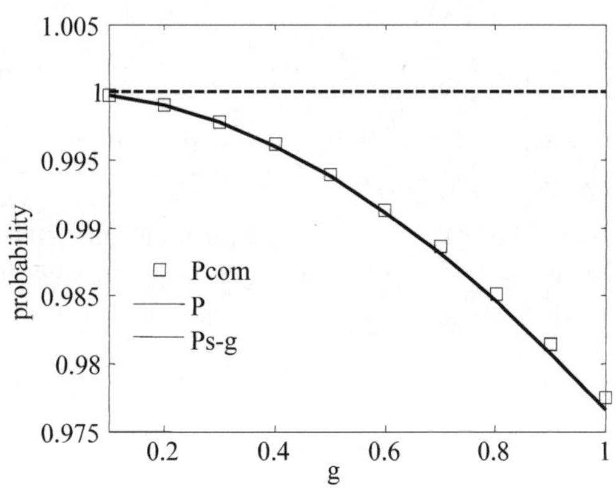


(a)

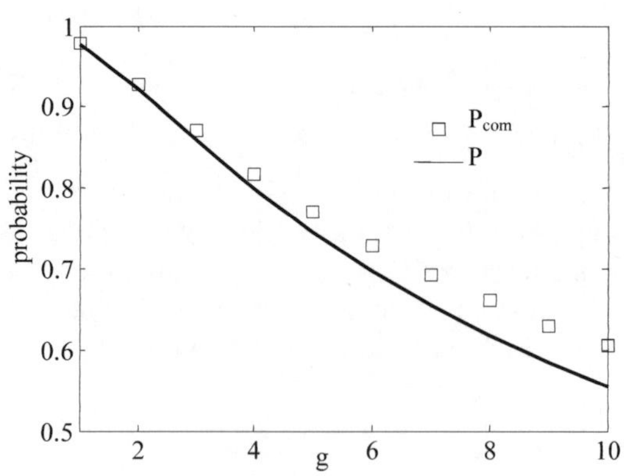


(b)

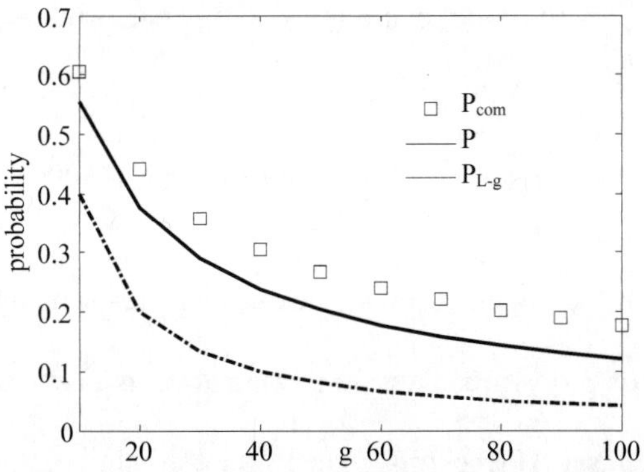


(c)

**Figure 5.** $P$ of different models at $g$ is 0.1-100

To validate the probability of molecules reflected through a hole, $P_{com}$ calculated by computer is used, then $P_{L\text{-}g}=4/g$、 $P_{S\text{-}g}=1$ 、$P$ and $P_{com}$ are discussed in figure 5. In figure 5(a), both values of $P_{com}$ and $P$ decrease as $g$ decreases. During the whole range of small $g$, values of $P$ can match those of $P_{com}$ very well with the largest description is less than 0.1%. Figure 5(b) shows that both values of $P_{com}$ and $P$ decrease as $g$ decreases. Although the description of $P_{com}$ and $P$ is growing slowly as g is growing, the largest description is less than 9%. In figure 5(c), when g is extremely large, the values of $P$ are still close to those of $P_{com}$. As the above analysis, all the values of $P$ can match $P_{com}$ well.

Combing equation(8) and equation(5), a new expression of quality factor of a rigid rectangular structures perforated with square holes is obtained:

$$Q_N=(Q_1^{-1}+Q_2^{-1})^{-1}=r[f(c_1)+f(c_2)N_xN_yA_{hole}(1-P_N^2)(2Ld)^{-1}]^{-1}Q_{Bao} \quad (9)$$

## 5. Validation of the improved model

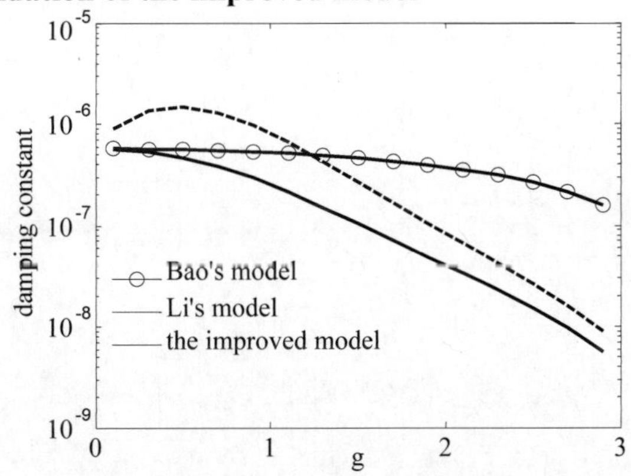


**Figure 6.**Damping constants of different models at $g$=0~3

To check the accuracy of the improved model, we use the improved model to predict the air damping constant of Kwork's gyroscopies[6]. Obviously, Bao's model[4] can only give the prediction of air damping of non-perforated structure. It is well known that with the same overall sizes, the air damping of structures perforated with holes is much small than that of non-perforated structures. The comparison of these models at different values of $g$ is shown in figure 6. At $g$=0-1.25, while Li's model[5] fails to give an acceptable prediction of air damping, the improved model can work well. The biggest description of Li's model and the improved model that occurs at $g$=0.9 is 7%, as $g$ is growing; the description becomes small and small. Consequently, the improved model can remove the limitation of Li's model, and it can give a better prediction of air damping of perforated structures.

## 6. Summary and conclusions

For perforated structures in rarefied air, Li's model[5] is valid only for large values of $g$. By researching on molecules going through holes, a better prediction model of the probability molecules entering the gap though the hole is found. Compared with the computer numerical simulation, the

IOP Conf. Series: Materials Science and Engineering **224** (2017) 012007 doi:10.1088/1757-899X/224/1/012007

improved model is validated at a wide range of *g*. The improved model can make a better performance for predicting the air damping of perforated structures than Li's model.

**Acknowledgments**
This work was supported by the National Natural Science Foundation of China (Grant No 51375091).

**References**

[1] Melvas P, Kalvesten E, Stemme G A. surface- micromachined resonant-beam pressure- sensing structure *J. Micromech. Syst.* **10** 498-502 (2001)
[2] Zook J D, Burns D W, Guckel H. Characteristics of plolysilicon resonant microbeams *Sensors Actuators A* **35** 51-9(1992)
[3] Schaaf S A, Chambre P L. Flow of Rarefied Gases (Princeton, NJ: Princeton University press)( 1961)
[4] Minhang Bao, Heng Yang, Hao Yin. Sun Energy transfer model for squeeze-film air damping in low vacuum *J. Micromeh. Microeng* **12** 341-346(2002)
[5] Pu Li, Rufu Hu. A model for squeeze-film damping of perforated MEMS devices in the free molecular regime. *J. Micromeh. Microeng* **21** 0256006(10pp) ( 2011)
[6] Younis M I, Nayfeh A H. Simulation of squeeze-film damping of microplates actuated by large electrostatic load *ASME J. Comput. Nonlinear Dyn.* **2** 232-41(2007)

IOP Conf. Series: Materials Science and Engineering **224** (2017) 012008 doi:10.1088/1757-899X/224/1/012008

# A Squeeze-film Damping Model for the Circular Torsion Micro-resonators

**Fan Yang**[1], **Pu Li**[2,a]

[1,2] School of Mechanical Engineering, Southeast University, Jiangning, NanJing211189, People's Republic of China

[a] seulp@seu.edu.cn

**Abstract.** In recent years, MEMS devices are widely used in many industries. The prediction of squeeze-film damping is very important for the research of high quality factor resonators. In the past, there have been many analytical models predicting the squeeze-film damping of the torsion micro-resonators. However, for the circular torsion micro-plate, the works over it is very rare. The only model presented by Xia et al[7] using the method of eigenfunction expansions. In this paper, The Bessel series solution is used to solve the Reynolds equation under the assumption of the incompressible gas of the gap, the pressure distribution of the gas between two micro-plates is obtained. Then the analytical expression for the damping constant of the device is derived. The result of the present model matches very well with the finite element method (FEM) solutions and the result of Xia's model, so the present models' accuracy is able to be validated.

## 1. Introduction

Squeeze-film damping, which is a main energy dissipation mechanism of MEMS resonators, plays an important part in the device's dynamic characteristics. There have been many studies over the torsion micro-mirrors in the past few years. The Reynolds equation is a common tool to calculate squeeze-film damping. In early year of 1962, Langlois[1] proposed a general equation for the compressible flow gas film, this equation is the traditional Reynolds equation. Pan et al[2] firstly linearized the nonlinear Reynolds equation under the assumption of small displacement between the fixed substrate and the movable mirror. By means of the solutions of Fourier series, they calculated the expressions of the damping pressure distribution by solving a simplified equation. Darling et al[3] derived the same expression based on the approach of Green's Function. Pandey et al[4-6] derived a modified Reynolds equation for perforated system. However, most of the previous works focused on the performance of the model of rectangular micro-plate, there are few models for the research of the circular micro-plate of a torsion micro-mirror, the one and only model proposed by Xia et al[7]. The modified Reynolds equation proposed by Bao[8] under the assumption of the incompressible gas is considered in this paper, and then extend this equation to the form of cylindrical coordinate. The analytical expression of the damping constant is developed using the Bessel series. Then we verified the accuracy of the model by comparing the present model with the FEM models as well as the result of Xia's model.

## 2. Problem formulation

The schematic view of the circular micro-plate is shown in Fig.1. $R$ is the radius of the micro-plate, $g_0$ is the thickness of the air gap. Under the incompressible gas condition, the governing equation proposed by Bao et al[8] is

$$\frac{\partial^2 p}{\partial x^2}+\frac{\partial^2 p}{\partial y^2}=\frac{12\mu}{g_0^{\ 3}}\frac{\mathrm{d}h}{\mathrm{d}t} \tag{1}$$

where $p(x,y,t)$ is the air pressure of squeeze-film gap, $p(x,y,t)=p_a+\Delta p(x,y,t)$ , $p_a$ is the ambient pressure, $\mu$ is the standard viscosity, $h(t)$ is the gap thickness, the expression of $h(t)$ is, $h(t)=g_0+x\theta_0 e^{\omega t}$ ,where $\theta_0 \mathrm{e}^{\omega t}$ is the angle of the torsion micro-plate.

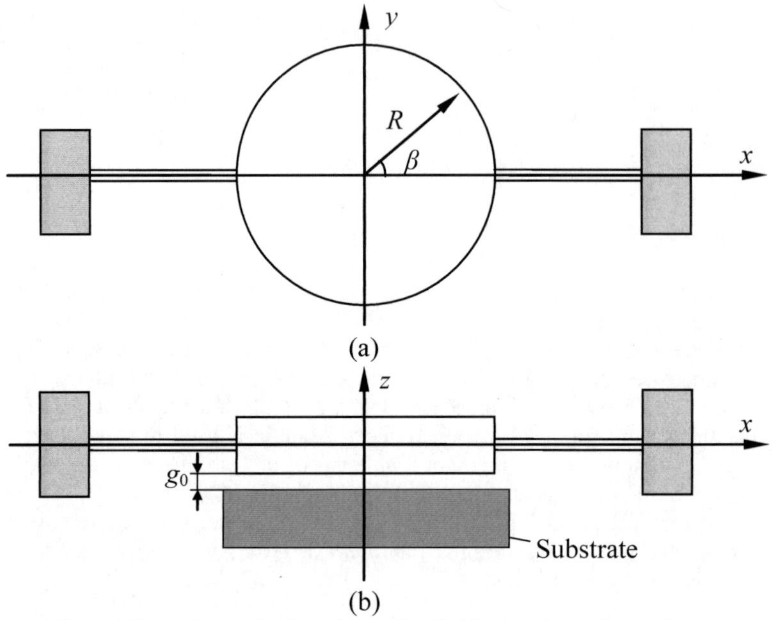


**Fig. 1** The schematic drawing of a rigid circular micro-plate.
(a)Top view of the circular micro-plate. (b) Side view of the circular micro-plate.

The following nondimensional variable is introduced for convenience:

$$P(x,y,t)=\frac{\Delta p(x,y,t)}{p_a} \tag{2}$$

Substituting Eq.(2) into Eq.(1), then change the cartesian coordinate to cylindrical coordinate, leads to

$$\frac{\partial^2 P}{\partial r^2}+\frac{1}{r}\frac{\partial P}{\partial r}+\frac{1}{r^2}\frac{\partial^2 P}{\partial \beta^2}=\frac{12\mu\theta_0}{h_0^{\ 3}p_a}\omega\cdot r\sin\beta\cdot e^{\omega t} \tag{3}$$

The boundary conditions is

$$P(R,\beta,t)=0 \tag{4}$$

The expression of $P(R,\beta,t)$ is able be expressed as a Bessel series

$$P(r,\beta,t)=\sum_{k=0}^{\infty}\sum_{q=1}^{\infty}J_k\left(\frac{a_{kq}}{R}r\right)[A_{kq}\sin k\beta+B_{kq}\cos k\beta]\mathrm{e}^{\omega t} \tag{5}$$

where $A_{kq}$ and $B_{kq}$ are complex amplitude to be determined, $J_k\left(a_{kq}/R\cdot r\right)$ is the first kind Bessel function of kth order. $a_{kq}$ is the qth zero of $J_k(x)$. Obviously, Eq. (5) meets the boundary condition above. Substituting Eq.(5) into Eq.(3), by comparing each side of the equation, we obtain

$$A_{kq}=\begin{cases}0, & k\neq 1\\ A_{1q}, & k=1\end{cases},\quad B_{kq}=0 \tag{6}$$

So, Eq.(5) is able to be simplified as follows

IOP Conf. Series: Materials Science and Engineering **224** (2017) 012008 doi:10.1088/1757-899X/224/1/012008

$$P(r,\beta,t)=\sum_{q=1}^{\infty}A_{1q}\cdot J_1\left(\frac{a_{1q}}{R}r\right)\cdot\sin\beta\cdot \mathrm{e}^{\omega t} \tag{7}$$

Substituting Eq.(7) into Eq.(3), leads to

$$\sum_{q=1}^{\infty}A_{1q}\left[J_1''\left(\frac{a_{1q}}{R}r\right)+\frac{1}{r}J_1'\left(\frac{a_{1q}}{R}r\right)-\frac{1}{r^2}J_1\left(\frac{a_{1q}}{R}r\right)\right]=\frac{12\mu\theta_0}{g_0^{\ 3}p_a}\cdot\omega r \tag{8}$$

Using the orthogonality of the Bessel series, both sides of Eq.(8) multiply $rJ_1\left(\frac{a_{1q}}{R}r\right)$, then integrating the outcome of both sides from $r$=0 to $R$, we obtain

$$A_{1q}=-\frac{24\mu\omega\theta_0 R^3}{g_0^{\ 3}p_a}\cdot\frac{1}{a_{1q}^{\ 3}J_2(a_{1q})} \tag{9}$$

The squeeze film damping torque is able to be calculated as the integration of the pressure distribution acting on the circular plate, the expression is

$$T_{\text{Squeeze}}=\int_0^R\int_0^{2\pi}(p-p_a)\cdot r\sin\beta\cdot r\mathrm{d}r\mathrm{d}\beta \tag{10}$$

So, the total damping torque is

$$\begin{aligned}T_{\text{damping}}&=\mathrm{e}^{\omega t}\cdot p_a\sum_{q=1}^{\infty}A_{1q}\int_0^R\int_0^{2\pi}r^2J_1\left(\frac{a_{1q}}{R}r\right)\sin^2\beta\mathrm{d}r\mathrm{d}\beta\\&=-24\pi\mu\omega\theta_0\cdot\frac{R^6\mathrm{e}^{\omega t}}{g_0^{\ 3}}\cdot\sum_{q=1}^{\infty}\frac{1}{a_{1q}^4}\end{aligned} \tag{11}$$

So the corresponding damping constant $C_\theta$ is

$$C_\theta=-\frac{T_{\text{damping}}}{\omega\theta_0\mathrm{e}^{\omega t}}=\frac{24\pi\mu R^6}{g_0^{\ 3}}\cdot\sum_{q=1}^{\infty}\frac{1}{a_{1q}^4} \tag{12}$$

## 3. Validation

Now, we verify the accuracy of the present model by comparing the present model with the FEM models and also the results of Xia's model. In ANSYS, FLUID136 elements are used for modeling the squeeze-film. The parameters and dimensions of the circular micro-plate are listed in Table1.

**Table1.** The parameters and dimensions of the circular micro-plate

| Parameters | Description | Values |
|---|---|---|
| R | Radius of the circular micro-plate | 500μm |
| $T_p$ | Thickness of the circular micro-plate | 10μm |
| $g_0$ | Gap spacing | 5μm |
| $p_a$ | Ambient pressure | 101325 Pa |
| ps | Density | $2330\text{kg}\cdot\text{m}^{-3}$ |
| $u$ | Viscosity coefficient | $1.85\times10^{-5}$ N· s·m$^{-2}$ |
| $f$ | frequency | 3000Hz |

Fig.2 shows the comparison of damping constants over the results of present model and FEM models and the results of Xia's model as a function of $R$. Other parameters of the model set as Table1 shows. We varied the radius of the circular micro-plate from 50μm to 1000μm. We find the theoretical model in this paper agrees very well with FEM results as well as Xia's results in a wide range of radius, the discrepancy of the results is very small. We also find the damping constant increasing as the increase of the radius of the circular plate. Because the contact area between the plate and the gas is increasing exponentially as the increase of the size of circular micro-plate, so the corresponding damping constant is increasing.

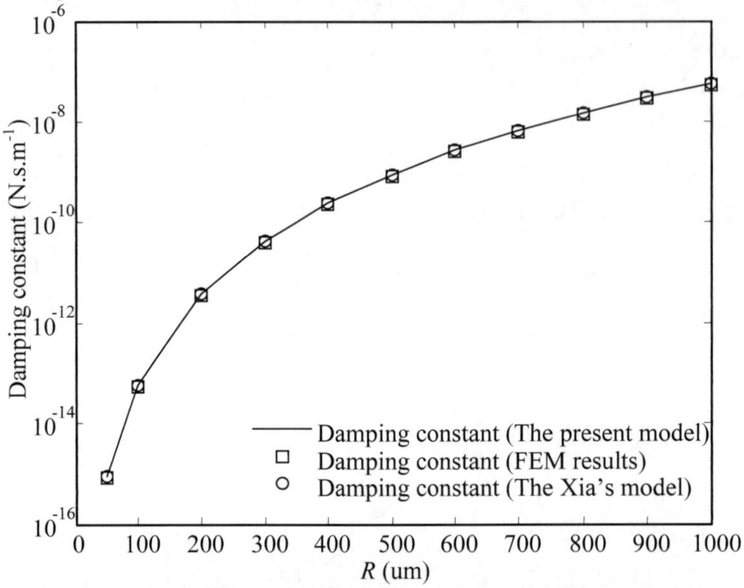


**Fig. 2** The comparison of damping constants over the results of present model and the FEM models and the Xia's model under different *R*.

Fig.3 shows the comparison of damping constants over the results of present model and the FEM models and the results of Xia's model as a function of gap spacing from 2μm to 20μm. Similarly, the theoretical model in this paper agrees very well with FEM results as well as Xia's results in a wide range of gap spacing. We also find the damping constant decreasing as the increase of the gap spacing. Because the gas is easier to escape from the gap between two plates as the gap spacing is increasing, so the corresponding damping constant is decreasing.

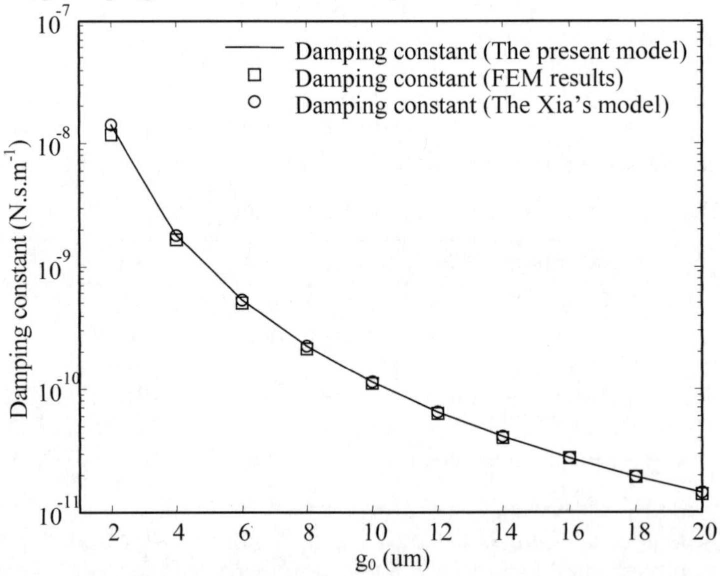


**Fig. 3** The comparison of damping constants over the results of present model and the FEM models and the Xia's model as a function of $g_0$.

Fig.4 gives the rate of convergence for the circular micro-plate. As it shows, the difference between N=5 and N=20 is very small. For those equations we proposed, the sum of 5 terms is enough to achieve convergence. Therefore, we use the sum of 20 terms to calculate the squeeze-film damping in this paper. This simplification of the series calculation not only meets the requirement of accuracy but also save time on the calculation.

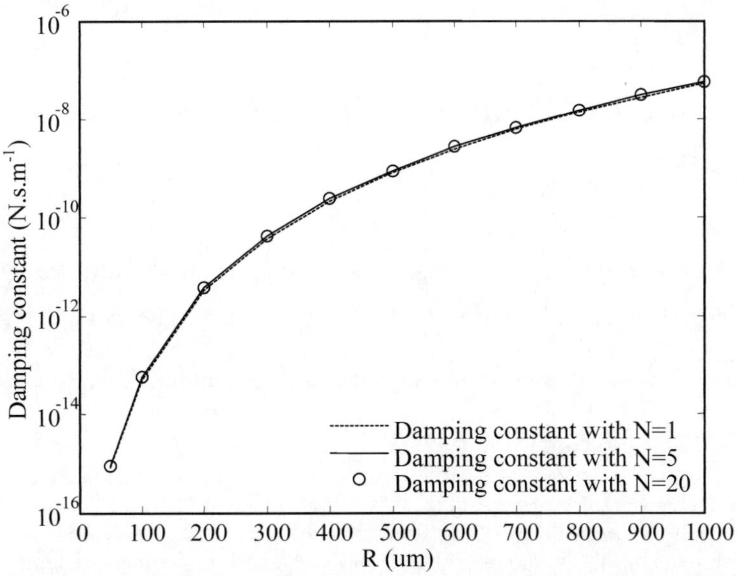


**Fig.4** The comparison of damping constant in the cases of N = 1, 5 and 10

## 4. Conclusions

In this paper, an analytical model for calculating the squeeze-film damping of the circular torsion micro-plate is given. The gas of the squeeze-film gap between two micro-plates is considered as incompressible. The pressure distribution of the gas is approximated through the Bessel series solutions. Then we obtained the analytical expression of the damping constant of the device. Through comparison of damping constants over the results of present model and the FEM models and the Xia's model as a function of the radius of the circular micro-plate and gap spacing, we find the analytical model in this paper agrees very well with FEM results as well as Xia's results, the discrepancy of the results is very small. In this paper, to simplify the calculation of the squeeze-film damping, the sum of 20 terms is enough to realize convergence and accuracy.

**Acknowledgment**

This project is supported by National Natural Science Foundation of China (Grant no. 51375091).

## References

[1] Langlois W E. Isothermal squeeze film Q. Appl. Math. 1962 20131-50

[2] Pan F, Kubby J, Peeters E, Tran AT, Mukherjee S Squeeze film damping effect on the dynamic response of a MEMS torsion mirror. J Micromech Microeng 8(1998): 200-208

[3] Chang K, Lee S, Li S Squeeze film damping effect on a mems torsion mirror. J Micromech Microeng 12(2002): 556-561

[4] A.K.Pandey, R.Pratap, F.S.Chau, Analytical solution of the modified Reynolds equations for squeeze film damping in perforated MEMS structures, Sensors Actuators A 135 (2007): 839–848.

[5] A.K.Pandey, R.Pratap, A comparative study of analytical squeeze film damping models in rigid rectangular perforated MEMS structures with experimental results, Microfluid Nanofluid 4 (2008):205–218.

[6] M. Bao, H.Yang, Y.Sun, Modified Reynolds' equation and analysis of squeeze-film air damping of perforated structures, Journal of Micromechanics and Microengineering 13 (2003):795 - 800.

[7] Xia C, Qiao D, Zeng Q, Yuan W, The squeeze-film air damping of circular and elliptical micro-torsion mirrors. Microfluid Nanofluid 19 (2015): 585-593

[8] M. Bao, H.Yang, Squeeze film air damping in MEMS, Sensors and Actuators A136 (2007): 3–27

IOP Conf. Series: Materials Science and Engineering **224** (2017) 012009 doi:10.1088/1757-899X/224/1/012009

# Static and Dynamic Performance Simulation of Direct-Acting Force Motor Valve

**Xinghai Ye[1], Jianjun Ding[2], GangZheng[1], Kunpeng Jiang[3] and Dongdong Chen[2]**

[1] School of Mechanical Engineering, Jiangsu University, Zhenjiang, Jiangsu, 212013, China

[2] Cold rolling General factory, Ma Steel, Maanshan, Anhui,243000

yesky791124@126.com

**Abstract:** This work focuses on static and dynamic characteristics of direct-acting force motor valve. First, we analyzed the structure features and operating principle of the Mitsubishi-Hitachi force motor valve (FMV) and the operating principle of its internal permanent-magnet moving-coil force motor magnetic circuit, determined the transfer function of the FMV force motor system, and established a mathematical model for the system. Secondly, we established a static performance analysis model using the AMESIM software and utilized the model in combination with experimental results to analyze the effects of electro-hydraulic servo valve structural parameters on static characteristics. Lastly, we deduced the trajectory equation of the system, established the relationship between dynamic characteristic indexes and structural parameters, and analyzed the effects of different parameter values on the dynamic characteristics of the system. This research can provide a theoretical guidance for designing and manufacturing the FMV body.

**Keywords:** Force motor valve; Mathematical model; Simulation model; Static and dynamic performance; Parameters design

## 1. Introduction

Compared with traditional mechanical transmission, hydraulic transmission uses hydraulic fluid as the working medium for energy transfer and control. With its unique advantages as strong anti-interference ability and high control accuracy, it has become one of the key technologies in modern transmission and control systems [1]. In the hydraulic system, a hydraulic valve, by adjusting the flow, direction and pressure of liquid, controls the actuator's motion direction, motion velocity, thrust, etc. [2]. If it malfunctions, the whole system can not work properly, resulting in incalculable losses. Hitachi force motor valves (FMV) are widely used in mill AGC systems [3] to control the mill work roll gap due to its characteristic of fast response, high control precision, and strong anti-pollution capacity, all which could affect the accuracy of machining precision of steel plates directly. Thus, real-time and effectively monitoring its various performance indicators is one of the important aspects to control the cost of steel production. Researchers [4-7] have performed some related researches on direct drive servo valves. Due to the limitations of theory and actual conditions, studies on FMV and relevant simulation tests were rarely reported. In addition, Mitsubishi-Hitachi hasn't made its related technical parameters open to the public [8-12]. In this study, we first illustrated the structure and principle of FMV servo valve, then established a simulation model and applied simulation software to obtain its optimal structure parameters, analyze its main structure parameters affecting FMV static and dynamic properties, set up a relevant database and last experimentally verified the simulation model.

IOP Conf. Series: Materials Science and Engineering **224** (2017) 012009 doi:10.1088/1757-899X/224/1/012009

This research is of guidance for the design, manufacture, and measurement of similar direct-acting servo valves.

## 2. FMV structural features and working principle

Figure 1 shows the structure and principle diagram of the FMV. The spool connecting to the force motor coil together with the hydraulic part lie in the same working space, as a wet motor. The displacement sensor consists of the anti-magnetic sensor shield, differential pressure transmitter (coil), differential pressure transmitter (mandrel) and other components. The valve body consists of the spool, sleeve, etc. The force motor consists of the coil, permanent magnet, magnet guide, centralizing spring, etc. The adjustment mechanism for centralization consists of the lock bolt, lock ring, adjusting bolt, lock nut, intermediate adjust seat, key, seat ring, position ring, etc..

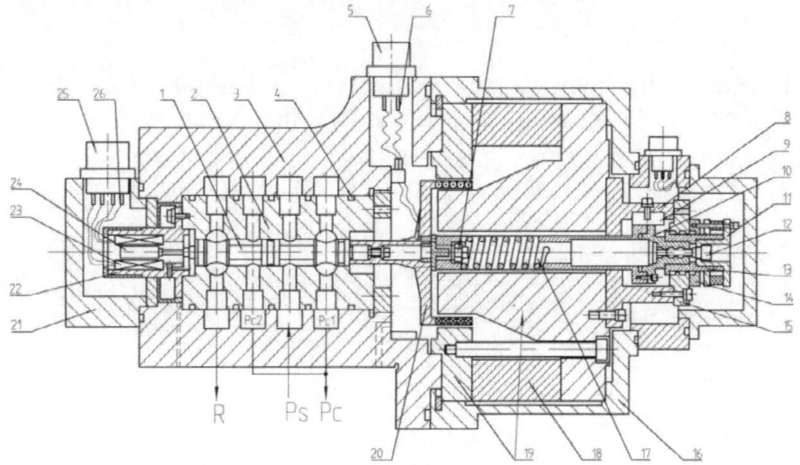


**Figure 1.** FMV structure principle diagram.

1,spool; 2, valve sleeve; 3, valve body; 4, O-ring; 5, connector; 6, wire; 7, bolt; 8, positioning ring; 9, key; 10, seat ring; 11, lock ring; 12, lock bolt; 13, adjust bolt; 14, lock nut; 15, median adjust seat; 16, lid; 17, spring; 18, permanent magnet; 19, guide magnet; 20, coil ; 21, lid; 22, anti-magnetic shield of displacement sensor; 23 differential pressure transmitter (coil); 24, differential pressure transmitter (mandrel); 25, connector; 26, conductor.

The working principle of the FMV is as follows: When the current signal is not input, the centralizing spring has no deformation, the spool is at the zero position, the oil-supply chamber (Ps) is not connected to the oil-control chamber ($P_{c1}$), the oil control chamber ($P_{c2}$) is not connected to the oil-return chamber (R), while mutually-connected Pc1 and Pc2 turn into the oil-control chamber ($P_c$). When the forward current flows, the coil acted by force drives spool to move to the left side. When the force reaches equilibrium with damping force from the centralizing spring, the spool will stop at some positions on the left side. At this time, Ps and Pc1 are closed, $P_{c2}$ connects with oil-return chamber (R), $P_{c1}$ connects with $P_{c2}$, oil-supply chamber (Ps) and oil-control chamber ($P_c$) are closed, oil-return chamber (R) connect with oil-control chamber (Pc). When the opposite current flows, the system state is just opposite.

## 3. Setup of FMV mathematical model

### *3.1. Force motor magnetic circuit of permanent magnet moving coil*

As shown in figure 2, the permanent magnet moving-coil force motor mainly consists of the working coil, guide magnet, permanent magnet, centralizing spring, etc. [13]. In the magnetic circuit, the ampere force is

$$F = B_a \pi D N_c I_C \sin\theta \tag{1}$$

where F is the ampere force, namely, the electromagnetic force applied on the moving coil, [N]; Ba is the magnetic induction intensity around the charged coil [T]; Ic is the current through the charged coil [A]; Nc is the number of turns of the charged coil; D is the average diameter of the charged coil [m]; and θ is the acute angle between the charged coil and the magnetic induction [°].

Figure 3 shows the magnetic path of the permanent magnet moving-coil force motor. Let e stands for the output voltage of the direct current (DC) servo amplifier at the front of the permanent magnet moving-coil force motor, $r_a$ stands for the internal resistance of the DC servo amplifier, and $r_c$ stands for the resistance of the control coil, the equation of electric circuit of the working coil is:

$$e = i(r_a + r_c) + L_c \frac{di}{dt} + K_d \frac{dX_v}{dt} \tag{2}$$

where Lc is the self-induction coefficient of the the coil; Kd is the counter electromotive force constant, $K_d = B_a \pi D N_c$. Taking Laplace transform of equation (2) finds:

$$E(s) = I(s)(r_a + r_c) + L_c sI(s) + K_d sX_v(s) \tag{3}$$

Further simplifying the equation obtains:

$$I(s) = \frac{E(s) - K_d S X_V}{(r_a + r_c) + L_{CS} S} = \frac{[E(s) - K_D S X_v(s)]/(r_a + r_c)}{1 + S/\omega_a} \tag{4}$$

Where $\omega_a = L_{cs}/(r_a + r_c)$ .

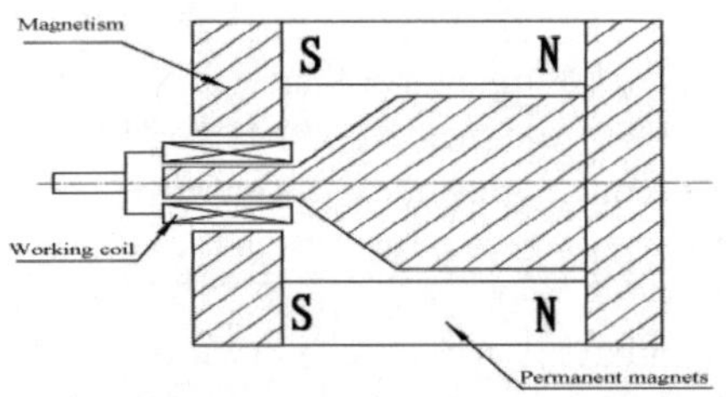


**Figure 2.** Schematic of charged coil structure.

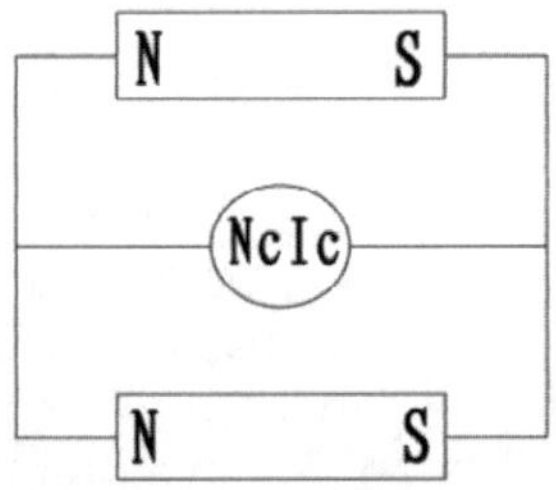


**Figure 3.** Schematic of magnetic circuit.

3.2. Setup of *mathematical* model for FMV force motor system

Inside the FMV spool, when the electromagnetic drive force, hydrodynamic force, elastic force, and frictional resistance reach an equilibrium, the relationship between the force generated by the force motor and the corresponding displacement is

$$F = M_V \frac{d^2 X_v}{dt^2} + (B_n + B_s)\frac{dX_v}{dt} + k_v X_v + k_f X_v \tag{5}$$

where Mv is the total mass of the FMV spool, charged coil, and other parts [kg]; Xv is the displacement of the FMV spool [m]; Bn is the viscous damping coefficient of the system; Bs is the damping coefficient generated by the transient hydrodynamic force; Kv is the stiffness of the FMV centralizing spring [N/m]; and Kf is the stiffness of the stable hydrodynamic force from which the FMV spool suffers [N/m].

Combining equation (1) with (5) finds:

$$B_a \pi D N_c I_c = M_v \frac{d^2 X_v}{dt^2} + (B_n + B_s) \frac{dX_v}{dt} + K_v X_v + K_f X_v, \tag{6}$$

$$B_a \pi D N_c I_c(s) = M_v X_v(s) S^2 + (B_n + B_s) X_v(s) S + K_v X_v(s) + K_f X_v(s), \tag{7}$$

and after simplification, one obtains the transfer function between the FMV spool displacement and the driving current:

$$G(s) = \frac{X_v(s)}{I_c(s)} = \frac{K \prod_{i=1}^{m} (s + Z_i)}{\prod_{j=1}^{n} (s + P_j)} = \frac{B_a \pi D N_c}{M_V S^2 + (B_n + B_s) S + K_v + K_f}$$

$$= \frac{\frac{B_a \pi D N_c}{M_v}}{S^2 + \frac{(B_n + B_s) S}{M_v} + \frac{K_v + K_f}{M_v}} \tag{8}$$

where K is the transfer function gain of the FMV force motor system; Zi (i = 1,2,3 ... m) and Pj (j = 1,2,3 ... n) represent all the zeros and all the poles of the system, respectively.

Through equation (8), one can separately find the FMV's damping ratio $\delta = \frac{B_n + B_s}{2\sqrt{M_v (K_v + K_f)}}$ and inherent frequency $W_n = \sqrt{\frac{K_v + K_f}{M_f}}$ with its the frequency characteristic equation of

$$G(j\omega) = \frac{\frac{B_a \pi D N_c}{M_v}}{-\omega^2 + \frac{(B_n + B_s) j\omega}{M_v} + \frac{K_v + K_f}{M_v}} \tag{9}$$

$$= \frac{\frac{B_a \pi D N_c}{M_v} \left[\frac{K_v + K_f}{M_v} - \omega^2 - \frac{(B_n + B_s)}{M_v} j\omega\right]}{\left(\frac{K_v + K_f}{M_v} - \omega^2\right)^2 + \left[\frac{(B_n + B_s)\omega}{M_v}\right]^2} \tag{10}$$

Through equation (10), one can respectively find the amplitude and the phase angle of the transfer function of the FMV force motor, A(ω) and φ(ω):

$$A(\omega) = \sqrt{\frac{(B_a \pi D N_c)^2 \left[\left(\frac{K_v + K_f}{M_v^2} - \frac{\omega^2}{M_v}\right)^2 + \left[\frac{\omega (B_n + B_s)}{M_v^2}\right]^2\right]}{\left[\left(\frac{K_v + K_f}{M_v} - \omega^2\right)^2 + \left[\frac{(B_n + B_s)}{M_v}\omega\right]^2\right]^2}} \tag{11}$$

$$\varphi(\omega) = \tan^{-1} \frac{-(B_n + B_s)\omega}{K_v + K_f - \omega^2 M_v} \tag{12}$$

Thus, the log amplitude-frequency characteristic of the FMV force motor system, L(ω) is

$$L(\omega) = 20\lg(A(\omega)) = 20\lg \sqrt{\frac{(B_a \pi D N_c)^2 \left[\left(\frac{K_v + K_f}{M_v^2} - \frac{\omega^2}{M_v}\right)^2 + \left[\frac{\omega (B_n + B_s)}{M_v^2}\right]^2\right]}{\left[\left(\frac{K_v + K_f}{M_v} - \omega^2\right)^2 + \left[\frac{(B_n + B_s)}{M_v}\omega\right]^2\right]^2}} \tag{13}$$

The dynamic characteristics of a second-order system are represented in a series of indexes under the unit step response including the overshoot, adjustment time, peak time, rising time, stable state error, etc. [14]. The dynamic characteristics of FMV depend on its damping ratio and the inherent frequency, both of which in turn depend on the viscous damping coefficient, the damping coefficient generated by the transient hydrodynamic force, the mass of FMV's moving parts, the stiffness of the centralizing spring, and others. Therefore, in order to find the parameter values making the dynamic characteristics of the FMV force motor be more optimal, it is necessary to analyze the effects of these factors on FMV's dynamic characteristics.

## 4. Simulation analysis of FMV static characteristics

The static characteristics of the servo valve can be assessed by the null bias, hysteresis loop, internal leakage, etc. under the unloaded state [15]. This paper applies AMESIM, the hydraulic system simulation software, to analyze the static characteristics of the FMV.

### *4.1 FMV static simulation model*

Tables 1 and 2 list the structure parameter values used in the AMESIM simulation.

**Table 1.** FMV's electric and hydraulic parameter values

| Hydraulic parameter | Value | Name |
|---|---|---|
| $P_s$ | 5 MPa | Oil-supply pressure |
| P | 850 kg/m$^3$ | Density of oil |
| M | 0.00432 N·s/m$^2$ | Dynamic viscosity |
| A | 69° | Jet angle |
| E | 2.1X10$^7$ N/m$^2$ | Elastic modulus of oil |
| f | 0.1 Hz | Frequency of sine wave |
| U | 0.002V | Amplitude of sine wave |

**Table 2.** Parameter values of FMV structure

| Structure parameter | Value | Name |
|---|---|---|
| d | 15 mm | Spool diameter |
| $u_{lap}$ | -0.001~0.001 | Cover amount of valve orifice |
| M | 0.06 kg | Spool mass |
| δ | 0.01~0.3 mm | Radial gap between spool and valve body |
| ε | 0.1~1 | Damping coefficient |
| w | 0.002 mm | Throttle orifice width |

Figures 4 and 5 show the simulation models for FMV flow and pressure characteristics established through AMESIM.

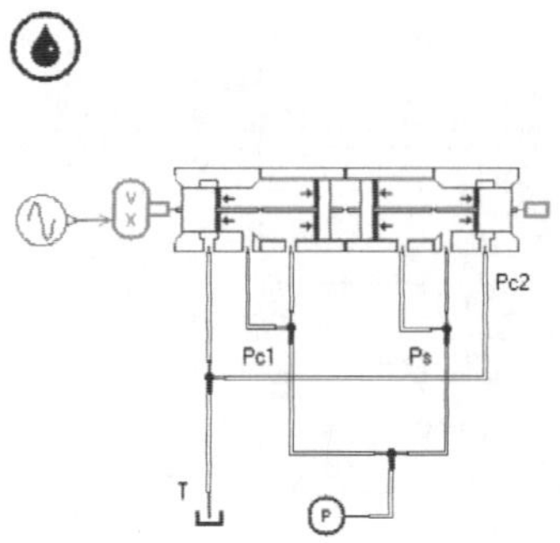


**Figure 4.** Simulation model for FMV flow characteristics.

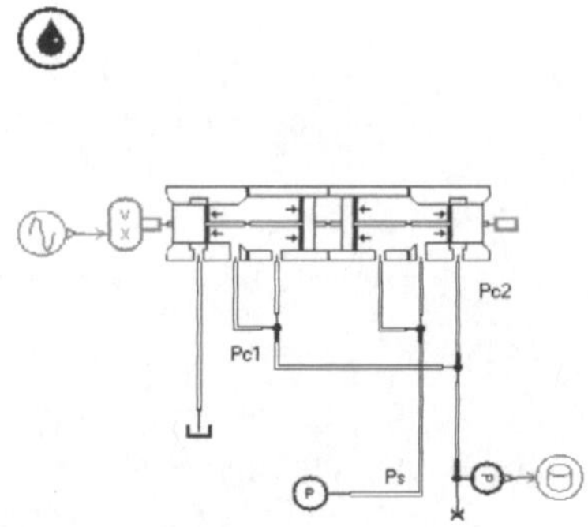


**Figure 5.** Simulation model for FMV pressure characteristics.

IOP Conf. Series: Materials Science and Engineering **224** (2017) 012009 doi:10.1088/1757-899X/224/1/012009

*4.2 Analysis of FMV static characteristics*

Figures 6 and 7 show the simulation curves of FMV flow and pressure characteristics, respectively.

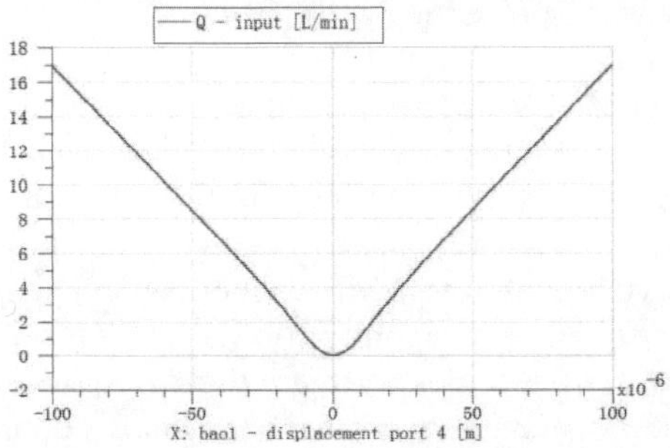


**Figure 6.** Simulation curves of FMV flow characteristics.

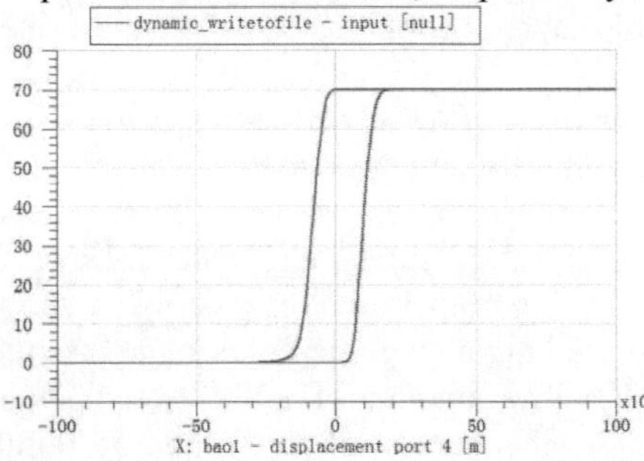


**Figure 7.** Simulation curves of FMV pressure characteristics.

In order to test the FMV's static state operation curve, a static state test system was established according to the hydraulic principle schematic of FMV static state test platform, as shown in figure 8.

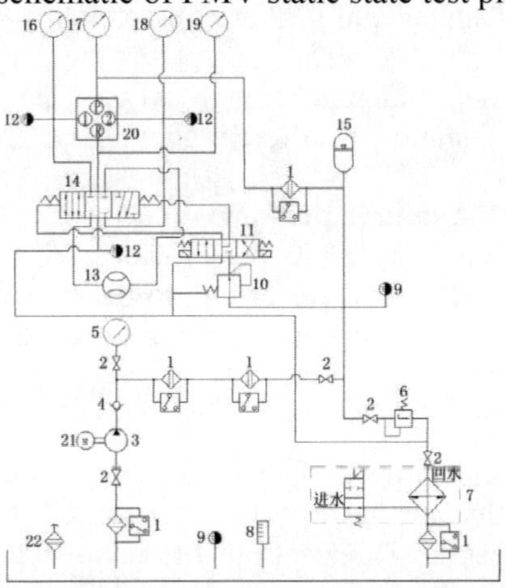


**Figure 8.** schematic of the hydraulic principle of FMV static test platform.

The current required for the FMV servo valve was 6 A, and the set pressure was 1.5 MPa. Figures 9 and 10 show the curves plotted based on the static tester-measured flow and pressure characteristics of the FMV servo valve.

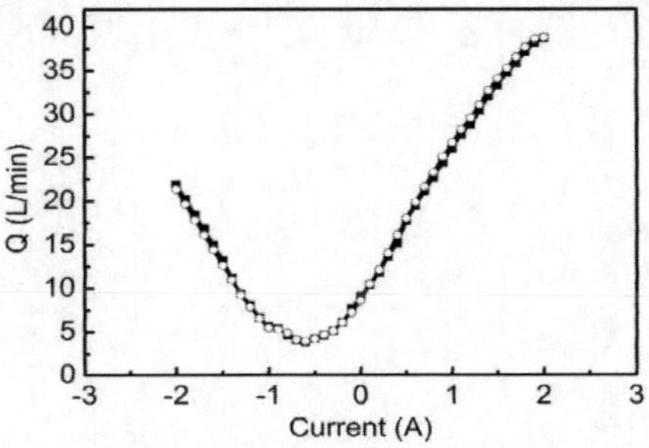


**Figure 9.** Characteristic curve of FMV flow.

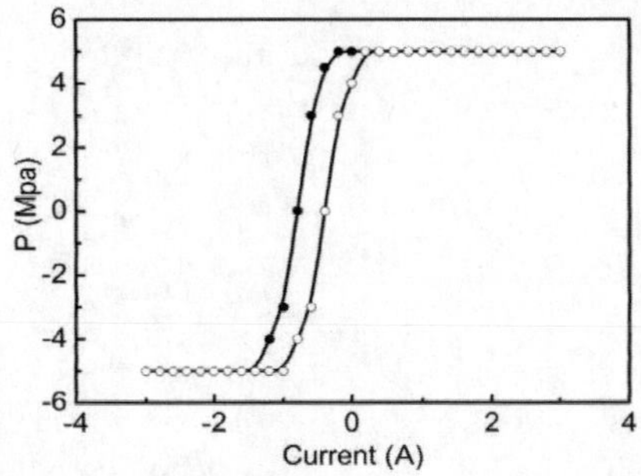


**Figure 10.** Characteristic curve of FMV pressure.

Simulation curves were obtained based on the relationships of the FMV's flow and pressure to the spool displacement and corresponded one by one to the coordinate parameters of experimentally tested curves. In practical engineering, the size of current signal is not necessarily given in accordance with the size of rated current, it is enough to meet applications only using the actually tested current range.

IOP Conf. Series: Materials Science and Engineering **224** (2017) 012009 doi:10.1088/1757-899X/224/1/012009

From figures 6, 7, 9 and 10, it is obvious that experimentally test curves are roughly consistent with the simulation curves, with a smaller difference, thus indicating that the static research result obtained in this paper is good enough and can be used to detect static properties of the FMV.

*4.3 Analysis of valve structure parameters impacting FMV static characteristics*

The valve structure parameters impacting FMV static characteristics include the radial gap between its spool and sleeve, the cover amounts of spool and sleeve, null bias, and the form of the valve orifice [16]. Analytic results are given as follows:

1) The gap radial gap between spool and sleeve affects the FMV's internal leakage, that is, increased radial gap will enlarge the internal leakage and reduce the control flow.

2) The amount of lap between spool and sleeve of the electro-hydraulic servo valve orifice includes the amount of zero, plus and minus laps. Their characteristics have been analyzed previously [17] and are not discussed in the study.

3) When no signal is inputted due to machining sizes, electromagnetic properties, and other causes, the output flow from the servo valve is not equal to zero. The FMV's zero bias is set at about -0.8 A to prevent sudden disruption of power supply-resulted interconnection of the oil-supply chamber with the oil-control chamber, and further mill activating-resulted roll damage.

4) The cross-section type of valve orifice comprises the cylindrical spool valve type, cone valve type, ball valve type, and axial triangular trough valve type. Different cross-section types have different section areas of flow, resulting in different gain characteristics of control flow.

## 5. Simulation analysis of FMV dynamic performance

The indexes used to measure the dynamic performance of the control system include stability, accuracy and rapidity of the system [18]. According to the transfer function of the FMV force motor system, i.e., equation (8), the root locus equations of the system can be found as follows:

$$\prod_j^n (s+P_j) = M_v S^2 + (B_n + B_s)S + K_v + K_f = 0 \tag{14}$$

After theoretical analysis, the simulation software Matlab was used to make codes, plot diagram of the FMV force motor system, calculate the dynamic performance indicators, and assess the dynamic performance of the control system. After setting the parameters, the variable method was used to analyze the effects of different parameter values on the dynamic characteristics of the FMV force motor system. Let $K_1=B_n+B_s$, $K_2=K_v+K_f$ and $K_3=B_a\pi DB_C$ and according to the above FMV structure parameter values, let $M_v$ = 0.02 kg, $K_3$ = 28800 N/mA, $K_2$ = 28800 N/m, and let $K_1$ be 4.8, 14.4, 24, 33.6, and 43.2. Simulation by bringing these values in the Matlab codes obtained the root locus plot (Figure 11), phase angle margin diagram (Figure 12), unit step response curve (Figure 13), unit pulse response curve (Figure 14) and Bode diagram (Figure 15) of the system at different $K_1$ values.

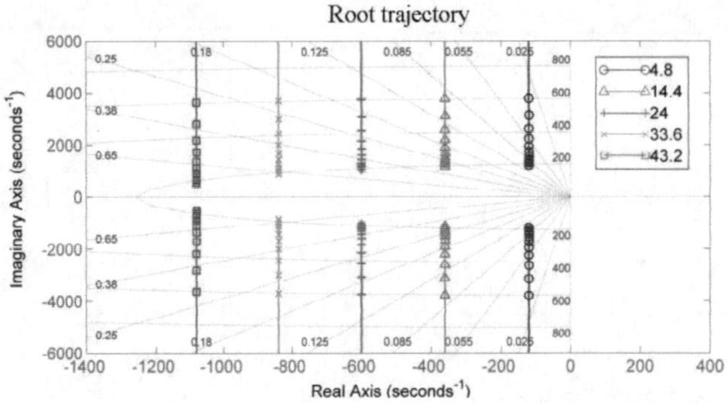


**Figure 11.** FMV's root locus plot at different $K_1$ values.

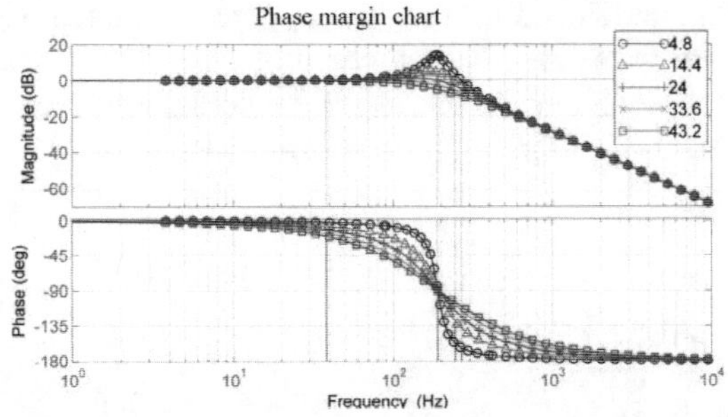


**Figure 12.** FMV's phase angle margin at different $K_1$ values.

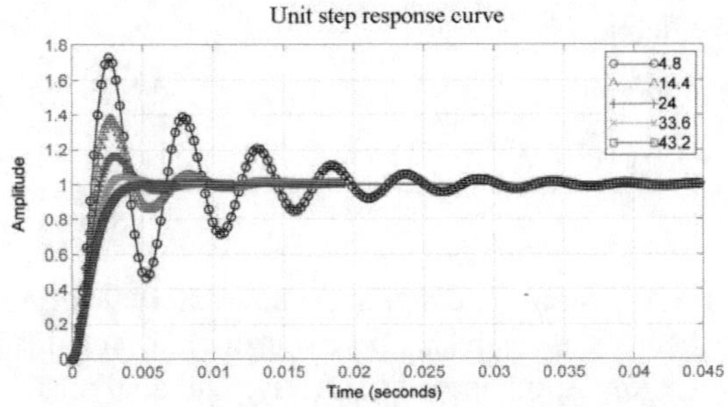


**Figure 13.** FMV's unit step response curve at different $K_1$ values.

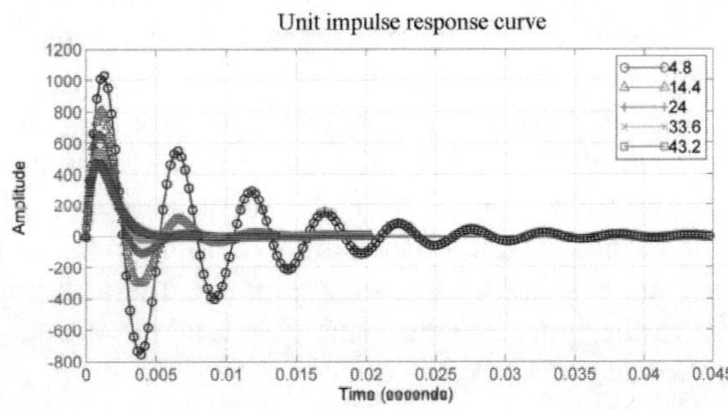


**Figure 14.** FMV's unit pulse response curve at different $K_1$ values.

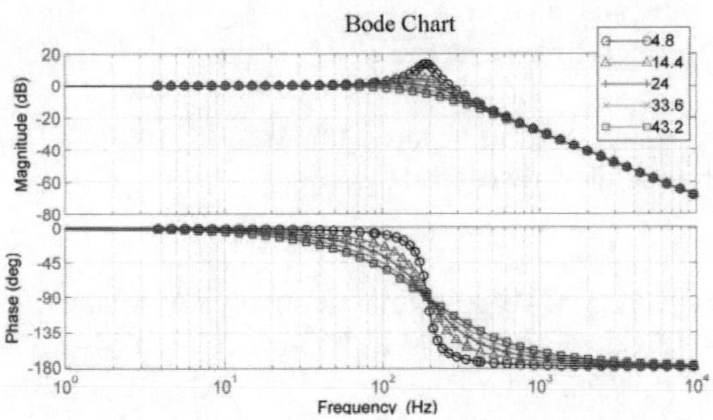


**Figure 15.** FMV's Bode plot at different $K_1$ values.

Figure 11 clearly shows that all the closed-loop poles of the system at different $K_I$ values lie in the left-half part of the complex plane, indicating the system is stable. From figure 12, figure 14 and figure 15, the stable phase angle margin, the peak time and the unit pulse response curve, and the frequency width of the FMV force motor system at different $K_I$ values can be obtained, respectively. From figure 13, the overshoot ($\sigma$), adjust time ($t_s$), peak time ($t_p$), rise time ($t_r$), etc. of the unit step signal can be found.

Table 3 lists all the data calculated based on the above-mentioned analysis. From the table, it can be seen that when other parameters of the FMV force motor system remain invariable, the dynamic performance of the system at $K_I$=24 or so best satisfies the actual requirement of production.

**Table 3.** Effect of different $K_I$ values on FMV system performance

| $K_I$ (unit step signal) | 4.8 | 14.4 | 24 | 33.6 | 43.2 |
|---|---|---|---|---|---|
| Phase angle margin (°) | 16.26 | 50.22 | 90 | 163.7 | 180 |
| Overshoot σ (%) | 72.9 | 36.56 | 15.83 | 4.81 | 0 |
| Adjust time $t_s$ (s) | 0.032 | 0.0094 | 0.0067 | 0.0049 | 0.0039 |
| Rise time $t_r$ (s) | 0.00094 | 0.0011 | 0.0014 | 0.0018 | 0.0024 |
| Frequency width (Hz) | 200 | 198 | 191 | 190 | 185 |
| Damping ratio δ | 0.1 | 0.3 | 0.5 | 0.7 | 0.9 |
| Inherent frequency $W_n$ (rad/s) | 1200 | 1200 | 1200 | 1200 | 1200 |

## 6. Conclusions

This study investigates the FMV's static and dynamic characteristics and optimizes its performance parameters using combined theoretical analysis and experimental verification approach as well as an integrated static and dynamic simulation method. Firstly, it analyzes the structural characteristics and working principle of Mitsubishi-Hitachi FMV, which laid a basis for establishing its theoretical model; Secondly, it explores the working principle of the internal permanent-magnet moving-coil force motor magnetic circuit, determined the transfer function based on the automatic control principle, established the mathematical model of the FMV force motor system, and revealed the influencing laws of these structural parameters on its performance, all of which provide references for the servo valve design; Thirdly, it establishes a simulation model of FMV's static characteristics using AMESIM. The simulation results are roughly consistent with the experimental results; and fourthly, it sets up the database of FMV's structure parameters, determines the dynamic characteristic indexes, and obtaines the curves of root locus, phase angle margin, unit step response and unit pulse response with variable parameter $K_I$ and Bode plots, further analyzes its dynamic characteristics and influencing factors, and provides a basis for the improvement of FMV.

## Acknowledgments

The study is supported by the National SME Innovation Fund (13C26213202029) and Zhenjiang Municipal Science and Technology Program (SGY20130037).

**Author:** Xinghai Ye, (1979-), male, born in Maanshan, Anhui Province, China. Ph.D. Candidate, major in hydraulic technology and its related researches.

## References

[1] Yuanfang Gu . Hydraulic AGC [J]. Hydraulics Pneumatics & Seals, 1997, 2: 34-35.
[2] Chunxin Wang. Hydraulic Control System [M]. Beijing: China Machine Press, 2013.
[3] Xianhui Chen. Research on Characteristics and Test System for FMV Electro-hydraulic Servo Valve [J]. Machine Tool and Hydraulics, 2013, 41(14): 59-62
[4] Qipeng Li. Research on Key Technologies of Direct Drive Servo Valve [D]. Zejiang: Zejiang University, 2005.
[5] Haibing Jiang, Jiajin Zheng, and Zhaozhong Zhou. Test Research and Failure Analysis of Force Motor Valve (FMV) Static Characteristics [J]. Machine Tool and Hydraulics, 2008, 36 (9): 276-277.
[6] Hailing Wang. Development of a New Torque Motor Direct-Drive Servo Valve [J]. Chinese Hydraulics and Pneumatics, 2014, 2:94-97.

[7] T. Minav, A. Virtanen, L. Laurila, J. Pyrhönen. Storage of energy recovered from an industrial forklift [J]. Automation in Construction, 2012, 22:506-515.
[8] D. Wang, C. Guan, S. Pan, M. Zhang, X. Lin, Performance analysis of hydraulic excavator powertrain hybridization [J].Automation in Construction, XXXX, 18(3):249-257
[9] T. Lin, Q. Wang, B. Hu, W. Gong. Research on the energy regeneration systems for hydraulic excavators [J]. Automation in Construction, 2010,19(8):1016-1026
[10] T.H. Ho, K.K. Ahn. Design and control of a closed-loop hydraulic energy regenerative system [J]. Automation in Construction, 2012, 22:444-458
[11] T. Lin, Q. Wang, B. Hu, W. Gong. Development of hybrid powered hydraulic construction machinery [J]. Automation in Construction, 2010,19(1):11-19
[12] Yu Mo. Mechanism of Torque Motor Demagnetization and Development of Demagnetizer [J]. Hydraulics and Pneumatics, 2006, 10:77-79
[13] Liyi Feng. Development of Electro-hydraulic Control System Test Device [D]. Beijing: Beijing University of Technology, 2008.
[14] Lian Jin. Electro-Hydraulic Servo Valve Optimization Design [D]. Xi'an: Northwestern Polytechnical University, 2007.
[15] Xin Xu. Theoretical and Experimental Researches on Electro-Hydraulic Servo Valve [D]. Zhengjian: Jiangsu University of Science and Technology, 2008.
[16] Jihai Jiang, Jinchun Song, Changshi Gao. Hydraulic and Pneumatic Transmission [M]. Beijing: Higher Education Press, 2011, 303-304.
[17] Rongjuan Li. Application of MATLAB in Stability Determination of Classic Control System [J]. Inner Mongolia Petrochemical Industry, 2011, 19: 32-33.
[18] Wang Chunxing. Hydraulic Control System [M]. Beijing: China Machine Press, 2013.

IOP Conf. Series: Materials Science and Engineering **224** (2017) 012010 doi:10.1088/1757-899X/224/1/012010

# Simulation Analysis of Helicopter Ground Resonance Nonlinear Dynamics

**Zhu Yan, Lu Yu-hui and Ling Ai-min**

Science and Technology on Rotorcraft Aeromechanics Laboratory, CHRDI, Jingdezhen, China

everyanyan@126.com

**Abstract.** In order to accurately predict the dynamic instability of helicopter ground resonance, a modeling and simulation method of helicopter ground resonance considering nonlinear dynamic characteristics of components (rotor lead-lag damper, landing gear wheel and absorber) is presented. The numerical integral method is used to calculate the transient responses of the body and rotor, simulating some disturbance. To obtain quantitative instabilities, Fast Fourier Transform (FFT) is conducted to estimate the modal frequencies, and the mobile rectangular window method is employed in the predictions of the modal damping in terms of the response time history. Simulation results show that ground resonance simulation test can exactly lead up the blade lead-lag regressing mode frequency, and the modal damping obtained according to attenuation curves are close to the test results. The simulation test results are in accordance with the actual accident situation, and prove the correctness of the simulation method. This analysis method used for ground resonance simulation test can give out the results according with real helicopter engineering tests.

## 1. Introduction

The aeromechanical instability of a helicopter on the ground, commonly referred to as ground resonance, is a complex phenomenon involving both the rotor and body degrees of freedom in which the rotor lead-lag regressing mode which includes substantial body pitch and roll motion may become unstable. The dynamics problem of ground resonance must be considered in the development of helicopter is more important for a new model. For the analysis method of helicopter ground resonance, the linearization hypothesis method based on balance position of small disturbance is often used in engineering practice. It is assuming that the dynamic characteristics of landing gear system and rotor dampers are linear. In a range of small vibration amplitude, from microscopic side, the results will be correct when the actual vibration amplitude is not beyond the range, sometimes the assumption can lead to some incorrect results when beyond the range. In fact, the dynamic characteristics of the landing gear system including absorber with wheel and the rotor dampers are nonlinear, especially the stiffness and damping of landing gear are high nonlinear characteristics and very sensitive to environmental. The linearization will cause remarkable error to the stability of ground resonance at the taking off and landing conditions with large amplitude movement.

Since the 1950s, the stability problem of helicopter ground resonance had been studied in the domestic and abroad, the studies conducted give an insight into the mechanism of coupling, the influence of design parameters, and the test verification, as described in references [1,2], and obtained a lot of research achievements. A nonlinear analysis model of rotor and fuselage coupled had been established in references [3~6], the nonlinear damping characteristics of rotor dampers effect on the stability of rotor and fuselage coupled was evaluated. The use of a wheeled landing gear simulation model, as described in reference [7], is one of considerable value in obtaining an understanding of the gear movement characteristics and the influence of gear design on ground resonance. Above studies only focus on the influence study of single component to ground resonance. The analysis method study of helicopter ground resonance considering the nonlinear dynamic characteristics of the landing gear wheel and absorber and rotor dampers has not been carried out.

The nonlinear dynamic characteristics of the landing gear wheel and absorber and rotor dampers were analyzed, referenced the modeling method of rotor dampers and landing gear components to the literatures [8~10], the calculation model of ground resonance considering the nonlinear dynamic characteristics of components was established, in this paper. The transient responses of the body and rotor blade were simulated under some disturbance using the numerical integral method. The frequency and damping of dynamic systems were obtained by the mobile rectangular window method. The nonlinear numerical simulation of ground resonance was carried out, and the stability margin or stability boundary of ground resonance was provided.

## 2. Nonlinear analytical model

### *2.1. Landing gear wheel nonlinear dynamics model*

The nonlinear relationship of the static stiffness and displacement at the station of vertical compression can be expressed as:

$$K(s_{Z0},\Delta s_z)=|\frac{k_{z1}}{w}+2k_{z2}\frac{s_{Z0}}{w^2}+k_{z2}\frac{\Delta s_z}{w^2}+3\frac{k_{Z3}s_{z0}^2}{w^3}+3k_{Z3}\frac{s_{Z0}\Delta s_z}{w^3}+k_{Z3}\frac{\Delta s_z^2}{w^3}|(1.08pw\sqrt{dw}) \tag{1}$$

The stiffness is the quadratic polynomial of compression displacement and the cubic polynomial of width w. The tire diameter and inflated-gas pressure are constant factors. When the wheel under static load compression, the static displacement is $S_{Z0}$ .

Use tire static compression characteristic test data for fitting, the coefficients $k_{z0}$, $k_{z1}$, $k_{Z2}$, $k_{z3}$ in above equations can be confirmed. When calculating the dynamic stiffness, the stiffness formula should be multiplied by a tire inflation pressure factor c , then, dynamic stiffness is given by $K'(s_{Z0},\Delta s_z)=c\cdot K(s_{Z0},\Delta s_z)$. Modified with the tire dynamic vertical stiffness test results, the inflation pressure factor c can be concluded.

For ground resonance analysis of helicopter, the lateral stiffness of wheel should be considered. In development of model, the tire lateral stiffness test should be done, and the rigidity and damping corresponding vibration amplitude and frequency can be obtained. Then, wheel lateral stiffness is not only the function of design parameters of tire, such as tire width w, diameter d and inflation pressure p , but also the function of displacement of static compression and lateral vibration, expressed as

$$K_Y(s_Z,s_Y)=1.24pw[a_{y0}+a_{y1}\frac{s_Z}{w}+a_{y2}(\frac{s_Z}{w})^2](1+c_y s_Y) \tag{2}$$

Tire provides heading stiffness while the mainwheel is braking. For the ground resonance analysis of braking condition, the heading stiffness test of wheel is needed. The expression of heading stiffness is similar to the lateral stiffness.

$$K_X(s_Z,s_X)=5pd[a_{x0}+a_{x1}\frac{s_Z}{d}+a_{x2}(\frac{s_Z}{d})^2](1+c_x s_X) \tag{3}$$

With wheel landing gear, whether nosewheel or tailwheel, or other configurations, the helicopter must be designed to landing gear steering wheel can be free. When a helicopter sliding or running, take-off or landing on the ground, the rotation function of the steering wheel can't be locked. The lateral stiffness will be reduced with the steering wheel, and the influence is big, especially the static compression displacement of tire is small, the lateral stiffness to reduce is the largest. The expression of torsional stiffness is similar to the heading stiffness.

IOP Conf. Series: Materials Science and Engineering **224** (2017) 012010 doi:10.1088/1757-899X/224/1/012010

$$K_\theta(s_Z,\theta)=1.8pw^3[a_{\theta 0}+a_{\theta 1}\frac{s_Z}{d}+a_{\theta 2}(\frac{s_Z}{d})^2] \tag{4}$$

Stiffness and damping of a wheel are the same as the elastomeric dampers, expressed in elastic stiffness and damping stiffness respectively. However, damping stiffness is usually expressed as the ratio of the elastic stiffness, known as the loss angle, the ratio of the damping stiffness and elastic stiffness is equal to loss tangent. In fact, tire stiffness is changing with frequency, but this change is small shown from same test results. The change is obvious for damping, however, is usually loss angle increases with the increase of frequency. From the results of experiment and processed into equivalent viscous damping, equivalent viscous damping has little change with frequency. Therefore, tire damping expression constructed as irrelevant to the frequency.

$$\begin{aligned} &F_{DT}(s_{Z0},\Delta s)=C(s_{Z0},\Delta s)\dot{\Delta s}\\ &C(s_{Z0},\Delta s)=|K(s_{Z0},\Delta s)C_d(s_{Z0},\Delta s)|\\ &C_d(s_{Z0},\Delta s)=(d_0+d_1 s_{Z0}+d_2 s_{Z0}{}^2)(1+c_{dd}\Delta s) \end{aligned} \tag{5}$$

The ratio coefficient changed with the quadratic of static compression displacement, and the linear relationship with Δs as corresponding vibration direction, are considered in the expression.

*2.2. Landing gear shock-absorber nonlinear dynamics model*

According to the literature [9], the axial force of the shock-absorber is expressed as, $F_S=F_a+F_f+F_h$, Where $F_a$ is the air spring force; $F_f$ is the internal friction force; $F_h$ is the damping force for oil.

Air spring force for the low-pressure chamber and high-pressure chamber:

$$F(s)=a\frac{1}{1-bs},\ 0\le S\le S_{\max}^L\ ;\quad F'(s)=a\frac{1}{1-bS_{\max}^{\mathrm{L}}}+a'\frac{1}{1-b'(s-S_{\max}^{\mathrm{L}})},\ S_{\max}^{\mathrm{L}}<S<S_{\max}^H \tag{6}$$

When calculate the dynamic stiffness, the force is multiplied by temperature coefficient c. The displacement integration method is to solve stiffness. Using the least square method, with the results of buffer static compression feature test and stiffness test, the coefficients a、b、a′、b′ and c can be obtained.

According to the fluid mechanics of the classic theory of local pressure loss, the main factors affecting oil damping force is speed $\dot{S}$. Using experimental modeling method, the mathematical model of oil damping force can be established, with the relationship of movement speed and static compression position, $F_h=a(s)\dot{S}^2$. According to the principle of equal work, the work of friction and oil damping force with steady-state forced vibration in a cycle is equal to the work of damping forces by test measured.

*2.3. Rotor lag damper nonlinear model*

There are two kinds of dampers used for helicopter rotor system usually, the elastomeric damper and the hydraulic shimmy damper. According to the types and the main design parameters, model for the two kinds of dampers respectively

The nonlinear physical model of elastomeric damper mentioned in the literature [6], is written as the following analytical form of the load and displacement.

$$\begin{aligned} &F_D(s,\dot{s})=G_S(s)s+G_D(s)\dot{s}\\ &G_S(s)=|k_{s0}+k_{s1}s+k_{s2}s^2+k_{s3}s^3+k_{s4}s^4+k_{s5}s^5+k_{s6}s^6+k_{s7}s^7|\\ &G_D(s)=|c_{d0}+c_{d1}s+c_{d2}s^2+c_{d3}s^3+c_{d4}s^4| \end{aligned} \tag{7}$$

Where $k_{s0}$, $k_{s1}$,…, $c_{s0}$, $c_{s1}$,…are model parameters. The damping force of elastomeric damper depends on only its displacement, in the opposite direction with the relative movement velocity, and the relevance with frequency is very small essentially, shown by experimental data. The damping modeling method of elastomeric lag damper is the same as wheel.

The nonlinear curve of damping force of hydraulic damper and the axial velocity, as shown in literature [3], is divided into three periods for modelling. Its formula is expressed as:

$$\begin{aligned} F_D(v) &= S_{D0} + | S_{D1} + S_{D2}v + S_{D3}v^2 | v \quad && v \le v_1 \\ F_D(v) &= S_{D4} + | S_{D5} + S_{D6}v + S_{D7}v^2 | v \quad && v_1 < v \le v_2 \\ F_D(v) &= S_{D8} + S_{D9}v && v_2 < v \le v_{\max} \end{aligned} \tag{8}$$

Experimental data show that the damping force of hydraulic damper related to displacement, speed and frequency also. The opening force and speed of the PORV of hydraulic damper can be determined according to the test data. With the experimental data of amplitude and frequency and sampling time, and other known quantity, the relationship between damping force and speed can be deduced.

### *2.4. Equations of ground resonance*

Using the spatial dynamic model, the helicopter dynamic system is divided into two parts of the fuselage and rotor system. The fuselage as a rigid body with six degrees of freedom: x, y, z, $\phi_x$, $\phi_y$, $\phi_z$. For each blade, the freedom degree of lead-lag motion is only considered, denote by $\xi_i$.

The movements of body and landing gear and rotor blade are described on basis of analysis of these dynamic models. The time domain nonlinear force of rotor dampers for blades lead-lag motion and landing gear for body motion can be determined, using the established nonlinear dynamic models of landing gear and rotor dampers, the inertial loads acting on the blade and rotor hub and body can also be obtained. Then, the equation of lead-lag motion of each blade in the rotor rotating-coordinate system and the equation of body motion in the fixed-coordinate system can be carry out respectively by Lagrange theorem. Finally, the analysis model of ground resonance considering the nonlinear dynamic characteristics of these parts can be set up.

The single gear balance equation:

$$K_{zn}Z_n + C_{zn}\dot{Z}_n + m_n\ddot{Z}_n = K_{hn}Z_{hn} + C_{hn}\dot{Z}_{hn} \quad (\mathrm{n} = 1,2,3) \tag{9}$$

The force and moment equilibrium equations of rotor and body coupling:

$$M_f\ddot{X} + \sum_{n=1}^{3}\left(K_{xn}X_n + C_{xn}\dot{X}_n\right) - S_b\sum_{k=1}^{N_b}\ddot{\xi}_k \sin\psi_k = F_X^a \tag{10}$$

$$M_f\ddot{Y} + \sum_{n=1}^{3}\left(K_{yn}Y_n + C_{yn}\dot{Y}_n\right) + S_b\sum_{k=1}^{N_b}\ddot{\xi}_k \cos\psi_k = F_Y^a \tag{11}$$

$$M_f\ddot{Z} + \sum_{n=1}^{3}\left(K_{zn}Z_n + C_{zn}\dot{Z}_n\right) + T - M\mathrm{g} = F_Z^a \tag{12}$$

$$I_X\ddot{\phi}_X + I_{XZ}\ddot{\phi}_Z + \sum_{n=1}^{3}\left[Y_n\left(K_{zn}Z_n + C_{zn}\dot{Z}_n\right) - Z_n\left(K_{yn}Y_n + C_{yn}\dot{Y}_n\right)\right] - S_bZ_f\sum_{k=1}^{N_b}\ddot{\xi}_k \cos\psi_k = M_{\phi X} \tag{13}$$

$$I_Y\ddot{\phi}_Y + \sum_{n=1}^{3}\left[Z_n\left(K_{zn}X_n + C_{zn}\dot{X}_n\right) - X_n\left(K_{yn}Z_n + C_{yn}\dot{Z}_n\right)\right] - S_bZ_f\sum_{k=1}^{N_b}\ddot{\xi}_k \sin\psi_k = M_{\phi Y} \tag{14}$$

$$I_Z\ddot{\phi}_Z + I_{XZ}\ddot{\phi}_X + \sum_{n=1}^{3}\left[X_n\left(K_{zn}Y_n + C_{zn}\dot{Y}_n\right) - Y_n\left(K_{yn}X_n + C_{yn}\dot{X}_n\right)\right] + S_b\sum_{k=1}^{N_b}\ddot{\xi}_k (Y_f \sin\psi_k + X_f \cos\psi_k) = M_{\phi Z} \tag{15}$$

The blade lead-lag behavior equation:

$$I_b\ddot{\xi}_k + F_{\xi_k}^a + M_d\left(\dot{\xi}_k, \xi_k\right) + S_b\Omega^2\xi_K e_H + S_b(\ddot{Y}_G - Z_H\ddot{\phi}_X)(\cos\varphi_k - \sin\varphi_k\xi_k) - S_b(\ddot{X}_G + Z_H\ddot{\phi}_Y)(\sin\varphi_k + \cos\varphi_k\xi_k) = M_{\zeta k} \tag{16}$$

$$(\mathrm{k} = 1,2,\ldots,\mathrm{Nb})$$

Where M, K and C, respectively, represent mass, stiffness and damping. The index n is corresponding to the numbered gear, the index h represents absorber, f represents the body center of gravity. $I_X, I_{XZ}, I_Y, I_Z, I_{XZ}$ are the body moments of inertia. $I_b, S_b$, respectively, represent the moment of inertia and static moment of blade around the vertical hinge. $F_X^a, F_Y^a, F_Z^a, M_{\phi X}, M_{\phi Y}, M_{\phi Z}$, respectively, represent aerodynamic force and moment. $F^a_{\xi_k}$ is aerodynamic drag. $M_d(\dot{\xi}_k, \xi_k)$ is restraint moment of blade root spring. $M_{\xi k}$ is Coriolis moment of blade lagging.

## 3. Analysis and discussions for simulation test

To investigate the ground resonance test simulation with the same conditions of a certain type helicopter. The calculation values of simulation are converted to close to the locations of test points, then to compare with the test curves. From the aspects such as frequency and reduction damping ratio, analysis the accuracy of the simulation calculation results, analysis the reason of the differences.

The vibration response time histories of the nose from the ground resonance test of a certain helicopter with a certain weight are shown in figure 1. Figure 1(a) is the nose lateral vibration response, in the condition of zero lift and in ground idle speed, the attenuation frequency is 1.67 Hz, the damping ratio is 4.8955%, and the frequency of the exciting period of the experimental curve is 1.67 Hz, by the experimental data analysis. It shows that the motion of blade lead-lag natural model has not been excited, and is not the body pitch frequency also. The nose vertical vibration response is shown in figure 1(b), 50% lift, in nominal rotor speed, the excitation frequency is 3.33 Hz, attenuation frequency is 3.33 Hz, and damping ratio is 1.307357%. The data results show that the motion of blade lead-lag natural model has not been excited, and is not the body yaw frequency also. Therefore, the accurate frequency excitation is hard to do in the actual model test.

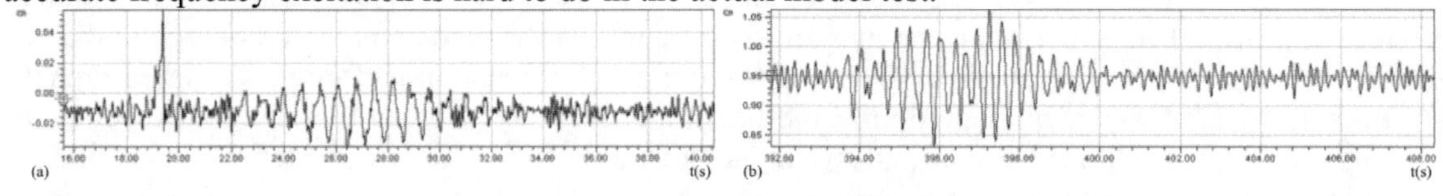


**Figure 1.** Vibration response time histories of the nose from the ground resonance test of a certain helicopter, (a) The nose lateral, zero lift, (b) The nose vertical, 50% lift.

The simulation results of ground resonance are shown in figure 2. For simulation calculation, the nose lateral vibration response is coupled by the lateral movement of body center of gravity and rolling motion, and the nose vertical vibration response is coupled by the vertical movement of body center of gravity and pitch motion. The nose lateral vibration response time histories of simulation test is shown in figure 2(a), in ground idle speed. Figure 2(c) is the nose vertical vibration response time histories of simulation test, in nominal rotor speed. At the same time, the response of blade lead-lag motion under the same conditions is simulated respectively, as shown in figures 2 (b) and 2(d).

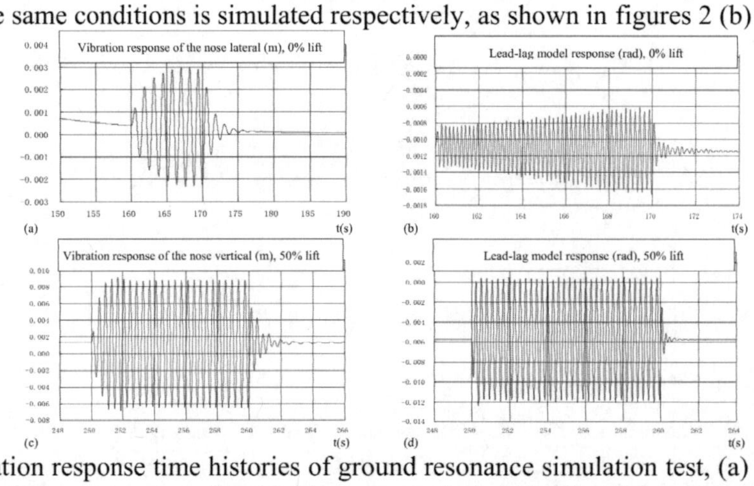


**Figure 2.** Vibration response time histories of ground resonance simulation test, (a) The nose lateral, zero lift, (b) Lead-lag motion, zero lift, (c) The nose vertical, 50% lift, (d) Lead-lag motion, 50% lift.

An exciting force was applied to the rotor at some time, lasting t seconds, in a simulation test of ground resonance. After the excitation had been removed, the system was free vibration. For the simulation test of ground resonance in figures 2(a) and 2(b), an exciting force was applied by simulated at 160s, and stopped the excitation after ten seconds, the rotor lift was zero, in ground idle speed. The attenuation frequency of blade was 1.2994 Hz, damping ratio was 4.21531%, calculated according to figure 2(b). Switched these to the fixed coordinate system, the frequency was 2.9 Hz, was just the frequency of lead-lag regressing mode in this state, and the damping ratio was 1.88875%.

For figures 2(c) and 2(d), an excitation was simulated at 250s, and stopped the excitation after ten seconds also, the rotor lift was 50% of the weight, in nominal speed. The attenuation frequency of blade was 2.9485 Hz, damping ratio was 2.41647%, calculated according to figure 2(d). Switched these to the fixed coordinate system, the frequency was 2.9682 Hz, was just the frequency of lead-lag regressing mode in this situation, and the damping ratio was 2.4%.

The comprehensive analysis of the frequency and damping in figures 1 and 2 shows that the excited frequency of simulation test is not the same as model experiment. The frequency of actual experiment is not accurate, is not the nature frequency of body, nor the rotor lead-lag regressing frequency, but the attenuation frequency is close to the simulation results, the corresponding attenuation damping ratio is similar. This indicates the simulation test results are in accordance with the practice. The large fluctuation of the simulation curve shows the influence of nonlinear dynamic characteristics of the components on the stability of the helicopter. The large blade attenuation damping of simulation test and the fast response attenuation indicate that the system has a high stability.

## 4. Conclusion

As the effects of nonlinearities existed in systems, there will probably be errors from engineering analysis when the linearized hypothesis is taken into account. In order to present an accurate analysis for helicopter ground resonance rotor/fuselage coupled dynamic systems, nonlinear differential equations are derived in this paper. Time-domain analyses is applied to predict dynamic instabilities and compared with test data of type helicopter. By comparing the frequency and damping ratio, concluded that the simulation test can exactly lead up the blade lead-lag regressing mode frequency, and the modal damping obtained according to attenuation curves closed to test results, the correctness of the analytical model had been validated. This analysis method used for ground resonance simulation test can give out the results according with real helicopter tests.

## References

[1] Robert P Coleman and Arnold M Feingold 1958 Theory of Self-excited Mechanical Oscillation of Helicopter Rotor with Hinged Blades *NACA TR*-1351

[2] Cheng Jin-song 2001 The Primary Analysis of a New Model Helicopter Ground Resonance *Chines Journal of Applied Mecihanics* 18(z1)

[3] Ling Ai-min 2009 Simulation Analysis of Helicopter Ground Resonance Test *Helicopter Technique* **vol 3**

[4] Robinson C S and Wood E R and King R L 1998 Full nonlinear simulation of coupled rotor/fuselage response using symbolically derived equations of motion *The American Helicopter Society 54th Annual Forum*

[5] Hu Guo-cai and Liu Quan 2010 Influence of Nonlinear Hydraulic Lag Damper on Scaled Rotor Ground Resonance *Acta Aeronautica et Astronautica Sinica* 31(11)

[6] Hu Guo-cai and Xiang Jin-wu 2005 Analysis of the Effect of Nonlinear Characteristics of Elastomeric Lag Damper on Helicopter Air Resonance in Forward Flight *Acta Aeronautica et Astronautica Sinica* 26(2)

[7] Monterrubio L and Sharf I 2002 Influence of Landing-Gear Design on Helicopter Ground Resonance *CASI*

[8] Felker F F and Lau B H 1987 Nonlinear Behavior of an elastomeric lag damper undergoing dual-frequency motion and its effect on rotor dynamics *Journal of the American Helicopter Socirty* 34(4):45-53

[9] Hu Guo-cai and Chen Jun-min 2007 Research on Dynamic Stiffness of Bi-Chambered Shock-Absorber of Helicopter Landing Gear *Journal of Naval Aeronautical Engineering Iinstitute* 22(6)

[10] Yuan Dong 2002 Establishment Method of a Landing-Gear Simulation Model *Flight Dynamics* 20(4)

# Research on high acceleration high precision air suspension control system performance evaluation algorithm

**Ling Xiang [1], Zhang Yu[1]**

[1]Guangdong Institute of Intelligent Manufacturing, Guangdong Provincial Key Laboratory of Modern Control Technology, Guangdong Modern control technology with optical and electrical Public Laboratory ,Guangzhou 510070, Guangdong, China

lingxiang33@163.com

**Abstract.** The motion performance of the high acceleration high precision air suspension platform is affected by the electromechanical characteristics of the drive system, the fluid characteristics of the air suspension guide rail and the load disturbance. The mathematical model of the system is established, and the controller performance is designed based on the optimal objective function value. The ideal controller performance is proposed as the evaluation benchmark. The performance of the system is evaluated by the ratio of the optimal objective value and the evaluation benchmark. Based on the established evaluation index, the influence of noise disturbance and estimation vector on the stability and robustness of the system is evaluated, and the system performance is further optimized.

## 1. Introduction

The precision air suspension platform is a complex kinematic system with high acceleration, high precision, high speed, no friction and fast stability. The comprehensive performance of the system is affected by the hardware structure, control system and driving mode. The motion performance of the high-acceleration and high-precision air suspension platform is affected by the electromechanical characteristics of the drive system, the fluid characteristics of the air suspension guide rail and the load disturbance, so it is necessary to analyze the factors that affect the running performance of the system.

There are many criteria for judging the operating performance of the air suspension system, in which the system stability and track accuracy are the most important. Stability is the basic requirement of the control system. Load disturbance is also a factor that can't be neglected in the control system. Erkorkmaz et al.[1] use the input shaping and state feedback controller to suppress the residual vibration of the air suspension platform and realize the precision control of the air suspension platform. Ham[2] and Viktorov[3] respectively in the application of fuzzy algorithm and neural network algorithm to suppress the residual vibration and vibration control system by input shaping technique. Although these control algorithms have certain compensation to the system, but the system performance also affected.

Dai Yi et al.[4] analyzed the influence of controller parameters on system performance by IP plus predictive feed-forward control method, and proposed the selection principle of controller. The dynamic performance of linear motor servo system with different load characteristics is compared and analyzed by Nanjing University of Aeronautics and Astronautics He Jiayuan[5] from the dynamic performance indicators and following error. Zhu Xiaogang[6] established a servo control system model with speed feed-forward and acceleration feed-forward, and analyzed the influence of different load on the control performance of air suspension platform. The above research on the precision air suspension platform focuses on the design of the control method and the influence of different load on the system, and does not propose a criterion which can be used as the evaluation standard to judge the performance of the control system. Therefore, a performance evaluation method of high acceleration high precision air suspension control system based on optimal objective function is proposed in this paper. The controller performance is designed based on the optimal objective function value according to the mathematical model of the system. The performance of the ideal controller is taken as the evaluation criterion, and the ratio of the optimal target value and the evaluation criterion of the precision air suspension system is taken as the performance evaluation index of the system, which is used to quantify the influence of noise disturbance and estimation vector on system stability and robustness.

## 2. System modeling

Due to external disturbances (such as the ripple of the linear motor, the electrical noise in the drive and the measurement noise, the cable power generated by the cable, etc.), changes in the environment and changes in system parameters will directly affect movement accuracy and fast stability of high acceleration high precision air suspension platform. Therefore, the control of the precision air suspension platform is simply in the expected minimum time through the rapid and optimal control strategy to make accurate movement to reach the ideal location.

For the high acceleration high precision air suspension platform shown in Fig. 1, both the X and Y movement directions are in the form of parallel-mounted two-motor drive, the thrust direction of the two motors is the same, and the push rod is pushed to move.

The mathematical model of the high acceleration high precision air suspension platform is described by a linear system with multiple inputs and outputs, namely,

$$\begin{cases} x(k+1)=A_c x(k)+B_c u_c(k)+B_{c,w} w(k) \\ y_c\ (k)=C_c x(k)+v(k) \end{cases} \tag{1}$$

The system is affected by the internal and external environment of the system. Form in, $k \in Z^+$ is the number of time steps, $x=[x_1,\cdots,x_n]^T \in R^n$, $u_c=[u_{c1},\cdots,u_{cm}]^T \in R^m$, $y_c=[y_{c1},\cdots,y_{cp}]^T \in R^p$, $w=[w_1,\cdots,w_n]^T \in R^n$, $v=[v_1,\cdots,v_p]^T \in R^p$ are the system state vector, control vector, output vector, system disturbance and random measurement noise vector.

The system state vector is set to the displacement and velocity of the center of the fader, $x(k+1)=[x_1,x_2,v_1,v_2]^T$. Form in, $x_1$ and $x_2$ are the center of the displacement data items, which can be measured directly through the precision grating; Speed items $v_1$ and $v_2$ cannot be directly measured, the need to estimate the value of the algorithm; $A_c \in R^{n*n}$, $B_c \in R^{n*m}$, $C_c \in R^{q*n}$, $B_{c,w} \in R^{n*1}$ are the constant matrices with appropriate dimensions, which are subject to external disturbances.

$$A_c=\begin{bmatrix} 0 & 0 & 1 & 0 \\ 0 & 0 & 0 & 1 \\ A_{31} & A_{32} & 0 & A_{34} \\ 0 & A_{42} & A_{43} & A_{44} \end{bmatrix}, B_c=\begin{bmatrix} 0 & 0 \\ 0 & 0 \\ 1 & 1 \\ B_{41} & B_{42} \end{bmatrix}, C_c=\begin{bmatrix} 1 & 0 & 0 & 0 \\ 0 & 1 & 0 & 0 \end{bmatrix}, B_{c,w}=\begin{bmatrix} 1 & 0 & 0 & 0 \\ 0 & 1 & 0 & 0 \\ 0 & 0 & 1 & 0 \\ 0 & 0 & 0 & 1 \end{bmatrix}.$$

According to the speed synchronization deviation, the control strategy is divided into two parts: the optimal estimate of the state vector and the optimal gain of the system. Based on the established system dynamics model, the optimal control rate $u_c(k)$ is designed,

$$u_c(k)=-L(k)\cdot\left[x_1(k), x_2(k), \hat{v}_1(k), \hat{v}_2(k)\right]^T \tag{2}$$

Form in, $\hat{v}_1(k)$ , $\hat{v}_2(k)$ respectively represent the estimated values of the velocity terms $v_1$ , $v_2$ . According to the optimal control rate to be further designed as the control system performance index analysis of the optimal objective function value to minimize its value:

$$J_{Aoofv}=E\left[x^T(N)Q_1x(N)+\sum_{k=0}^{N-1}\left(x^T(k)Q_2x(k)+u_c{}^T(k)Ru_c(k)\right)\right] \tag{3}$$

Form in, $Q_1$, $Q_2$ are the state vector weighting matrix, which are the symmetric nonnegative matrix. $R$ is the input vector weighting matrix, which are the symmetric positive definite matrix.

## 3. Control performance evaluation

In order to evaluate the current high acceleration high precision air suspension motion platform controller performance deviates from the optimal performance of the controller, the random performance indicator is adopt to evaluate the performance of the control system. According to the control loop and input and output data to get the performance measurement of the control system, which is used to quantify the degree of performance of the evaluation control system under this control.

### *3.1. Control performance evaluation benchmark*

Since the velocity term of the system state vector is estimated from the initial value, the actual calculated velocity theoretical value should be the ideal value. Set the instantaneous velocity $v$ of time $t$: $v=\lim_{\Delta t\to 0}\frac{\Delta x}{\Delta t}$.

When $\Delta t$ is smaller, the average speed $\bar{v}$ is closer to the instantaneous velocity. Therefore, the optimal objective function of the ideal controller is:

$$J_{Ioofv}=E\left[x^T(N)Q_1x(N)+\sum_{k=0}^{N-1}\left(x^T(k)Q_2x(k)+u_c{}^T(k)Ru_c(k)\right)\right] \tag{4}$$

Form in, $x(k+1)=[x_1,x_2,\bar{v}_1,\bar{v}_2]^T$.

### *3.2. Control system evaluation index*

The optimal target value of the high acceleration high precision air suspension system is compared with the evaluation standard. The ratio is taken as the scale and criterion for evaluating the performance of the system:

$$\eta=\frac{J_{Aoofv}}{J_{Ioofv}} \tag{5}$$

The effect of noise perturbation and estimation vectors on the stability and robustness of the system are evaluated by this performance index. $J_{Aoofv}$ represent the optimal objective function value of the system; $J_{Ioofv}$ represent the system theory optimal objective function value.

The significance of the evaluation index $\eta$ is: the closer the value to the value 1 indicates that the control performance of the control system is less affected by the noise disturbance and the estimated vector, and the performance is better than the control performance without ideal. Specifically, the value of the evaluation index is expressed as a value between 0 and 1. When it is greater than or equal to 0.9, it indicates that the performance of the controller is excellent and when the evaluation value is between 0.6 and 0.9, the performance of the controller is good and when the evaluation value is less than 0.6, it means that the performance of the controller is low and needs improvement.

## 4. Simulation analysis

Based on the above mentioned above, a model of the air suspension system is established, considering a double input and double output discrete time system with a shape like (1), the initial state of each parameter and system is as follows:

IOP Conf. Series: Materials Science and Engineering **224** (2017) 012011 doi:10.1088/1757-899X/224/1/012011

$$A_c = \begin{bmatrix} 0 & 0 & 1 & 0 \\ 0 & 0 & 0 & 1 \\ 0.1 & -0.2 & 0 & -0.15 \\ 0 & -0.56 & 0.1 & -0.23 \end{bmatrix}, B_c = \begin{bmatrix} 0 & 0 \\ 0 & 0 \\ 1 & 1 \\ 321 & 348 \end{bmatrix}, C_c = \begin{bmatrix} 1 & 0 & 0 & 0 \\ 0 & 1 & 0 & 0 \end{bmatrix}, B_{c,w} = \begin{bmatrix} 1 & 0 & 0 & 0 \\ 0 & 1 & 0 & 0 \\ 0 & 0 & 1 & 0 \\ 0 & 0 & 0 & 1 \end{bmatrix}.$$

System disturbance: $w(k) \sim N(0, I_{4*4})$ ; Random measurement noise: $v(k) \sim N(0, I_{2*2})$ ; Initial state vector: $x(0) = [100,100,0,0]^T$ ;State estimate initial value: $\hat{x}(0) = [97.12 \quad 49.3 \quad 0.05 \quad 0.05]^T$ ; Filtering estimation error variance initial value: $p(0) = 4I_{4*4}$ ; State vector weighted matrix: $Q_1 = Q_2 = 25I_{4*4}$; Input vector weighted matrix: $R = I_{2*2}$ .

According to the system parameters and initial conditions given above, the kalman filter gain matrix $K(k)$ is calculated: $K = \begin{bmatrix} 0.7525 & -0.2238 \\ -0.2475 & 0.7762 \\ 0.0198 & -0.0587 \\ 0.0715 & -0.0970 \end{bmatrix}$. At the optimal gain for linear quadratic regulator is calculated: $L = \begin{bmatrix} 1.2885 & -2.5370 & -0.0071 & -1.9164 \\ -1.1905 & 2.3410 & 0.0071 & 1.7693 \end{bmatrix}$.

Based on the filter gain and optimal gain of linear quadratic regulator, the optimal state estimation can be further calculated, which is shown in Figure 2,and the output of the system corresponding is shown in Figure 3. Based on the proposed control performance evaluation index calculation method, the value is calculated, which is shown as Figure 4.

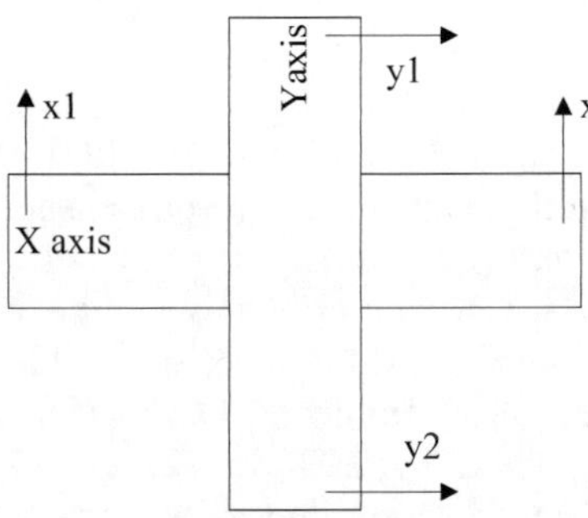


**Figure 1.** Structure schematic of high acceleration high precision air suspension platform.

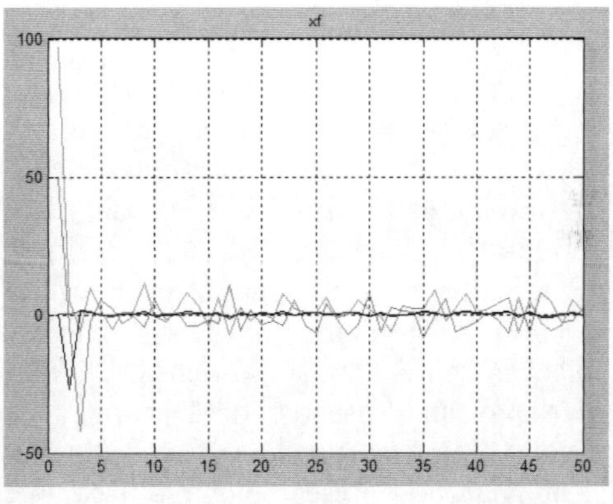


**Figure 2.** Optimal state vector estimation.

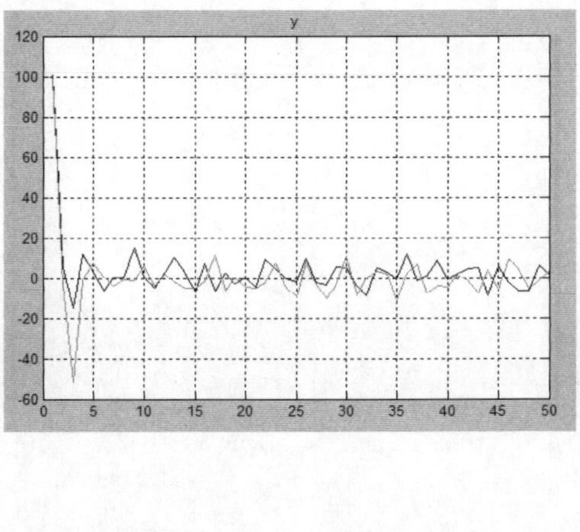


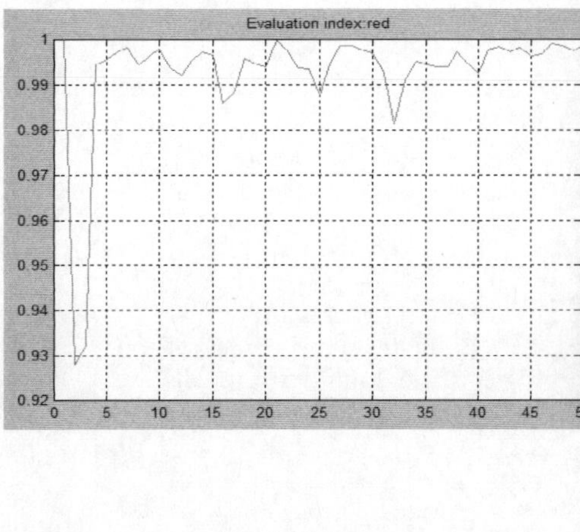

**Figure 3.** System output response.

**Figure 4.** High acceleration high precision air suspension system control performance index value.

It can be seen from the figure 4, high acceleration high precision air suspension system control performance index value eventually tends to 1, which means good control performance. This means that although the noise and disturbance have some influence on the system control performance, the system can maintain the stable output and have high control performance through the compensation effect of the designed control algorithm.

## 5. Conclusion

In this paper, the mathematical model of the system is established based on the characteristics of the high acceleration high precision air suspension system, and the current controller performance is calculated based on the optimal objective function value. The performance of the ideal controller is taken as the evaluation criterion, and the ratio of the optimal target value and the evaluation criterion of the precision air suspension system is taken as the performance evaluation index of the system. The performance index is used to quantify the influence of the performance of the air suspension system under this control strategy and the control performance of the system, and to evaluate the influence of the noise disturbance and the estimation vector on the stability and robustness of the system. Then the system performance is optimal.

## Acknowledgments

This paper is supported by Guangzhou Science and Technology Program under Grant 2016201604030045 and Guangdong academy of Sciences Funds of Young Scholar under Grant qnjj201604.

## References

[1] Erkorkmaz K, Gorniak J and Gordon D. Precision machine tool XY stage utilizing a planar air bearing arrangement[J]. *CIRP Annals-Manufacturing Technology*, 2010, 59(1): p 425–428.

[2] B.K. Yoo, W.C. Ham. Adaptive control of robot manipulator using fuzzy compensator, *IEEE Transactions on Fuzzy Systems,* 2000, vol 8(2), p 186-199.

[3] Viktorov V, Belforte G and Raparelli T. Modeling and identification of gas journal bearings: Externally pressurized gas bearing results[J]. *Journal of tribology,* 2005, 127(3): p 548–556.

[4] Dai Yi, Zhou Yunfei and Han Aiguo et al. Optimal Design of Controller Parameters in High-Speed/High-Accuracy Motions [J]. *China Mechanical Engineering*, 2004,15(12) : p 1045–1047.

[5] He Jiayuan. Research on Dynamic Performance of Linear Motor Servo System Based on PMAC[D]. *Nanjing University of Aeronautics and Astronautics*, 2013.

[6] Zhu Xiaogang. Research on multi-parameter control and trajectory accuracy of two-dimensional precision aerostatic bearing motion stage[D]. *Guangdong University of Technology*, 2016.

IOP Conf. Series: Materials Science and Engineering **224** (2017) 012012 doi:10.1088/1757-899X/224/1/012012

# Analysis of Design Parameters Effects on Vibration Characteristics of Fluidlastic Isolators

**DENG Jing-hui, CHENG Qi-you**

Science and Technology on Rotorcraft Aeromechanics Laboratory, China Helicopter Research and Development Institute, Jingdezhen, 333001, China

qy_cheng@163.com

**Abstract.** The control of vibration in helicopters which consists of reducing vibration levels below the acceptable limit is one of the key problems. The fluidlastic isolators become more and more widely used because the fluids are non-toxic, non-corrosive, nonflammable, and compatible with most elastomers and adhesives. In the field of the fluidlastic isolators design, the selection of design parameters is very important to obtain efficient vibration-suppressed. Aiming at getting the effect of design parameters on the property of fluidlastic isolator, a dynamic equation is set up based on the theory of dynamics. And the dynamic analysis is carried out. The influences of design parameters on the property of fluidlastic isolator are calculated. Dynamic analysis results have shown that fluidlastic isolator can reduce the vibration effectively. Analysis results also showed that the design parameters such as the fluid density, viscosity coefficient, stiffness (K1 and K2) and loss coefficient have obvious influence on the performance of isolator. The efficient vibration-suppressed can be obtained by the design optimization of parameters.

## 1. Introduction

The control of vibration in helicopters which consists of reducing vibration levels below the acceptable limit is one of the key problems. Increasing demands for expanding the flight envelop of helicopters, such as nap of earth flying, high speed, high maneuvers, coupled with the need to improved system reliability and reduce maintenance costs have resulted in more stringent vibration specifications.

Various methods have been applied to vibration control in the engineering field [1-5]. Traditionally, passive isolators and dampers are used to attenuate mechanical vibrations. The traditional approach to passive vibration isolation is to install relatively soft springs or elastomeric isolators to provide a low primary natural frequency [6-9]. These isolators would also incorporate sufficient damping to control resonant response. Soft systems with primary natural frequencies well below the N/rev exciting frequency are required to achieve isolation. Such systems result in large relative motion between the pylon and the airframe due to static loads. Natural frequencies low enough to isolate N/rev vibration would have static (1G) deflections up to 0.50 inches. Since flight controls and power transmission drive shafts cross this interface it is advantageous to keep the relative motion as small as practical. An effective method for isolating the N/rev vibration that did not allow large relative motions between the pylon and airframe was needed.

Fluidlastic products can also provide the long, predictable service life typical of elastomeric rotor bearings, because all of the relative motion across a Fludlastic device is accommodated by shear of the elastomer. There are no sliding seals, bushings, or bearings exposed to the environment [9]. Over a long service life, they typically show gradual degradation in appearance, and benign failure modes that allow for replacement long before their performance has been compromised. These characteristics also allow for on-condition replacement eliminating fixed service life intervals [10, 11]. Many fluidlastic products cannot be distinguished from a conventional bonged rubber part by their external appearance.

The present study is aimed at analyze the influence of design parameters on property of fluidlastic isolator. The dynamic equation is set up based on the theory of dynamics. And the dynamic analysis is carried out. The influences of design parameters on the property of fluidlastic isolator are calculated. The method is used to analyze the actuator optimization selection of helicopter.

## 2. Mathematical Model

Figure 1 shows the schematic diagram of fluidlastic isolator. The fluidlastic isolator consists of an inner cylinder and an outer cylinder concentrically bonded together with elastomers to form two chambers that are joined by a tuning port through the inner cylinder. The elastomer serves both as the seal for the chamber and the compliant strain member in the isolator. The chambers and the tunning port are completely filled with an inviscid fluid. As the inner cylinder moves up or down, producing an inertial force that cancels the elastomeric spring force at a discrete frequency.

A mechanical model used to analyze the dynamic performance of fluidlastic device is shown in Figure 2.

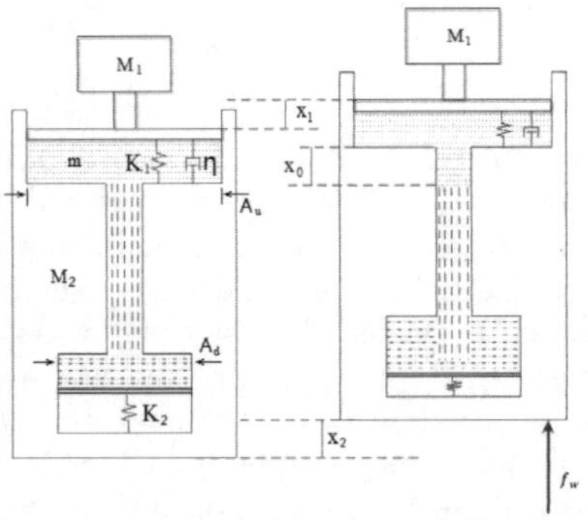

**Figure 1.** Cross-section of a fluidlastic Isolator

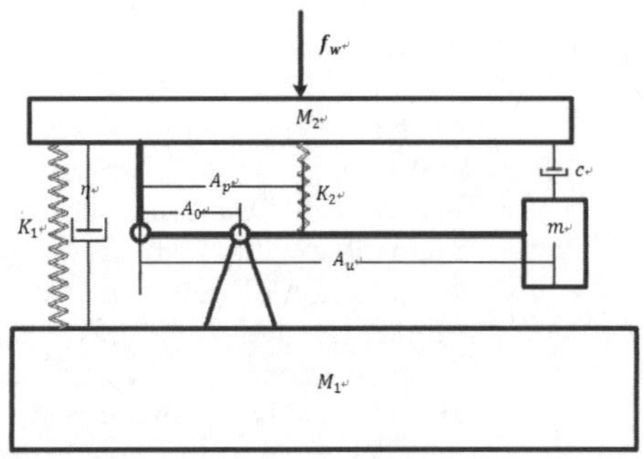

**Figure 2.** Mechanical Model of Isolator

According to the figure 1 and figure 2, the equations of motions (EOM) for the fluidlastic tuned isolator can be expressed as:

$$M_1\ddot{x}_1 = -K_1(1+i\eta)(x_1-x_2)-f-[m(\ddot{x}_0+\ddot{x}_2)+c\dot{x}_0]-K_2x_d \quad (1)$$

$$M_2\ddot{x}_2 = K_1(1+i\eta)(x_1-x_2)+f+K_2x_d+c\dot{x}_0+f_w \quad (2)$$

$$[m(\ddot{x}_0+\ddot{x}_2)+c\dot{x}_0](A_u-A_0)+K_2x_d(A_p-A_0)=fA_0 \quad (3)$$

Where, $f$ is the load that the M2 imposed on the lever mechanism. And Ap is defined as:

$$A_p = A_uA_0/A_d \quad (4)$$

The factors are defined as:

$$R_1 = \frac{A_u}{A_0} = \frac{x_0}{x_1-x_2} \quad (5)$$

$$R_2 = \frac{A_d}{A_0} = \frac{x_0}{x_d} \quad (6)$$

Taking Eq.(5) and Eq.(6) into the Eq.(3) ,We obtain the equation of $f$ :

$$f=[mR_1\ddot{x}_1-m(R_1-1)\ddot{x}_2+cR_1(\dot{x}_1-\dot{x}_2)](R_1-1)+K_2\frac{R_1}{R_2}(x_1-x_2)\left(\frac{R_1}{R_2}-1\right) \quad (7)$$

Taking Eq.(7) into the Eq.(1) and Eq.(2) ,We obtain the following relations:

$$M_1\ddot{x}_1+m\ddot{x}_1R_1^2+c\dot{x}_1R_1^2+K_1(1+i\eta)x_1+K_2\frac{R_1^2}{R_2^2}x_1=mR_1(R_1-1)\ddot{x}_2+c\dot{x}_2R_1^2+K_1(1+i\eta)x_2+K_2\frac{R_1^2}{R_2^2}x_2 \quad (8)$$

$$M_2\ddot{x}_2 + m(R_1-1)^2\ddot{x}_2 + c\dot{x}_2R_1^2 + K_1(1+i\eta)x_2 + K_2\frac{R_1^2}{R_2^2}x_2 = mR_1(R_1-1)\ddot{x}_1 + c\dot{x}_1R_1^2 + K_1(1+i\eta)x_1 + K_2\frac{R_1^2}{R_2^2}x_1 \quad (9)$$

The Eq.(8) and Eq.(9) can be expressed as:

$$\begin{bmatrix} R_2^2M_1 + R_1^2R_2^2m & -R_2^2mR_1(R_1-1) \\ -mR_1(R_1-1)R_2^2 & R_2^2M_2 + R_2^2m(R_1-1)^2 \end{bmatrix}\begin{Bmatrix}\ddot{x}_1\\ \ddot{x}_2\end{Bmatrix} + \begin{bmatrix} R_1^2R_2^2c & -R_1^2R_2^2c \\ -R_1^2R_2^2c & R_1^2R_2^2c \end{bmatrix}\begin{Bmatrix}\dot{x}_1\\ \dot{x}_2\end{Bmatrix}$$
$$+\begin{bmatrix} R_2^2K_1(1+i\eta)+R_1^2K_2 & -R_2^2K_1(1+i\eta)-R_1^2K_2 \\ -R_2^2K_1(1+i\eta)-R_1^2K_2 & R_2^2K_1(1+i\eta)+R_1^2K_2 \end{bmatrix}\begin{Bmatrix}x_1\\ x_2\end{Bmatrix} = \begin{Bmatrix}0\\ f_w\end{Bmatrix} \quad (10)$$

When the vibration load is a harmonic excitation, the load and displacement can be written as follows:

$$\left.\begin{aligned} f_w &= Fe^{j\omega t} \\ x_1 &= X_1e^{j\omega t} \\ x_2 &= X_2e^{j\omega t} \end{aligned}\right\} \quad (11)$$

The Eq.(11) is substituted into the Eqs.(10). We obtain the following relations:

$$\frac{X_1}{X_2} = \frac{\Delta_1}{\Delta_2} = \frac{\omega^2m(R_1-1)R_1R_2^2 - K_2R_1^2 - jc\omega R_1^2R_2^2 - K_1(1+i\eta)R_2^2}{\omega^2mR_1^2R_2^2 - K_2R_1^2 - jc\omega R_1^2R_2^2 + \omega^2M_1R_2^2 - K_1(1+i\eta)R_2^2} \quad (12)$$

From the equation (12), the Transmissibility of vibration can be written as:

$$T = \left|\frac{X_1}{X_2}\right| = \left|\frac{\Delta_1}{\Delta_2}\right| \quad (13)$$

When the $X_1, \Delta_1, c, \eta$ are all equal to zero, the frequency of undamped isolation system can be obtained:

$$f = \frac{1}{2\pi}\sqrt{\frac{K_1R_2^2 + K_2R_1^2}{(R_1-1)R_1R_2^2m}} \quad (14)$$

The amplitude of vibration can be written as:

$$X = \frac{R_1R_2^2F}{\left[K_1(1+i\eta)R_2^2 + K_2R_1^2\right] - m(R_1-1)^2R_2^2\omega^2 + jcR_1^2R_2^2\omega} \quad (15)$$

From the equation (15), the dynamic stiffness of isolation system can be obtained:

$$K^* = \frac{F}{X} = \frac{\left[K_1(1+i\eta)R_2^2 + K_2R_1^2\right] - m(R_1-1)^2R_2^2\omega^2 + jcR_1^2R_2^2\omega}{R_1R_2^2} \quad (16)$$

According to the static equation, the static stiffness of isolation system can be also obtained:

$$K = K_1 + K_2\frac{R_1^2}{R_2^2} \quad (17)$$

## 3. Results and Discussion

Main rotor speeds depend on the size of the helicopter but are typically low (160 to 450 RPM) because they are limited by tip speed and noise constraints. Since the number of rotor blades is generally between two and seven, N/rev frequencies are usually between 12 and 30 Hz. The frequency be focused in this paper is about 25 Hz.

The transmissibility versus stiffness (K1) curve is shown in Figure 3. It is observed that the K1 has obvious influence on the transmissibility. When the K1 increase to some value, the transmissibility will be attain to minimum value. This is means that the selection for K1 is very important to get a good effect of vibration-suppressed.

The influence of K2 on the Transmissibility of vibration is shown in Figure 4. The results show that there is the same trend of transmissibility curves at different K2. The transmissibilities at the three stations are all less than 0.1 between 20Hz and 30Hz. It is means that the fluidlastic isolators can provide efficient vibration-suppressed.

IOP Conf. Series: Materials Science and Engineering **224** (2017) 012012 doi:10.1088/1757-899X/224/1/012012

It is also observed that the K2 has obvious influence on the resonance frequency. The resonance frequency of isolator increased with K2. So we can adjust the K2 to make the frequency of isolator suit the excitation. This is an advantage of the fludilastic isolators.

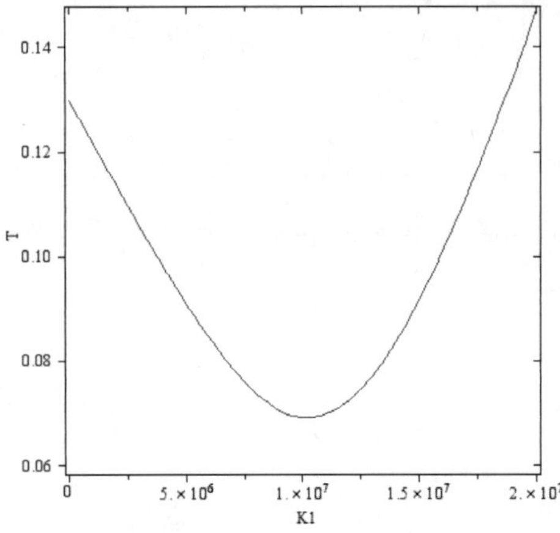


**Figure 3.** The influence of K1 on the Transmissibility

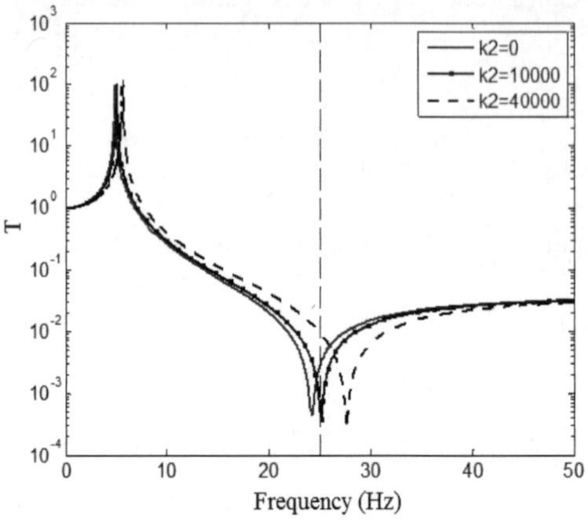


**Figure 4.** The influence of K2 on the Transmissibility

The influence of loss coefficient on the Transmissibility of vibration is shown in Figure 5. The results show that the loss coefficient has obvious influence on the transmissibility of vibration. With the increase of the loss coefficient , the transmissibility of vibration decreased. It is means that the efficiency of vibration-suppressed is improved. However, the result is also show that the loss coefficient has no influence on the resonance frequency of isolator.

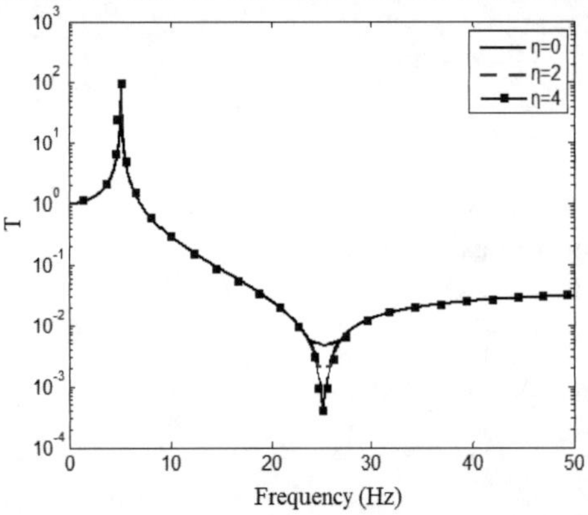


**Figure 5.** Loss coefficient vs transmissibility

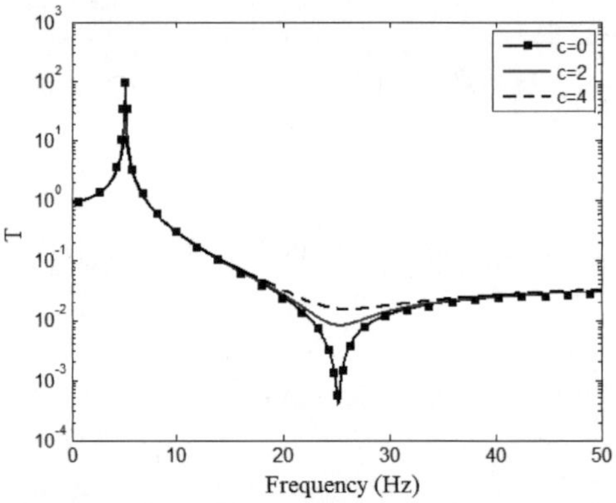


**Figure 6.** Viscosity coefficient vs transmissibility

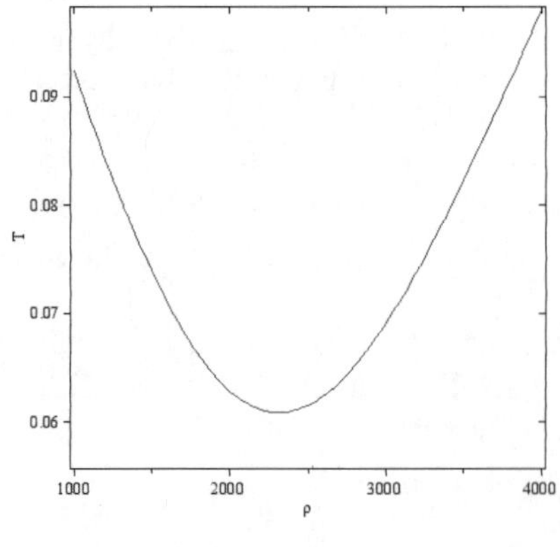


**Figure 7.** The density versus transmissibility

The influence of fluid viscosity coefficient on the Transmissibility of vibration is shown in Figure 6. The results show that the viscosity coefficient has obvious influence on the transmissibility of vibration. With the decrease of the viscosity coefficient, the transmissibility of vibration decreased. It is means that the efficiency of vibration-suppressed is improved.
The transmissibility versus fluid density curve is shown in Figure 7. The results show that the alteration trend of transmissibility influenced by fluid density is similar to that affected by K1. There is also an optimal value exists for fluid density to get a good effect of vibration-suppressed.
According to the equation (14), the nature frequency of isolator is decreased with the mass of fluid. When the nature frequency of isolator is varied from excitation frequency, the Transmissibility will be increased.

## 4. Conclusions

The dynamic equation is set up based on the theory of dynamics. The dynamic analysis is carried out. The influences of design parameters on the property of fluidlastic isolator are calculated. Findings are listed below:
1) Dynamic analysis results have shown that fluidlastic isolator can reduce the vibration effectively. The transmissibility can reach 0.1 between 20Hz and 30Hz. The performance of isolator can be designed by the design parameters.
2) The K1, K2 and loss coefficient have obvious influence on the performance of isolator. When the K1 increase to some value, the transmissibility will be attain to minimum value. The transmissibility of vibration decreased with the increase loss coefficient. So the selection for K1, K2 and loss coefficient is very important to get a good effect of vibration-suppressed.
3) The nature frequency and transmissibility of vibration are all affect by the fluid performance. To design a fluidlastic isolator with higher static stiffness and well vibration-suppressed efficiency, the density of fluid should be higher value. However the fluid viscosity coefficient should be lower value.

## References

[1] Strehlow, H., Mehlhose, R., Znika, P., Roth, D., "Review of MBB's Passive and Active Vibration Control Activities", The Royal Aeronautical Society in London, January 1990.
[2] Narayana Rao, K.S., Pranesh, B., Ravindranath, R., "Design, Development, and Validation of an Anti-Resonant Isolation System for Advanced Light Helicopter", 24th European Rotorcraft Forum, Marseille, France, September, 1998.
[3] John Heilmann, Doug Swanson, Askari Badre-Alam. VIBRATION ATTENUATION THROUGH THE USE OF ACTIVE FRAHMS. American Helicopter Society 59th Annual Forum, Phoenix Arizona, May 6-8, 2003.
[4] C.T. Hugin, C. Hatch. Global control of helicopter vibrations using a semi-active vibration control system. 47th AIAA/ASME/ASCE/AHS/ASC Structures, Structural Dynamics, and Materials Con, 1-4 May 2006, Newport, Rhode Island.
[5] Brown T J, Mccloud J L. Multicyclic control of a helicopter rotor considering the influence of vibration, load, and control motion. In: 36th annual National Forum of the AHS. Washington. D C, 1980, Washington. D C: AHS ,1980, 72.
[6] Gupta B P, Logan A H, Wood E R. Higher harmonic control for rotary wing aircraft. In: AIAA/AHS/ASE Aircraft Design Systems and Operations Metting, San Diego, California, 1984. New York: AIAA, 1984, 2484.
[7] Halwes, Dennis R, "Nodal Beam Inertial Flexure", U.S. Patent 3, 858,831, January 7, 1975.
[8] Flannelly, W. G., "Dynamic Anti-resonant Vibration Isolator", U.S. Patent 3,322,379, November 3, 1964.
[9] Smith, M. R., Redinger, W. S., "The Model 427 Pylon Isolation System", American Helicopter Society 55th Annual Forum, Montreal, Quebec, Canada, May 1999.
[10] McGuire, Dennis P, "Fluidlastic Dampers and Isolators for Vibration Control in Helicopters". American Helicopter Society 50th Annual Forum, Washington, DC, May, 1994.
[11] Dennis P. McGuire. Active Vibration Control Using Fluidlastic Pylon Struts. American Helicopter Society 62nd Annual Forum, Phoenix, Arizona May 9-11,2006.

IOP Conf. Series: Materials Science and Engineering **224** (2017) 012013 doi:10.1088/1757-899X/224/1/012013

# Improving Stiffness-to-weight Ratio of Spot-welded Structures based upon Nonlinear Finite Element Modelling

**Shengyong Zhang**

Department of Mechanical and Civil Engineering, College of Engineering and Sciences, Purdue University Northwest, USA

syzhang@pnw.edu

**Abstract.** Spot welding has been widely used for vehicle body construction due to its advantages of high speed and adaptability for automation. An effort to increase the stiffness-to-weight ratio of spot-welded structures is investigated based upon nonlinear finite element analysis. Topology optimization is conducted for reducing weight in the overlapping regions by choosing an appropriate topology. Three spot-welded models (lap, doubt-hat and T-shape) that approximate "typical" vehicle body components are studied for validating and illustrating the proposed method. It is concluded that removing underutilized material from overlapping regions can result in a significant increase in structural stiffness-to-weight ratio.

## 1. Introduction

Most automobiles and light trucks have their body-in-white (BIW) structures comprised of stamped or bonded panels connected with combinations of mechanical fasteners, resistance welds and bonding. Due to its advantages of high speed and adaptability for automation, resistance spot welding is conventionally the dominant joining technology for assembled vehicle construction. A modern automobile may contain 3000 to 4000 spot welds in its sheet metal structure. Over the last few years low density material (e.g. aluminium and magnesium alloys) has been suggested to substitute carbon steels for fabricating vehicle BIW. Again, it is noted that assembly of aluminium sheets is still dominated by resistance spot welding [1]. The integration of spot welds determines the overall body assembly stiffness and has an influence on the associated noise, vibration and harshness (NVH) characteristics.

Much of the recent research in the automotive industry has been motivated by the need to improve fuel economy and dynamic performance. One way to achieve such an improvement is to reduce BIW weight. The BIW typically represents about 30 percent of the total weight of a vehicle and is therefore a prime target for weight reduction. Among the strategies employed for lightweight BIW design are (1) low-density material substitution (aluminium, magnesium and thermoplastics are the most common candidates)[2, 3]; (2) manufacturing process improvement ( an array of new manufacturing techniques are being adopted by the automotive industry, such as hydro-forming, stamping of Tailor-welded blankets, etc.)[4, 5]; and (3) design architecture optimization (FE and sensitivity based optimization provides a promising way to reduce BIW weight, for example)[6-9]. The availability of advanced high-strength steel (AHSS) is emphasized as an emerging technology that would help reduce vehicle weight by allowing the use of reduced-gauge underbody parts. Reference [10] studies the microstructure and mechanical properties of spot weldments in various grades of AHSS and proposes a formula for measuring the effect of carbon and alloying elements on weld fusion zone hardness.

This study focuses on improving the effectiveness of material in the overlap regions of spot-welded structures (e.g. automotive BIW). Nonlinear FEA and topology optimization are employed to locate and remove underutilized material for maximizing the structural stiffness-to-weight ratio. Three spot-welded models with sections found commonly in vehicle body structures are examined as validation case studies to evaluate the proposed weight-reducing design.

## 2. Nonlinear FE model and topology optimization

### *2.1 Nonlinear FE model of a welded specimen*

Figure 1 shows the configurations and dimensions of the considered specimen. The specimen is composed of two sheets assembled using resistance spot welding. The sheets are made of A366 steel (Young's modulus E=190 GPa, Poisson's ratio ν=0.25). Each sheet is 120 mm long, 15 mm wide and 1 mm thick. The spot weld is at the centre of the overlap zone and has a diameter of 6 mm. Elastic material properties of the spot nugget are set to be E=200 GPa and ν=0.2. A gap of 0.02 mm between the two sheets is assumed in the overlap zone. Finite element model of the specimen is shown in Figure 2. Three-dimensional body elements are used in the discretization. The spot nugget in the specimen is meshed with layers of hexahedral elements by using volume sweeping, a technique with which one can fill an existing unmeshed volume with elements by sweeping the mesh from a source area to a target area throughout the volume. One end of the specimen is subjected to uniform loading, while the other end is clamped. Due to the symmetry in the structural geometry, external loading and boundary conditions about the specimen's longitudinal axis, only half of the specimen is modelled. Symmetric boundary conditions are applied to the plane of symmetry for restricting out-of-plane displacements and in-plane rotations.

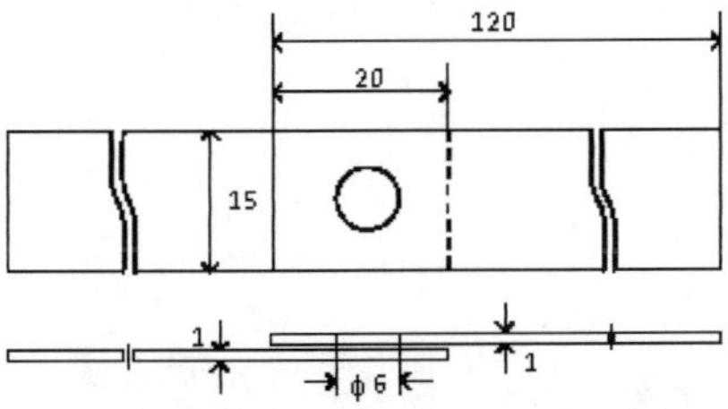


**Figure 1**. Spot welded specimen. All dimensions are in millimeters.

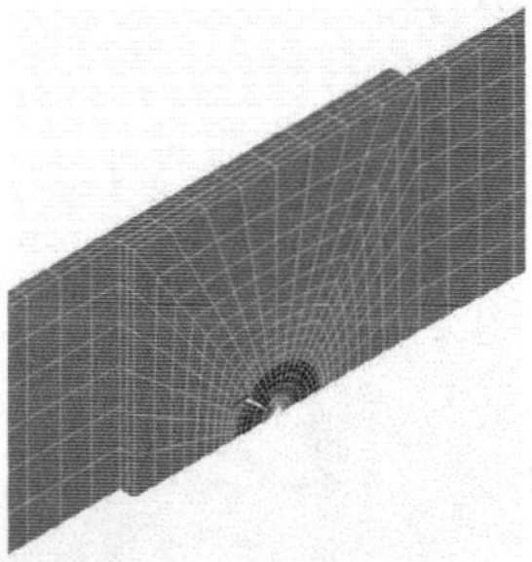

**Figure 2**. FE model with symmetric constraints.

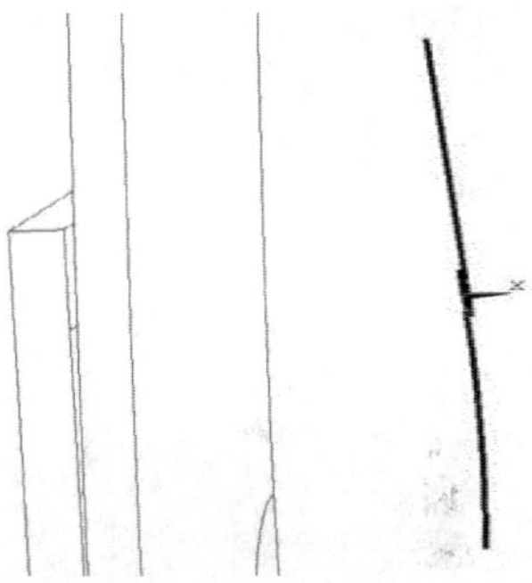

**Figure 3**. Deformations and nonlinear large displacement.

Linear FE analysis is conducted to investigate the specimen responses to the above prescribed loadings and constraints. It is found that contact and interference have occurred between the two sheets in the overlap zone. Ignoring the effects of contact and interference in the FE modelling will lead to incorrect predictions of the specimen compliance and other structural parameters. Contact problem has nonlinear behaviour involved and is regarded as a boundary nonlinear process. Because of the inherent complexities, it is difficult to develop analytical models to assess the contacting process. Most researches on this nonlinear behaviour focus on FE-based numerical approaches. In FE models various contact elements are utilized to simulate the contacting surfaces. Elements TARGE170 and CONTA173, for example, are a contact pair developed in ANSYS: element CONT173 describes the boundary of a deformable body and are potentially in contact with target surface defined by TARGE170. In this study 3-D surface-to-surface contact pairs are applied to the contact zone between the two sheets. Both the contact elements and target surfaces are defined as flexible bodies in the FE model. Figure 3 shows the deformed specimen in which surface contact is avoided due to the application of the contact pairs.

Geometric nonlinearity is another consideration in the modelling of this specimen for more accurate predictions. Geometric nonlinearity primarily refers to stiffness changes related to geometry constraints or the magnitude of strains. Unlike material nonlinearity, geometric stiffening has no hard transition marker. However, large displacement is an indicative of the transition from linear to nonlinear behaviour. In this study, the onset of geometric nonlinearity is assumed when displacement exceeds one-third of the sheet thickness, see Figure 3.

*2.2 Material removal for improving stiffness-to-weight ratio*

From material saving perspective, spot welded structures should be designed to have as less material as possible in the overlap zone. Stiffness-to-weight ratio is a measure of the effects of weight reduction efforts on the assembly stiffness of spot welded structures. Assembly stiffness is defined to be the quotient of an applied force and corresponding deformation of the structure at the same location. The weight used in calculating the stiffness-to-weight ratio refers to the mass of material contained in the overlap zone in this study. One objective of this research focuses on maximizing the assembly stiffness under specified amount of weight reduction in the overlap zone. Nonlinear FE analysis is conducted. Figure 4 shows the resulting von Mises stress distribution in the overlap zone when the lap-welded specimen is subjected to the above prescribed external loading and boundary constraints. The nonuniformities of the stress distributions are a result of the inherently no uniform internal load distributions required for static equilibrium constraints, providing guidance for evaluating the effectiveness of material in a loaded structure. Low stressed material is deemed to be underutilized. Removing or eliminating regions that contain relatively low levels of stress in the structure results in a near uniform stress distribution by reducing the stress variation. Fully stressed design, for example, is one optimum design methodology based on this criterion to distribute a limited amount of material in the design domain. Figure 5 shows the path plots of von Mises stresses calculated on the symmetry plane across the overlap zone. The two peaks on the plots correspond to the maximum stresses at either side of the nugget along the path. The stress concentrations shown in Figure 5 suggest that failures of spot welded structures most likely occur at the interception of the nugget boundary with the interface of the welded sheets.

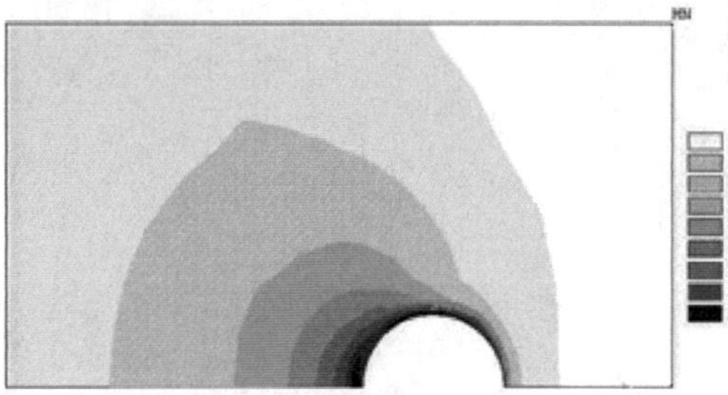

**Figure 4**. Nonuniform von Mises stress distribution in the overlap zone.

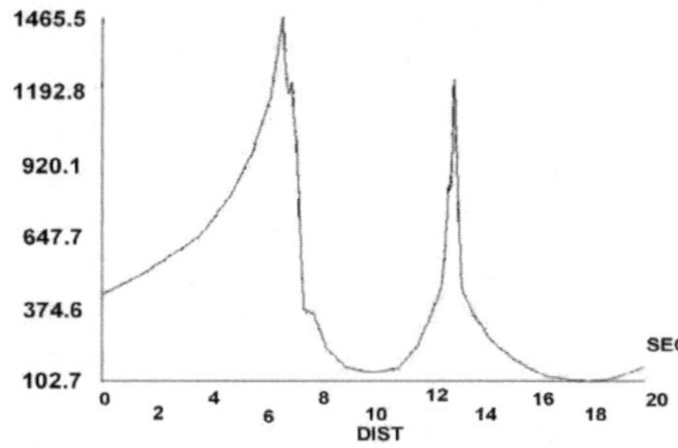


**Figure 5**. Path plot of von Mises stress across the overlap zone.

*2.3 Topology optimization*

Topology optimization is a form of shape optimization. The goal of topology optimization is to find the best use of material for a body such that an objective criterion (structural stiffness in this study) takes on a maximum value subject to given constraints (weight reduction in this study). Existing topology optimization routines work by removing materials based upon strain energy density or von Mises stress distributions in the design domain. In this study stiffness-based optimization is performed in order to reduce the specimen's weight and increase material effectiveness. Reducing weight is achieved by removing material from the sheets in the overlap zone. However, such material removal will result in a decrease in the structural stiffness. Optimization analysis aims to minimize the adversely influence of this effort. The objective function in this study is defined to be the specimen's compliance, because minimizing structural compliance is equivalent to maximizing the structural stiffness. The overlap zone of the welded sheets is to be optimized for improving material effectiveness. A 20% reduction of the volume in the optimized region is specified as the constraint, which means that 20% of the material in the overlap zone is to be removed in a manner that maximizes the stiffness with the given load configuration.

Figure 6 plots the material density of each finite element in the design domain, ranging from 0 to 1. Material density distributions provide an approach for determining material configuration in the most

IOP Conf. Series: Materials Science and Engineering **224** (2017) 012013 doi:10.1088/1757-899X/224/1/012013

effective way: if the density of an element is lower than user-specified threshold, the element is identified to be less stressed and can be eliminated from the design domain with negligible influence on the objective function. Figure 7 shows the optimum result with rough surfaces. Less stressed elements are invisible. The rough surfaces can be smoothed using appropriate algorithm based on geometries and results for better visualization. Optimum results provide the designer an insight from which regions the material can be removed for improving structural stiffness-to-weight ratio.

**Figure 6**. Distribution of density functions.

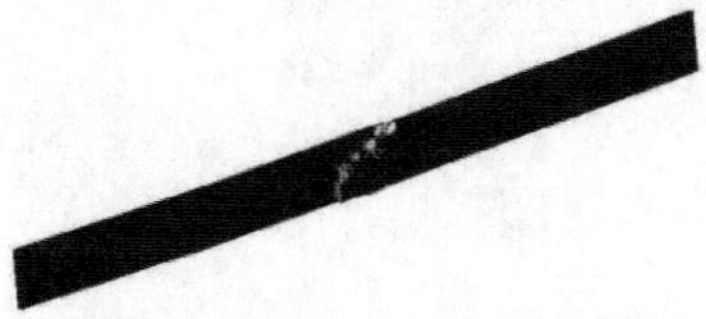

**Figure 7**. Elimination of less stressed material

## 3. Case studies

Three spot-welded models (lap, double-hat and T-shaped) are studied for validating the above-mentioned optimum design. The optimum process is illustrated by taking a lap model as an example. The baseline lap model is composed of two flat sheets assembled using multiple resistance spot welds with a pitch of 30 mm. Each sheet is 60 mm in length, 90 mm in width and 1 mm in thickness. Both contact nonlinearity and geometric nonlinearity are involved in the FE modeling. Topology optimization is performed when the lap model is subjected to a tensile load shown in Figure 8(a). The volume of material in the overlap zone is to be optimized for maximizing the model's stiffness-to-weight ratio. Figures 8(b) and 8(c) show the optimization results and demonstrate the regions where the material is not efficiently utilized. Modified lap model shown in Figure 8(d) is formed by removing material from these regions. Figures 8(e) and 8(f) plot the respective deformations of the baseline and modified lap models subject to a given set of forces and boundary constraints. The applied forces and corresponding deformations permit calculations of the assembly stiffness, which in turn are used to calculate the stiffness-to-weight ratio. Comparison results are reported in Table 1. The above optimization is based upon structural stiffness. The considered lap model can also be optimized for minimizing maximum stress. The overlapping width (W) and spot weld spacing (S) are defined as two design variables in this study. W is assumed to range from 15 to 30 mm and S from 20 to 40 mm. Optimization of these design variables is performed for maximum von Mises stress minimization.
Similar analysis is conducted for the double-hat model and T-shaped model separately. Results are shown and compared in Figures 9 and 10 and in Table 1. It is found that the assembly stiffness-to-weight ratio of each modified model has been improved significantly compared to that of the corresponding baseline model.

## 4. Conclusions

Most BIW structures are comprised of panels assembled by resistance spot welding. Reducing BIW weight has been an effective way for improving fuel economy and dynamic performance. Design architecture optimization has been employed to achieve such a weight reduction because it can provide internal force management that yields an optimized structure with equivalent stiffness at low BIW weight.
A FEA-based method is investigated to reduce the weight of spot-welded structures without a significant decrease in structural stiffness. Contact and geometric nonlinearities involved in the structure responses are simulated in the FE models for more accurate numerical predictions. Topology optimization is conducted to identify less stressed material in the overlapping regions. The identified material has low effectiveness and can be eliminated without significant influence on the structural stiffness. Finally, three types of spot-welded models with sections found commonly in vehicle body structures are analyzed and optimized to validate the proposed method for designing architecture with optimum structural stiffness-to-weight ratios.

**Table 1.** Increases in the stiffness-to-weight ratio of the modified models

| | Lap | Doubt-hat | T-shape |
|---|---|---|---|
| Stiffness-to-weight ratio | 1.24 | 1.25 | 1.22 |

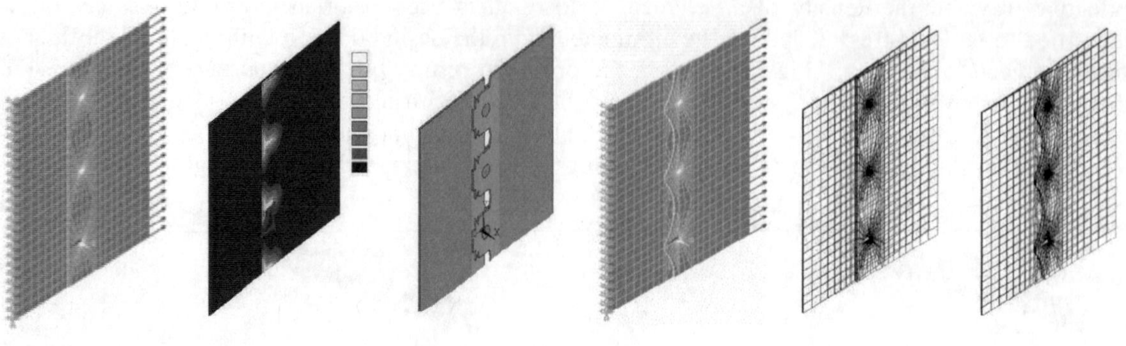

(a) (b) (c) (d) (e) (f)

**Figure 8. FE analysis and optimization of lap model. (a) baseline lap model; (b) distribution of variable density; (c) optimum results after removal of underutilized material; (d) modified lap model; (e) deformation of the baseline lap model; and (f) deformation of the modified lap model.**

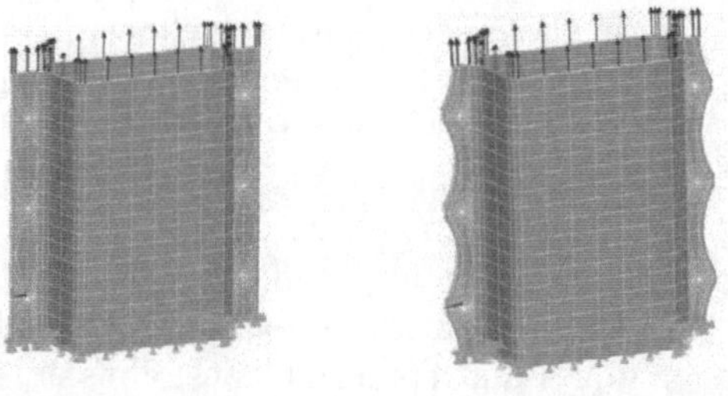

**Figure 9**. FE analysis and optimization of double-hat model. Baseline one and modified one

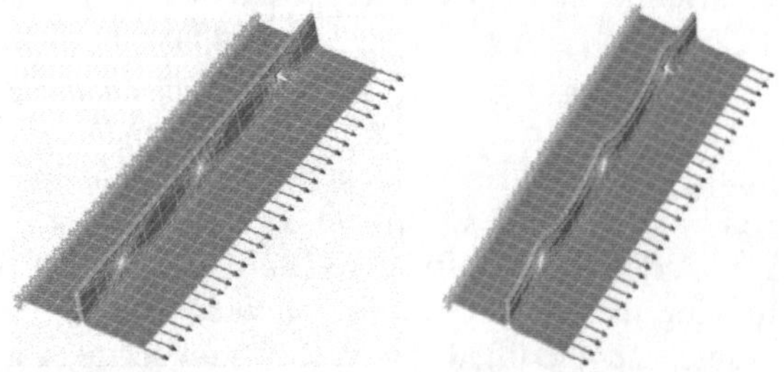

**Figure 10.** FE analysis and optimization of T-shaped model. Baseline one and modified one.

## Reference

[1] Gould, J.E. 2012 *Joining Aluminum Sheet in the Automotive Industry–A 30 Year History* Welding Journal, 91, 23s–34s.

[2] Chung, Y-D., Kang, H., and Cho, W-S. 2000 *The Development of Lightweight Vehicle using Aluminum Space Frame Body. Proceedings of the FISITA World Automotive Congress* Seoul, Korea, 12-15 Jane 2000, paper number F2000G361.

[3] Hirsch, J. 2011 *Aluminum in Innovative Light-Weight Car Design* Materials Transactions 52, pp 818–824.

[4] Kreis, O. and Hein, P. 2001 *Manufacturing System for the Integrated Hydroforming, Trimming and Welding of Sheet Metal Pairs* Journal of Materials Processing Technology, 115, pp 49-54.

[5] Köklü, U. 2013 *Investigation into the Formability of Al-1050 Tailor-Welded Blanks with Antilock Braking System* International Journal of Advanced Manufacturing Technology. 66, pp 221–229.

[6] Patton, R., Brehob, E., State, M., Furman, V., Geck, P. and Cummins, M. 2000 *Advanced Material Technologies for 21st Century Trucks* SAE paper no. 2000-01-3124.

[7] Baskin, D., Reed, D., Seel, M., Takacs, Z. and Vollmer, A.B.2008 *A Case Study in Structural Optimization of an Automotive Body-in-White Design* SAE Technical Paper 2008-01-0880.

[8] Christensen, J., Bastien, C., Blundel, M.V., etc. 2012 *Generation of Optimized Hybrid Electric Vehicle Body in White Architecture from a Styling Envelope*. Global Journal of Researchs in Engineering Automotive Engineering, 12(1), Version 1.0, pp 1–7.

[9] Li, Z. and Mei, J. 2012 *A Lightweight Optimization Method of Vehicle Body Structure Design* Proc. of FISITA 2012 World Automotive Congress, paper number F2012-E09-026, pp 1063–1074.

[10] Khan, M.I., Kuntz, M.L. etc. 2008 *Microstructure and Mechanical Properties of Resistance Spot Welded Advanced High Strength Steels* Materials Transactions. 49(7), pp 1629–1637.

IOP Conf. Series: Materials Science and Engineering **224** (2017) 012014 doi:10.1088/1757-899X/224/1/012014

# Thermoelastic Damping in Cone Microcantilever Resonator

**Pu Li[a] and Hongyue Zhou**

School of Mechanical Engineering, Southeast University, Jiangsu, China

[a]seulp@seu.edu.cn

**Abstract.** Microbeams with continuous or discontinuous variable cross-section have been applied in Microelectromechanical Systems (MEMS) resonators, such as tapered microbeam, torsion microbeam and stepped microbeam. Thermoelastic damping (TED), which is verified as a fundamental energy lost mechanism for microresonators, is calculated by the Zener's model and Lifshits and Roukes's (LR) model in general. However, for non-uniform microbeam resonators, these two classical models are not suitable in some cases. On the basis of Zener's theory, a TED model for cone microcantilever with rectangular cross-section has been derived in this study. The comparison of results obtained by the present model and Finite Element Method (FEM) model proves that the proposed model is able to predict TED value for cone microcantilever. In addition, TED in cone microcantilever is nearly same as TED in wedge microcantilever. The results show that quality factors (*Q*-factors) of cone microcantilever and wedge microcantilever are larger than *Q*-factor of uniform microbeam at low frequencies. The Debye peak value of a uniform microcantilever is equal to $0.5\Delta_E$, while those of cone microcantilever and wedge microcantilever are about $0.438\Delta_E$ and $0.428\Delta_E$, respectively.

## 1. Introduction

With a fast growth of advanced manufacture technologies of Microelectromechanical Systems (MEMS), microbeam resonators with discontinuously or continuously variable cross-section have already been applied in real applications [1-3]. Non-uniform microbeam resonators are considered seriously due to their enhanced properties. A permanent topic for MEMS researchers and designers is to obtain a higher quality factor (*Q*-factor). Thermoelastic damping (TED) is a significant and fundamental mechanism of energy lost in non-uniform microbeam resonators. Especially, TED determines the *Q*-factor of microresonators. TED is caused by an irreversible heat flux produced by temperature gradients in the vibrating microbeam. It is unfortunate that TED cannot be entirely avoided by the improved design and fabrication. Hence, to accurately predict the TED value for the design and optimization of non-uniform microbeam resonators is of importance. Zener [4-5] firstly provided a 1D model of TED in a uniform reed (beam) in 1937 and 1938. And the TED formula of Zener's model for the beam with thickness $h$ is expressed as follow

$$Q_{\text{Zener}}^{-1} = \frac{1}{2\pi}\frac{\Delta W}{W} = \frac{E\alpha^2 T_0}{C_V}\frac{\omega\tau}{1+(\omega\tau)^2} \tag{1}$$

where $\Delta W$ represents the work lost per cycle, $W$ represents the maximum stored energy per cycle. The thermal relax time $\tau = h^2 C_V / \pi^2 \kappa$, $E$ is the elastic modulus (Young's modulus), $T_0$ denotes the equilibrium temperature, $\kappa$ is the thermal conductivity, $\alpha$ denotes the coefficient of thermal expansion, $C_V$ denotes the specific heat, and $\omega$ is the vibrational frequency. Lifshitz and Roukes (LR) [6] proposed an accurate formula for TED in the uniform beam in 2000, which is given by

$$Q_{\mathrm{LR}}^{-1}=\frac{E\alpha^2T_0}{C_V}\frac{6}{\xi^2}\left[1-\frac{1}{\xi}\left(\frac{\sinh\ \xi+\sin\ \xi}{\cos\ \xi+\cosh\ \xi}\right)\right] \tag{2}$$

here the dimensionless variable $\xi=h\sqrt{\frac{\omega C_V}{2\kappa}}$.

Zener's model and LR's model are widely used to calculate TED for uniform microbeam resonators. However, Zener's model and LR's model are failed to provide an appropriate estimation in some cases. In 2006, Candler et al [7] studied TED in the slotted microbeam. They compared their experimental $Q$-factors with the theoretical $Q$-factors obtained from Zener's model, but found that Zener's model is not able to give an appropriate estimation. In 2011, Li et al [8] showed that LR's model is effective for microresonators with added proof masses when all supported cantilevers have the same and constant thickness. In 2013, Tunvir [9] investigated TED in double fixed stepped microbeams, and studied the effect of step position on thermoelastic dissipation. In 2016, Zhou et al [10] investigated TED in the wedge microcantilever resonator with rectangular cross-section. As papers reported above, Li et al [8] and Tunvir [9] both suggested that LR's model and Zener's model are not valid to evaluate TED in non-uniform microbeam resonators. The cone microcantilever is another typical microbeam with discontinuous variable cross-section, so it is meaningful to develop the corresponding TED model. In this study, TED model for the cone microcantilever whose width and thickness changes linearly is derived on the basis of Zener's theory [4].

## 2. TED definition of cone microcantilever

Figure 1 shows three types of microcantilevers with rectangular cross-section. Figure 1(a) shows a simple uniform cantilever, Figure 1(b) shows a wedge cantilever, and Figure 1(c) shows a cone cantilever. The cantilever is continuous, homogeneous and otropic. The $x$-axis is along the microcantilever length direction, and the $y$-axis is along the microcantilever thickness direction. The $h$, $b$ and $l$ represent right-end thickness, width, and length of microcantilever, respectively. And the thickness function $h(x)$ and cross-sectional area function $A(x)$ of the cone microcantilever shown in Figure 1(c) are $h(x)=hx/l$ and $A(x)=bh(x)$, respectively.

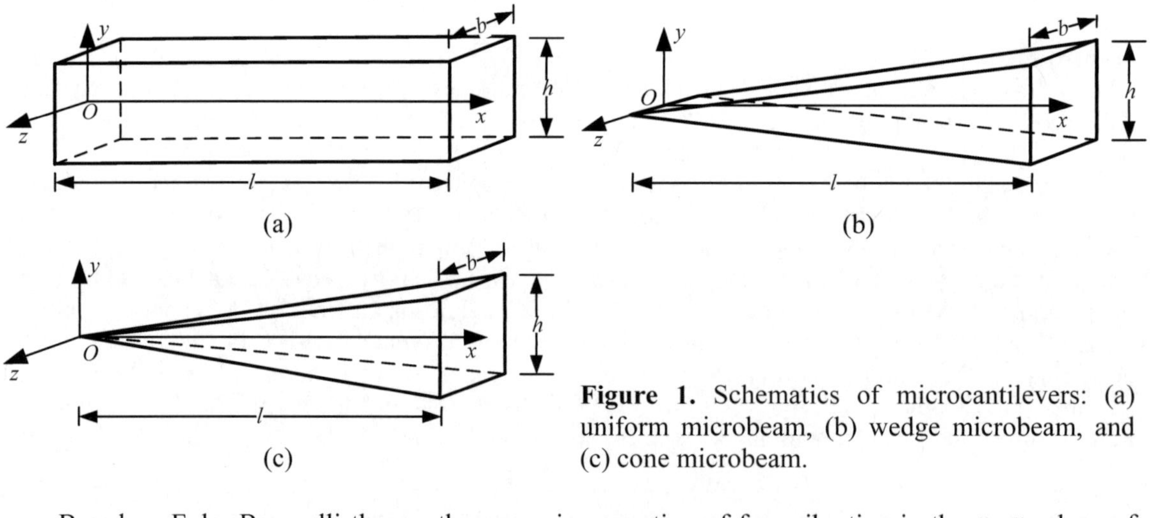


**Figure 1.** Schematics of microcantilevers: (a) uniform microbeam, (b) wedge microbeam, and (c) cone microbeam.

Based on Euler-Bernoulli theory, the governing equation of free vibration in the $x-y$ plane of cone microcantilever can be expressed as

$$\frac{\partial^2}{\partial x^2}\left[M_M(x)+M_T(x)\right]+\rho A(x)\frac{\partial^2 y(x,t)}{\partial t^2}=0 \tag{3}$$

here $\rho$ denotes the mass density, $y(x,t)$ represents the transverse vibrational displacement, $M_T(x)$ represents the thermoelastic moment , and $M_M(x)$ represents the mechanical moment.

The governing equation of 1D ($y$-direction) heat conduction is based on the Principle of Fourier heat conduction, which can be expressed as

$$\frac{\partial\theta(x,y,t)}{\partial t}=\chi\frac{\partial^2\theta(x,y,t)}{\partial y^2}+y\frac{\Delta_E}{\alpha}\frac{\partial}{\partial t}\frac{\partial^2 y(x,t)}{\partial x^2} \tag{4}$$

here $\theta(x,y,t)$ represents the temperature field function, $\Delta_E$ is called Zener's modulus and $\Delta_E=E\alpha^2T_0/C_V$.

In the paper [4], $\theta(x,y,t)$ is written as the total summation of thermal modes. It is worth noting that the first-order thermal mode affects thermoelastic energy dissipation seriously. Therefore, it is viable that $\theta(x,y,t)$ of cone microcantilever can be written as the first-order thermal mode. Additionally, the thermal condition of the cone microcantilever surface is assumed to be adiabatic. At last, one can get the formula of $\theta_0(x,y,t)$

$$\theta_0(x,y,t)=\hat{\theta}_0(x,y)\mathrm{e}^{\mathrm{i}\omega t} \tag{5}$$

here $\hat{\theta}_0$ is the amplitude of $\theta_0$ and written as

$$\begin{cases}\hat{\theta}_0(x,y)=\dfrac{E\alpha T_0}{C_V}Y_{xx}(x)\dfrac{\mathrm{i}\omega}{\mathrm{i}\omega+\mu_0}S_0\sin\left(\dfrac{1}{h(x)}\pi y\right)\\ \mu_0(x)=\dfrac{\chi\pi^2}{h^2(x)}\\ S_0(x,y)=\dfrac{2}{h(x)}\displaystyle\int_{-h(x)/2}^{h(x)/2}y\sin\left(\dfrac{1}{h(x)}\pi y\right)\mathrm{d}y=\dfrac{4h(x)}{\pi^2}\end{cases} \tag{6}$$

here $y(x,t)=Y(x)\mathrm{e}^{\mathrm{i}\omega t}$.

The total energy dissipation over one period due to TED can be calculated by [9]

$$\Delta W=\pi\iiint_V\hat{\sigma}\,\mathrm{Im}\left(\hat{\varepsilon}_{\mathrm{thermal}}\right)\mathrm{d}V=\pi\frac{bE^2\alpha^2T_0}{12C_V}\int_L\frac{\omega\chi\pi^2h^5(x)}{\omega^2h^4(x)+\chi^2\pi^4}Y_{xx}^2\mathrm{d}x \tag{7}$$

here the amplitude of thermal strain in cone microcantilever $\hat{\varepsilon}_{\mathrm{thermal}}=\alpha\hat{\theta}$, the amplitude of mechanical stress $\hat{\sigma}=EyY_{xx}$, and Im represents the imaginary part.

The maximum energy stored $W$ in the microcantilever can be calculated by[9]

$$W=\frac{1}{2}\iiint_V\sigma\varepsilon_{\mathrm{elastic}}\mathrm{d}V=\frac{bE}{24}\int_L h^3(x)Y_{xx}^2\mathrm{d}x \tag{8}$$

According to the energy definition [11], TED model in the cone microcantilever is given by

$$Q_{\mathrm{TED}}^{-1}=\frac{1}{2\pi}\frac{\Delta W}{W}=\chi\pi^2\Delta_E\frac{\displaystyle\int_L\frac{\omega b(x)h^5(x)}{\omega^2h^4(x)+\chi^2\pi^4}Y_{xx}^2\mathrm{d}x}{\displaystyle\int_L b(x)h^3(x)Y_{xx}^2\mathrm{d}x} \tag{9}$$

Without losing generality, the non-dimensional form of equation (9) can be

$$Q_{\mathrm{TED}}^{-1}=\Delta_E\frac{\displaystyle\int_0^1\frac{\eta X^6}{\eta^2X^4+1}Y_{XX}^2(X)\mathrm{d}X}{\displaystyle\int_0^1 X^4Y_{XX}^2(X)\mathrm{d}X} \tag{10}$$

where $\eta=\dfrac{\omega h^2}{\chi\pi^2}$. It is obvious that $Y(X)$ is an important step to calculate thermoelastic damping, and the formula of $Y(X)$ will be derived in Section 3.

## 3. Modal parameters

### *3.1. Modal shape function*

The non-dimensional thickness function, non-dimensional width function, and non-dimensional cross-sectional area function are $h(X)=hX$, $b(X)=bX$, and $A(X)=b(X)h(X)$, respectively. Here, the non-dimensional variable $X=x/l$. Because $M_T(x)$ is neglected compared to $M_M(x)$, the non-dimensional form of equation (3) is simplified as following:

$$\frac{\mathrm{d}^4Y(X)}{\mathrm{d}X^4}+\frac{8}{X}\frac{\mathrm{d}^3Y(X)}{\mathrm{d}X^3}+\frac{12}{X^2}\frac{\mathrm{d}^2Y(X)}{\mathrm{d}X^2}-\frac{\lambda^2}{X^2}Y(X)=0 \tag{11}$$

where $A_0=bh$, $I_0=bh^3/12$, $\lambda^2=\omega^2\rho A_0l^4/EI_0$.

The solution of equation (11) is given by [12]

$$Y(X)=\frac{1}{X}\left[C_1J_2\left(2\sqrt{\lambda X}\right)-C_2I_2\left(2\sqrt{\lambda X}\right)\right] \tag{12}$$

here $J_2$ is the 2nd order Bessel function of the first kind with real argument, and $I_2$ is the 2nd order Bessel function of the first kind with imaginary argument. The $C_1$ and $C_2$ are undetermined coefficients.

### 3.2. Modal eigen frequency

Considering boundary conditions of the cone microcantilever, the rotation and displacement of the right-end are zero, $Y(1)=Y'(1)=0$, then the following equations are

$$\begin{cases} J_2\left(2\sqrt{\lambda}\right)C_1-I_2\left(2\sqrt{\lambda}\right)C_2=0 \\ \left[\sqrt{\lambda}J_1\left(2\sqrt{\lambda}\right)-2J_2\left(2\sqrt{\lambda}\right)\right]C_1+\left[-\sqrt{\lambda}I_1\left(2\sqrt{\lambda}\right)+2I_2\left(2\sqrt{\lambda}\right)\right]C_2=0 \end{cases} \tag{13}$$

For meaningful solutions of $C_1$ and $C_2$, the eigen frequency equation is given by

$$\begin{vmatrix} J_2\left(2\sqrt{\lambda}\right) & -I_2\left(2\sqrt{\lambda}\right) \\ \sqrt{\lambda}J_1\left(2\sqrt{\lambda}\right)-2J_2\left(2\sqrt{\lambda}\right) & -\sqrt{\lambda}I_1\left(2\sqrt{\lambda}\right)+2I_2\left(2\sqrt{\lambda}\right) \end{vmatrix}=0 \tag{14}$$

And the equation (14) can be expanded to

$$J_1\left(2\sqrt{\lambda}\right)I_2\left(2\sqrt{\lambda}\right)-J_2\left(2\sqrt{\lambda}\right)I_1\left(2\sqrt{\lambda}\right)=0 \tag{15}$$

Solve the equation (15) in MATLAB and obtain the $\lambda$. By setting $C_2=1$ and $C_1=\gamma=I_2\left(2\sqrt{\lambda}\right)/J_2\left(2\sqrt{\lambda}\right)$, then the square of $Y_{XX}(X)$ is obtained by

$$Y_{XX}^2(X)=\frac{1}{X^7}\left[\lambda X^{1.5}(\gamma J_0-I_0)+4\sqrt{\lambda}X(-\gamma J_1+I_1)-6\sqrt{X}(\gamma J_0+I_0)+\frac{6}{\sqrt{\lambda}}(\gamma J_1+I_1)\right]^2 \tag{16}$$

$J_0$, $J_1$, $I_0$ and $I_1$ are all Bessel functions of the first kind with respect to $2\sqrt{\lambda X}$.

By substituting equation (16) and different structural geometry parameters into equation (10), then the correspongding value of TED can be computed by integration.

## 4. Validation and discussion

In this part, the results computed by the present model for cone microcantilevers are compared with the FEM results obtained by ANSYS. Solid226, which is a 3D 20-Node element, is applied in coupled thermal-structural analysis. A harmonic force $p\sin(\omega t)$ excites the cone microcantilever. The single-crystal silicon is applied in simulation. And the material properties of silicon [13] are listed in Table 1.

Firstly, the values of TED in first-ten order vibrational modes of wedge microcantilever [10] and cone microcantilever are investigated. For instance, these two kinds of microcantilevers have the same dimensions: $l=300\mu\text{m}$ and $h=b=10\mu\text{m}$. TED values against the modal order are shown in Figure 2.

As observed in Figure 2, results based on the present model match well with those based on the FEM model. The relative errors in two cases are smaller than 5% over the range of modal order. In reality, most of microbeam resonators only operate at their fundamental frequencies. TED of wedge microcantilever [10] and cone microcantilever in the first flexural mode as a function of virational frequencies are calculated as shown in Figure 3. As expected, the differences between the results obtained from the current model and the FEM model are also no more than 5%, and can be neglected. Figure 3 and Figure 4 show the perfect agreement proving that the present model is effective.

**Table 1.** Values of material properties of single-crystal silicon

| Symbols | Units | Values | Symbols | Units | Values |
|---|---|---|---|---|---|
| $E$ | GPa | 160 | $\kappa$ | W/m/K | 120 |
| $\rho$ | Kg/m$^3$ | 2330 | $\alpha$ | K$^{-1}$ | $2.6\times10^{-6}$ |
| $\upsilon$ | / | 0.22 | $\chi$ | m$^2$/s | $7.4033\times10^{-5}$ |
| $C_V$ | J/m$^3$/K | $1.6\times10^6$ | $T_0$ | K | 300 |

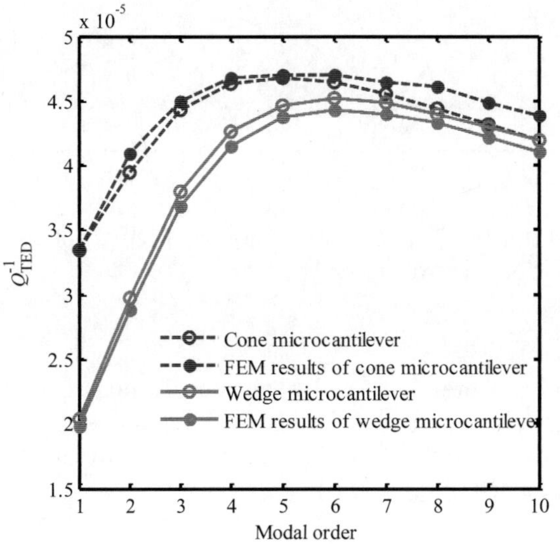


**Figure 2.** Comparison of TED values in first-ten order modes

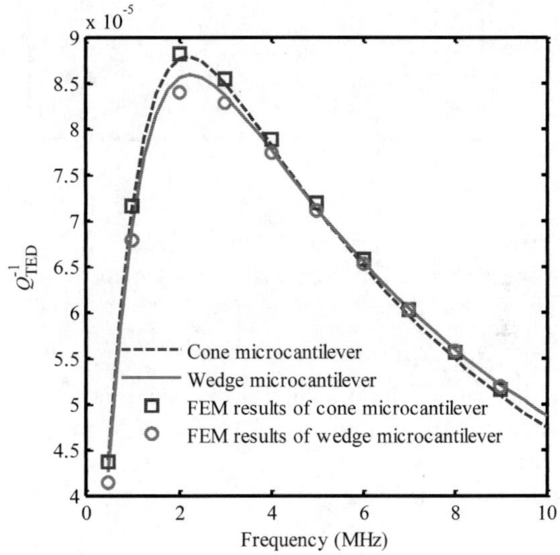


**Figure 3.** Comparison of TED values against frequencies. The two types of microcantilever resonators have the same dimension.

Figure 4 shows the normalized TED ( $Q^{-1}_{TED}/\Delta_E$ ) as a function of the variable $\eta$. There is almost no difference between TED curves of wedge microcantilever [10] and cone microcantilever. Moreover, this point suggests that TED in cone microcantilever does not depend on the width $b$, which is the same as the characteristic of wedge microcantilever [10]. However, the performance of wedge microcantilever and cone microcantilever is different from that the uniform microbeam. The small value of $\eta$ stands for a low frequency but a large aspect ratio $l/h$. For $\eta$ less than 1, TED values of the uniform microbeam are larger than those of two non-uniform microcantilevers, which means that two types of non-uniform microcantilevers own better performances and higher $Q$-factors. Thereby, for longer and slender microresonaters, the cone or wedge microcantilever should be chose in real application instead of the uniform microcantilever. On the other hand, it is found that the Debye peak value of cone microcantilever is about $0.438\Delta_E$, while that of Zener's model is $0.5\Delta_E$. In the paper [10], the Debye peak value of wedge microcantilever is about $0.428\Delta_E$. Moreover, the results suggest that for non-uniform microbeam resonators, TED is dependent of frequencies and the mode shape.

Figure 5 shows the imaginary region of temperature profile in a typical cone microcantilever. One can observe that tensional regions of the cone microcantilever lose temperature while compressed regions generate temperature. This phenomenon of interest is right caused by TED.

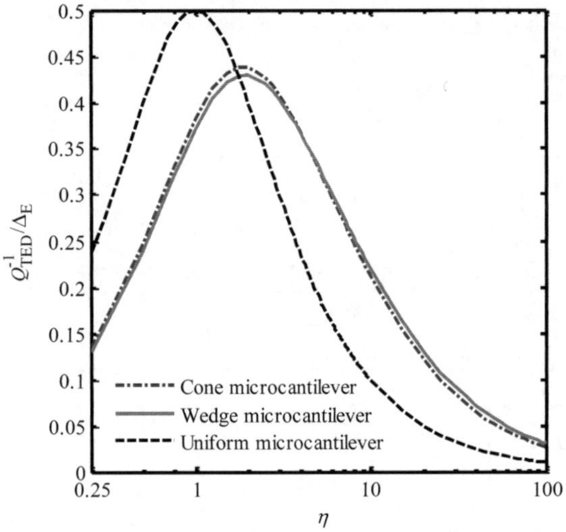


**Figure 4.** Comparison of normalized TED curves for three types of microcantilever resonators

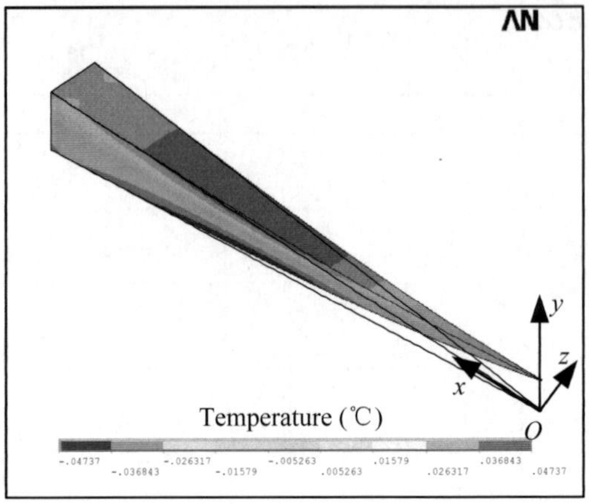


**Figure 5.** Imaginary region of a representative cone microcantilever

## 5. Conclusions

In this study, TED model for cone microcantilevers are presented and analyzed. Compared with numerical results obtained from the FEM model, and perfect agreements imply that the proposed TED model is correct and reasonable. The current work could help MEMS designers to achieve better performance (higher *Q*-factor) by reducing thermoelastic damping.

## Acknowledgements

We acknowledge the financial support from National Natural Science Foundation of China (Grant No. 51375091).

## References

[1] Cleland A N Pophristic M and Ferguson I 2001 Appl Phys Lett **79**(13) p 2070-2072.
[2] Liu T Li W and Huang Q 2009 *8th IEEE Conference on Sensors* p 234-237.
[3] May A L Ramirez P J Cortes L A et al 2011 IEEE Sens J **11**(6) p 1361-1374.
[4] Zener C 1937 Phys Rev **52**(3) p 230-236.
[5] Zener C 1938 Phys Rev **53**(1) p 90-99.
[6] Lifshitz R and Roukes M L 2000 Phys Rev B **61**(8) p 5600-5609.
[7] Candler R N Duwel A Varghese M et al 2006 J Microelectromech Syst **15**(4) p 927-934.
[8] Li P and Hu R F 2011 Jpn J Appl Phys **50**(7) p 077202.
[9] Tunvir K 2013 Microsyst Technol **19**(5) p 721-731.
[10] Zhou H Y Li P and Zuo W L 2016 *2016 IEEE International Conference on Mechatronics and Automation* p 1590-1595.
[11] Nowick A S and Berry B S 1972 *Anelastic relaxation in crystalline solids* (New York: Academic Press).
[12] Kamke E 1977 *Ordinary differential equations manual* (Beijing: Science Press).
[13] Yoneoka S Roper C S Candler R N et al 2010 J Microelectromech Syst **19**(2) p 357-366.

IOP Conf. Series: Materials Science and Engineering **224** (2017) 012015 doi:10.1088/1757-899X/224/1/012015

# Structure Design and Performance Analysis of High-Speed Miniature Ball Bearing

**Songsheng Li, Guoye Zhang, Bin Chen and Yuan Shen**

School of Mechatronic Engineering and Automation, Shanghai Key Laboratory of Intelligent Manufacturing and Robotics, Shanghai University, Shanghai, China, 200072

03810020@shu.edu.cn

**Abstract.** The working performances of miniature ball bearings are obviously affected by its' geometric structure parameters. In this paper, quasi-static analysis theory is applied in the design of miniature ball bearings. Firstly, it is studied the influence of geometry structure, preload and rotating speed on the dynamic performance of bearing. Secondly, bearing dynamic characteristics are analyzed which include the bearing stiffness and Spin to roll Ratio. Lastly, the contact stress and bearing life are calculated. The results indicate that structure parameters play an importance role in bearing's dynamic performances. Miniature ball bearings which have lager ball number, bigger ball diameter and smaller inner race groove radius can get better performances while velocity and preload have great impact on the bearing life. So that parameters of miniature bearing should be chosen cautiously.

## 1. Introduction

High speed miniature bearings are widely used in high speed instruments, which require high precision and long service life. In order to adapt to a variety of working conditions and satisfy interchangeability, the angular contact miniature ball bearings are often chose in these applications. The design of its geometric structure parameters and the selection of axial preload have close relationship with contract load, contract stress, service life and stiffness. The bearing performance can be affected by the subtle change of structure greatly due to its small geometry [1-2].

In structure design, machining method, performance analysis and measurement method of miniature bearing, many scholars have done a lot of researches. For calculating the critical speed of rotating machinery supported by miniature bearing, the stiffness of miniature bearings under various loads must be predicted first [3-5]. The friction torque is a very important performance of the miniature bearing, and the measurement of the torque is a difficult task in bearing test field Chen [6] has studied the miniature bearing and proposed that the better test data gathered must use more sophisticated test platform. In Gao's [7] paper the factors which cause the friction torque and the measuring methods for the friction torque of the miniature bearing are summarized and analyzed by him. Li [8] established an analysis modal for calculating high speed miniature ball bearing friction torque based on thermal elastohydrodynamic lubrication (TEHL), which has proved by using high precision miniature bearing testing machine. Zhang [9] and his team mainly analysis the design and manufacture of miniature air bearing, and explore the influence of microscale effect and slip flow effect on lubrication characteristics respectively. A new type of grinding process is proposed in Gui's [10] paper, which can effectively improve the wear resistant property of miniature bearing. A new kind of high strength alloy material is introduced by Tomasello [11], which can greatly improve the corrosion resistance and low temperature properties of miniature bearing. The damages due to contamination and high handling forces and preload options which are used to remove axial play and boost axial and radial stiffness are also presented [12]. According to Zhu's [13] paper the structure and design scheme of a closed–type miniature bearing are introduced, and the causes of the vibration of the bearing are also analyzed. Based on the quasi-dynamics of high-speed miniature ball bearing and the principle of energy conservation and taking the characteristics of lubrication oil and friction mechanism into account, a mathematic model of friction torque is established in Li's [14] paper. In this paper, based on the quasi-static model established by A B Jones [15], the calculation equations of miniature bearing parameters such as stiffness, life and contact stress are established.

In conclusion, it's difficult to design and test the performance of miniature bearing, and it is of great value to study the influence of miniature bearing geometry parameters on the contact load, contact stress and bearing life, stiffness. In this paper, the effects of miniature ball bearing geometric structure parameters such as ball number, ball diameter, contact angle, groove radius on bearing working performances are analyzed. Moreover, some suggestions are put forward on bearing parameter selection. Undoubtedly, it can provide guidance for the design of miniature bearings.

## 2. Analytical calculation model

According to the quasi-static model established by A B Jones, it is known that combine bearing's quasi-static equilibrium equation with displacement deformation compatibility equation, then iterate those equations can get many dynamic performance parameters, such as contact deformation, contact load, contact stress, ratio of revolution and rolling and bearing stiffness.

### *2.1. Contact stress and contact load of ball with ring*

Using the iterative method to solve the quasi-static equilibrium equation and displacement-deformation compatibility equation of the bearing can get the contact deformation $\delta_{ij}$ of every ball. According to Hertz contact theory, the contact load between ball and bearing ring is

$$Q_{ij} = K\delta_{ij}^{\frac{3}{2}} \tag{1}$$

$$K = \frac{1}{\left(1/K_1\right)^{\frac{2}{3}} + \left(1/K_2\right)^{\frac{2}{3}}} \tag{2}$$

where $Q_{ij}$ is contact load between the ball number j and raceway, for inner ring i =1, for outer ring i =2; $K$ is Load displacement coefficient; $K_1$ is outer ring stiffness; $K_2$ is inner ring stiffness; $\delta_{ij}$ is contact deformation between the ball number j and rings.

The max contact stress between ball and bearing ring is

$$\sigma_{ij} = \frac{3Q_{ij}}{2\pi ab} \tag{3}$$

where $a$, $b$ is long and short radius of Hertz contact ellipse respectively, the calculation process provided by Wang [16] and Harris [17].

### *2.2. Bearing stiffness*

Putting the contact load $Q_{ij}$, gyro torque $M_{gi}$ of balls and centrifugal force $F_{ci}$ into force balance equation (forma4-6), can get the total bearing loads, which include axial load $F_a$,, radial load $F_r$ and angular bending moment load $M$.

$$F_a = \sum_{j=1}^{z}\left[Q_{2j}\sin\beta_{2j} - \frac{2(1-\lambda)M_{gi}}{d}\cos\beta_{2j}\right] \tag{4}$$

$$F_r = \sum_{j=1}^{z}\left[Q_{2j}\cos\beta_{2j} + \frac{2(1-\lambda)M_{gi}}{d}\sin\beta_{2j}\right] \tag{5}$$

$$M = \sum_{j=1}^{z} \begin{bmatrix} Q_{2j} \sin \beta_{2j} \\ -\frac{2(1-\lambda) M_{gi}}{d} \left( R \cos \beta_{2j} - f_2 D_b \right) \cos \phi_j \end{bmatrix} \tag{6}$$

where $F_a$, $F_r$ is axial and radial load of bearing respectively; $M_{gj}$ is gyro torque of j ball; $\beta_{2j}$ is the contact angle between the j ball and the inner ring; $\varphi_j$ is position angle of j ball; $D_b$ is ball diameter; for outer ring $\lambda$=1, for inner ring $\lambda$=2.

$$[K] = \begin{bmatrix} \frac{\partial F_x}{\partial x} & \frac{\partial F_x}{\partial y} & \frac{\partial F_x}{\partial z} & \frac{\partial F_x}{\partial \theta_x} & \frac{\partial F_x}{\partial \theta y} \\ \frac{\partial F_y}{\partial x} & \frac{\partial F_y}{\partial y} & \frac{\partial F_y}{\partial z} & \frac{\partial F_y}{\partial \theta_x} & \frac{\partial F_y}{\partial \theta y} \\ \frac{\partial F_z}{\partial x} & \frac{\partial F_z}{\partial y} & \frac{\partial F_z}{\partial z} & \frac{\partial F_z}{\partial \theta_x} & \frac{\partial F_z}{\partial \theta y} \\ \frac{\partial M_x}{\partial x} & \frac{\partial M_x}{\partial y} & \frac{\partial M_x}{\partial z} & \frac{\partial M_x}{\partial \theta_x} & \frac{\partial M_x}{\partial \theta y} \\ \frac{\partial M_y}{\partial x} & \frac{\partial M_y}{\partial y} & \frac{\partial M_y}{\partial z} & \frac{\partial M_y}{\partial \theta_x} & \frac{\partial M_y}{\partial \theta y} \end{bmatrix} \tag{7}$$

where $x$, $y$, $z$ is the three coordinate axes of the bearing.

Using the variables in formula (3-5) to taking the derivative of the displacement ($\delta_x$、$\delta_y$、$\delta_z$、$\theta_x$、$\theta_y$) of the bearings on the five degrees of freedom, can get the relationship between bearing loads and displacement. The stiffness matrix of bearing on five degrees of freedom can be transformed into the above equation.

### *2.3. Spin to roll ratio*

Spin to roll ratio on behalf the size of spin motion of ball on the ring. With the increasing of the ratio, the spin becomes more violent, the bearing generates more thermal, and also deteriorate the lubrication. According to A B Jones raceway control theory [18], the spin motion of the ball occurs only in the non-controlled raceway, so spin to rolling ratio of the ball in controlled raceway is 0, and in non-controlled raceway the ratio can be got through formula (8) and (9):

$$\omega_{si} = \left[ -\omega_i \sin \beta_i + \omega_b \sin(\beta_i - \alpha) \right] \tag{8}$$

$$\omega_{roll} = -\omega_i \frac{D_m}{d} \tag{9}$$

where $\alpha$ is the angle between the ball spin axis and the central axis of the bearing; $\beta_i$ is the contact angle of the ball; $D_m$ is bearing pitch circle diameter; $\omega_i$ is the relative rotating speed between ring and cage, if the control ring is inner ring $i$ = 1, if not $i$ = 2.

### *2.4. Bearing fatigue life*

The fatigue life of bearing is discrete, according to literature [3], it can be got from the following formula:

$$L_i = \left( \frac{Q_{ci}}{Q_{ei}} \right)^3 \tag{10}$$

$$Q_{ci} = 98.1 \left( \frac{f_i}{f_i - 1} \right)^{0.41} \frac{(1 \mp}{(1 \mp} \quad , \quad , \quad \left. \frac{\gamma}{\text{)s} \beta_i} \right)^{0.3} D_b^{1.8} Z^{-\frac{1}{3}} \tag{11}$$

$$Q_{ei} = \left( \frac{1}{z} \sum_{j=1}^{j=z} Q_j^{\frac{1}{e}} \right)^e \tag{12}$$

where $Q_{ci}$ is the rated static load of the ring; $Q_{ei}$ is working load of the ring; $f_i$ is curvature radius of the ring; $Z$ is the number of the ball; for the rotating ring $e$=1/3, for static ring $e$=0.3.

### *2.5. Model verification*

In order to verify the correctness of the model, it is used to calculate the bearing parameters in Li's paper and compare the result with it. Bearing parameters shown in table 1, bearing structure shows in figure 1 and the results are as table 2.

**Table 1.** Bearing parameters in Li's [19] paper

| Parameters | Value |
|---|---|
| Number of balls | 11 |
| Diameter of ball ($D_b$/mm) | 3.175 |
| Pitch diameter of bearing ($d_m$/mm) | 16 |
| Contact angle ($\beta$/°) | 15 |
| Outer curvature coefficient ($f_1$[a]) | 0.55 |
| Inner curvature coefficient ($f_2$[b]) | 0.53 |
| Material | GCr15 |
| Axil force $P_r$ (N) | 70 |

[a] $f_1$=R$_1$/D$_b$, [b] $f_2$=R$_2$/D$_b$
where R$_1$, R$_2$ is the bearing outer and inner raceway curvature radius respectively.

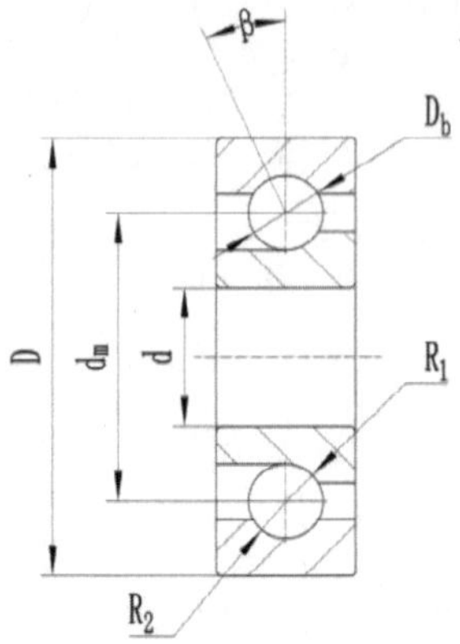

**Figure 1.** Schematic diagram of parameters

**Table 2.** Compare results of two papers

| Parameters | Result of Li's paper | Result of this paper |
|---|---|---|
| Contact stress $\sigma$/ (MPa) | 1200 | 1372 |
| Stiffness $K$/ (N·mm) | 22 | 21 |
| Spin to roll Ratio | 0.27 | 0.25 |
| Life $L$/ (h) | 6000 | 6600 |

It can be seen from table 2 that the results of this modal are close to Li's study. So, it indicates that the modal in this paper is credibility.

## 3. Design calculation and result analysis

Using computer programmes above formula, take a miniature bearing for example whose parameters are shown in table 3. Material of the bearing is GCr15. When it's working, the rotate speed of outer ring is 36000r/min and the load is 4N, using constant pressure preload. Because of the radial load the stress condition of each ball is inconsistent, and the contact load, contact deformation, contact stress, Spin to roll ratio changes with ball position angle, so it is necessary to analyze and calculate the force of all balls. According to paper [20] the reasonable range of bearing parameters is shown in table 4.

**Table 3.** Geometric structures parameters of this example.

| Parameter | Value |
|---|---|
| Number of balls | 6 |
| Diameter of bal ($D_b$/mm) | 1.2 |
| Pitch diameter of bearing ($d_m$/mm) | 4.17 |
| Contact angle ($\beta$/°) | 15 |
| Outer curvature coefficient ($f_1$) | 0.5337 |
| Inner curvature coefficient ($f_2$) | 0.5667 |

**Table 4.** Reasonable range of each bearing parameter

| Parameter | Value |
|---|---|
| Number of balls | 5~9 |
| Diameter of bal ($D_b$/mm) | 1.0~1.3 |
| Contact angle ($\beta$/°) | 0~45 |
| Outer curvature coefficient ($f_1$, $f_2$) | 0.52~0.59 |
| Rotational sp6eed ($n$/rpm) | $10^3$~$10^4$ |
| Axial preload ($F_a$/N) | 2~7 |

### *3.1. The max contact stress between ball and outer inner raceway*

When the bearing in a high speed, centrifugal force makes the contact stress in outer raceway much bigger then it in inner raceway. But because the curvature coefficient in inner and outer raceway is different, the contact stress is related to rotate speed and geometric construction. Figure 2 shows the max contact stress change trend, out ring of the bearing rotate speed is 36000r/min while inner ring is static. As shown in Figure 1a and Figure 1b, when the pitch circle of bearing is constant, increasing the number of ball properly can reduce the contact stress, but if too many balls increased will decrease the ball diameter. It is not conducive to reducing ring contact stress. According to Figure 2c, the contact stress decreases with the increasing of contact angle. According to Figure 2d and Figure 2e, contact stress in raceway is increased with the increase of its curvature radius. As shown in Figure 2f and Figure 2g, contact stress is increased with the increasing of axial preload. Meanwhile, as the rotate speed increasing, the contact stress in inner ring is reduced.

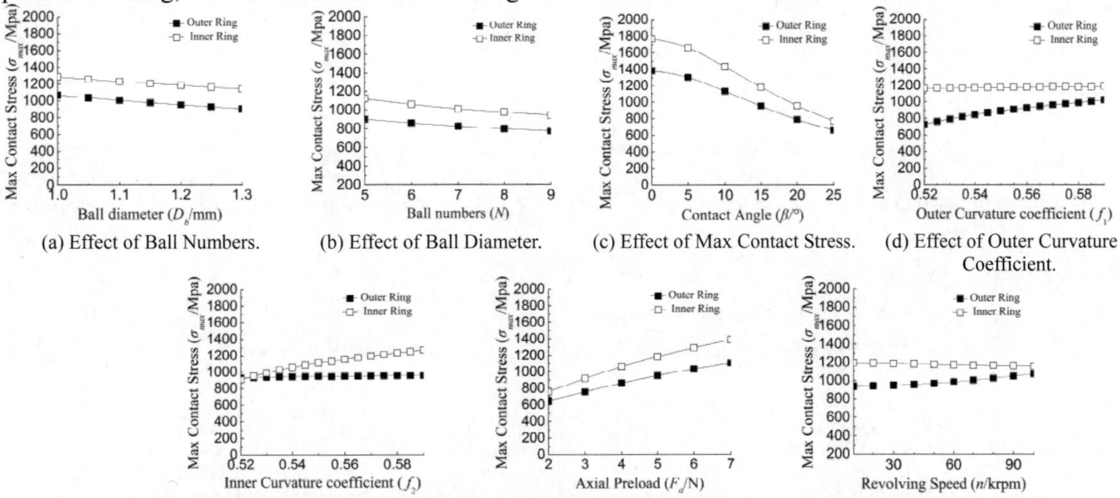


(a) Effect of Ball Numbers. (b) Effect of Ball Diameter. (c) Effect of Max Contact Stress. (d) Effect of Outer Curvature Coefficient.

(e) Effect of Inner Curvature Coefficient. (f) Effect of Axial Preload. (g) Effect of Revolving Speed.

**Figure 2.** The Max Contact Stress Change Trend Between Ball and Rings.

## 3.2. Bearing stiffness

According to Figure 3a and Figure 3b, stiffness increases with the increasing of ball numbers and diameter, especially radius stiffness. That because the more balls, the stronger bearing capacity. From Figure 3c, it can be known that with the increasing of contact angle, $K_a$ and $K_\theta$ increase obviously, but $K_r$ showed a downward trend. According to Figure 3d and Figure 3e, the bigger curvature radius coefficient is, the smaller the bearing stiffness becomes. According to Figure 3f and 3g, the bearing stiffness increases with the increasing of axial preload and the stiffness decreases with the increasing of rotational speed.

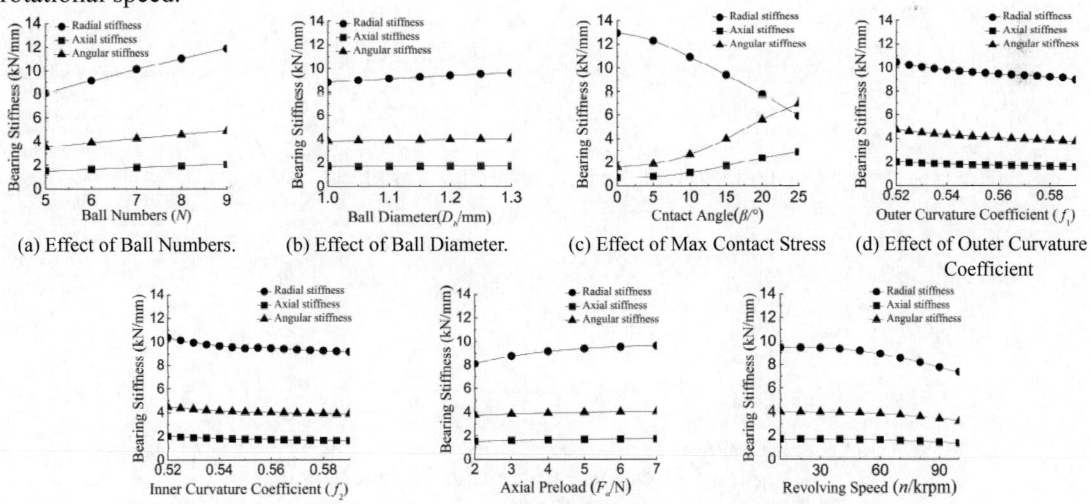


(a) Effect of Ball Numbers. (b) Effect of Ball Diameter. (c) Effect of Max Contact Stress (d) Effect of Outer Curvature Coefficient

(e) Effect of Inner Curvature Coefficient. (f) Effect of Axial Preload (g) Effect of Revolving Speed

**Figure 3.** Stiffness Change Trend.

## 3.3. Spin to roll ratio

The influence of bearing's parameters to spin to roll ratio is showed in Figure 4. According to raceway control theory, spin exists only on the non-control raceway. Figure 4a and Figure 4b shows that ratio increases with ball numbers and ball diameter. According to the Figure 4c, ratio increases with the contact angle significantly. According to the Figure 4d and 4e, $f_1$ has little effect on the ratio, and with the increasing of $f_2$ ratio will decrease. According to the Figure 4f and 4g, decrease radial force can lower the ratio.

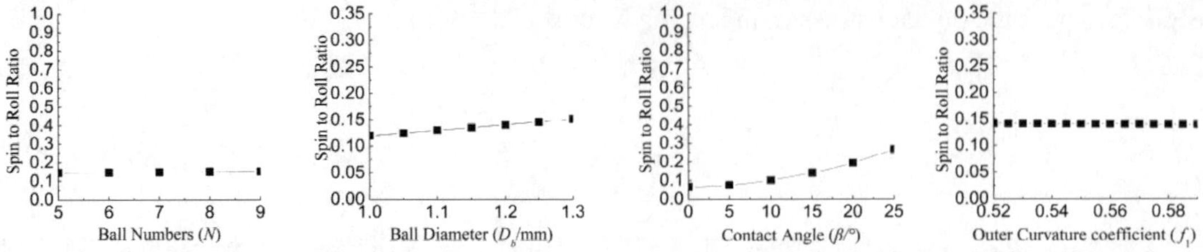


(a) Effect of Ball Numbers. (b) Effect of Ball Diameter. (c) Effect of Max Contact Stress. (d) Effect of Outer Curvature Coefficient.

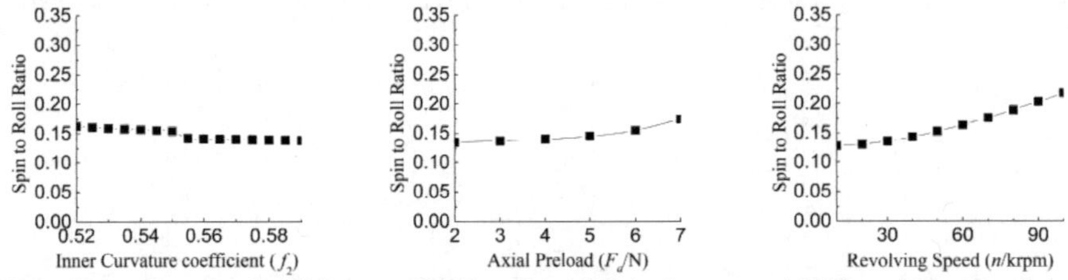


(e) Effect of Inner Curvature Coefficient. (f) Effect of Axial Preload. (g) Effect of Revolving Speed.

**Figure 4.** Spin to Rolling Ratio Change Trend.

### *3.4. Bearing fatigue life*

Figure 5 is the influence of the variation of parameters on the fatigue life of bearing *L*. According to Figure 5a and 5b, increasing ball numbers and ball diameter can prolong the fatigue life of the bearing. According to Figure 5c, increasing contact angle appropriately bearing life will be extended. According to Figure 5d and 5e, increasing $f_2$ will lead to the reduction of bearing life, while $f_1$ has little effect on fatigue life. According to Figure 5f and 5g, increasing axial preload and rotate speed can reduce the fatigue life of bearing.

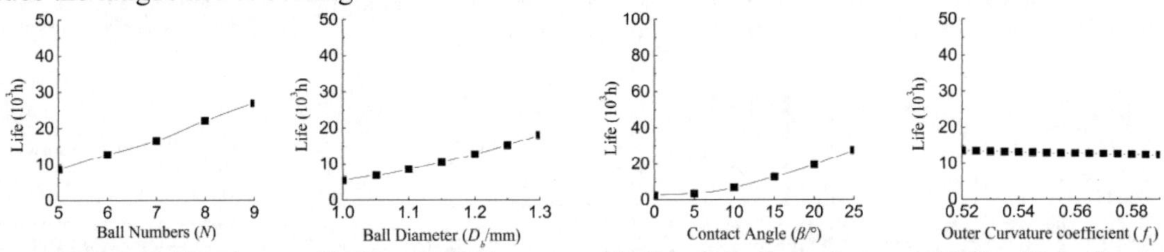


(a) Effect of Ball Numbers. (b) Effect of Ball Diameter. (c) Effect of Max Contact Stress. (d) Effect of Outer Curvature Coefficient.

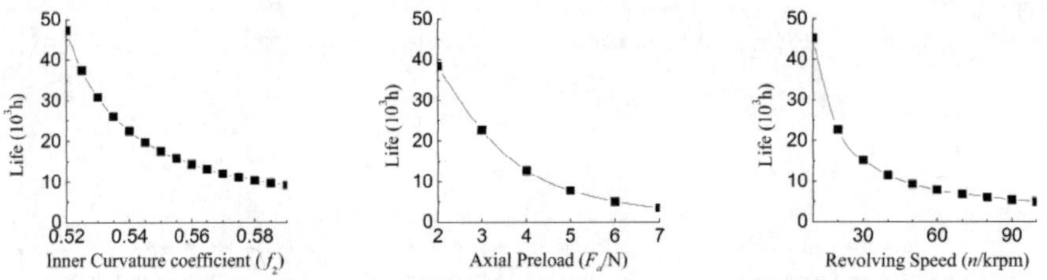


(e) Effect of Inner Curvature Coefficient. (f) Effect of Axial Preload. (g) Effect Of Revolving Speed.

**Figure 5.** Bearing Fatigue Life Change Trend.

## 4. Bearing design

(1) It is hoped that a working bearing has small contact load and contact stress, low Spin to roll ratio, high stiffness and longer service life. According to the above analysis, to achieve those targets the ball number and the diameter should be larger, however, Spin to roll ratio will be higher and heat generation will be intensify. Because of the significant influence of ball numbers on the stiffness and life, as well as the ball diameter's significant influence on Spin to roll ratio, it should be appropriate when increases the number of balls and decreases ball diameter.

(2) The bigger contact angle and the smaller contact stress, the longer bearing life will be. But small angle can reduce the bearing ratio and increase the radial stiffness. According to Shun's [21] paper, when the contact angle is 15 degrees, the friction torque of the bearing is the smallest. According to the study of Yang [22], when the miniature bearing is loaded, the contact angle will increase significantly compared with the normal bearing, so bearing contact angle should be a small value, the range from 10 degrees to 25 degrees will be better.

(3) Raceway curvature radius coefficient changes will lead to the change of control ring, and the increase of the curvature radius will enlarge the contact stress, reduces the stiffness and shortens the life. According to Deng [23] and Wang's [24] research, a smaller radius of curvature of the inner raceway contributes to the formation of a thicker lubricant film, and bearing's friction is small, which contributes to the formation of elastohydrodynamic lubricating film in its interior. Therefore, it is necessary to select a relatively small raceway curvature radius, and curvature radius of the inner raceway should be smaller than the outer.

Based on the above design principle, the improved parameters are shown in table 5. The table 6 shows the comparison of bearing performance parameters before and after improvement. Despite the slight increase of the contact stress and the Spin to roll ratio, the other parameters have been improved considerably.

**Table 5.** Parameters of improved bearing

| **Parameters** | **Value** | |
|---|---|---|
| | **Scheme1** | **Scheme2** |
| Number of balls | 7 | 8 |
| Diameter of ball ($D_b$/mm) | 1.2 | 1.0 |
| Pitch diameter of bearing ($d_m$/mm) | 4.16 | 4.16 |
| Contact angle ($\beta$/°) | 15 | 15 |
| Outer curvature coefficient ($f_1$) | 0.58 | 0.58 |
| Inner curvature coefficient ($f_2$) | 0.53 | 0.53 |

**Table 6.** Bearing performance comparison before and after improved

| **Parameters** | **Before improved** | **After improved** | |
|---|---|---|---|
| | | **Scheme 1** | **Scheme 2** |
| Contact stress $\sigma$/ (MPa) | Outer 859.9<br>Inner 1060.5 | Outer 940.6<br>Inner 952.8 | Outer 1013<br>Inner 993.9 |
| Stiffness $K$/ (N·mm) | Axial 1662.0<br>Radial 9166.5<br>Angle 3916.9 | Axial 1972.1<br>Radial 10832<br>Angle 4457.7 | Axial 2113.7<br>Radial 11313<br>Angle 4733.9 |
| Spin to roll Ratio | 0.15 | 0.16 | 0.13 |
| Life $L$/ (h) | 12751 | 14033 | 7953.7 |

## 5. Conclusions

(1) The performance of miniature bearing is greatly influenced by its internal geometry parameters, so the geometric parameters of each part should be selected reasonably. When the outer ring rotates, the appropriate increase of the number of balls can be considered, and the appropriate contact angle should be chosen.

(2) The stiffness of bearing can be improved by increasing the axial preload. At the same time, the contact load and contact stress will increase, bearing life will reduce. So when design bearings the requirement of bearing stiffness and life must be taken into account.

(3) Because of the ball centrifugal force, high rotating speed makes the bearing contact stress in outer ring increases obviously, and decreases in the inner ring which reduces bearing stiffness and life. The spin motion of the ball on the raceway is aggravated and the heating is serious, which should pay attention to when design bearings.

**Acknowledgements**
The study was supported by the National 'the twelfth five-year' Projects of Science and Technology of China (No. D.50-0109-15-001).

**References**
[1] Yu L M, Chen X Y and Gu J M 2011 *J. Machinery* **5** 71
[2] Beatty R F and Rowan B F 1982 *Determination of Ball Bearing Dynamic Stiffness*
[3] Okayasu A, Ohta T, Azuma T and Fujita T 1990 *Vibration Problem in the LE-7 LH2 Turbo Pump*. AIAA **90** 2250
[4] Butner M F, Murphy B T and Akian R A 1991 *J ASME Rotating Mach*. Veh. Dyn **35** 155
[5] Chen C Y, C S Liu and Y C Li 2015 *J. Microsystem Technologies* **21** 1
[6] Bugra H and Vance J M 2015 *J. Journal of Propulsion & Power* **20** 634
[7] Gao X 2003 *J. Shanghai Measurement & Testing* **6** 13
[8] Li S S, Chen J and Lin J 2013 *J. Lubrication Engineering* **8** 32
[9] Zhang W M, Meng G and Di C 2008 *J. Journal of Vibration and Shock* **5** 27
[10] Gui L F, Tang R J and Zhu Z L 1986 *Failure Analysis of Miniature Bearings J. ASM* 167
[11] Tomasello C M, Maloney J L and Ward P C. 1998 *J. ASTM Special Technical Publication* **1327** 437
[12] Vol N 2003 *J. Machine Design* **75** 54
[13] C D Zhu 2002 *J. Bearing* **12** 7
[14] Li S S, Lin J and Chen J 2014 *J. Journal of Shanghai University(Natural Science Edition)* **4** 429
[15] Jones A B 1978 *Rotor Bearing Dynamic Technology Design Guide, Part II: Ball Bearings* (New York: Shaker Research Corporation Ballston Lake) pp 96-115
[16] Wan C S 1978 *Rolling bearing analysis method* (Beijing: China Machine Press) p 113
[17] Harris T A 1991 *Rolling Bearing Analysis* (New York: A Wiley-Interscience Publication) p 231
[18] Jiang X Q 2001 *D.* Study on thermal characteristics of spindle bearing and its effect on speed and dynamics *Zhejiang University*
[19] Li S S 2006 *D.* Study on dynamic characteristics of ball bearing –rotor system in ultra high speed spindle *Shanghai University*
[20] Liu C H, Wei M Q and Chen L 2000 *J. Bearing* **10** 6
[21] Shun Z C and Z K Wang 1979 *J. Bearing* **1** 28
[22] Yang C G, Jia H G and B Liu 2010 *J. Computer Simulation* **27** 291
[23] Deng S E, Li X L and J G Wang 2011 *J. Journal of Mechanical Engineering* **47** 114
[24] Wang J G and Hong Y F 2002 *J. Machinery Design & Manufacture* **1** 59

IOP Conf. Series: Materials Science and Engineering **224** (2017) 012016 doi:10.1088/1757-899X/224/1/012016

# Innovation Design of Persimmon Processing Equipment Driven by Future Scenarios

**DUAN Xiao-fei [2], SU Xiu-juan [3], GUAN Lei [4] and ZHANG Wei-she [1,5]**

[1] School of Engineering Machinery, Chang' an University, Xi'an 710064, Shan Xi, China.

[2] sharfling@foxmail.com, [3] 739618547@qq.com, [4] 956461303@qq.com,
[5] zwshe@chd.edu.cn

**Abstract.** This article aims to discuss the methods of innovative by future scenarios design ,to help the designers be more effective of the design of persimmon processing machinery.By analyzing the persimmon traditional processing process, conceiving persimmon processing future scenarios and using the UXD and Morphological matrix , it can get the comprehensive function schemes.It Select the most optimal schemes which match the future scenarios best by illustrating the schematic design of the rotary-light Dried-persimmon Processing Machinery.It is feasible and effective to carry out the scenario design research and construct the reasonable future scenario, and combine the function analysis method to carry on the product plan innovation and the development.

## 1. Introduction

With the progress of industrial society, product replacement cycle is shorter and shorter, and effective product innovation design method can accelerate the development of new products. In the development of the product, to build the future scenario, the product concept can be gradually specific, conducive to finding unexpected discoveries(UXD) for innovative design[1] [2]. Based on the future scenarios, use functional analysis and morphological matrix for functional scheme integration, conducive to do a more comprehensive and more targeted design.

Persimmon processing in our country has a long history, mainly made to be dried persimmon. The existing persimmon processing equipment, have the problems of labor intensity, time-consuming and low-quality[3]. In this paper, the persimmon processing equipment scheme is innovatively designed by the combination of future scenarios, UXD and morphological matrices to improve the quality and production efficiency of persimmon and provide an effective method for the development and innovation of persimmon processing equipment.

## 2. The construction of persimmon processing in future scenarios

### *2.1 Build future scenarios of persimmon processing*

(1) Persimmon traditional processing needs to achieve the main functions are: cleaning, peeled, up-shelf, drying or baking, kneading, frosting, down-shelf and storage, etc.[4]. The processing and equipment we designed need to ensure the persimmon's taste and texture, as far as possible in accordance with the traditional process to achieve the function, do not merge in this deletion.

(2) Persimmon up-shelf and down-shelf are reciprocal behavior,which could combine into an action.

(3) The environmental space of "drying" scenario and the "frosting&storage"scenario are

IOP Conf. Series: Materials Science and Engineering **224** (2017) 012016 doi:10.1088/1757-899X/224/1/012016

similar,which can be combined and share a set of environmental space equipment; "up-shelf" and "down-shelf"are the mutual action, which can share a set of structures.

Among them, in the "persimmon peeling" scenario , the current semi-automatic peeled machine to meet the basic requirements, so follow the original technology and equipment. In the process of artificial kneading, the times of press is lesser, and the operation is sophisticated; on the contrary, the cost of the robot is expensive and the frequency of use is low, the effect is less than the manual. Figure 1 to Figure 3 shows the series of scenes, depicting the future of persimmon formation.

## 3. Design and optimization of persimmon processing equipment

In fact, the function of the product is often complex and diverse, the scheme design is the process of defining the product features and functional solving of the product[5]. In the process of design, the morphological matrix is constructed by the functional definition, functional decomposition and functional element solving of the product, and the possibilities of the design scheme are obtained. On this basis, the scheme design will be brought back to the future scenario that has been built, select the optimal solution which meet the design requirements.

### *3.1. Functional analysis*

Functional decomposition is an important step in the design of innovation[6], according to Figure 1 to Figure 3 the three scenarios in the future scene of the use of equipment, persimmon forming process, to get the functional decomposition results shown in Figure 4. The "persimmon up-shelf" and "persimmon down-shelf" share a set of equipment, the functions both are fixed persimmon. The three processes of "Drying", "frosting" and "storage" are mainly affected by the environment , which is related to the temperature, humidity, wind speed and other factors, collectively referred to as drying function.The "kneading" process need human intervention, defined as kneading processing function. The final shape of persimmon are shaped with the space and shape constraints, defined as shaping function.

### *3.2. Scheme integration*

According to the results of the functional analysis in Figure 4, the feasible realization of the corresponding functional scene is extracted by analyzing the unexpected discovery (UXD) and future scenarios of persimmon processing, resulting in possible sub-functional solution. According to the morphological matrix synthesis method to get a large number of possible design[7], as shown in Table 1, theoretically available persimmon processing and forming equipment, the number of integrated scheme N = 192.

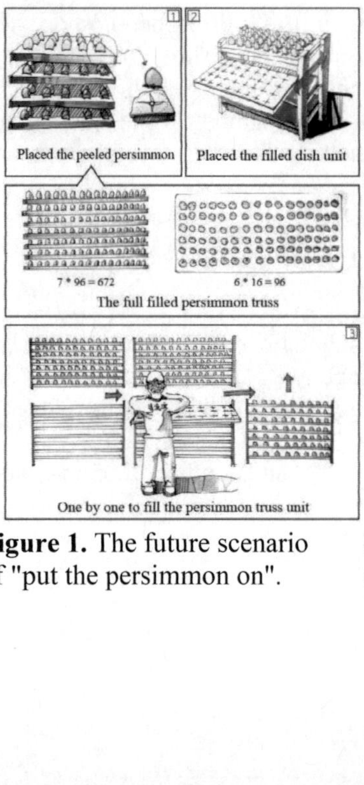


**Figure 1.** The future scenario of "put the persimmon on".

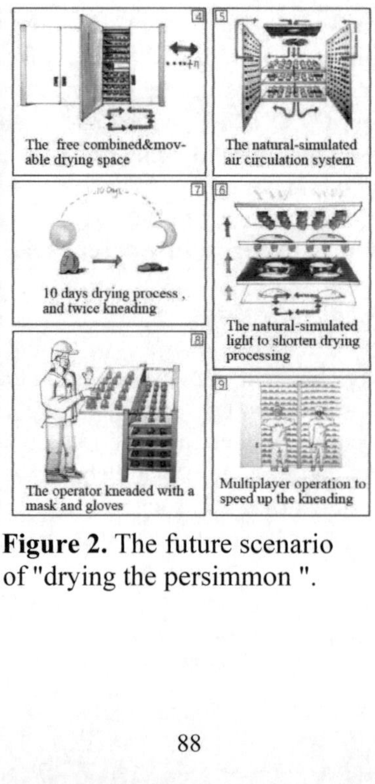


**Figure 2.** The future scenario of "drying the persimmon ".

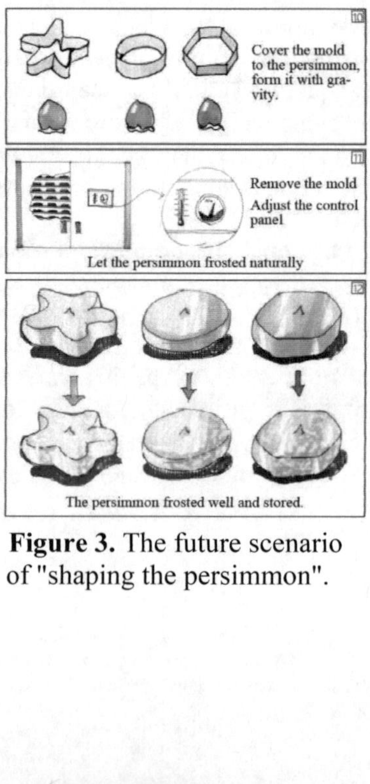


**Figure 3.** The future scenario of "shaping the persimmon".

IOP Conf. Series: Materials Science and Engineering **224** (2017) 012016 doi:10.1088/1757-899X/224/1/012016

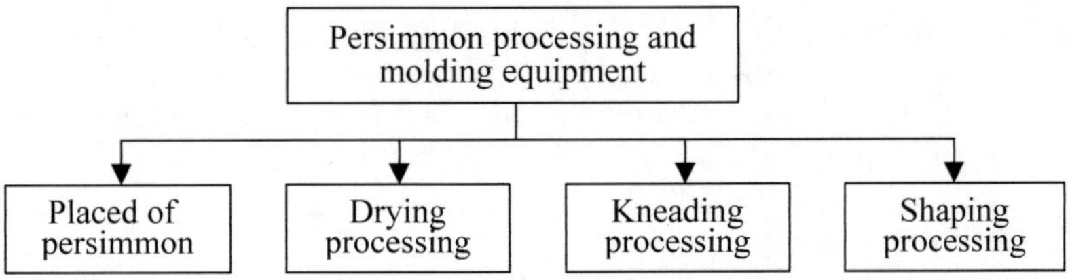


**Figure 4.** Persimmon processing equipment functional decomposition.

**Table 1.** The solving of morphological matrix.

| Sub-function | Possibility scheme 1 | 2 | 3 | 4 |
|---|---|---|---|---|
| A placed of Persimmon | 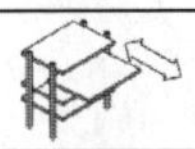 pulling plate | 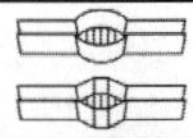 Mold fixed | 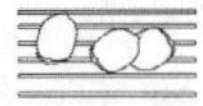 Free placement | 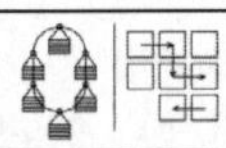 Space mobiling |
| B Drying processing | 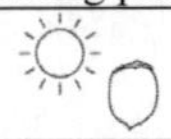 Natural light |  phototheral hot air | 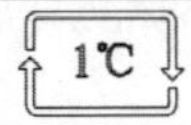 phototheral cold air |  Naturally dry |
| C kneading processing | 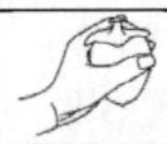 Hand kneading | 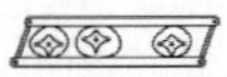 Four-bar mechanism | 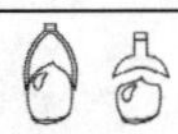 Manipulator kneading | |
| D shaping processing | 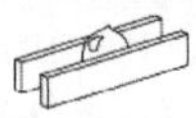 Horizontal squeeze | 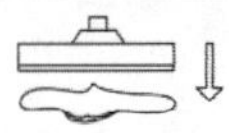 Vertical squeeze | 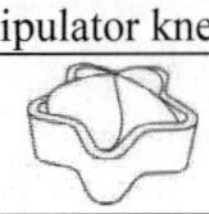 Mold shaping | 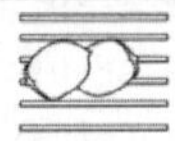 Natural forming |

### *3.3. Determine the feasible scheme*

Morphological matrix can provide a large number of solutions,and it is difficult to select, so according to the functional compatibility and the need gratification the two principles to screen, to get the following four feasible solutions.

Scheme 1: Semi-automatic Simple Persimmon Molding Equipment (A1 + B2 + C1 + D4). The dry time and the ambient temperature of this scheme are controllable. The use of artificial kneading and natural forming, effectively reduce the cost of equipment, but the labor workload. The pulling plate placing device can improve the equipment space utilization , and according to the site, the placing device can expand in both vertical and horizontal.

Scheme 2: Natural Light Rotary Persimmon Forming Equipment (A4 + B1 + C1 + D3). This scheme simulates the natural environment, to ensure persimmon taste, and the dry time and ambient temperature are controllable. Processing process transparent, there is a certain ornamental. But need manually kneading, the workload is greater. Due to the simulation of traditional processing methods, the production cycle is longer.

Scheme 3: Automatic Persimmon Forming Equipment (A2 + B2 + C2 + D1). This scheme has a high degree of automation, shorter processing and forming cycle. But the equipment structure is more complex, and the higher production costs.It is applicable to a single small amount of persimmon fast processing and shaping processing.

Scheme 4: Persimmon processing and rapid forming equipment (A3 + B3 + C3 + D3). This scheme has a high degree of automation,and the kneading process is unity, fine, and the results of the consistency of good. The machine and mold control forming, shaping faster, modeling more accurate.

### *3.4 Scheme Optimization*

The above four schemes are matched with the proposed future scenario model, and the application of the cost analysis method and the fuzzy evaluation method are used to optimize the scheme[8] [9], and

IOP Conf. Series: Materials Science and Engineering **224** (2017) 012016 doi:10.1088/1757-899X/224/1/012016

select the optimal scheme. Scheme 2 Compared with other programs, its advantages are: ① mobile truss components transform from horizontal to multi-layer rotation, making full use of space, at the same time, facilitating the flow of air in the equipment ; ② the equipment appearance transform from the opaque box to the transparent shell, using ambient light to drying, which is more energy efficient.

## 4. Functional design of the scheme

Design the structure of the optimal functional design scheme, that is, natural light rotary persimmon forming equipment, and considering the human-machine-environment constraints to optimize the scheme[10], the final model is expressed by the Computer three - dimensional model. The result is shown in Figure 5.

### *4.1. Persimmon placement structure design*

Persimmon "kneading area" and "mold forming area" have the door with 3 × 1.2m , as shown in Figure 6. The staff can fix the persimmon on the persimmon dish from the both sides ,after the completion of the persimmon formation ,also from this area off the shelf.

### *4.2 .Drying processing structure design*

Equipment drying processing mainly through air flow and natural light to exchange the heat produced by persimmon, improve the efficiency of persimmon processing. In the drying process ,the truss module is in a rotating state, powered by the motor, and assisted by chain and orbital to rotate, as shown in Figure 7. The fan provides the air source of the internal space, and the airflow discharges moisture from the outlet along the path shown in Figure 8 (a), and can be in full contact with the persimmon surface. In the truss, the persimmon is placed staggered between the upper and lower layers, increased the airflow and persimmon contact opportunities, and the air flow has a good guiding effect because of the truss trapezoidal mouth design, as shown in Figure 8 (b) below.

**Figure 5.** Construction Scheme of Light Rotary Persimmon Forming Equipment.

**Figure 6.** Double door structure scheme.

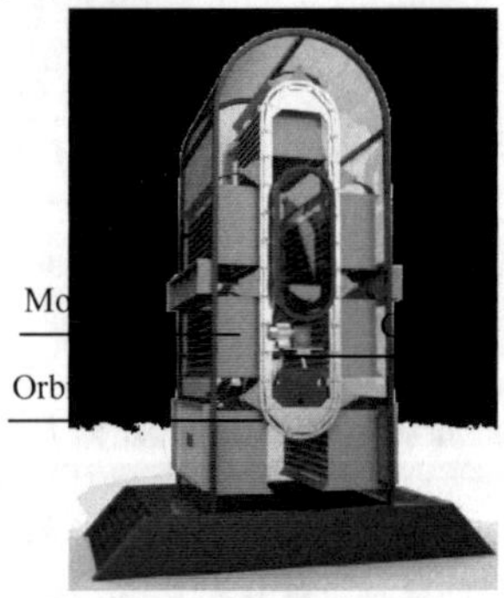

**Figure 7.** Rotation structure scheme.

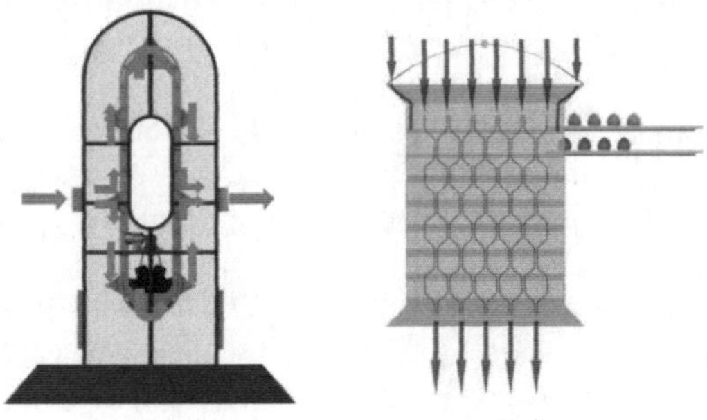

(a) (b)

**Figure 8.** Closed space airflow path

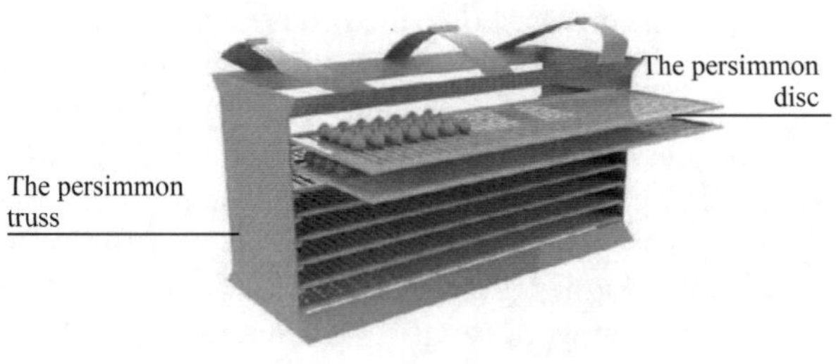


**Figure 9.** Structure of frame and persimmon disc

### *4.3. Kneading processing structure design*

In the process of persimmon formation, the persimmon truss is fixed on the rotating structure. When the persimmon truss unit is rotated to the kneading area, the persimmon is pressed . In the persimmon truss unit, the persimmon disc adopts the pull type, as shown in Fig. 9, to facilitate the operation and speed up the kneading speed. Here can use the external equipment to pinch or manual press pinch in two ways, which is high flexibility.

*4.4 Persimmon shaping structure design*

The persimmon shape is implemented in the mold forming area, in the same time the other side of the device can knead the persimmon. It use mold to forming, as shown in Figure 10, put the corresponding mold of persimmon plate just on the persimmon for a period of time, and use the force of gravity to shape the persimmon. For the kneaded persimmon, do the forming operation.The the force of gravity of the mold forming, the standing time and the pressing times are all from experiments. "Five-pointed star" persimmon shaping results shown in Figure 11.

## 5. Conclusion

In this paper, based on the construction of future persimmon forming scenario model, the functional matrix of the persimmon shape forming equipment is solved and the morphological matrix is used to synthesize multiple functional schemes. The optimal four functional schemes are matched with the future situation model. Finally, a functional scheme of persimmon shape forming equipment with high matching degree is obtained: the light rotary persimmon shape forming equipment,and expressed and illustrated by the three-dimensional entity. It verified that construct the reasonable future scenario and combine the functional analysis method to innovate and develop the product scheme is feasible and effective .

**Figure 10.** Persimmon shape forming mold.

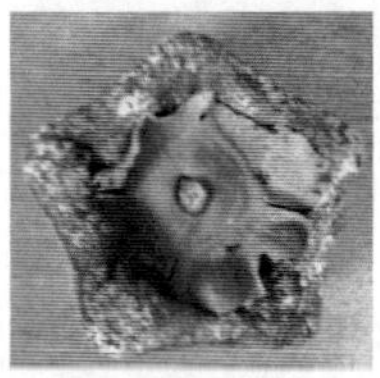

**Figure 11.** "Five-pointed star" persimmon shape.

## References

[1] Liu X M and Tan R H 2006 *Product Innovation Driven by Constructive Model of Unexpected Discoveries for Situated Cognition of Function-Behaviour-Structure. Chinese Journal of Mechanical Engineering.* **42(12)** 186-91.

[2] Suwa M, Gero J S and Purcell T 2000 *Unexpected discoveries and S-invention of design requirements; Important vehicles for a design process.Design Studies.* **21(6)** 539-67.

[3] Chen H 2016 *5HD-35 type persimmon dryer and its technology research. Fujian Agricultural Machinery.* **(04)** 23-6.

[4] Ma T T, Hao G, Yan H L, Jia R F, Tian C R and Fu C C 2012 *The Application of Food Quality and Safety Management System in the Processing of Dried Persimmon. Academic Periodical of Farm Products Processing.* **(09)** 145-9.

[5] Chen J F, Cheng S Y and Yang X L 2017 *A Metho- dology of Innovation Design Based on Function Expansion. Modular Machine Tool & Automatic Manufacturing Technique.* **(01)** 28-34.

[6] Zhang P, Dong Y F and Zhang H G 2016 *Functional Decomposition Process Model for Complex Electrome- chanical Based on Design-centric Complexity and TRIZ.Journal of Mechani- cal Engineering.* **52 (23)** 17-24.

[7] Qu A L and Yu Z B 2010 *The Design and Simulation Analysis of Separater of Grait and Soil Based on Function Elements Methods. Ningxia Engineering Technology.* **9(2)** 125-7.

[8] Song M, Jiang S F and Chai G Z 2015 *Schemes Econo- mic Evaluation Methods of Product Conceptual Design Based on Cost Unit. Journal of Zhejiang University of Technology.* **43(03)** 265-9.

[9] Lei X, Li D, Ma Y M, Ni M and Zhao K L 2011 *Comprehensive Evaluation System and Fuzzy Evaluation on the Design Proposal of Energy-Saving Hydraulic System.Journal of Inner Mongolia University of Technology (Natural Science).* **30(03)** 309-14.

[10] Xiang Z R and Liang G Y 2016 *The Method of Product Design Based on Constrained Man-Machine-Environment Compa- tibilitles. Decoration.* **(02)** 136-7.

IOP Conf. Series: Materials Science and Engineering **224** (2017) 012017 doi:10.1088/1757-899X/224/1/012017

# Reliability Analysis for the Gear of Power Assisting Cycle with Multi-Extremum Response Surface Method

**Chun-Yi ZHANG[1,a], Ai-Hua WANG[1,b], Tian Sun[1], Hui-Zhe JING[1] and Cheng-Wei LI[1]**

[1]School of Mechanical and Power Engineering, Harbin University of Science and Technology, Harbin 150080, China

[a]zhangchunyi@hrbust.edu.cn, [b]1512863091@qq.com

**Abstract.** To more accurately calculate the dynamic reliability for the center gear of power assisting cycle with various failure modes, the multi-extremum response surface method is adopted. The first, the gear's torque, material density, elastic modulus and poisson's ratio are sampled in small batches as the random input samples, the dynamic extremum of deformation, stress and strain are obtained as output response by finite element model. Then, the multi-extremum response surface equation is established. Finally, multitudinous sample points are obtained by using Monte Carlo method and linkage sampling to multi-extremum response surface equation, which can used to calculate the reliability of the deformation, stress and strain of the gear in the comprehensive failure mode. The results show that the comprehensive reliability degree of gear is 0.9949 when the allowable deformation, stress and strain are 0.47 mm, 540 Mpa and 0.003 , respectively.

## 1. Introduction

The gears support complex loads during the operation of the power assisting cycle operation. Once the gears fail, the energy stored in the spring can't be released and the whole system does not run. So, the structural reliability analysis of gear is very necessary.

In recent years, a structural analysis method considering the tooth contact of internal gear system is introduced through the comparison with the simplified gear system model and applied to the structural analysis of a 2-stage differential-type gearbox for wind turbine by Cho et al [1]. The key meshing states of the gear pair for the contact fatigue and the bending fatigue was investigated based on the cumulative fatigue criterion and the stress-life equation by Deng et al [2]. Although the problem of gear contact was studied in different degrees, they were based on deterministic analysis method, and it is not discussed further from the angle of probability.

Currently, scientists have presented a lot of research on the gear reliability analysis method, response surface method is the most widely used [3-6]. The reliability analysis method based on response surface and the Markov chain Monte Carlo (MCMC) were proposed to improve the precision of the response function, and the dependability sensitivity analysis was conducted to quantify the effects of assembly errors, machining errors and stochastic external loads on gear transmission reliability by Tong et al [7], then the reliability was calculated using the founded mathematical model by Monte Carlo method (MCM). However, these methods only consider the reliability of single failure mode analysis without considering the failure correlation of dynamic reliability problems.

In order to solve the problem of multiple failure mode correlation, the multi-extremum response surface method was introduced in the reliability analysis of gear. The gear's torque, material density, elastic modulus and poisson's ratio were sampled as random input variables to calculate the overall maximum deformation and maximum stress and maximum strain as output response. Then, these dates were used to fit out multi-extremum response surface mathematical model, and to calculation reliability with the Monte Carlo method and sampling linkage technology [8]. Finally, Monte Carlo method [9], [10] was used to verify the accuracy and effectiveness of this method.

## 2. Multi-extremum response surface method

In order to solve the problem of structural reliability analysis under multiple failure modes, the multi-extremum response surface method (MERSM) is adopted. The principle of MERSM is as follows: Firstly, Monte Carlo method is used to sample random of input parameters, the dynamic extremum of deformation, stress and strain were obtained as output response by finite element model. The extremum response surface function was established to reflect the relationship between the input parameters and the extremum output. Then, a multi-extremum response surface mathematical model is established on the basis of the extremum response surface model, and the multi-extremum response surface model is used to replace the finite element model (FEM) for structural reliability analysis.
Assume that $m$ as output objects in the structural reliability analysis($m \in Z$ ). If $X^{(i)}$ is an input at the $i$-th output objects, and term "i" varies from 1 to m, then the corresponding output response is $y^{(i)}$, the functions is shown as:

$$y^{(i)}(X) = a^{(i)} + \sum_{j=1}^{n} b_j^{(i)} X_j^{(i)} + \sum_{j=1}^{n} c_j^{(i)} (X_j^{(i)})^2 \quad (j = 1, 2, \cdots, n) \tag{1}$$

Where $x_j^{(i)}$ is the $j$-th component of the input variable $x$ in the $i$-th output response (object) and $a^{(i)}$, $b_j^{(i)}$, and $c_j^{(i)}$ are the undetermined coefficients of constant term, linear term, and quadratic term, respectively. The number of undetermined coefficients is $m(2n+1)$.

The undetermined coefficients are gained based on least square method when the number of samples is enough, because the vector $\boldsymbol{D}^{(i)}$ is formed by

$$\boldsymbol{D}^{(i)} = [a^{(i)} \; b_1^{(i)} \; b_2^{(i)} \cdots b_k^{(i)} \; c_{11}^{(i)} \; c_{22}^{(i)} \cdots c_{kk}^{(i)}] \tag{2}$$

From (2), we can gain the undetermined coefficients of (1) and further the mathematical model of MERSM.

## 3. Certainty analysis of bicycle gear

The static analysis of the gear is carried out under the finite element simulation platform. The material of gear is 40Cr, and the boundary condition gear's torque is 9152N / mm. 47768 nodes and 27651 elements are generated by using tetrahedral meshes, and the finite element mesh model of gears is shown in Figure 1.
The distribution of stress, deformation and strain of the center gear are shown in Figure 2, Figure 3 and Figure 4, respectively. It can be seen that the maximum stress and maximum strain of the gear occur at the tooth root line of the contact gear. The maximum displacement is tooth top position. The maximum stress, maximum displacement and maximum strain of the center gear are 506.44Mpa, 0.44673 mm, 0.0028407, respectively.

IOP Conf. Series: Materials Science and Engineering **224** (2017) 012017 doi:10.1088/1757-899X/224/1/012017

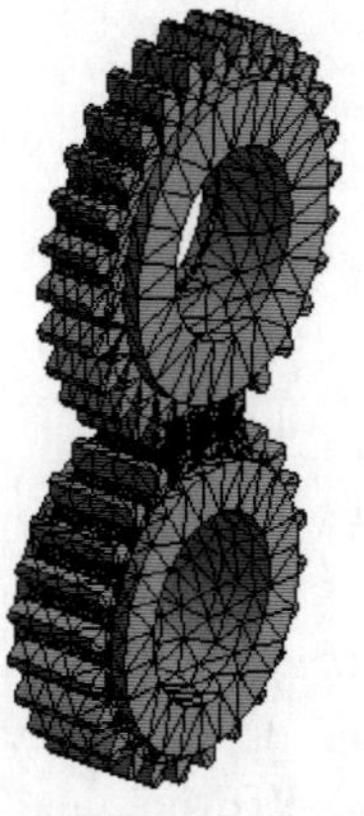

**Figure 1.** Grid of gear

**Figure 2.** Stress distribution on gear

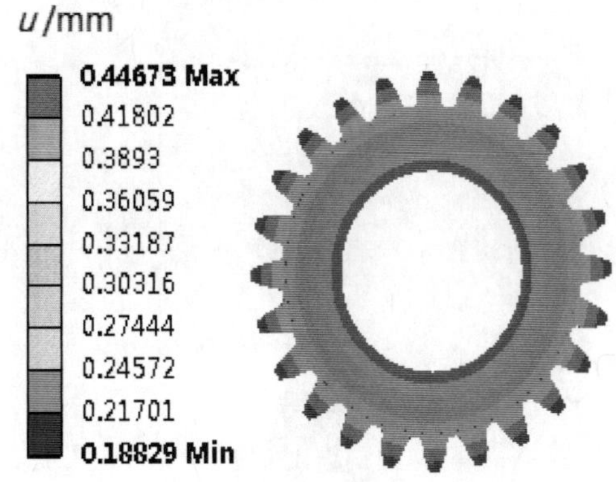


**Figure 3.** Deformation distribution on gear

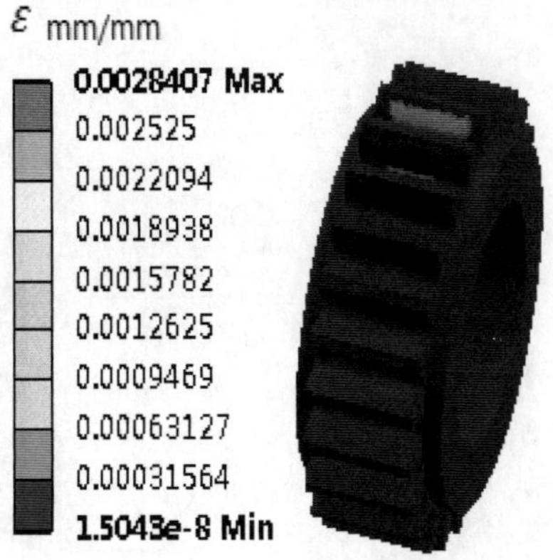


**Figure 4.** Strain distribution on gear

## 4. Reliability analysis of bicycle gear

### *4.1. Reliability analysis*

In the light of the uncertainties of material parameters mentioned in the handbook, the torque, material density, Young's modulus, and Poisson's ratio are chosen as input random variables, which were assumed to follow normal distribution and be independent mutually. The distributions of input random variables are listed in Table 1.

**Table 1.** Input random variables of MERSM reliability analysis.

| Input random variables | Young's modulus, E/ (Mpa) | Material density, $\rho$/ ($kg/m^3$) | Torque, $T$/ (N • mm) | Poisson's ratio, $\mu$ |
|---|---|---|---|---|
| Mean | 211000 | 7870 | 9152 | 0.277 |
| Standard deviation | 6330 | 236.1 | 274.56 | 0.00831 |

Using the MC method and Latin Hypercube Sampling technology simultaneously extract random input variables and output response at the location of the maximum deformation, maximum stress and maximum strain. Then, we can establish extreme response surface equation with these sample points by least square, finally the Multi-extremum response surface equation is shown as:

$$\begin{cases} \hat{y}_1 = -67.8892 - 4.3026\times10^{-4}\rho + 1.1124\times10^{-10}E - 6.5839\mu + 0.0865T \\ \quad + 3.1464\times10^{-8}\rho^2 - 3.1481\times10^{-22}E^2 + 18.8054\mu^2 - 2.6825\times10^{-6}T^2 \\ \hat{y}_2 = -2.4011\times10^{-4} - 2.7785\times10^{-9}\rho + 6.5854\times10^{-16}E - 1.7401\times10^{-6}\mu + \\ \quad 4.3186\times10^{-7}T + 1.9956\times10^{-13}\rho^2 - 1.8523\times10^{-27}E^2 + 3.9624\times10^{-5}\mu^2 - 1.0927\times10^{-11}T^2 \\ \hat{y}_3 = -0.0839 - 3.0803\times10^{-6}\rho + 1.3390\times10^{-13}E + 0.0798\mu + 8.3817\times10^{-5}T \\ \quad + 2.0600\times10^{-10}\rho^2 - 3.8401\times10^{-25}E^2 - 0.1269\mu^2 - 2.9531\times10^{-9}T^2 \end{cases} \tag{3}$$

### *4.2. The results of reliability analysis*

Through 10000 simulations on three extremum response surfaces by MC simulation method, and based on material handbook and some related material tests, it can be found that the allowable deformation, allowable stress and allowable strain are $u = 0.47$ mm, $\sigma = 540$ Mpa and $\varepsilon = 0.003$, respectively. Referencing these material parameters, the results of gear reliability analysis are listed in Figure 5.

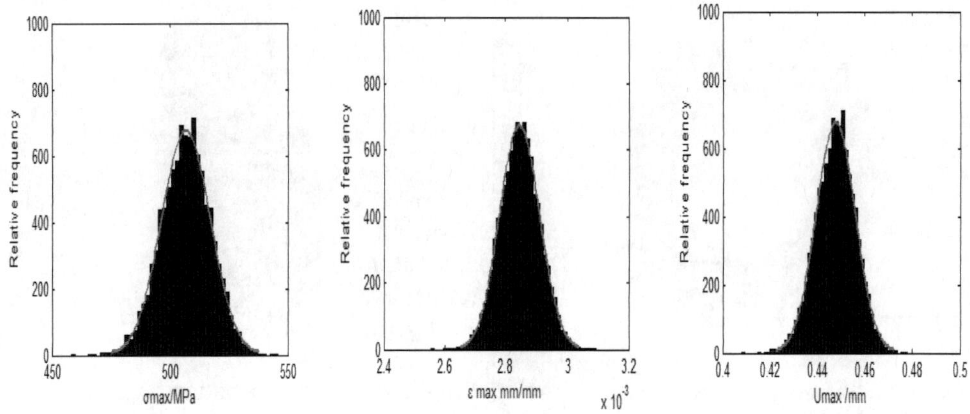


(a)Distribution of gear stress (b)Distribution of gear strain (c)Distribution of gear deformation

**Figure 5.** Results of reliability analysis.

It can be seen in Figure 5 that the frequency distribution of the stress, deformation and strain of the center gear is obeying normal distribution, and the mean were 506.5062 MPa, 0.44766 mm and 2.845 $\times 10^{-3}$, the standard deviation was 102.7062, $6.5125\times10^{-5}$ and $3.9449\times10^{-9}$. The overall failure number is 51, and the comprehensive reliability degree of the gear is 0.9949 and the calculation time is 1.44s.

## 5. Methods validation

In order to verify the validity of the multi-extremum response surface method based on the input variables in table 1, the reliability of MCM, MRSM and MERSM were compared at the same condition. Under different times of simulation, the method of calculating time and reliability are shown in table 2. The results show that MERSM is a kind of high precision and high efficiency in reliability analysis.

**Table 2.** The results of reliability analysis of gear with three methods

| Samples | Computational time/s | | | Computational Precision | | |
|---|---|---|---|---|---|---|
| | MCM | MRSM | MERSM | MCM | MRSM | MERSM |
| $10^2$ | 6300 | 0.97 | 0.62 | 0.99 | 0.98 | 0.99 |
| $10^3$ | 59160 | 1.63 | 0.824 | 0.992 | 0.988 | 0.991 |
| $10^4$ | - | 5.36 | 1.44 | - | 0.9925 | 0.9949 |
| $10^5$ | - | 15.28 | 2.53 | - | 0.99573 | 0.99642 |

IOP Conf. Series: Materials Science and Engineering **224** (2017) 012017 doi:10.1088/1757-899X/224/1/012017

## 6. Conclusions

(1) The location and value of maximum deformation, maximum stress and maximum strain are obtained by the deterministic analysis of gear contact. The comprehensive reliability of gear is 0.9949 when the allowable maximum of deformation, stress and strain are 0.47 mm, 540 Mpa and 0.003, respectively.

(2) The comparison shows that the calculation precision used MERSM is consistent with MCM, and superior than MRSM, the advantage of its high computational efficiency is more obvious with the increase of the number of simulation.

## Acknowledgments

This paper is co-supported by the National Natural Science Foundation of China (No. 51275138), the Science Foundation of Heilongjiang Provincial Department of Education (No. 12531109).

## References

[1] Cho J R, Jung K Y, Park M H, et al. Finite element structural analysis of wind turbine gearbox considering tooth contact of internal gear system. *J. Journal of Mechanical Science and Technology*, 2013, **27**(7):2053-2059.

[2] Song D , Lin H , X H H, et al. Finite element analysis of contact fatigue and bending fatigue of a theoretical assembling straight bevel gear pair. *J. Journal of Central South University*, 2013, **20**(2):279-292.

[3] Guan T M, Li J B, Lei L. Tooth Surface Contact Fatigue Reliability Analysis of Cycloidal Gear Based on Monte-Carlo. *J. Advanced Materials Research*, 2013, 605-607:811-814.

[4] Li G, Mi W, Lu K, et al. Reliability analysis on gear contact fatigue strength considering the effect of tolerance. *J. Open Mechanical Engineering Journal*, 2014, **8**(1):630-635.

[5] Liu Z L, Zhao M M, Peng-Peng M A, et al. Large Helical Gears Contact Analysis and Contact Strength Reliability Analysis Based on ANSYS. *J. Journal of Zhengzhou University*, 2015.

[6] Zhao H, Ru Z, Chang X, et al. Reliability Analysis Using Chaotic Particle Swarm Optimization. *J. Quality & Reliability Engineering International*, 2014, **31**(8):1537 - 1552.

[7] Tong C, Sun Z L, Chai X D, et al. Gear Contact Fatigue Reliability Based on Response Surface and MCMC. *J. Journal of Northeastern University*, 2016.

[8] Zhang C Y,Lu Cheng,Fei C W,et al.Linked Analysis of reliability of Aeroengine Blisk with dual extremum response surface method. *J.Propulsion Technology*, 2016, **37**(6):1158-1164.

[9] Seila A F. Simulation and the Monte Carlo Method. *J. Technometrics*, 2012, **24**(2):167-168.

[10] Cheng Q, Zhao H, Zhao Y, et al. Machining accuracy reliability analysis of multi-axis machine tool based on Monte Carlo simulation. *J. Journal of Intelligent Manufacturing*, 2015:1-19.

# Design and Performance Analysis of a new Rotary Hydraulic Joint

**Yong Feng, Junhong Yang, Jianzhong Shang, Zhuo Wang and Delei Fang**

Department of Mechanical Engineering, National University of Defense Technology, Changsha, 410073, China

yangjunhong@nudt.edu.cn

**Abstract.** To improve the driving torque of the robots joint, a wobble plate hydraulic joint is proposed, and the structure and working principle are described. Then mathematical models of kinematics and dynamics was established. On the basis of this, dynamic simulation and characteristic analysis are carried out. Results show that the motion curve of the joint is continuous and the impact is small. Moreover the output torque of the joint characterized by simple structure and easy processing is large and can be rotated continuously.

## 1. Introduction

As an important part of robots, robots joint has a great effect on the performance of robots. At present the drive device of robots with drive of servo motor and reducer, is heavy and limited in load capacity, which has seriously restricts on the practicability of robots [1].Therefore robot joints with the advantages of small size, light weight and strong load capacity etc. have become a main studying interest.

There have been a few approaches to improve the load capacity of the robot joint, without increasing the mass. The one which enhance the joint performance is to use high power density motor and reducer. For example the robot Cheetah uses a planetary gear reducer and an EM motor whose power mass density is 7kw/kg [2]. The motor is small, high precision and no pollution, but they are expensive.

Compared with the motor drive, hydraulic driving have much predominance, such as big driving power. Therefore many high-load hydraulic actuators are designed to use on robots. Professor Liang [3] has proposed a rotary actuator based on screw device, which has an advantage of bigger output torque ($\geq 2.5kN\ m$) , but the corner is only 60 degrees. In the literature [4], a straight hydraulic cylinder is used to drive the hinge joint, the structure of the joint is simple, but the angle range of the joint is limited. Ref. [5] shows a kind of joint which adopts rope to drive pulley. It's easy to arrange this joint on robots, but the flexible wire rope is easy to deformation. In addition, there are other types of hydraulic joints, such as gear and rack rotary hydraulic cylinder [6], screw rotary cylinder [7] etc. These joints are not good choice for robots, due to leakage, larger size, and other reasons.

According to the characteristics of the robot joints described above, a new kind of hydraulic joint is proposed and can convert the reciprocating linear motion of the piston into the rotary motion of the output shaft. With a compact structure, driving capacity strongly, and being rotated continuously, this new joint can be applied especially for mobile robot which need large load driving capacity.

## 2. Design of the wobble plate hydraulic joint

In this section, the mechanism of the new hydraulic joint is presented, and the structure of the joint is designed.

### *2.1. Structure design of the hydraulic joint*

The structure of the wobble plate hydraulic is shown in figure 1, which is mainly composed of the base, the cylinder, the wobble shaft, the piston, the rod, and the output shaft (Z axis). The two bases are connected by connecting rods, the linear hydraulic cylinders are evenly distributed on the circular base, two symmetrical rods and the fixing device are connected by screw threads. The linear bearing is installed on the linear bearing seat, there is a rotation pair between the linear bearing and the fixing device. Two perpendicular swing rod which constitute wobble plate are mounted on the output shaft through the bearing, the swing rod and the output shaft form a rotating pair, and a cylindrical pair are formed between swing rod and linear bearing simultaneously.

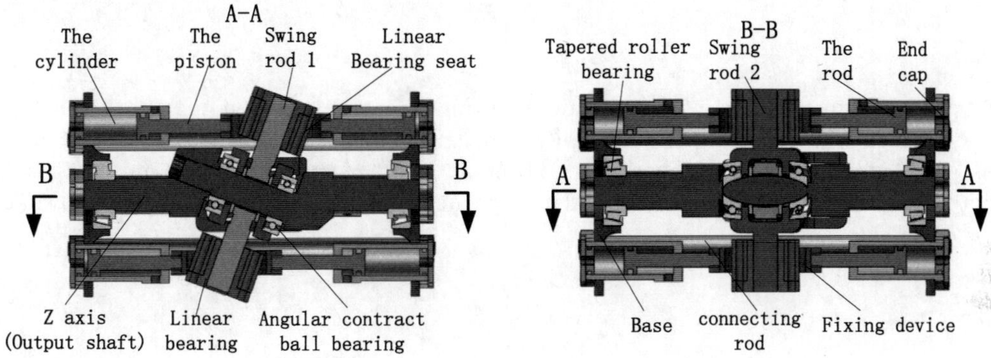


**Figure 1.** The structure of the joint.

### *2.2. Principle of the hydraulic joint*

As shown in the figure 2, the servo valve causes the rods on both sides of the swing rod to produce a linear motion in the opposite direction by controlling the flow direction of the oil. Under the action of the rod, the swing rods are rotated in the plane A-A and B-B, which also drives Z axis to rotate. At the same time, the swing rod is also rotated relative to its own axis. In this way, the reciprocating linear motion of the piston is converted into rotary motion of the output shaft. However, when the piston is moved to the limit position, the corresponding swing rod is in the same plane with the bent portion of the output shaft. At this point, no matter how big the piston thrust is, the output shaft can't be driven, this point is called the dead point of the mechanism. In order to remove dead point, similar structures are arranged on the plane A-A and B-B, when the swing axis of the A-A plane is at the dead point, the swing axis of the B-B plane can rotates the output shaft. Thus the continuous rotation of the hydraulic joint is achieved.

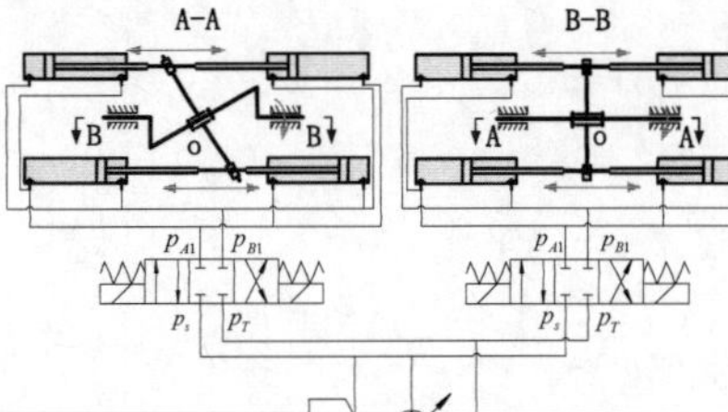


**Figure 2.** The principle of the joint.

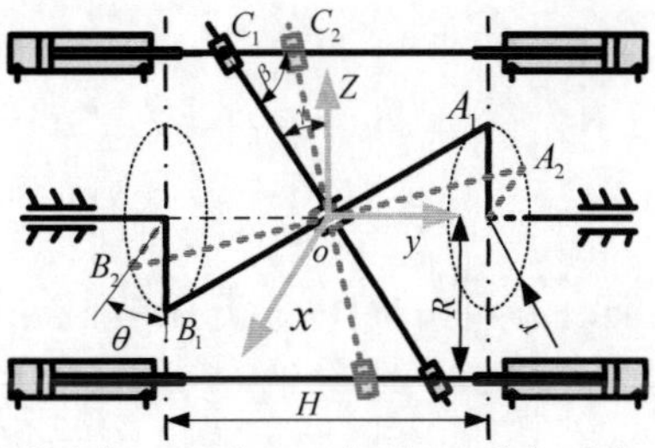


**Figure 3.** The mathematical model of the joint.

## 3. Mathematics model of the new joint

### *3.1. Kinematics model*

The geometric relationship and some parameters of the internal structure of wobble plate hydraulic joint are shown in figure 3. The relationship between displacement of the ith piston $y_i (i=1,2,3,4)$ and angle $\theta$ can be expressed as follows. The meanings of $R$, $r$, $\beta$, $\gamma$ and $H$ are shown in figure 3.

$$\begin{cases} y_1 = -\dfrac{2Rr}{H}\cos\theta \\ y_2 = -\dfrac{2Rr}{H}\sin\theta \\ y_3 = \dfrac{2Rr}{H}\cos\theta \\ y_4 = \dfrac{2Rr}{H}\sin\theta \end{cases} \tag{1}$$

*3.2. Dynamic model*

Before the establishment of the dynamic model by using Lagrange equation, the swing plate hydraulic joints are divided into four modules. Module 1 includes piston, rod, and fixing device. Module 2 contains linear bearings and linear bearing seat. Module 3 comprises swing rod 1 and swing rod 2. The last module is the Z axis. Respectively, calculate the kinetic energy of each module as follows.

$$T_1 = \sum_{i=1}^{4} \frac{1}{2} m_1 \dot{y}_i^2 \tag{2}$$

$$T_2 = \sum_{i=1}^{4} \frac{1}{2} m_2 \dot{y}_i^2 + \sum_{i=1}^{4} \frac{1}{2} I_{2x} \dot{\beta}_i^2 \tag{3}$$

$$T_3 = \frac{1}{2} I_{31x} \gamma_1^2 + \frac{1}{2} I_{31z} \gamma_2^2 + \frac{1}{2} I_{32x} \gamma_2^2 + \frac{1}{2} I_{32z} \gamma_1^2 \tag{4}$$

$$T_4 = \frac{1}{2} I_{4y} \dot{\theta}^2 \tag{5}$$

Where $I_{2x}$ is the moment of inertia of module 2 relative to the x-axis, $I_{4y}$ is the moment of inertia of Z axis relative to the y-axis, $I_{3ij}$ $(i=1,2; j=x,z)$ is the moment of inertia of swing rod $i$ relative to the $j$-axis. $\gamma_i$ is the angle between the swing rod $i$ and the Z-axis, $\beta_i$ is the angle between the rod $i$ and swing rod, and the expressions of $\gamma_i$ and $\beta_i$ are as follows.

$$\gamma_i = \arctan\frac{y_i}{R} (i=1,2) \tag{6}$$

$$\begin{cases} \beta_i = \dfrac{\pi}{2} + \gamma_i (i=1,2) \\ \beta_i = \beta_{i-2} (i=3,4) \end{cases} \tag{7}$$

The kinetic energy of system $T$ is obtained.

$$T = \sum_{i=1}^{4} T_i \tag{8}$$

Because the joints are symmetrical about the origin of the coordinate system, the potential energy of the system $V$ can be obtained.

$$V = 0 \tag{9}$$

From equations (3) to (11), The Lagrange is obtained as:

IOP Conf. Series: Materials Science and Engineering **224** (2017) 012018 doi:10.1088/1757-899X/224/1/012018

$$L = T + V = \sum_{i=1}^{4} T_i \tag{10}$$

$$\frac{d}{dt}\frac{\partial L}{\partial \dot{\theta}} - \frac{\partial L}{\partial \theta} = Q \tag{11}$$

Assuming that the joint is affected only by the fluid force $F_i$ and torque $\tau$, the power generated by $F_i$ and $\tau$ is equal to the power produced by the generalized force $Q$.

$$\tau\theta + \sum_{i=1}^{N} F_i \dot{y}_i = Q\theta \tag{12}$$

From equations (2) and (14), the generalized force $Q$ .can be obtained.

$$Q = M + F_1\frac{2Rr}{H}\sin\theta - F_2\frac{2Rr}{H}\cos\theta - F_3\frac{2Rr}{H}\sin\theta + F_4\frac{2Rr}{H}\cos\theta \tag{13}$$

The meaning of $F_i$ $(i = 1,2,3,4)$ is the output force of the ith cylinder.

## 4. Simulation

The new hydraulic joint is modeled in MATLAB/Simulink. In order to verify the wobble plate hydraulic joint can output a large torque, the servo valve is fully opened, and the external force of each swing rod is shown in the figure 4. Meanwhile the displacement of Z axis is $\theta = \pi t/6$. Shown in Figure 5, the piston motion curve is continuous, which is consistent with the mathematical model.

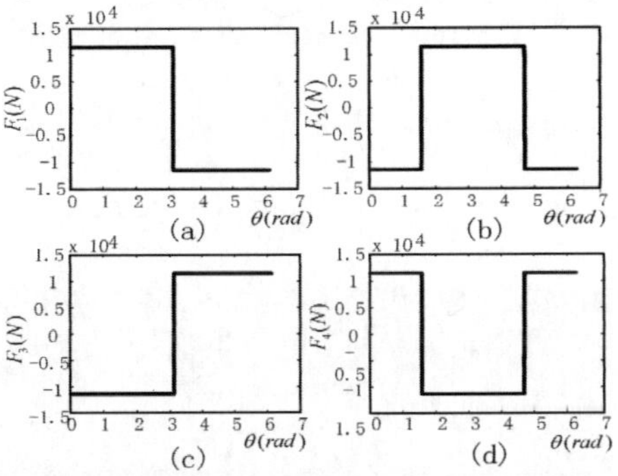


Figure 4. The external force of swing rod.

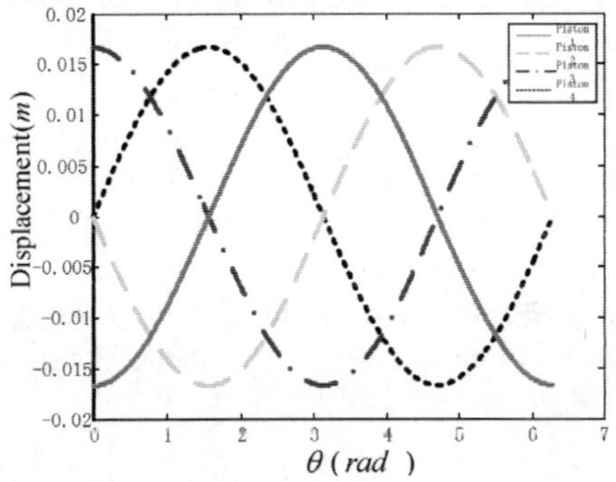


Figure 5. The displacement of each piston.

Figure 6 shows the output torque curve of the joint. It can be seen that the output torque of the joint is at least 380 $N\bullet$ , and the output torque varies with the rotation angle of the output shaft. In order to verify the validity of the mathematical model, Adams virtual prototype simulation is carried out after MATLAB simulation. The results of Adams simulation is shown in Figure. 7.

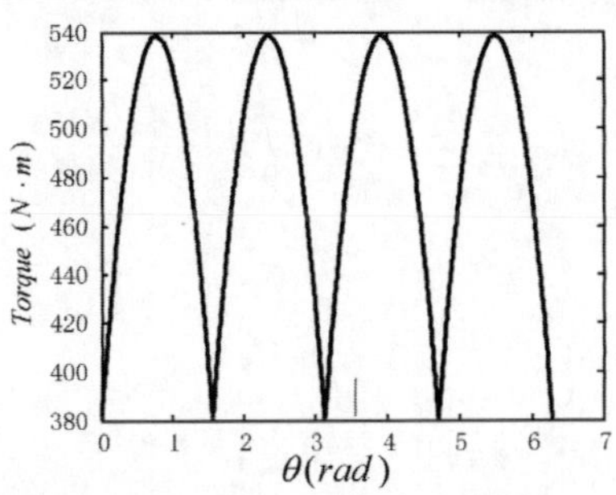


**Figure 6.** The output torque of the joint.

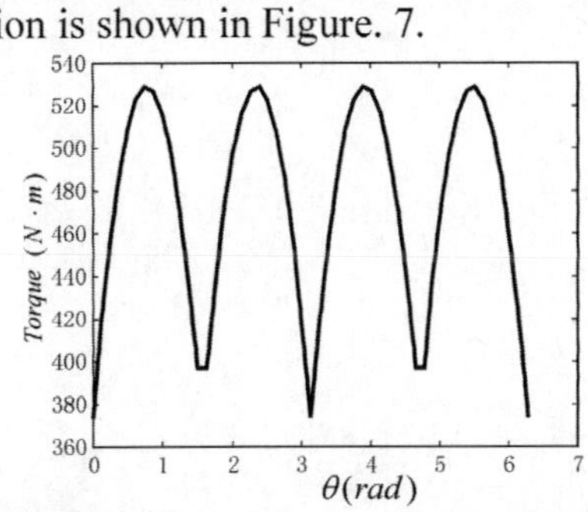


**Figure 7.** Adams Simulation Result.

By comparison, it is found that the output torque in Figure 7, is smaller than that in Figure 6, at any point. The reason is that the influence of frictional resistance and the inertial force of eccentric mass are ignored in the mathematical model. On the whole, two sets of curves are basically consistent, which proves that the mathematical model and simulation are correct.

## 5. Conclusion

In the present paper, a new joint is proposed. With advantages of compact structure, continuous rotation, and small mass, this joint, whose maximum output torque can be more than 380 $N\bullet$ , is just about 4.75kg. In addition, through the movement and dynamic simulation of the joint, it's found that the joint motion curve was continuous and the maximum output torque is related to the rotation angle of the joint.

## Acknowledgement

This work was financially supported by the National Natural Science Foundation of China with Grant No: 51675522.

## References

[1] Hegao Cai, Robot Will be a Hot Spot of Technological Development in the Twenty First Century, China Mechanical Engineering, 2000, 11(1):58-60

[2] Seok S, Wang A, Otten D, et al. Actuator design for high force proprioceptive control in fast legged locomotion, International Conference on Intelligent Robots and Systems. IEEE, 2012:1970-1975.

[3] Xichang Liang, Guangjian Wang, Xiaoguang Zheng, Research on Rotary Actuator Based on Screw Device, Chinese Journal of Aeronautics, 2003, 24(3):282-285.

[4] Focchi M, Guglielmino E, Semini C, et al. Control of a Hydraulically-Actuated Quadruped Robot Leg, 2010, 58(8):4182-4188.

[5] Xiong C, Jiang X, Sun R, et al. Control methods for exoskeleton rehabilitation robot driven with pneumatic muscles, Industrial Robot, 2009, 36(3):210-220.

[6] Xinwei Luo Operation Principle Fault Analysis of Gera and Rack Rotary Hydraulic Cylinder, Baosteel Meishan, 2012(3):52-53.

[7] Peng Kong, Xiankun Pu, Optimized Design for Lead Angle of Output Shaft of the Double Helix Rotary Cylinder, Hydromechatronics Engineering, 2014(14): 51-53.

IOP Conf. Series: Materials Science and Engineering **224** (2017) 012019 doi:10.1088/1757-899X/224/1/012019

# Walking Posture Control of Transmission Line Single Arm Inspection Robot

**Yan Yu[1,2] , Liu Xiaqing [1,2] , Li Jinliang[1,2]and Ou Yuexiong[1,2]**

[1]State Grid Hunan Electric Power Company Live Working Center,Changsha 410100,China

[2]Hunan Province Key Laboratory of Intelligent live working Technology and Equipment (Robot) ,Changsha 410100,China

380147536@qq.com

**Abstract.** To control the walking posture according to transmission line single arm inspection robot, the robot is divided into normal walking and climbing walking two state, and gives the definition, then based on the state space method of state variable feedback and PD control method is used to control the two states, two kinds of control method of simulation by using Matlab,in the end, the two control methods proposed is validated in the actual circuit structures. The results show that, the proposed control method is rapid and effective, and can meet the needs of practical application.

## 1. Introduction

The electricity sector in order to grasp the running state of line, find fault equipment in time, At present, they mainly used to monitor the line on the line of the manual inspection line, This paper describes the design of a compact, high reliability, double arm inspection robot for the single span line between two towers.

Single arm inspection robot to large span wire for driving path. Under the influence of gravity, the line will present an opening up of a parabolic shape, Therefore, in the process of continuous walking, the robot will inevitably encounter the situation of downhill and uphill, which will show an opening to the parabolic shape. In addition, affected by the structure of the robot arm, robot in walking, center of gravity can not always maintain in the perpendicular bisector of the two wheel shaft connection, leading to the robot due to the instability and attitude in uncontrolled phenomenon.

The problem for the analysis of the attitude control of robot, the domestic and foreign scholars have made some research. The literature on the control of the state of the body is rarely seen. Literature[1-4]about the line tracking robot is analyzed and found robot in the process of obstacle, line grasping arm would swing along the ground, Therefore, a PD control method based on the minimum energy consumption is proposed.

The method can effectively control the swing angle of the robot to return to the equilibrium point quickly and smoothly. In this paper, the PD control method is simulated and analyzed, but the practical application is not given. Literature [5] has carried out some research on the deflection of the designed three arm line inspection robot, by adding a center of gravity adjustment mechanism on the robot to eliminate the robot's overall deflection caused by the moving of the center of gravity, this way has achieved certain results, but increase the burden of the system but also reduces the efficiency of robot. Literature [6] patrol robot for an additional obstacle resistant tension tower bridge, the real-time positioning of the sensor based control, but does not take into account the deflection between the arm and the ground line grasping, cannot reflect the attitude of robot arm at the end of the execution.

Literature [7]on the two arms and three arms of the robot obstacle navigation process, just made some simulation research. In the literature [8], the phenomenon of posture instability appeared in the application of a wide range of boom type inspection robot in the cable accessories is studied, from the perspective of institution and developed the corresponding pitch instability self balancing mechanism, lateral buckling self balancing mechanism, and analyzes its working principle.

In this paper, a power transmission line inspection robot walking wheel arm charged on attitude control. Firstly, the climbing ability of the robot is analyzed, and the maximum climbing angle of the robot is given. Secondly, the dynamic model of the robot's walking is established, taking into account that the robot can be divided into normal walking posture and walking posture. After the analysis of normal walking posture, the robot is designed by using the control method based on state feedback to design the attitude controller of the robot. After the analysis of the climbing walking posture, the system has a strong nonlinear, simple linear processing can not meet the needs of climbing attitude control, So the attitude controller of the robot climbing walking is designed by using the PD control method. Finally, the effectiveness of the two attitude controllers is verified by experiments.

## 2. Static analysis of robot climbing

Two wheeled robot arm as shown in Figure 5, Figure 1 static model for climbing, Among them, Ⅰ is the robot's walking wheel, Ⅱ is the pressing wheel, Ⅲ is the control box of the robot. In order to increase the climbing ability of the robot, a pressing wheel mechanism is designed on the arm, which is used to increase the positive pressure of the walking wheel, thereby increasing the friction force.

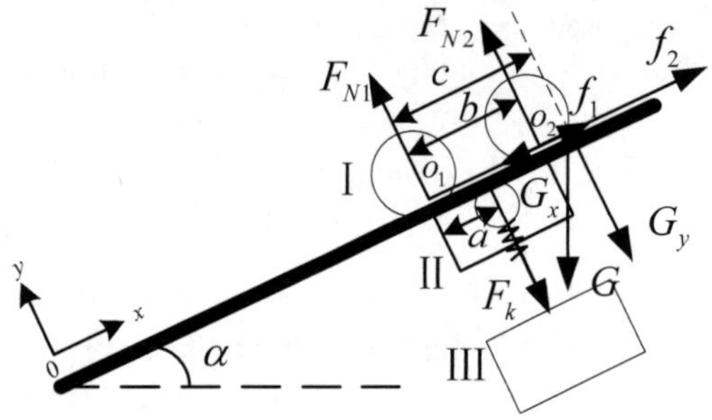

Figure 1. Static model of robot climbing.

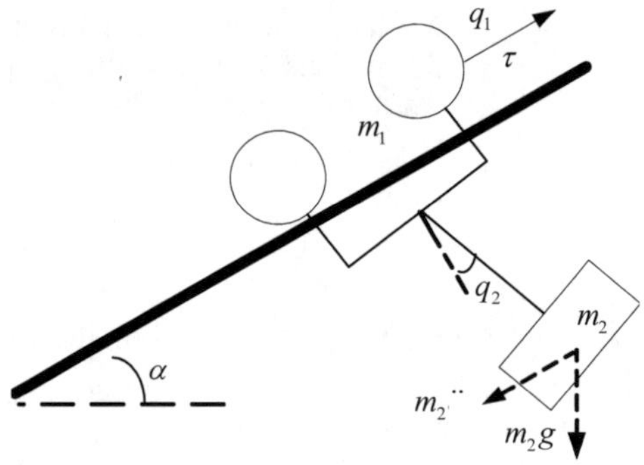

Figure 2. Dynamic model of robot climbing.

Static equation of robot climbing:

$$G_x = G\sin\alpha \tag{1}$$

$$G_y = G\cos\alpha \tag{2}$$

$$f = f_1 + f_2 \tag{3}$$

$$f - G_x \geq 0 \tag{4}$$

$$f = \mu(F_{N1} + F_{N2} + F_k) \tag{5}$$

$$F_k = -kx \tag{6}$$

$$F_k a - F_{N2} b + G_y c = 0 \tag{7}$$

$$F_{N1} b - F_k (b-a) + G_y (c-b) = 0 \tag{8}$$

Above the formula, $\alpha$ stands for slope; $G$ is the gravity of the robot, $N$; $F_{N1}$ for the support of the line on the left wheel, $N$; $F_{N2}$ for the support of the line on the right wheel, $N$; $F_k$ for pressing wheel by the line pressing force, $N$; $f_1$ for the left running wheel by the friction, $N$; $f_2$ for the right running wheel by the friction, $N$。 $G_x$ is a component of Gravity $G$ acting on the $x$ axis, $N$; $G_y$ is a component of Gravity $G$ acting on the $y$ axis, $N$; $f$ for the robot to be the total friction, $N$; $\mu$ for friction factor; $k$ as elastic, $x$ for spring deformation, $m$; the maximum deformation of the spring is a $x_m$ ,Above the formula $x \in (0, x_m)$ ; $a$ is $F_k$ and $o_1$ between the force arm, $m$; $b$ is $F_{N2}$ and $o_1$ between the force arm, $m$; $c$ is $G_y$ and $o_1$ between the force arm, $m$。 From formula (7) and (8) can be seen, $F_{N1}$ and $F_{N2}$ are not equal, combined (1) to (8) can be obtained (9):

$$\cos(\alpha + \arctan(1/\mu)) \le \frac{2\mu kx}{G\sqrt{1+\mu^2}} \tag{9}$$

From the formula (9) can be seen, under the condition of not considering the driving ability of the driving wheel, the biggest factors influencing the climbing of the robot are the positive pressure exerted by the pressing wheel. Taking into account the maximum deformation of the compression spring is a $x_m$ , we can get the maximum climbing angle of the robot $\alpha_{\max}$ is:

$$\alpha_{\max} = \arccos\left(\frac{2\mu kx_m}{G\sqrt{1+\mu^2}}\right) - \arctan(1/\mu) \tag{10}$$

## 3. Dynamic analysis of robot walking

The robot dynamics model is shown in Figure 2, the formula (11) can be obtained by analyzing the force of the walking direction of the system,and the torque of the pendulum is obtained by the formula (12).

$$m_1\ddot{q}_1 + m_2 l\ddot{q}_2 + c_1\dot{q}_1 + (m_1 + m_2)g\sin\alpha = \tau \tag{11}$$

$$m_2 l\ddot{q}_1 + m_2 l^2\ddot{q}_2 + c_2\dot{q}_2 - m_2 gl\sin(q_2 + \alpha) = 0 \tag{12}$$

The quality of $m_1$ for two walking wheels in the formula, $kg$ ; $m_2$ is the quality of the robot arm and control box, $kg$ ; $l$ is the length of the robot arm, $m$; $\tau$ is the driving wheel drive power, $N$ ; $c = [c_1 \;\; c_2]^T$ is the coefficient of friction; $q_1$ is the displacement of the robot, $m$ ,is the input variable of the system; $\dot{q}_1$ is the robot moving speed, $m/s$ ; $\ddot{q}_1$ is the acceleration of robot motion, $m/s^2$ ; $q_2$ is the robot arm swing angle, but also the output variable of the system moving acceleration; $\dot{q}_2$ is the angular velocity of the robot arm swing, $rad/s$ ; $\ddot{q}_2$ is the robot arm swing angular acceleration, $rad/s^2$ ; $g$ is gravity constant coefficient。 $|q_2 + \alpha| \le \alpha_d$ , $\alpha_d$ takes 15° in this paper; $\alpha$ is for line slope.

According to the line slope size, in the case of slope $|\alpha| \le 8°$ , the dynamic equations of the robot can be partially linearized, and the equations can be controlled by modern control theory; in the case of slope $|\alpha| > 8°$ ,control using PD control method.Therefore, in order to facilitate the discussion, the two controllers are designed to facilitate the discussion, and the following two definitions are given for the convenience of the discussion:

**Definition 1 Normal walking:** The robot walked on the ground which is less than or equal to 8 degrees.

**Definition 2 Climbing walking:** The robot walked on the ground which is greater than or equal to 8 degrees.

### *3.1. Normal walking controller design*

When the robot walked on the ground, $c \approx 0$ , $\alpha \approx 0$ , because the swing angle $q_2$ is very small, so we can deal with the linear approximation of the system equation. Besides, $x = [q_1 \quad \dot{q}_1 \quad q_2 \quad \dot{q}_2]^T$ , $x = [\dot{q}_1 \quad \ddot{q}_1 \quad \dot{q}_2 \quad \ddot{q}_2]^T$ .

Combined (11) to (12) can be obtained:

$$m_1\dot{x}_2 + m_2 l\dot{x}_4 = \tau \tag{13}$$

$$\dot{x}_2 + l\dot{x}_4 - gx_3 = 0 \tag{14}$$

Combined formula (13) formula (14), elimination of $\dot{x}_4$ , can be obtained:

$$(m_1 - m_2)\dot{x}_2 + m_2 g x_3 = \tau \tag{15}$$

Combined formula (13) formula (14), elimination of $\dot{x}_2$ , can be obtained:

$$(m_1 - m_2)l\dot{x}_4 - m_1 g x_3 + \tau = 0 \tag{16}$$

As a result, we can get the following 4 first order differential equations.

$$\dot{x}_1 = x_2 \tag{17}$$

$$\dot{x}_2 = -\frac{m_2 g}{m_1 - m_2}x_3 + \frac{1}{m_1 - m_2}\tau \tag{18}$$

$$\dot{x}_3 = x_4 \tag{19}$$

$$\dot{x}_4 = \frac{m_1 g}{(m_1 - m_2)l}x_3 - \frac{1}{(m_1 - m_2)l}\tau \tag{20}$$

Taking into account that when the robot is walking, only the displacement of $q_1$ can be measured directly by the encoder, $q_1$ as the only output of the system $y = x_1$ . Combined formula (17) - (20) the state space equation of the robot when the robot is walking:

$$\dot{\boldsymbol{x}} = \boldsymbol{A}\boldsymbol{x} + \boldsymbol{B}\tau \tag{21}$$

$$y = \boldsymbol{C}\boldsymbol{x} \tag{22}$$

Where $A = \begin{bmatrix} 0 & 1 & 0 & 0 \\ 0 & 0 & -\dfrac{m_2 g}{m_1 - m_2} & 0 \\ 0 & 0 & 0 & 1 \\ 0 & 0 & \dfrac{m_1 g}{(m_1 - m_2)l} & 0 \end{bmatrix}$, $B = \begin{bmatrix} 0 \\ \dfrac{1}{m_1 - m_2} \\ 0 \\ -\dfrac{1}{(m_1 - m_2)l} \end{bmatrix}$, $C = [1 \quad 0 \quad 0 \quad 0]$.

After actual measurement, the parameters of the system are $m_1 = 2kg$ , $m_2 = 10kg$ , $l = 0.4m$ . Therefore, the determinant value of the controllability matrix $P_c$ and the observability matrix $P_o$ of the system are respectively given by formula (23) and formula (24):

$$|P_c| = |A \quad AB \quad A^2B \quad A^3B| = 20.6670 \neq 0 \tag{23}$$

$$|P_o| = |\begin{bmatrix} C & CA & CA^2 & CA^3 \end{bmatrix}^T| = 2401 \neq 0 \tag{24}$$

By the formula (23) and (24) we can know that the system is controllable and observable.

Using the following control law:

$$\tau = Kx \tag{25}$$

Through the analysis, we can know that there are two zeros point in the open loop poles of the 4 order system and the critical is very stable. The closed-loop poles of the system can be configured on the $\lambda = [-8+6j \quad -8-6j \quad -0.4+0.3j \quad -0.4+0.3j]$, to meet the requirements of the system regulating time and absolute stability, the feedback gain matrix K should be able to satisfy the following formula:

$$\det(\lambda I - (A - BK)) = (\lambda + 8 \pm 6j)(\lambda + 0.4 \pm 0.3j) \tag{26}$$

We can first calculate the $K = [1.7 \quad 5.8 \quad 342.9 \quad 56.1]$. In our practical system, the state variable is only measured by displacement $x_1$, so it is needed to design the observer for the system to provide the estimated value of the state variable which can not be measured directly. The actual control law is given by formula (27):

$$\tau = K\hat{x} \tag{27}$$

The state variable estimate $\hat{x}$ is given by the formula (28):

$$\dot{\hat{x}} = A\hat{x} + B\tau + L(y - C\hat{x}) \tag{28}$$

L is the observer gain, which is related to the pole of the observer, and satisfies the formula:

$$\det(\lambda I - (A - LC)) = ((\lambda + \lambda_1)(\lambda + \lambda_2))^2 \tag{29}$$

In application, the closed-loop poles of the system and the poles of the observer should be guaranteed to be 10 to 2 times. After repeated testing, the observer poles are determined to be $-16 \pm 21.3j$.

The observer gain matrix of this system is determined by using Ackerman's formula $L = [64 \quad 2437.3 \quad 918.9 \quad 9974.4]^T$.

*3.2. Design of climbing walking controller*

The dynamic equations of the robot can be expressed as follows:

$$M\ddot{q} + G(q) = \begin{bmatrix} \tau - c_1\dot{q}_1 \\ -c_2\dot{q}_2 \end{bmatrix} \tag{30}$$

Above the formula, $M = \begin{bmatrix} m_1 & m_2 l \\ m_2 l & m_2 l^2 \end{bmatrix}$ is a constant positive definite matrix. $G(q) = \begin{bmatrix} m_1 g q_1 \sin\alpha \\ -m_2 g l \sin(q_2 + \alpha) \end{bmatrix}$.

The formula (31) to construct auxiliary function:

$$P(q) = \frac{1}{2} m_1 g q_1^2 \sin\alpha + m_2 g l \cos(q_2 + \alpha) \tag{31}$$

Lyapunov function is defined as follows:

$$V(q,\dot{q}) = \frac{1}{2}\dot{q}^T M \dot{q} + P(q) + \frac{1}{2} k_p q_1^2 \tag{32}$$

In the formula, the position gain $k_p > 0$, we know that $V(q, \dot{q}$ is globally positive definite.

The formula (11) and the formula (12) describe the system in $|\alpha| > 8°$ when using the following control law:

$$\tau = -k_p q_1 - k_v q_1 \tag{33}$$

In the formula, the velocity gain is $k_v > 0$.

Derivative of the formula (31):

$$\dot{V}(\boldsymbol{q},\dot{\boldsymbol{q}}) = \dot{\boldsymbol{q}}^T M\ddot{\boldsymbol{q}} + \frac{1}{2}\dot{\boldsymbol{q}}^T \dot{M}\dot{\boldsymbol{q}} + \dot{P}(\boldsymbol{q}) + k_p q_1 \dot{q}_1 \quad (34)$$

Into the formula (30) can be obtained:

$$\dot{V}(\boldsymbol{q},\dot{\boldsymbol{q}}) = \dot{q}_1(\tau + k_p q_1) - c_1\dot{q}_1^2 - c_2\dot{q}_2^2 \quad (35)$$

$\tau$ in the formula (33) into the formula (35) can be obtained:

$$\dot{V}(\boldsymbol{q},\dot{\boldsymbol{q}}) = -(k_v + c_1)\dot{q}_1^2 - c_2\dot{q}_2^2 \le 0 \quad (36)$$

The global stability of the control law type (30) is proved.

## 4. Simulation analysis and experiment

According to the preamble design of two kinds of controller, the actual parameters of measuring robot, using Matlab simulation platform is built, respectively, given two kinds of controller system initial parameters, using the designed control law, the robot in different conditions of walking posture control simulation.

### 4.1. Normal walking simulation analysis

The simulation conditions are as follows: the initial value of the given state variable $x_0$ and the state error $e_0$, $x_0 = [1\ \ 0\ \ 0\ \ 0]$, $e_0 = [0.5\ \ 0\ \ 0\ \ 0]$. The simulation platform is Matlab, which provides the basis for the actual control of the robot. The simulation results are shown in figure 3.

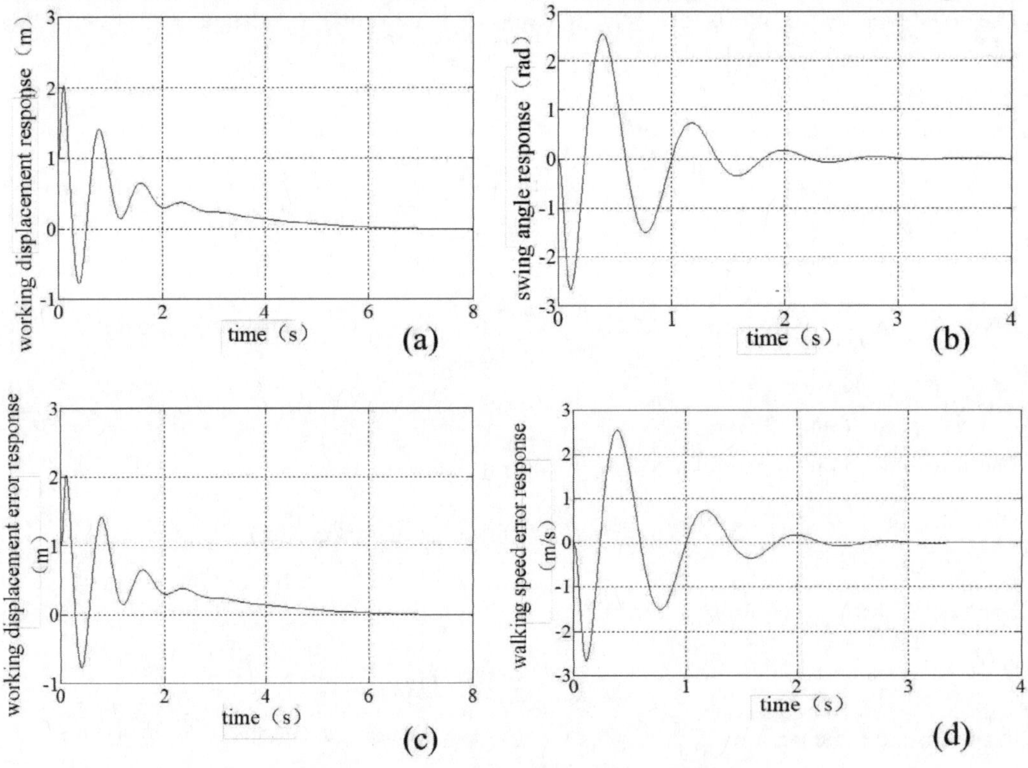


**Figure 3.** Robot normal walking simulation.

From figure 3 (a), we can see that the robot in the start stage of the displacement response to the time will be relatively large fluctuations, but after the 6S, the response tends to be stable, the displacement of the robot from the initial value of 1 to the target value of 0.

Figure 3 (b) shows the response of the robot to the time, after 2 seconds, the robot's swing angle convergence to the target value of 0.

Figure 3 (c) reflects the time response of the robot's walking displacement error, and the time is rapidly converging to 0 in a second or so.

Figure 3 (d) reveals the time response curve of the robot's walking speed error. Can be seen from the figure, in a second or so the time that convergence to 0.

*4.2. Simulation analysis and experiment of hill climbing*

The simulation conditions of robot climbing are: the slope is 10° , the target displacement distance is 12 *m* , After repeated adjustment, determine the control system of the PD parameters: $k_p$=60000 , $k_v$=10000 . Swing angle response as in Figure 3 as shown, figure with solid lines mark the robot swing angle of the target value of 0° , because there is no for swinging angle control design special joint, but controlled in the joint of the walking robot, so robot swing angle response shows an amplitude cosine function rule. As can be seen from the trend of the attenuation, the robot's swing angle will eventually converge to the target value of 0° .

The experiment was carried out on the set of analog circuits, as shown in figure 4, to verify the effect of this designed controller.

The experimental conditions are consistent with the simulation conditions. The robot from the same starting point for 10 consecutive times climbing walking experiment, using inclination angle sensor for robot body swing angle acquisition as per second frequency, and average of 10 times in the experimental data plotted as shown in figure 5 shows the angular response.

We can see that the inconsistent with the several intermediate response and simulation results, mainly because some parts of the line have rust, interfering with the machine swing angle response rules, but the robot actual placed the swing angle of the horn rules and simulation rules are basically the same. The practical application results show that the proposed controller can control the swing angle quickly and effectively.

**Figure 4.** Robot attitude control experiment.

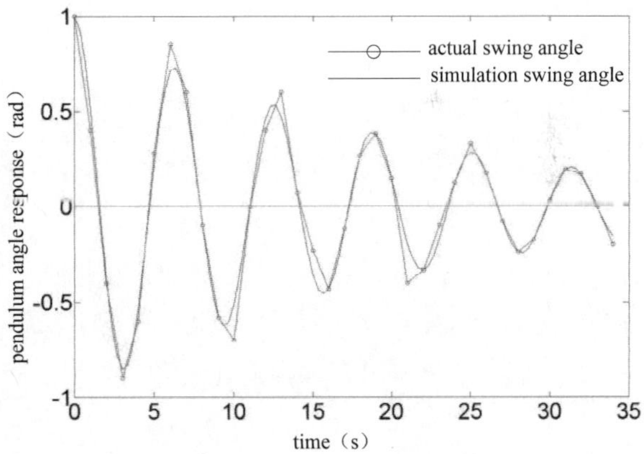


**Figure 5.** Pendulum angle response.

## 5. conclusion

We establish the static model of the single arm robot transmission line, and get the maximum climbing angle of robot. Based on the analysis of the robot's walking state, the two concepts of normal walking and climbing walking are proposed, which provide guidance for the correct and effective walking control strategy. According to the two walking strategies, the corresponding controller is designed, and the simulation is carried out by using the MATLAB tool. Finally, the effectiveness of the proposed two kinds of attitude controllers is verified by experiments, which can effectively solve the problem of robot's walking attitude control.

IOP Conf. Series: Materials Science and Engineering **224** (2017) 012019 doi:10.1088/1757-899X/224/1/012019

**References**

[1] Li E, Liang Z.-Z and Hou Z.-G 2009 *Energy-based balance control approach to the ball and beam system, International Journal of Control* **82** 981

[2] Yang D.-W,Feng Z.-R and Zhang X 2012 *Mechanism design and kinematic analysis of a novel tribrachiation robot for transmission line inspection, Journal of Xi'an Jiaotong University,* **9** 43

[3] Li Z, Ruan Y 2010 *Autonomous inspection robot for power transmission lines maintenance while operating on the overhead ground wires, International Journal of Advanced Robotic Systems,* **7** 107

[4] Wang J.-J,Su Z.-M and Wu J.-H 2014 *Live eliminate tools development for drainage board on220 kV transmission lines , Journal of Electric Power Science and Technology* **4** 93

[5] He Y,Wu G.-P and Wang W 2015 *Advanced Design and motion plan of an improved inspection robot running across obstacles ,Journal of Sichuan University(Engineering Science Edition)* **6** 157

[6] Farinelli, A, Iocchi, L and Nardi, D 2016 *Distributed on-line dynamic task assignment for multi-robot patrolling. Autonomous Robots,* 1

[7] Huang Q, Shao L and Li, N 2016 *Dynamic detection of transmission line outages using hidden markov models. IEEE Transactions on Power Systems* **31** 2026

[8] Cheng J, Qiu J, Ji H, Wang E, Takagi T and Uchimoto T 2017 *High precision ultrasonic guided wave technique for inspection of power transmission line. Chinese Journal of Mechanical Engineering* **30** 170

IOP Conf. Series: Materials Science and Engineering **224** (2017) 012020 doi:10.1088/1757-899X/224/1/012020

# Automatic Verification System for Flatness Error of Large Flat

**Minjian XING**

No. 1500, Zhang Heng Rd. Pu Dong New District, 201203 Shanghai

xingminjian@126.com

**Abstract**. This paper introduces the detection method of flatness error of flat plate, designs a large-scale flatness error automatic verification system. The structure, working principle and flatness error of least square method are described in detail, the uncertainty is analyzed. The advantages of the system are compared and analyzed by experiments, and it has a good application value.

## 1. Introduction

Large flat is used in a wide range of applications such as military, heavy industry and other enterprises. On-site testing methods are usually used for verification due to its large size, weight, and extremely inconvenient to move. The traditional detection method uses bridge plate and electronic level to carry on the segmentation examination, and finally obtains its flatness error through the computation[1]. The method is not only inefficient, laborious, but also requires a high degree of skill in the operation and calculation. Sometimes, the result of the test is not reliable and needs to be re-tested. Therefore, it is necessary to find a convenient, accurate and reliable method to replace the manual operation.

## 2. System structure and working principle

### *2.1 system structure*

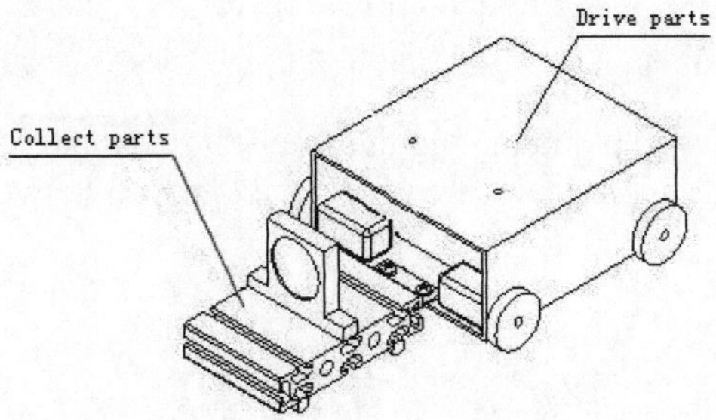


**Figure 1.** Structure of the Intelligent Car

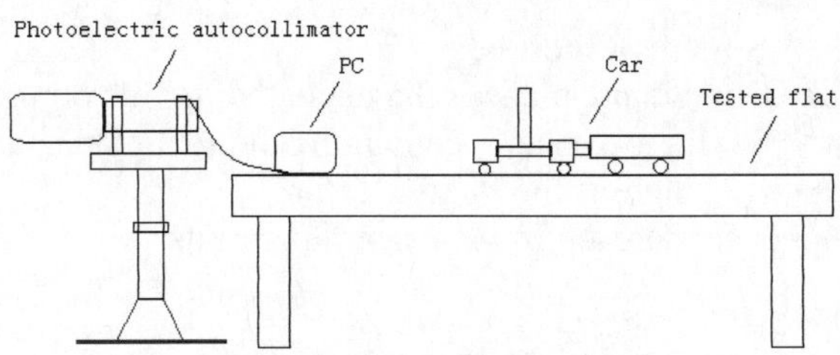


**Figure 2.** System Works

Since auto-collimator is easy to connect with the computer for data transmission, collimator is chosen in the system4. An intelligent car is designed to automatically collect measurement data to realize automatic data acquisition in the whole process. The collected data are input into computer and calculated by flatness error software. The whole flatness measurement system consists of the autocollimator, the intelligent car, the flatness error evaluation software and etc. The intelligent car is designed as two parts: drive part and collection part, as shown in Figure 1. The drive section uses two parallel stepping motors on the same side of the collector to control the intelligent trolley. The collecting section is as follows: two aluminum alloy profiles are connected by a long guide rail, two measuring optical axes are arranged at the lower part of the aluminum alloy profile, and the threaded holes are arranged on the long guide rail to adjust the distance between two aluminum profiles to achieve span adjustment.

*2.2 The system works*

The detection process is as follows. First of all, determine the distribution of measurement points according to the size of the flat plate, adjust the span of the aluminum profile and then open the auto-collimator and computer, start the smart car. The drive part of the intelligent car takes the collection part run along the established route. The plane mirror installed on the collection part produce a small inclination with movement. The autocollimator sends the small tilt signal collected into the computer for its storage. After the completion of the whole measurement, the flatness error of the flat plate is calculated by the calculation software, and the measurement is completed. The measurement process is shown in picture 2.

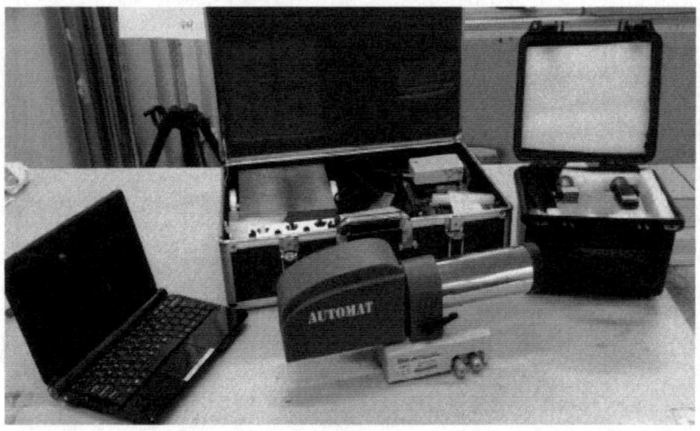


**Figure 3.** Physical Picture

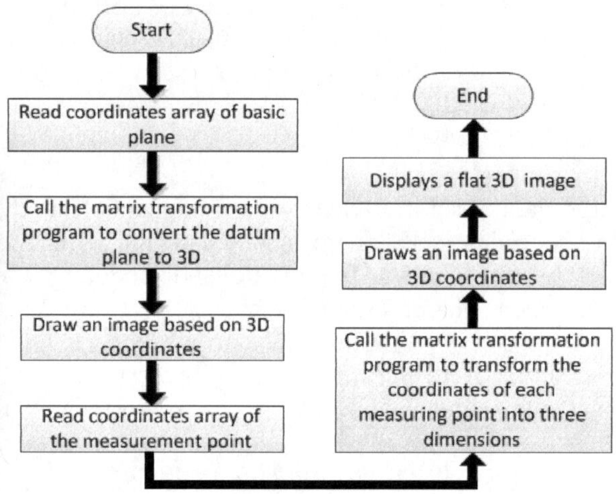


**Figure 4.** Flow chart

## 3. Assessment methods

The least square plane is the plane that points on the measured plane having the smallest distance to it[2]. Taking two planes containing the real plane parallel to the least square plane, the distance between the two planes is the flatness error[3].

Suppose there are some measuring points

$$P(X_i, Y, Z_{ij})\ (i = 0,1...k; j = 0,1...m)$$

The equation of the least squares ideal plane in the three-dimensional coordinate system is

$$\overline{Z_{ij}} = aX_i + bY_j + c \tag{1}$$

The mathematical model of the actual surface of the three-dimensional space can be expressed as:

$$Z_{ij} = aX_i + bY_j + c + f_{ij} \tag{2}$$

According to the basic idea of the least square method

$$\varphi(a, b, c) = \sum (z_{ij} - \overline{z_{ij}})^2 = \sum (z_{ij} - aX_i - bY_j - c)^2 \tag{3}$$

Where: $\varphi(a, b, c)$ is the sum of the squared distances of the actual measured value to the ideal plane.

Using the method of extreme value in mathematical analysis to get $\varphi(a, b, c)$, respectively, calculate the partial differential of $a, b, c$, the equation is:

$$\begin{cases} \frac{\partial \varphi}{\partial a} = -2\sum (z_{ij} - aX_i - bY_j - c)X_i = 0 \\ \frac{\partial \varphi}{\partial b} = -2\sum (z_{ij} - aX_i - bY_j - c)Y_j = 0 \\ \frac{\partial \varphi}{\partial c} = -2\sum (z_{ij} - aX_i - bY_j - c) = 0 \end{cases} \tag{4}$$

Simplifying this equation:

$$\begin{cases} \left(\sum X_i^2\right)a + \left(\sum X_iY_j\right)b + \left(\sum X_i\right)c = \sum X_iz_{ij} \\ \left(\sum X_iY_j\right)a + \left(\sum Y_j^2\right)b + \left(\sum Y_j\right)c = \sum Y_jZ_{ij} \\ \left(\sum X_i\right)a + \left(\sum Y_j\right)b + Nc = \sum z_{ij} \end{cases} \tag{5}$$

The formula 5 is rewritten in the form of a matrix:

$$\begin{pmatrix} \sum X_iZ_{ij} \\ \sum Y_jZ_{ij} \\ \sum Z_{ij} \end{pmatrix} = \begin{pmatrix} \sum X_i^2 & \sum X_iY_j & \sum X_i \\ \sum X_iY_j & \sum Y_j^2 & \sum Y_j \\ \sum X_i & \sum Y_j & \sum N \end{pmatrix} \begin{pmatrix} a \\ b \\ c \end{pmatrix} \tag{6}$$

It can be seen from equation 6, the least square ideal plane equation can be determined as long as the parameters a, b, c be gotten. The corresponding coordinate values $Z_{ij}$ can be gotten after corresponding coordinate of each measuring point is substituted into the plane equation[4]. The flatness error is:

$$f = \max(Z_{ij} - \overline{z_{ij}}) - \min(Z_{ij} - \overline{z_{ij}}) \tag{7}$$

## 4. Evaluation of uncertainty

1 Measurement repeatability in the standard uncertainty $u_1(a)$:

The standard deviation is the standard deviation of the experimental results obtained by measuring the straightness of the same cross-section of the same plate with the pitch method and the self-collimator with the same bridge span, and repeating the measurement ten times[5-6].

It can be measured by continuous measurement column and use A-type assessment method for assessment. First of all, determine the section measuring span of 500mm according to the measured section length. The same measurement point takes continuous measurement for 10 times in the same repeatability measurement condition. The intelligent car takes the bridge plate to move in turn.

$$u_1(a) = \sqrt{\frac{\sum_{i=1}^{10} (a_i - \bar{a})^2}{n - 1}} = 0.426'' \tag{8}$$

2 Components of estimation error $u_2(a)$:

According to the experimental analysis, the mechanical intelligent drive the bridge plate to move. When the span of the bridge is 500mm, the influence on the numerical value is 0.002μm / m. The input Δ can be considered as uniform distribution.

$$u_2(a) = 0.002 / (0.005 \times \sqrt{3}) = 0.2301'' \tag{9}$$

3Evaluation of the Uncertainty Component of the Effect of Indication Error of Photoelectric Autocollimator $u_3(a)$:

From the calibration certificate, photoelectric autocollimator indication error is 0.3 ". It can be considered that within ± 0.3" is uniform distribution, so:

$$u_3(a) = 0.3 / \sqrt{3} = 0.1732'' \tag{10}$$

**Table 1.** Standard Uncertainty Schedule

| Standard Uncertainty Component $u(x_i)$ | Source of Uncertainty | Standard Uncertainty Value $u(x_i)('')$ | Sensitivity coefficient $c_i = \frac{\partial f}{\partial x_i}$ | $\lvert c_i \rvert \times u(x_i)(\mu m)$ |
|---|---|---|---|---|
| $u(a)$ | The error associated with the measurement reading | 0.6649 | | |
| $u_1(a)$ | Measurement repeatability | 0.4265 | $\frac{\sqrt{n_1+n_2}}{2}L$ | $0.6649'' \times \frac{\sqrt{n_1+n_2}}{2} \times \frac{L}{1000}$ |
| $u_2(a)$ | Positioning error | 0.2301 | | |
| $u_3(a)$ | Self - Collimation Indication Error of Photoelectricity | 0.1732 | | |

4 Composite Standard Uncertainty $u(a)$:

$$u^2(a) = u_1^2(a) + u_2^2(a) + u_3^2(a) = 0.4265^2 + 0.2301^2 + 0.1732^2 = 0.6649'' \tag{11}$$

5 Extended Standard Uncertainty:

$$u_c^2 = c^2(a) \cdot u^2(a)$$

$$u_c = 0.00167\sqrt{n_1 + n_2}L\mu m$$

Extended Uncertainty: Taking k = 2, the extended uncertainty U is:

$$U = 2u_c = 0.0034\sqrt{n_1 + n_2}L\mu m$$

Indication of measurement uncertainty:

$$U = 0.0034\sqrt{n_1 + n_2}L\mu m \qquad k=2$$

Where: $L$- bridge span, mm;, $n_1$、$n_2$-the measurement of the long side and short side of the measurement section number.

Large plate of (4000 × 1000) mm, take the span of L = 250mm, $n_1$ = 16, $n_2$ = 4, the flatness of the expansion of uncertainty:

$$U = 0.0034\sqrt{n_1 + n_2}L = 3.8\mu m \qquad k=2$$

## 5. Experiments

Experiments were carried out on a large cast iron plate (4000 × 1000) mm of Shanghai Aircraft Factory. The two methods of plate flatness error automatic test system and manual test were used to carry on the experiment. Measurements: diagonal method, diagonal span: 258mm, long side, short side span: 250mm, diagonal measurement section n0 = 16, long side measuring section n1 = 16, short side measuring section n2 = 4. The flatness error obtained by manual test was 18.6μm. The final result of the automatic detection system was a flatness of 17.9μm. The comparison found that the difference between the two flatness error of only 0.7um, which has little affection of the level of plate discrimination. Manually test took 2 hours in the factory, 1 hour for manual calculation in the laboratory, and 3 hours in all. The automatic detection system got the results in 2 hours in the factory which saving time about 33 %. The more prominent advantage is that it can be directly re-measurement when in doubt.

## 6. Conclusion

Automatic verification system of planar flatness error can automatically collect the data, the test conclusion can be reached in the factory, and plane flatness convex state can visually seen. It improves the detection efficiency and avoids the introduction of human error in the measurement process, improves the accuracy and reliability. It can play an important role in the detection of large flat panel and has a good practical application value.

## References

[1] Shan Zhongde, Zhang Fei, Nie Jungang, Ren Yongxin. Study on the Detection Method and Measurement System of Head Flatness Error on Non-contact Detection [J]. China Academy of Machinery Science Technology,2016, 52(20):1-7

[2] Lei Xianqing, Li Fei, Tu Xianping, Wang Shi feng, Geometry searching approximation algorithm for flatness error evaluation [J]. Optics and Precision Engineering, 2013, 21(5):1312-1317

[3] Lei Kai, Qin Zhe. Evaluation algorithm of flatness error and program design [J]. Machine Tool and Hydraulics, 2011.02

[4] Shuyao Tian, Weina Lu, Chunxia Zhao, Ning Cai, Qianqian Chen. The research of Flatness error measure sampling program [C]. Iita International Conference on Geoscience & Remote Sensing 2010, 2:594-597

[5] Zhang Hao. Research on digital inspection and assessment algorithm of flatness errors [D]. Taiyuan, Taiyuan University of Science and Technoloty,2013

[6] Huang Hailin, Yang Xue, Liu Yingjie. Dimensional modeling of virtual experiment of flatness error measurement [J] .Production Technology and Management, 2009,26 (7)

IOP Conf. Series: Materials Science and Engineering **224** (2017) 012021 doi:10.1088/1757-899X/224/1/012021

# Lattice Boltzmann Simulation of Flow Past a Finite Cylinder

**Yi Liu, Deming Nie**

Institute of Fluid Mechanics, China Jiliang University, China

nieinhz@cjlu.edu.cn

**Abstract.** In this work the lattice Boltzmann method was adopted to simulate the flow past a finite cylinder which is mounted on a wall. The Reynolds number based on the height of the cylinder is ranging from 10 to 160. The effect of the Reynolds number on the flow pattern as well as the drag and lift coefficient of the cylinder was discussed. Results show that the flow becomes unsteady when the Reynolds number is larger than 60. Moreover, the flow pattern becomes more complex for larger Reynolds number.

## 1. Introduction

Flow past a cylinder is a topic of great academic and practical importance [1]. It is well known that Karman vortex street generally takes place when flow past a fixed circular cylinder, which leads to the oscillation of the drag and lift force experienced by the cylinder. Therefore, the investigation of the flow past a cylinder helps to understand the fundamental physical phenomena of bluff body aerodynamics. Things may become much more complex when the cylinder is mounted on a wall, since the effect of the wall is significant in most cases. The flow around a cylinder of finite length is representative of flows in various applications, such as high-rise buildings, chimneys and tube banks in heat exchangers.

Due to its importance a few experimental and numerical work have been devoted to investigate the flow past a finite cylinder which is mounted on a wall. Liu et al. [2] numerically investigated the flow past a finite circular cylinder. They revealed that the tip vortex is generated at the free-end while the horse-shoe vortex is generated at the plate-body junction for the Reynolds number ranging from 100 to 200. Wang and Zhou [3] studied the wake of flow around a finite-length square cylinder at Reynolds number of 221 and 9300 through an experimental work. They observed two types of arch type vortices, which are depending on the aspect ratio of cylinder. Bourgeois et al. [4] conducted an experimental study on the turbulent flow past a finite square cylinder which is mounted on a thin boundary layer. They presented the coherent and incoherent flow fields and investigated the effect of incoherent turbulence on the turbulent characteristics. Recently, Saha [5] performed direct numerical simulation to study flow past a finite square cylinder at a Reynolds number of 250. It was found that the effect of the cylinder aspect ratio on the flow characteristics is significant. Moreover, his work revealed that the Strouhal number and drag coefficient increases as the cylinder aspect ratio increases.

So far the largest and maybe the most realistic suspension simulations have been performed with the lattice Boltzmann method (LBM) [6, 7] that appears to be advocated as an effective computational tool for the simulation of particle suspensions, multiphase flow, microfluidics, and turbulence for its several remarkable advantages, such as easy coding, no requirement of re-meshing procedure, and high computational efficiency.

In comparison with the research on the flow past infinite cylinder, the number of research on finite length cylinder is limited. As mentioned, there are only a few numerical simulations of flows around tall cylinders and there are still open questions about parts of this flow. For example, the flow pattern before the cylinder is not clear. In addition, the lower part of the near wake is not fully understood. The main objective of the present study is to numerically investigate the flow past a finite cylinder which is placed on a wall in a direct numerical simulation framework for the Reynolds number $10 \le Re \le 160$. This range of Re is consistent with most of previous research work on flow past an infinite cylinder. This work mainly focuses on the effects of the Reynolds number on the flow pattern and drag and lift coefficient of the cylinder.

## 2. Numerical Model

We solve the motion of a fluid using the LBM. The discrete lattice Boltzmann equations of a single-relaxation-time model are expressed as [6]

$$f_i(\boldsymbol{x}+\boldsymbol{e}_i\Delta t, t+\Delta t) - f_i(\boldsymbol{x},t) = -\frac{1}{\tau}\left[f_i(\boldsymbol{x},t) - f_i^{(0)}(\boldsymbol{x},t)\right] \tag{1}$$

where $f_i(\boldsymbol{x}, t)$ is the distribution function for the microscopic velocity $\boldsymbol{e}_i$ in the ith direction, $f_i^{(0)}(\boldsymbol{x}, t)$ is the equilibrium distribution function, $\Delta t$ is the time step of the simulation, | is the relaxation time, $c_s$ is the speed of sound, and $w_i$ are weights related to the lattice model.

The fluid density $\rho_f$ and velocity u are determined by the distribution function

$$\rho_f = \sum_i f_i, \qquad \rho_f \boldsymbol{u} = \sum_i f_i \boldsymbol{e}_i \tag{2}$$

For the two-dimensional D2Q9 lattice model used here, the discrete velocity vectors are

$$\boldsymbol{e}_i = \begin{cases} (0,0), & \text{for } i = 0, \\ (\pm 1,0)c,\ (0,\pm 1)c, & \text{for } i = 1 \text{ to } 4, \\ (\pm 1,\pm 1)c, & \text{for } i = 5 \text{ to } 8, \end{cases} \tag{3}$$

where $c = \Delta\boldsymbol{x}/\Delta t$, and $\Delta\boldsymbol{x}$ is the lattice spacing. The speed of sound $c_s^2 = c^2/3$. Following Qian et al. [11], the equilibrium distribution function is chosen as

$$f_i^{(0)}(\boldsymbol{x},t) = w_i \rho_f \left[1 + \frac{3\boldsymbol{e}_i \cdot \boldsymbol{u}}{c^2} + \frac{9(\boldsymbol{e}_i \cdot \boldsymbol{u})^2}{2c^4} - \frac{3\boldsymbol{u}^2}{2c^2}\right] \tag{4}$$

where the $w_i$ are set to $w_0 = 4/9$, $w_{1\text{-}4} = 1/9$, and $w_{5\text{-}8} = 1/36$. In this model, the fluid viscosity is computed using the equation $\nu = c_s^2(\tau - 0.5)\Delta t$.

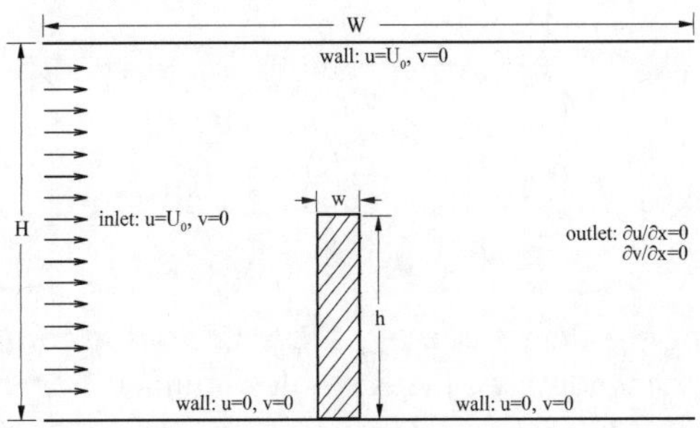


**Figure 1.** Two-dimensional physical model.

The physical model considered for the analysis is schematically shown in Fig. 1. $w$ is the width of the cylinder and $h$ is the height of the cylinder. In the simulations, we set $w$=20 (in lattice unit) and $h$=60 (in lattice unit). The computational domain is $W \times H = 30h \times 20h$ in this work. The cylinder is $10h$ away from the inlet. The boundary conditions employed for the present investigation are shown in Fig. 1. The Reynolds number is defined as $Re = U_0 h/\nu$, where $\nu$ is the kinematic viscosity of the fluid.

## 3. Results

Fig. 2 shows the computational results in the vicinity of the cylinder by the streamlines plots at $Re$ = 10, 20, 40 and 60. All the cases shown in Fig. 2 are steady state. Fig.2 shows that the flow separates at the corner of the cylinder for all Reynolds numbers. As a result, a closed steady recirculating region consisting of one vortex forms behind the cylinder. This recirculating region increases in size with the increase in Reynolds number.

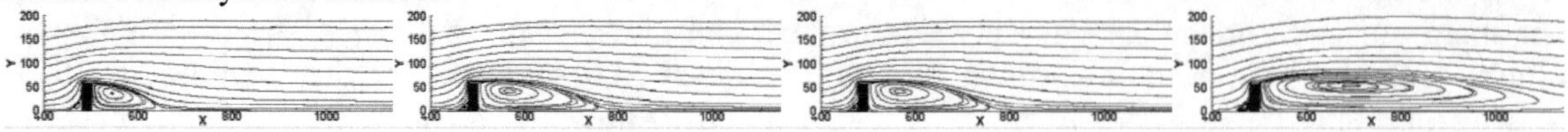

(a) (b) (c) (d)

**Figure 2.** Streamline for different Reynolds number: (a) $Re = 10$, (b) $Re = 20$, (c) $Re = 40$ and (d) $Re = 60$.

The instantaneous streamlines in Fig. 3 show the detailed views of the laminar vortex near the cylinder at different Reynolds number. In addition to the primary separation region, there is a secondary separation region behind the cylinder, which is much smaller than the primary one. Furthermore, another separation region is observed before the cylinder for all cases, which is an unusual phenomenon. It is also clear that the size of the separation region is not sensitive to the Reynolds number.

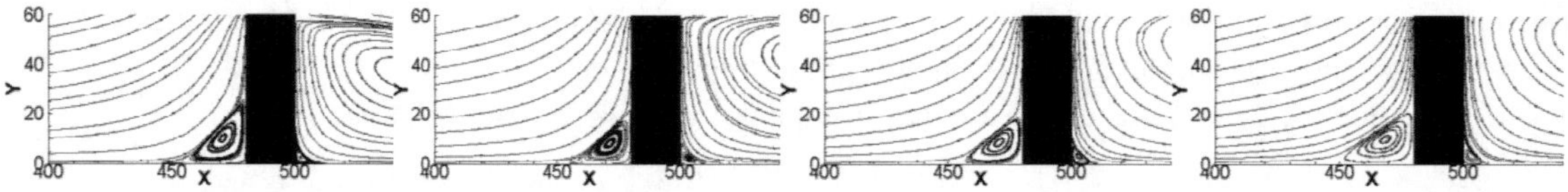

(a) (b) (c) (d)

**Figure 3.** Recirculation regions for different Reynolds number: (a) $Re = 10$, (b) $Re = 20$, (c) $Re = 40$ and (d) $Re = 60$.

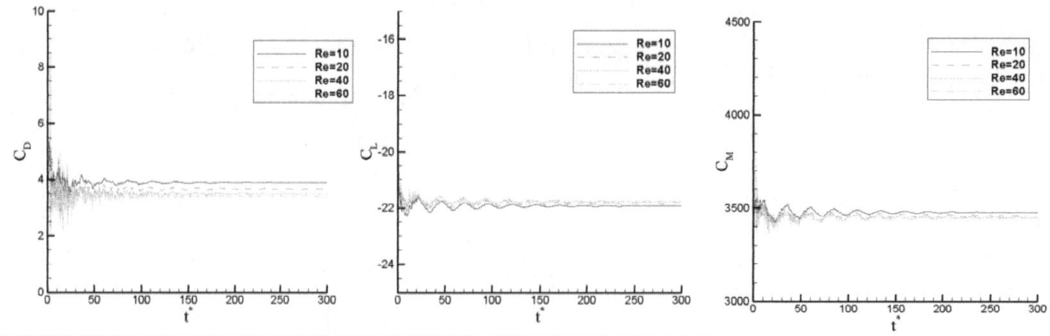


(a) (b) (c)

**Figure 4.** Normalized force experienced by the cylinder for different Reynolds numbers: (a) drag, (b) lift and (c) torque.

The cylinder is experiencing the drag force $F_D$, lift force $F_L$ and torque $F_M$ resulted from the flow past on it, which are generally depending on the Reynolds number. For the sake of comparison, all the results are normalized, i.e. $C_D=F_D/0.5\rho_f U_0 2h$, $C_L=F_L/0.5\rho_f U_0 2h$ and $C_M=F_M/0.5\rho_f U_0 2h^2$. As shown in Fig. 4, all results eventually reach a steady state after a damped oscillation of initial transients for all cases. The drag coefficient $C_D$ decreases as $Re$ increases. The same is true for the torque coefficient $C_M$, as shown in Fig. 4. However, the lift coefficient $C_L$ increases as $Re$ increases. In addition, the lift coefficient is much larger than the drag coefficient.

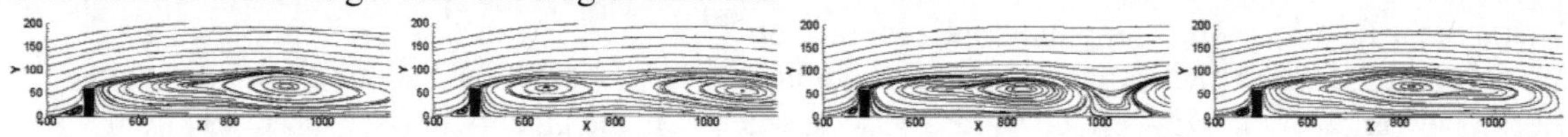

(a) (b) (c) (d)

**Figure 5.** Streamline at different times for $Re = 100$: (a) $t = 10000$, (b) $t = 14000$, (c) $t = 18000$ and (d) $t = 22000$.

The instantaneous streamlines in Fig. 5 show the detailed views of the laminar vortex shedding near the cylinder at $Re = 100$, for four successive moments of time which span over the whole period.

The results shown in Fig. 5 indicate that the flow does not reach a steady state. The vortex shedding is observed at this Reynolds number. In addition, two successive vortices will come to contact and merge into a larger vortex, as shown in Fig. 5. Fig. 6 shows the instantaneous vortex contour at different times at *Re* = 100, which clearly display the entire evolution of vortex behind the cylinder. Each time the vortex shedding takes place, the vortex will move to the downstream, as show in Fig. 6.

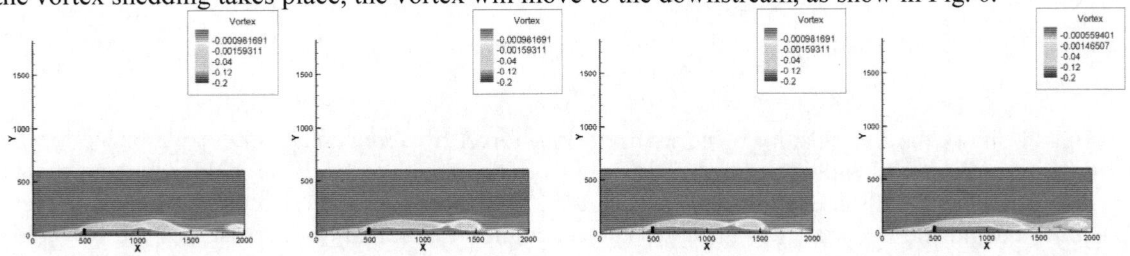


(a) (b) (c) (d)

**Figure 6.** Instantaneous vortex contour at different times for *Re*=100: (a) $t$ = 10000, (b) $t$ = 14000, (c) $t$ = 18000 and (d) $t$ = 22000.

Things may become more complex if increasing the Reynolds number. As shown in Fig. 7, the instantaneous streamline patterns of *Re* = 120 are different from those of *Re* = 100 shown in Fig. 5. A small vortex attached on the bottom wall is observed in the figure, which is located between two primary shedding vortices. Another significant difference is that the period is about 12000 time step for *Re* = 100, while it is about 28000 time step for *Re*=120. Moreover, the unsteadiness of flow can be demonstrated by showing the recirculation regions near the cylinder, as one can see in Fig. 8. During a single period, the size of the recirculation region before the cylinder decreases firstly, then increases. This process is repeated. In addition, this is also true for the secondary recirculation region behind the cylinder.

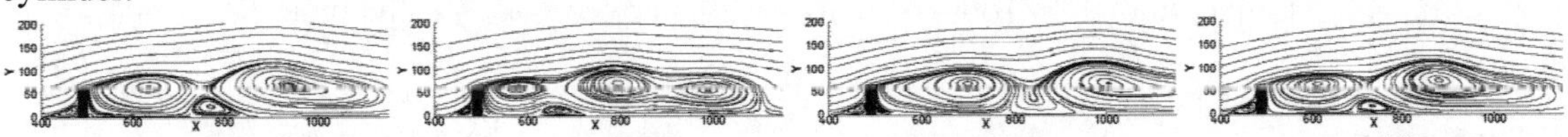

(a) (b) (c) (d)

**Figure 7.** Streamline at different times for *Re* = 120: (a) $t$ = 10000, (b) $t$ = 18000, (c) $t$ = 28000 and (d) $t$ = 38000.

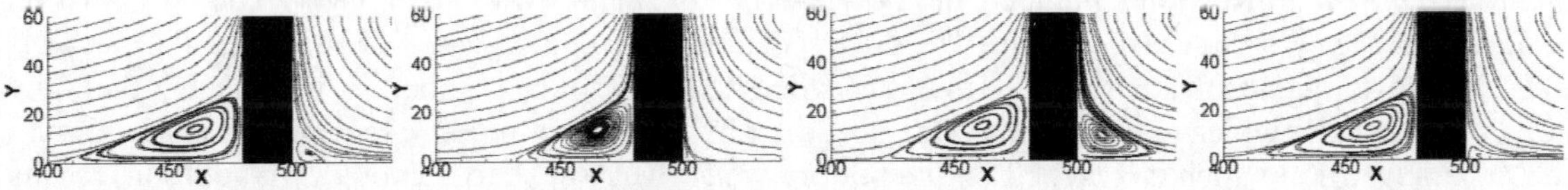

(a) (b) (c) (d)

**Figure 8.** Recirculation regions at different times for *Re* = 120: (a) $t$ = 10000, (b) $t$ = 18000, (c) $t$ = 28000 and (d) $t$ = 38000.

The flow pattern becomes much more complex when the Reynolds number is *Re* = 160, which is the largest Reynolds number studied in this work. The streamlines are shown in Fig. 9 and Fig. 10. In comparison with the results of *Re* = 120, more recirculation regions are generated behind the cylinder. Furthermore, the recirculation regions before and behind the cylinder are much larger, as shown in Fig. 10, which may lead to higher pressure drop and larger oscillations in drag and lift force experienced by the cylinder.

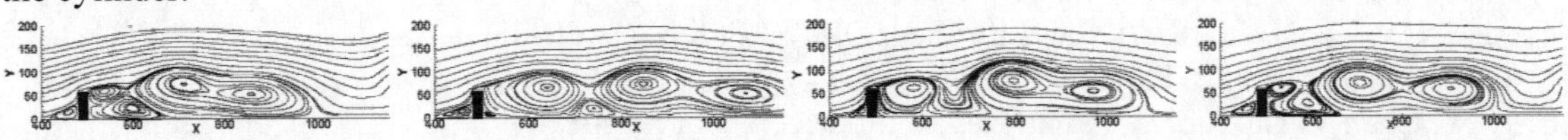

(a) (b) (c) (d)

**Figure 9.** Streamline at different times for *Re* = 160: (a) $t$ = 10000, (b) $t$ = 18000, (c) $t$ = 28000 and (d) $t$ = 38000.

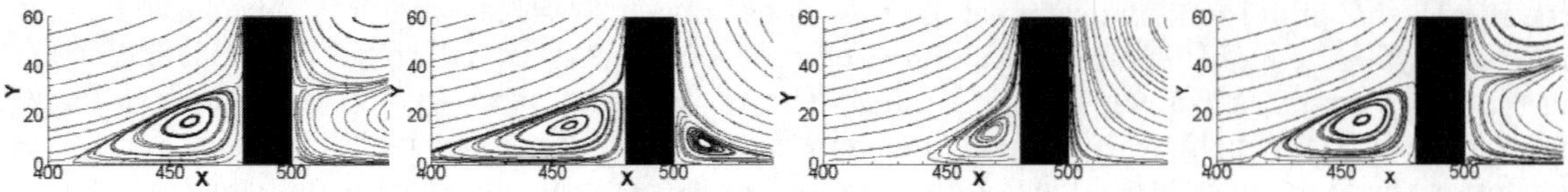


(a) (b) (c) (d)

**Figure 10.** Recirculation regions at different times for $Re$ = 160: (a) $t$ = 10000, (b) $t$ = 18000, (c) $t$ = 28000 and (d) $t$ = 38000.

Fig. 11 shows the time history of normalized drag force, lift force and torque experienced by the cylinder for different Reynolds numbers. Different from the results of small Reynolds numbers shown in Fig. 4, all the results are oscillating even after a long time, which is resulted from the vortex shedding behind the cylinder. For all cases the magnitude of oscillation increases as the Reynolds number increases. Similarly, the lift coefficient is much larger than the drag coefficient. And the normalized torque is larger than $C_D$ and $C_M$ for several orders of magnitude.

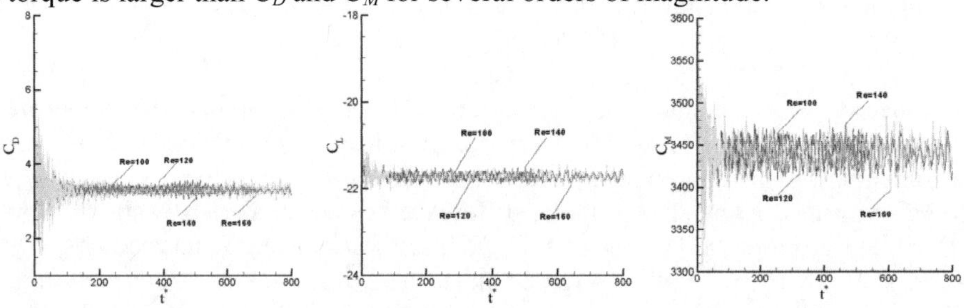


(a) (b) (c)

**Figure 11.** Normalized force experienced by the cylinder for different Reynolds numbers: (a) drag, (b) lift and (c) torque. Red: $Re$ = 100, green: $Re$ = 120, purple: $Re$ = 140 and blue: $Re$ = 160.

## 4. Conclusion

In this work, the lattice Boltzmann method was adopted to simulate the flow past a finite cylinder for the Reynolds number $10 \leq Re \leq 160$. The streamline pattern as well as the drag and lift force was presented. For small Reynolds number, the flow can reach a steady state eventually. The recirculation regions near the cylinder are displayed. The drag coefficient $C_D$ decreases as $Re$ increases. The same is true for the torque coefficient $C_M$. However, when the Reynolds number is increasing, the flow becomes unsteady, which leads to the vortex shedding behind the cylinder. Especially, there is a small recirculation region which is attached on the bottom wall when $Re$ is large enough. In addition, results also show that the drag and lift coefficients are oscillating with large magnitude.

## Acknowledgments

This work was supported by the Zhejiang Provincial Natural Science Foundation of China (Grant LY15A020004)

## References

[1] S. Kang, Phys. Fluids **15** (2003).
[2] Y. Liu Y, R.MC. So, Z.X. Cui, J. Fluids Struct. **20** (2005).
[3] H.F. Wang, Y. Zhou, J. Fluid Mech. **638** (2009).
[4] J.A. Bourgeois, P. Sattari, R.J. Martinuzzi, Phys. Fluids **23** (2011).
[5] A. K. Saha, Comput. Fluids **88** (2013).
[6] S.Y. Chen, G.D. Doolen, Annu. Rev. Fluid Mech. **30** (1998).
[7] C.K. Aidun, J.R. Clausen, Annu. Rev. Fluid Mech. **42**(2010).

IOP Conf. Series: Materials Science and Engineering **224** (2017) 012022 doi:10.1088/1757-899X/224/1/012022

# Thermal equilibrium of a Brownian particle in a fluctuating fluid: a numerical study

**Yi Liu, Deming Nie**

Institute of Fluid Mechanics, China Jiliang University, China

nieinhz@cjlu.edu.cn

**Abstract.** In this work the fluctuating lattice Boltzmann method was adopted to simulate the motion of a Brownian particle in a fluid in two dimensions. The temperatures characterizing the translation motion and rotational motion of the particle were calculated to evaluate the thermal equilibrium between the particle and the fluid. Furthermore, the effects of the fluid temperature and viscosity on the fluid pressure fluctuation were investigated. The linear relationships were observed in a log-log coordinate. Besides, the slopes of the linear relation were obtained, which keeps constant for all cases studied.

## 1. Introduction

Particles suspended in fluids experience a random force due to the thermal fluctuations in the fluid around them in addition to the average hydrodynamic force. Brownian motion may take place for those sub-micron/nanoscale particles. For many applications in microsystems, the ability to control and measure temperature inside microfluidic devices is critical since temperature often affects biological or chemical processes. Recent developments [1,2] demonstrate that the well-defined temperature dependence of the Brownian motion of nanoparticles could be used to present a temperature measurement technique which offers several benefits over existing methodologies. Brownian particle can be adopted to measure the local viscoelastic response of soft materials [3] or the topography of a surrounding polymer network [4]. The motion of a Brownian probe can also be used to characterize mechanical properties of molecular motors by analyzing the particle's trajectory [5]. Moreover, the biased Brownian motions or rectified Brownian motions, induced by an energy source [6] or by broken spatial reflection symmetry [7], provide a very effective technique for particle separation. Furthermore, it has been demonstrated [8,9] that nano-particles in a conventional base fluid, known as nanofluids, tremendously enhance the heat transfer characteristics of the original fluid. At the same time, study [10] has declared that Brownian motion is a key mechanism governing the thermal behavior of nanofluids. Due to its importance in engineering applications, there has always been a great deal of interest in developing algorithms that can provide a better understanding of particle's Brownian motion.

The most important and accurate approach to simulate particle Brownian motion may be the fluctuating hydrodynamics, which was proposed by Landau and Lifshitz [11]. In this approach, the thermal fluctuations in the fluid, which result in the Brownian motion of particles, are modeled by adding a random stress tensor to Navier - Stokes equations. Solving the fluctuating hydrodynamic equations coupled with the particle equations of motion result in the Brownian motion of particles. As a direct numerical simulation scheme, there is no need to add a random force term in the particles'equations since random fluctuations are applied directly to the particles.

The thermal equilibrium between the Brownian particles and the fluids is quite fundamental to the modelling particle Brownian motion, which has been paid little attention in the past. The research work on this subject is very limited. Therefore, in this work the fluctuating dynamics was used to simulate the motion of a Brownian particle in two dimensions. The present work aims to study the thermal equilibrium between the particle and the fluid, as well as the pressure fluctuation.

## 2. Numerical Model

The fluid flow is solved by the LB method. The discrete LB equations of a single-relaxation-time model are described as [12],

$$f_i(\boldsymbol{x}, c_i\Delta t, t+\Delta t) - f_i(\boldsymbol{x},t) = -\frac{1}{\tau}\left[f_i(\boldsymbol{x},t) - f_i^{(0)}(\boldsymbol{x},t)\right] + f_i^f(\boldsymbol{x},t) \tag{1}$$

where $f_i$ ($\boldsymbol{x}$, $t$) is the distribution function on the i-direction microscopic velocity $e_i$, $f_i(0)(\boldsymbol{x}, t)$ is the equilibrium distribution function, Δt is the time step of the simulation, $f_i^{\,f}(\boldsymbol{x}, t)$ is a stochastic term representing the thermal fluctuations, which is related to the fluctuating stress in the Navier-Stokes equation [12]. The fluid density $\rho$ and velocity $\boldsymbol{u}$ are determined by the distribution function,

$$\rho = \sum_i f_i, \qquad \rho_f \boldsymbol{u} = \sum_i f_i \boldsymbol{e}_i \tag{2}$$

For the two-dimensional D2Q9 lattice model used here, the discrete velocity vectors are,

$$\boldsymbol{e}_i = \begin{cases} (0,0), & \text{for } i = 0, \\ (\pm 1,0)c,\ (0,\pm 1)c, & \text{for } i = 1 \text{ to } 4, \\ (\pm 1,\pm 1)c, & \text{for } i = 5 \text{ to } 8, \end{cases}$$

where $c = \Delta x/\Delta t$, $\Delta x$ is the lattice spacing. The equilibrium distribution function is chosen as,

$$f_i^{(0)}(\boldsymbol{x},t) = \omega_i \rho\left[1 + \frac{3\boldsymbol{e}_i\cdot\boldsymbol{u}}{c^2} + \frac{9(\boldsymbol{e}_i\cdot\boldsymbol{u})^2}{2c^4} - \frac{3\boldsymbol{u}^2}{2c^2}\right] \tag{3}$$

As shown by Nie and Lin [12], the stochastic term is related to the fluctuating stress,

$$\sigma_{\alpha\beta}^f = -\tau\sum_i f_i^f e_{i\alpha} e_{i\beta} \tag{4}$$

According to the fluctuation-dissipation theorem, has the following property [11],

$$\begin{gathered} \langle \sigma_{\alpha\beta}^f \rangle = 0 \\ \langle \sigma_{\alpha\beta}^f(\boldsymbol{x}_1,t_1)\sigma_{\gamma\delta}^f(\boldsymbol{x}_2,t_2)\rangle = 2k_BT\mu\left(\delta_{\alpha\gamma}\delta_{\beta\delta} + \delta_{\alpha\delta}\delta_{\beta\gamma} - \frac{2}{3}\delta_{\alpha\beta}\delta_{\gamma\delta}\right)\delta(\boldsymbol{x}_1-\boldsymbol{x}_2)\delta(t_1-t_2) \end{gathered} \tag{5}$$

where <> denotes averaging over an ensemble, $k_B$ is the Boltzmann constant, $T$ is temperature of the fluid, $\mu$ is the dynamic viscosity of the fluid. The fluctuating stress is sampled from a Gaussian distribution with zero mean and a given variance of $2k_BT\mu$.

In this work we assume the stochastic term $f_i^f(\boldsymbol{x}, t)$ to be the following form to make sure of the conservation of mass and momentum [12],

$$\begin{aligned} f_0^f &= 0 \\ f_1^f &= f_3^f = \frac{1}{2\tau}\sigma_{yy}^f \\ f_2^f &= f_4^f = \frac{1}{2\tau}\sigma_{xx}^f \\ f_5^f &= f_7^f = -\frac{1}{4\tau}\left(\sigma_{xx}^f + \sigma_{yy}^f + \sigma_{xy}^f\right) \\ f_6^f &= f_8^f = -\frac{1}{4\tau}\left(\sigma_{xx}^f + \sigma_{yy}^f - \sigma_{xy}^f\right) \end{aligned} \tag{6}$$

## 3. Results and Discussion

In this work, the Brownian motion of a single particle in a periodic domain was numerically investigated. Only the hydrodynamic force was considered. The periodic domain is 256×256 in the simulations, which is large enough to eliminate the effect of boundary. The density of the fluid is $\rho = 1$ and the non-dimensional relaxation time $\tau = 0.8$, which leads to the fact that the viscosity of the fluid is $\nu = 0.1$. The radius of particle is $a = 3.5$. The solid/fluid density ratio is $\rho_s/\rho = 11$. In order to determine the magnitude of the fluid fluctuation, the variance of fluctuating stress is chosen as $2k_BT\mu = 10^{-4}$. It should be stated that all the above-mentioned parameters are in lattice unit.

In Fig. 1, the instantaneous velocity field of fluid is shown at different times, along with the Brownian particle. As shown in the figure, the velocity vectors are disorderly and disorganized, which represents the random molecular motion of the fluid. This is the origin of the Brownian motion of particle in a fluid, which is resulted from the essence of the present fluctuating lattice Boltzmann method. As one can see in Fig. 1, the particle moves randomly due to the fluid molecular collision. This suggests that it is impossible to predict the motion of particle, such as the velocity or the trajectory of particle. In addition, the rotation of particle is also attained in the simulation, which is different from the Langevin dynamics.

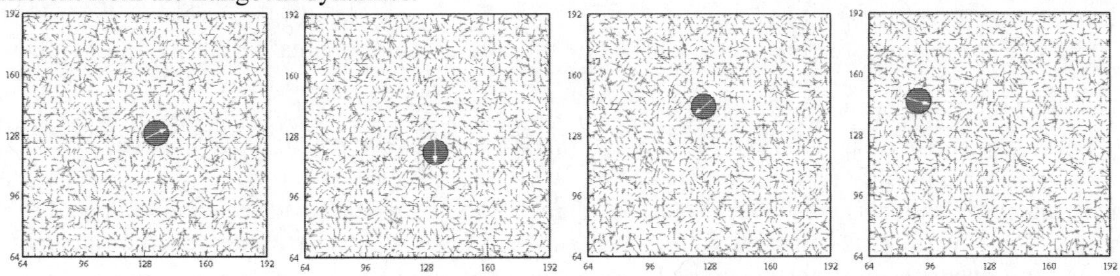

(a) (b) (c) (d)

**Figure 1.** Instantaneous flow for a Brownian particle suspended in a fluctuating fluid: (a) $t = 10000$, (b) $t = 50000$, (c) $t = 100000$, (d) $t = 200000$

Fig. 2 shows the pressure fluctuation of fluid at different times, which also presents the disorderly and disorganized results. As shown in the figure, the pressure fluctuation is randomly but homogeneously distributed in the whole domain. Undoubtedly, the results shown in Fig. 1 and Fig. 2 reveal the microstructure of fluid in a very simple way.

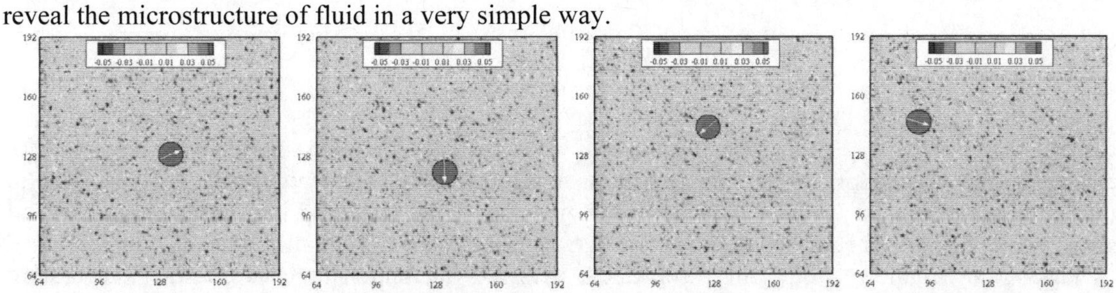

(a) (b) (c) (d)

**Figure 2.** Instantaneous pressure for a Brownian particle suspended in a fluctuating fluid: (a) $t = 10000$, (b) $t = 50000$, (c) $t = 100000$, (d) $t = 200000$

It has been shown that thermal equilibrium between the Brownian particle and the surrounding fluid molecular will reach and that an equipartition of energy for each degree of freedom will be observed, which can be described as

$$\langle U^2 \rangle = \langle V^2 \rangle = k_B T / M, \langle \Omega^2 \rangle = k_B T / J \tag{7}$$

where $U$ and $V$ refer to the translational velocity of $x$ and $y$ coordinate, respectively, and $\Omega$ refers to the rotational velocity of particle. $M$ and $J$ are the mass and inertial of the particle, respectively. An equipartition of energy for $x$ and $y$ translational motion can be observed, as well as the rotational motion, as displayed in Fig. 3, which shows the thermal equilibrium of $Re = 0.035$ and $Re = 0.15$, respectively.

The temperatures characterizing translational motion of $x$ and $y$ coordinate, and the rotational motion agree with each other after a certain time, but all about 5% less than or more than the effective temperature of the fluid, as on can see in Fig. 3. The similar results were also obtained by Ladd [13], which are however 15% less than the effective temperature of the fluid.

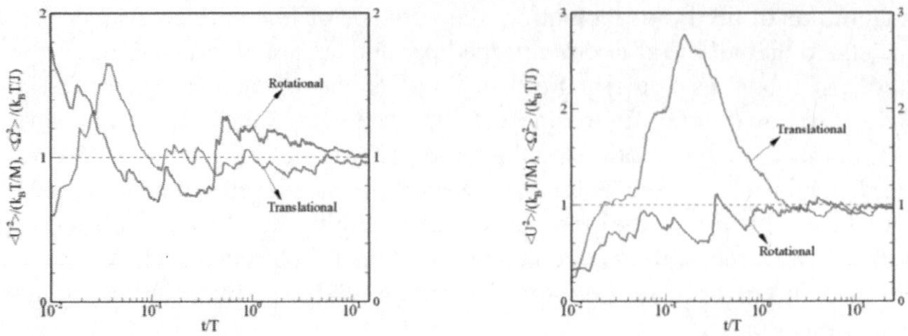


(a) (b)

**Figure 3.** Time history of normalized mean square of translational/rotational velocity: (a) $Re$=0.035, (b) $Re$=0.15

Fig. 4 summarizes the temperatures representing the translational and rotational velocity of a Brownian particle for different Reynolds number. All the results shown in the figure are normalized, which demonstrates that the difference between the particle temperature and the fluid temperature is within 5%. As a result, the thermal equilibrium between the Brownian particle and the fluctuating fluid is almost achieved irrespective of the Reynolds number.

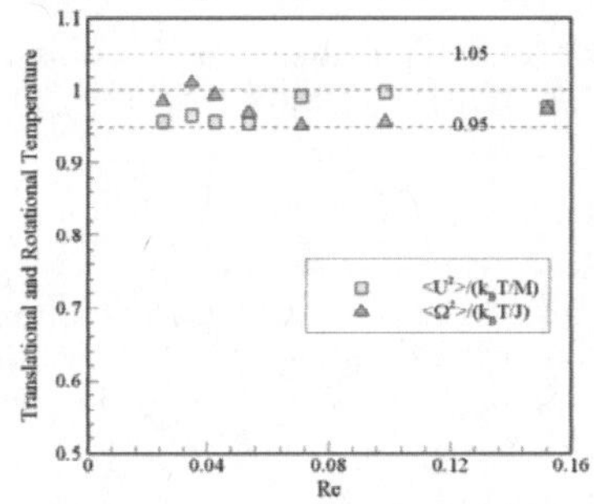


**Figure 4.** Summary of thermal equilibrium of a Brownian particle for different Reynolds number

The pressure fluctuation is a significant factor to determine the Brownian motion of particles in fluids. In order to present more insight into this, Fig. 5 shows the effects of the fluid temperature and viscosity on the pressure fluctuation. To better visualize the relationship, all the results are displayed in a log-log coordinate. As one can observe, the pressure fluctuation increases as the temperature increases, showing a linear relationship. More importantly, the slope of -0.5 is observed for all the results shown in Fig. 5(a), which is independent of the fluid viscosity. Similarly, the linear relationship between the pressure fluctuation and the fluid viscosity is also obtained, as shown in Fig. 5(b). However, the pressure fluctuation decreases as the viscosity increases. In addition, the slope of -0.08 is observed for all results.

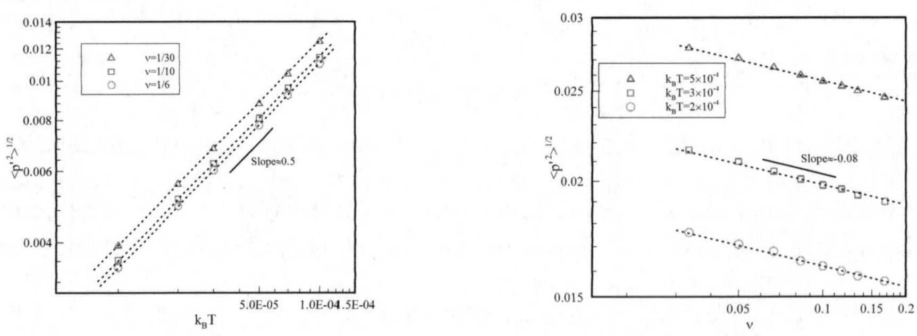


(a) (b)

**Figure 5.** The pressure fluctuation as a function of: (a) the fluid temperature and (b) viscosity

Fig. 6 shows the normalized translational and rotational velocity correlation functions, which are the fundamental results of a Brownian particle. As shown in the figure, all the results are decaying to zero, indicating that the velocity of Brownian particle become more unrelated for larger time interval. In addition, the similar rate of decay is observed for both horizontal and vertical components. The rotational velocity correlation function is decaying faster than that of translational velocity.

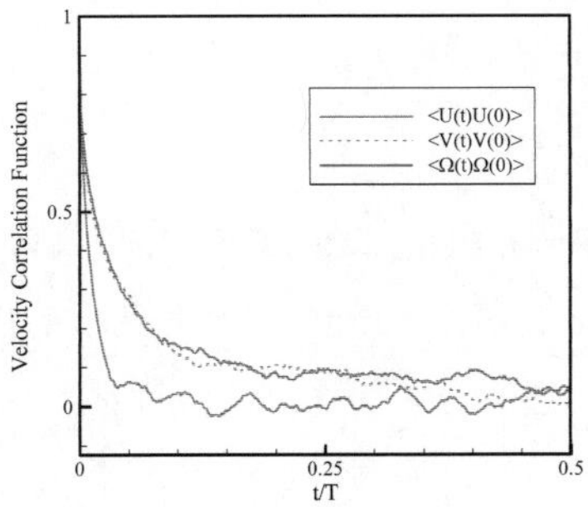


**Figure 6.** The translational and rotational velocity correlation functions

## 4. Conclusion

In this work a fluctuating lattice Boltzmann model was adopted to simulate the Brownian motion of a particle in two dimensions. Results show that the difference between the particle temperature and the fluid temperature is within 5% irrespective of the Reynolds number, indicating the thermal equilibrium is almost achieved. Moreover, it has been shown that the pressure fluctuation has a linear relationship with the fluid temperature in a log-log coordinate. The slope is shown to be about 0.5 for all cases. Similar feature is observed for the relationship between the pressure fluctuation and the fluid viscosity.

**Acknowledgments**

This work was supported by the National Natural Science Foundation of China (Grant 11272302) and the Zhejiang Provincial Natural Science Foundation of China (Grant LY15A020004)

**References**

[1] J. S. Park, C. K. Choi, K. D. Kihm, Meas. Sci. Technol. **16**, 1418(2005).
[2] K. Chung, J. K. Cho, E. S. Park, V. Breedveld, H. Lu, Anal. Chem. **81**, 991(2009).
[3] F. C. MacKintosh, C. F. Schmidt, Curr. Opin. Colloid Interface Sci. **4**, 300 (1999).
[4] C. Tischer, S. Altmann, S. Fisinger, J. K. H. Ho¨rber, E. H. K. Stelzer, E.-L. Florin, Appl. Phys. Lett. **79**, 3878 (2001).
[5] S. Jeney, E. H. K. Stelzer, H. Grubmu¨ller, E.-L. Florin, Chem. Phys. Chem. **5**, 1150 (2004).
[6] R. Dean Astumian, Science **276**, 917 (1997).
[7] I. Dere´nyi, R. D. Astumian, Phys. Rev. E **58**, 7781 (1998).
[8] P. Keblinski, S. R. Phillpot, S.U.S. Choi, J.A. Eastman, Int. J. Heat Mass Transfer **45**, 855 (2002).
[9] S. Lee, S. U. S. Choi, S. Li, J. A. Eastman, J. Heat Transfer **121**, 280 (1999).
[10] R. Prasher, P. Bhattacharya, P. E.Phelan, Phys. Rev. Lett. **94**, 025901 (2005).
[11] L.D. Landau, E.M. Lifshitz, *Fluid Mechanics* ( Pergamon Press, London, 1959).
[12] D. Nie, J. Lin, Particuology **7**, 501 (2009).
[13] A.J.C. Ladd, J. Fluid Mechanics **271**, 311 (1994).

IOP Conf. Series: Materials Science and Engineering **224** (2017) 012023 doi:10.1088/1757-899X/224/1/012023

# A Quadtree-gridding LBM with Immersed Boundary for Two-dimension Viscous Flows

**Jieke Yao[1, a], Wenliang Feng[1], Bin Chen[1], Wei Zhou[1] and Shikun Cao[1]**

[1]Chengdu Aircraft Industrial (group) CO.LTD., 610092, Chengdu, China

[a] yjk1106@163.com

**Abstract.** An un-uniform quadtree grids lattice Boltzmann method (LBM) with immersed boundary is presented in this paper. In overlapping for different level grids, temporal and spatial interpolation are necessary to ensure the continuity of physical quantity. In order to take advantage of the equation for temporal and spatial step in the same level grids, equal interval interpolation, which is simple to apply to any refined boundary grids in the LBM, is adopted in temporal and spatial aspects to obtain second-order accuracy. The velocity correction, which can guarantee more preferably no-slip boundary condition than the direct forcing method and the momentum exchange method in the traditional immersed-boundary LBM, is used for solid boundary to make the best of Cartesian grid. In present quadtree-gridding immersed-boundary LBM, large eddy simulation (LES) is adopted to simulate the flows over obstacle in higher Reynolds number (Re). The incompressible viscous flows over circular cylinder are carried out, and a great agreement is obtained.

## 1. Introduction

Lattice Boltzmann method (LBM) is a mesoscopic method, which springs from the later of 1980s. It is used mainly in the low-speed flows, porous media flows, multiphase flows and so on. However, traditional LBM has serious limitations for uniform grids. Some flow regions for large gradient need more refined grids, so numerical efficiency for traditional LBM is limited. In order to improve computational accuracy, the application of un-uniform grids has been mentioned gradually.

For the un-uniform grids, Yu et al proposed multi-block grids which are non-overlapped each other exchange message by interface [1]. The special treatment is carried out for different blocks to ensure conservation of the mass and momentum. The spline interpolation is adopted in the information exchange for coarse grids to refined grids to eliminate spatial asymmetry. However, it is very difficult to apply the spline interpolation for irregular boundary in the overlapping between coarse grids and refined grids. Therefore, un-uniform quadtree grid is used generally in the LBM. The quadtree grid is a hierarchical data structure, and its algorithm is very simple. In the irregular refined-region, information processing in the overlapping is very easy.

Immersed boundary method (IBM) is proposed by Peskin in 1972, and it mainly simulates mutual deformation between fluid and object boundary [2]. The IBM is based on macroscopic equation, and it is used in LBM as a method treating boundary in 2002 [3]. However, force density is carried out by Hooke's law, direct forcing method or momentum exchange method in the traditional immersed boundary-lattice Boltzmann method (IB-LBM). Therefore, no-slip boundary flowing over obstacle condition can be satisfied approximately. The velocity correction method proposed by Shu et al in 2010 contains unknown force density, and it is obtained by corrected velocity [4]. No-slip boundary condition is applied in this equation solved, so the velocity correction method can guarantee no-slip boundary condition better than traditional IB-LBM. For the IB-LBM, the flow in higher Reynolds number (Re) can be simulated hardly. If the flow for higher Re is simulated by refining grids persistently based on enough grids, this will cause a large number of grids, low computing speed, unstable numerical calculation, and so on. And at the same time, the flow will transform from laminar flow to turbulence when the Re is higher. To solve turbulent flow, large eddy simulation (LES) is adopted in IB-LBM.

An un-uniform quadtree grids IB-LBM is proposed in present paper. Boundary processing technique and LES are introduced to carry out the flows over circle cylinder in this paper.

## 2. The briefing introduction of IB-LBM

Being different from traditional numerical model, LBM is a mesoscopic method, and it can describe basic physical properties of microscopic equation and keep macroscopic equation. Lattice Boltzmann equation is acquired by dispersion of continuous Boltzmann equation in temporal and spatial variation [5]. For the simulation of obstacle in the flow field, lattice Boltzmann equation containing force proposed by Guo et al [6] is used. The equation is listed as following:

$$f_a\left(x+e_a\delta t,t+\delta t\right)-f_a\left(x,t\right)=-\frac{1}{\tau}\left(f_a\left(x,t\right)-f_a^{eq}\left(x,t\right)\right)+F_a\delta t \tag{1}$$

$$F_a=\left(1-\frac{1}{2\tau}\right)w_a\left(\frac{\vec{e}_a-\vec{u}}{c_s^2}+\frac{\vec{e}_a\cdot\vec{u}}{c_s^4}\vec{e}_a\right)\cdot\vec{f} \tag{2}$$

$$\rho\vec{u}=\sum_a\vec{e}_af_a+\frac{1}{2}\vec{f}\delta t \tag{3}$$

Where $\vec{x}$ is coordinate of Euler point in the flow field, $\vec{u}$ is the speed of fluid, $\rho$ is the density of fluid, $f_a$ is distribution function of particle, $f_a^{eq}$ is corresponding equilibrium distribution function, $\tau$ is single relaxation time, $\vec{e}_a$ is discrete velocity of particle, $w_a$ is weight coefficient, $\vec{f}$ is external force density acting on the fluid and $F_a$ is the external force acting on the wall. In the D2Q9 model [7], the velocity of particle is shown in Figure 1.

$$\left[\vec{e}_0\ \vec{e}_1\ \vec{e}_2\ \vec{e}_3\ \vec{e}_4\ \vec{e}_5\ \vec{e}_6\ \vec{e}_7\ \vec{e}_8\right]=c\begin{bmatrix}0&1&1&0&-1&-1&-1&0&1\\0&0&1&1&1&0&-1&-1&-1\end{bmatrix} \tag{4}$$

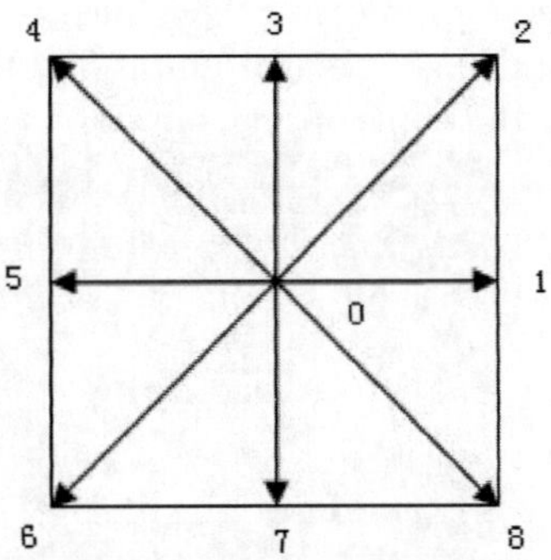


**Figure 1.** D2Q9 lattice model

Where $c=\delta x/\delta t$ (if $\delta x=\delta y=\delta t$, $c=1$), $\delta x$ and $\delta t$ are temporal step and spatial step respectively. Corresponding equilibrium distribution function is:

$$f_a^{eq}\left(\vec{x},t\right)=\rho w_a\left[1+\frac{e_a\cdot u}{c_s^2}+\frac{\left(e_a\cdot u\right)^2-\left(c_s\left|u\right|\right)^2}{2c_s^4}\right] \tag{5}$$

Where $c_s$ is the sound speed of model, $c_s = c/\sqrt{3}$ and weight coefficient is $w_0 = 4/9$, $w_1 = w_2 = w_3 = w_4 = 1/9$ and $w_5 = w_6 = w_7 = w_8 = 1/36$. In this model, the relationship of $\tau$ and kinematic viscosity coefficient for the fluid is $\upsilon = (\tau - 0.5)c_s^2 \delta t$.

## 3. The generation and refinement of quadtree grids

Traditional LBM requires uniform and regular grids [8], but high-density grids are demanded in the large gradient flow region of or near the curved solid. Therefore, the application of traditional LBM is limited. In order to improve numerical efficiency and accuracy, Filippov et al proposed grid refinement technique [9, 10], so the use of un-uniform grid makes LBM get wide application.

Quadtree grid based on recursion and decomposition principle is a hierarchical data structure [11, 12]. Its data storage and adaptive grid refinement are very simple. Main unit of quadtree grid is quadtree unit. Very quadtree unit contains four grid nodes. These grids nodes store geometry information and physical quantity in the flow field, and all computing processes are done by these grid nodes. Quadtree grid has two types: root unit and leaf unit. Root unit is divided into four equal leaf unit. Very tree unit stores pointers of four grid nodes, leaf units and neighboring units. The process of grid refinement is shown in Figure 2.

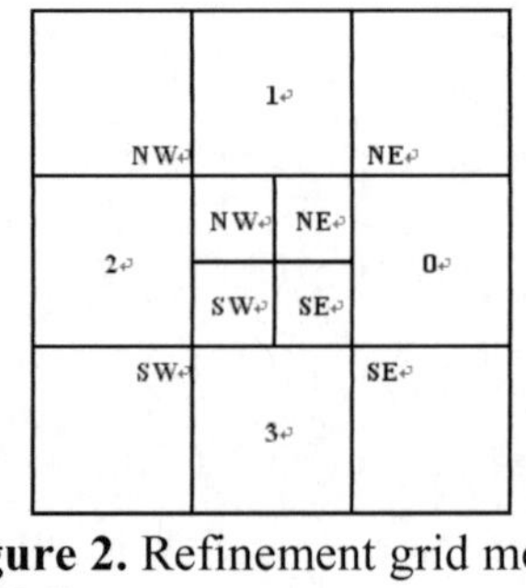


**Figure 2.** Refinement grid model

**The process of grid generation is as follows:**

- Firstly, rectangular region is used as bottom element, and it is divided into many quadtree grid units. Grid step of Very tree unit is unit length, and four vertexes are four grid nodes of quadtree unit. Four quadtree units on the left of it, on the right of it, above it and under it are neighboring units.
- Secondly, grid unit is divided by the controlling function or gradient of physical quantity for the nodes in the flow flied, and very quadtree unit is divided into four identical leaf units.

**The grids refined to three levels by the controlling function is shown in Figure 3.**

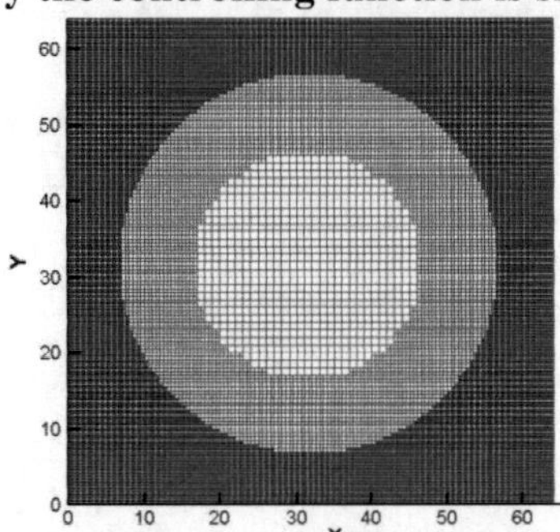


**Figure 3.** Quadtree grid in square domain with high grid density near the edge

In order to apply the quadtree grids refined at any boundary to the LBM, the present paper used the method in the reference [1] which presented that coarse grid and refined grid didn't have overlapping domain, and a smooth connection for the flow was guaranteed by information transfer, spatial interpolation and temporal interpolation on the overlap. However, the application of three spline interpolation in space is very difficult for the irregular refined region.

Because the four leaf units divided are identical in the quadtree grids, the spatial and temporal steps are equal in the same level grid. Equal interval interpolation in spatial and temporal aspects are adopted to replace Lagrange interpolation and three spline interpolation. This makes the LBM get more extensive application, and two order accuracy is guaranteed in spatial and temporal aspects.

Detailed implement steps of multi-levels grids refined for any boundary in the LBM is listed as follows: firstly, neighboring units are arranged by recursive algorithm for very leaf unit based on bottom unit. And the same time, the nodes of streaming and collision are marked in very level grid. Secondly, because the nodes for neighboring units are used in streaming, the streaming is done for grid units. In order to avoid repeating streaming for the same node, new marker is done for the node completing streaming until the streaming is finished in the same level grid. Then original markers are recovered for these. Thirdly, macroscopic calculation and collision are carried out at the nodes, so the two processes are done by the pointer in very level grids. Fourthly, according to the demand of multi-block grid, the streaming and collision for coarse grid are once, this process for refined grid is twice. And this process is circulated in the LBM computation.

## 4. The treatment of boundary condition

### *4.1. The boundary condition on the wall*

For the treatment of the boundary on the wall, the velocity correction proposed by Wu and Shu [4] is adopted. In the velocity correction, $\vec{u}^*$ is intermediate velocity, and it is obtained by:

$$\rho\vec{u}^* = \sum_a \vec{e}_a f_a \ , \ \rho\delta\vec{u} = \frac{1}{2} f\delta t \tag{6}$$

then equation (3) can be written as $\vec{u} = \vec{u}^* + \delta\vec{u}$. Where $\delta\vec{u}$ is the correction velocity at Eulerian points. In the IB-LBM, the boundary of the object is represented by a set of Lagrangian points $\vec{X}_B(s_l,t)(l = 1,2,\cdots,m)$. Here, we can set an unknown velocity correction vector $\delta U_B^l$ at very Lagrangian points. The velocity correction $\delta\vec{u}$ at the Eulerian point can be obtained by the following Dirac delta function interpolation:

$$\delta\vec{u} = \int_\Gamma \delta\vec{U}_B\left(\vec{X}_B,t\right)\delta\left(\vec{x} - \vec{X}_B(s,t)\right)ds \tag{7}$$

The Dirac delta function is $\delta\left(\vec{x} - \vec{X}_B(s,t)\right) = D_{ij}\left(\vec{x}_{ij} - \vec{X}_B^l\right) = \delta\left(x_{ij} - X_B^l\right)\left(y_{ij} - Y_B^l\right)$. Where $\delta(r)$ is proposed by Peskin [13] as $\delta(r) = \frac{1}{4}\left(1 + \cos\left(\frac{\pi|r|}{2}\right)\right)(if\,|r| \le 2)$ or $\delta(r) = 0(if\,|r| > 2)$. Using the Dirac delta function, the velocity correction at Eulerian points can be expressed as:

$$\delta\vec{u}\left(\vec{x}_{ij},t\right) = \sum_l \delta\vec{U}_B^l\left(\vec{X}_B^l,t\right)D_{ij}\left(\vec{x}_{ij} - \vec{X}_B^l\right)\Delta s_l \quad (l = 1,2,\ldots,m) \tag{8}$$

And at the same time, the velocity correction at Lagrangian points can be obtained:

$$\vec{U}_B^l\left(\vec{X}_B^l,t\right) = \sum_{i,j}\vec{u}\left(\vec{x},t\right)D_{ij}\left(\vec{x}_{ij} - \vec{X}_B^l\right)\Delta x\Delta y \tag{9}$$

Substituting equation (8) and $\vec{u} = \vec{u}^* + \delta\vec{u}$ into equation (9) gives:

$$\vec{U}_B^l\left(\vec{X}_B^l,t\right) = \sum_{i,j}\vec{u}^*\left(\vec{x},t\right)D_{ij}\left(\vec{x}_{ij} - \vec{X}_B^l\right)\Delta x\Delta y + \sum_{ij}\sum_n\left[\delta\vec{U}_B^n\left(\vec{X}_B^n,t\right)D_{ij}\left(\vec{x}_{ij} - \vec{X}_B^n\right)\Delta s\right]D_{ij}\left(\vec{x}_{ij} - \vec{X}_B^l\right)\Delta x\Delta y \tag{10}$$

Equation system (10) can be further rewritten as the following matrix form:

$$AX = B \tag{11}$$

where $X=\left\{\delta\vec{U}_B^1,\delta\vec{U}_B^2,\ldots,\delta\vec{U}_B^m\right\}^T$ , $B=\{\Delta u_1,\Delta u_2,\ldots,\Delta u_m\}^T$ with $\Delta u_l=U_B^l\left(X_B^l,t\right)-\sum_{i,j}u^*\left(x,t\right)\cdot D_{ij}\left(\vec{x}_{ij}-\vec{X}_B^l\right)\Delta x\Delta y \quad (l=1,2,\ldots,m)$.

Note that the elements of matrix $A$ are only related to the boundary points and their nearby Eulerian points. The pressure at the Lagrangian points is calculated by $p=c_s^2\rho$ . Therefore, the correction velocity at Eulerian points can be calculated by solving equation (11), the correction velocity at Lagrangian points can be obtained by equation (8).

### *4.2 The boundary condition in the far field*

Because the far boundary has a smaller influence on the flow nearby the wall, the treatment for the far field boundary can use initial condition. However, the wake has a greater influence on the flow, so the treatment of right boundary in the $x$ direction can use convective boundary condition proposed by Kim [14]:

$$\frac{\partial u}{\partial t}+U_0\frac{\partial u}{\partial x}=0 \tag{12}$$

where $U_0$ is the initial velocity of field, $u$ is the velocity of field in the $x$ direction.

## 5. The primary process of LBM and numerical experiments

### *5.1 Large eddy model*

Because IB-LBM simulates mainly the flow at high Re, kinematic viscosity coefficient is very small (kinematic viscosity coefficient is inversely proportional to Re). According the relationship of $\tau$ and $\upsilon$, relaxation time $\tau$ is c lose to 0.5. Therefore, distribution function of particle may be negative value. This may result in numerical instability. In order to solve this problem, LES proposed Hou and Sterling [15] is adopted. That combines IB-LBM with LES to solve the flow at higher Re. This method is based on standard Smagorinsky subgrid model [16], and its principle is that single relaxation time (SRT) for any node is different when they are colliding. Finally, the SRT can be obtained by formulation derivation as follows:

$$\tau=\frac{1}{2}\left(\sqrt{\tau_0^2+\frac{18}{\rho\delta t^2}C\Delta^2\sqrt{2Q}}+\tau_0\right) \tag{13}$$

Where $\tau_0$ is the SRT without the LES Smagorinsky model, $C$ is Smagorinsky constant and it is usually set to 0.1, $\Delta$ is filter scale and this value is usually spatial step, $Q=\overline{\Pi}_{ij}\overline{\Pi}_{ij}$ , $\overline{\Pi}_{ij}=\sum\vec{e}_{ai}\vec{e}_{aj}\left(f_a-f_a^{eq}\right)$ is the non-equilibrium stress tensor and $i$ , $j$ are $x$ and $y$ direction respectively.

### *5.2 The numerical experiments*

In this paper, the flows over circular cylinder are carried out to validate the methods of quadtree grids, the treatment of boundary and LES in the LBM. In the present simulations, the Re is defined as $\mathrm{Re}=(U_\infty D)/\upsilon$ , the drag coefficient $C_d$ and the Strouhal number $St$ are defined as $C_d=F_d/(0.5\rho_\infty U_\infty^2 D)$ , $St=(f_q D)/U_\infty$ . Where $U_\infty$ is the free stream velocity, $\rho_\infty$ is the free stream density, $D$ is the diameter of cylinder, $F_d$ is the drag and $f_q$ is the vortex shedding frequency. The drag is obtained by $F_d=-\frac{d}{dt}\left(\int_\Omega u_1 d\Omega\right)-\int_s\left[u_1\vec{u}\cdot\vec{n}+pn_1-\mu\left(\frac{\partial u_1}{\partial x_i}+\frac{\partial u_i}{\partial x_1}\right)\right]ds$ . Where $s$ the control surface, $\vec{n}$ the normal vector of control surface, and the subscripts 1 and 2 stand for the $x$ direction and $y$ direction respectively. The instantaneous vorticity is shown in Figure 4.

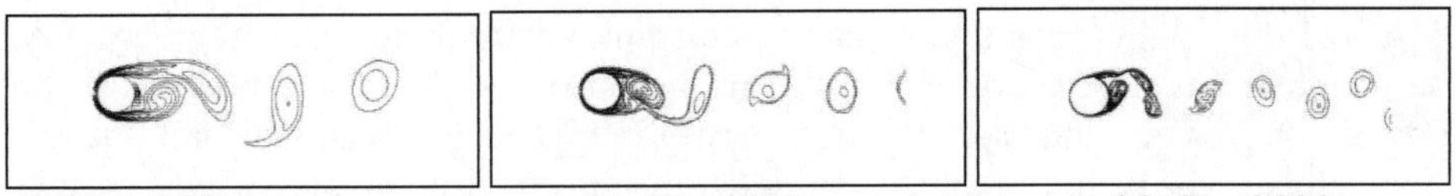

(a) Re=200 (b) Re=1000 (c) Re=3900

**Figure 4.** Vorticity contours for different Reynolds number

From seen the figure 4, the stable symmetric vortex behind the cylinder appears when the Re is much lower. However, the vortex shedding occurs when the Re is much higher. Because may lead to the periodic oscillation for the drag, the figure 5 gives the variation of drag coefficient according to time. Figure 6 gives instantaneous streamline for Re=3900.

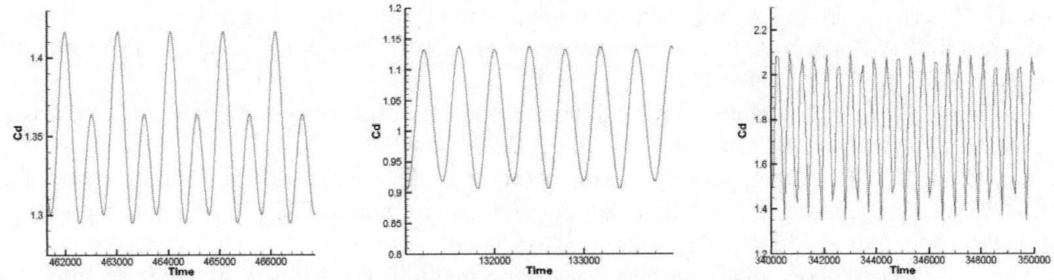


(a) Re=200 (b) Re=1000 (c) Re=3900

**Figure 5.** The variation of drag coefficient according to time

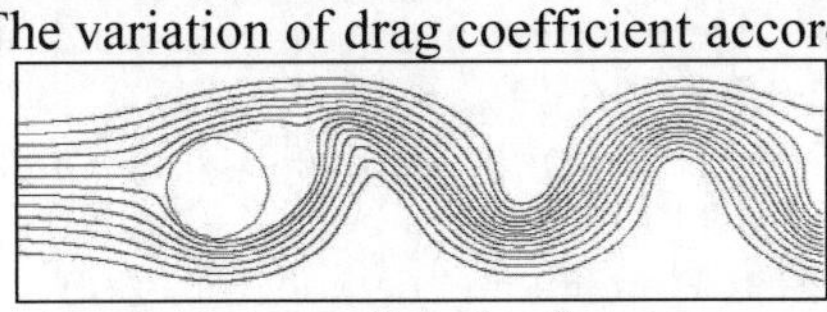

**Figure 6.** The instantaneous streamline for flow over a cylinder at Re=3900

The comparisons of mean drag coefficient $\overline{C}_d$ and Strouhal number $St$ are listed in table 1.

**Table 1.** The mean drag coefficient and Strouhal numbers for different Reynolds number

| Re | Method | $\overline{C}_d$ | $St$ |
|---|---|---|---|
| | Finite-element method and N-S equation [17] | 1.35 | 0.196 |
| 200 | A Fractional Step Method[18] | 1.3 | 0.196 |
| | The Present method | 1.36 | 0.196 |
| | Stanley K.Jordan al[19] | 1.24 | 0.206 |
| 1000 | Finite-element method and N-S equation [17] | 1.47 | 0.234 |
| | The Present method | 1.03 | 0.227 |
| | LBM-LES2D[20] | 1.827 | 0.225 |
| 3900 | Finite-element method[21] | 1.74 | 0.263 |
| | The Present method | 1.74 | 0.226 |

In Figure 6, because no-slip boundary based on velocity correction method stops fluid penetration, the streamlines around the cylinder round cylindrical surface. And at the same time, the streamline can smooth transition in different level, which is because information processing for interface between coarse mesh and refined mesh is appropriate to guarantee smooth transition of fluid. However, the essential cause is that numerical error is much smaller in the interpolation, which shows that the precision for equal interval interpolation in present paper dovetail nicely with the precision of LBM. Table 1 shows that the higher for Re, the greater for Strouhal number, which indicates the faster vorticity is shedding. A great agreement with literature is obtained in the table 1, but a small error is going on. That is due to the section of parameters in the model, such as the section of control domain, the section of computational domain, the method of interpolation, and so on.

## 6. Conclusion

Firstly, the quadtree-gridding refinement which allots the neighbors of leaf units and marks the nodes of streaming and collision in different level mesh is applied in the LBM for multi-level refinement of arbitrary boundary. Equal interval interpolation in temporal and spatial space is used in the different grid interface, and this method is simpler than three spline interpolation and Lagrange interpolation. Because the temporal and spatial space is equal in the same grid level, the present method has the second order accuracy which matches the accuracy of LBM in the temporal and spatial direction. Secondly, the velocity correction method is adopted in immersed boundary, which ensures the non-slip boundary compared to the boundary treatment technique in the traditional IB-LBM. Thirdly, the LES is applied to the present IB-LBM to simulate the flows with high Re around the obstacle. Lastly, the flows is simulated for the different Re to validate the present method, and the results agree well with those of other methods.

**References**

[1] D. Z. Yu, R. W. Mei, W. Shyy, A multi-block lattice Boltzmann method for viscous fluid flows[J]. International journal for numerical methods in fluids, 2002, 39(2): 99-120.
[2] C. S. Peskin, Flow patterns around heart valves: a numerical method[J]. Journal of Computational Physics, 1972, 10(2): 252-271.
[3] Z. G. Feng, E. E. Michaelides, Hydrodynamic force on spheres in cylindrical and prismatic enclosure[J]. Journal of Multiphase Flow, 2002, 28(3): 479-496.
[4] J. Wu, C. Shu, Y. H. Zhang, Simulation of incompressible viscous flows around moving objects by a variant of immersed boundary-lattice Boltzmann method[J]. International Journal for Numerical Methods in Fluids, 2010, 62(3): 327-354.
[5] X. He, L. S. Luo, Theory of the lattice Boltzmann method: from the Boltzmann equation to the lattice Boltzmann equation[J]. Physical Review E, 1997, 56(6): 6811-6817.
[6] Z. Guo, C. Zheng, B. Shi, Discrete lattice effects on the forcing term in the lattice Boltzmann method[J]. Physical Review E., 2002, 65(4): 046308.
[7] Y. H. Qian, D. d'Humieres, P. Lallemand, Lattice BGK model for Navier-Stokes equation[J]. Europhysic Letters, 1992, 17(6): 479-484.
[8] S. Chen, G. D. Doolen, Lattice Boltzmann method for fluid flows[J]. Annual review of fluid mechanics, 1998, 30(1): 329-364.
[9] O. Filippova, D. Hanel, Grid refine for lattice-BGK model[J]. Journal of Computational Physics, 1998, 147: 219–228.
[10] C. L. Lin, Y. G. Lai, Lattice Boltzmann method on composite grids[J]. Physical Review E, 2000, 62(2): 2219-2225.
[11] H. Samet, Applications of Spatial Data Structures [M]. Addison-Wesley, MA, 1990.
[12] H. Samet, Neighbor finding techniques for images represented by quadtrees [J]. Computer Graphics and Image Processing, 1982, 18(1): 37-57.
[13] C. S. Peskin, The immersed boundary method[J]. Acta Numerica, 2001, 11: 479-517.
[14] D. H. Kim, K. S. Yang, M. Senda, Large eddy simulation of turbulent flow past a square cylinder confined in a channel[J]. Journal of Computational Physics, 2004, 33(1): 81-96.
[15] S. Hou, J. Sterling and S. Chen et al, A lattice Boltzmann subgrid model for high Reynolds number flows[J]. Pattern formation and lattice gas automata, 1994, 9401004(6): 151-166.
[16] Y. L. He, Y. Wang, Q. Li, Lattice Boltzmann method and its application[M]. Beijing: Science Press, 2008.
[17] K. Herfjord, A study of two-dimensional separated flow by a combination of the finite element method and N-S equations[D]. Norwegian Institute of Technology, 1996.
[18] J. R. Meneghini, F. Saltara, C. L. R. Siqueira, J. A. Ferrari. Numerical simulation of flow interference between two circular cylinders in tandem and side-by-side arrangements[J]. Journal of Fluids and Structures 2001, 15(2): 327-350.
[19] S. K. Jordan, J. E. Fromm, Oscillatory drag, lift, and torque on a circular cylinder in a uniform flow[J]. Physics of Fluids, 1972, 15(3): 371-376.
[20] A. G. Kravchenko, P. Moin, Numerical studies flow over a circular cylinder at ReD=3900[J]. Phycics of Fluids, 2000, 12(2): 403-417.
[21] P. Beaudan, P. Moin, Numerical experiments on the flow past a circular cylinder at sub-critical Reynolds number[R], Report No. TF-62, Department of Mechanical Engineering, Stanford University, 1994.

IOP Conf. Series: Materials Science and Engineering **224** (2017) 012024 doi:10.1088/1757-899X/224/1/012024

# Parametric Design and Mechanical Analysis of Beams based on SINOVATION

**Z G Xu[1,a] , W D Shen[2] , D Y Yang [2] , W M Liu[1]**

[1]Shenzhen Research Institue of Shandong University, 518057, High-tech Zone A301, Virtual University Part, Shenzhen, China.
[2]School of Mechanical Eng, Shandong University, 250061, No 17923, Jingshilu, Lixiaqu, Jinan, Shandong, China

zhgxu@sdu.edu.cn

**Abstract**. In engineering practice, engineer needs to carry out complicated calculation when the loads on the beam are complex. The processes of analysis and calculation take a lot of time and the results are unreliable. So VS2005 and ADK are used to develop a software for beams design based on the 3D CAD software SINOVATION with C ++ programming language. The software can realize the mechanical analysis and parameterized design of various types of beams and output the report of design in HTML format. Efficiency and reliability of design of beams are improved.

## 1. Introduction

Beam structure is a common component in engineering structure, such as the beam of the crane, the lathe spindle,etc, which can be simplified as a beam structure. To analyze the design of beams under external load, such as, shear force, bending moment, deflection, rotation, bending stress, bending shear stress, torsional shear stress and tensile stress etc, large amount of calculations are needed. For these reasons, software systems are developed, for examples, Xu Y 2013, developed one parametric design system of crane girder based on UG[1]; LU J J. et al, developed a crane beam CAD system[2]. Fu K W. et al, developed a bridge parametric design and analysis system for main beam machine based on Pro/E [3]; Han Ling developed a bridge crane design system based on AutoCAD [4]. In this paper, a beam design software is developed based on SINOVATION, which can be used to analyze the mechanical properties of the beam under various loads in various forms, and the parametric design of the beam, the results of the analysis can be put forward in the form of report.

## 2. Mechanical Analysis of Statically Determinate Beam

In this paper, the parametric properties of the different beams are analyzed, for the the simply supported beam, cantilever beam, fixed end and movable hinge support beam and beams with two fixed ends, etc. A freely supported beam with three radial force is shown in fig. 1. In order to get the shear value at the point A of the beam, the superposition method is utilized. Firstly, the single shear value at A point, with a single load is calculated[5], then, the supposition shear value at A point, with supposition loads are calculated one by one, shear at A with all the individual load values are obtained, i.e. the actual shear A value. Similarly, at any position on the beam, the moment, angle, deflection, bending stress, bending shear stress, torsional shear stress and tensile stress value can be calculated.

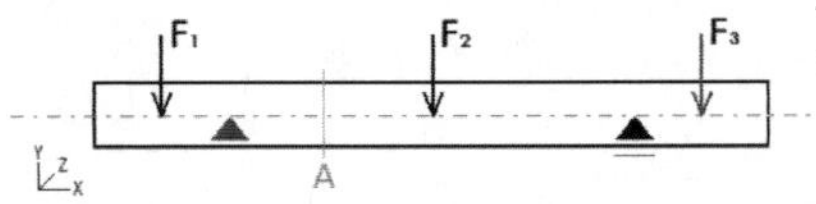


**Figure 1.** Simply supported beam

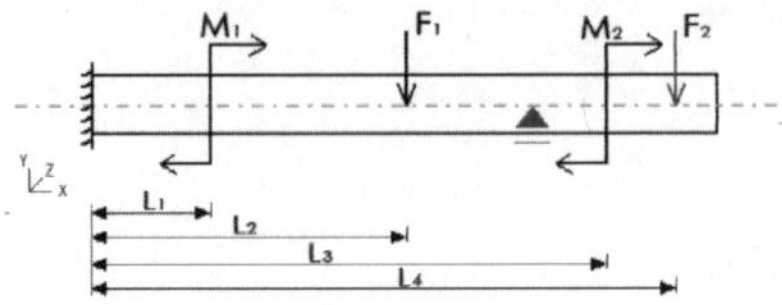


**Figure 2.** Fixed/movable supports

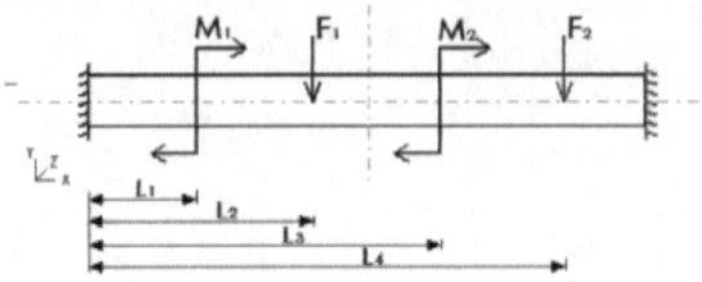


**Figure 3.** Three fixed ends

The parametric calculation is simple, for the simply supported beam and the cantilever beam. Calculation of parameters of beams supported by a fixed end and movable hinge , or by two fixed ends, are complex.The following is one calculation example for beam deflection. As shown in fig. 2, the beam is supported by a fixed end and a movable hinge,The beam is under one uniform loading due to self-weight q, at the same time, loaded by the radial force F1, F2, and the concentrated couple M1 and M2. The concentrated couple M1 has the end distance of L1, M2 has the end distance of L3. The radial force F1 has the distance L2, the radial force F2 has the fixed distance L4.

The calculation of the deflection of the beam with a distance of *x* from the fixed end is as follows. If *S* is the distance between the movable hinge and the fixed end, the length of the beam is L, the elastic modulus of the beam is E, the moment of inertia of the section to the *Z* axis is I., and the deflection value caused by the uniform load Q is as the follows.

When $0 \le x < S$,

$$\omega_q(x) = -\frac{qx^2}{24EI}\left(x^2 - 4Lx + 6L^2\right) + \frac{qx^2}{48SEI}\left(S^2 - 4LS + 6L^2\right)\left(3S - x\right) \tag{1}$$

When $S \le x \le L$,

$$\omega_q(x) = -\frac{qx^2}{24EI}\left(x^2 - 4Lx + 6L^2\right) + \frac{qS}{8EI}\left(S^2 - 4LS + 6L^2\right)\left(\frac{x}{2} - \frac{S}{6}\right) \tag{2}$$

When $0 \le x < L_2$,

$$\omega_{F_1}(x) = \frac{F_1x^2}{6EI}\left[\frac{L_2^2}{2S^3}\left(3S - L_2\right)\left(3S - x\right) - 3L_2 + x\right] \tag{3}$$

When $L_2 \le x < S$,

$$\omega_{F_1}(x) = \frac{F_1L_2^2}{12EI}\left[\frac{x^2}{2S^3}\left(3S - L_2\right)\left(3S - x\right) + 2L_2 - 6x\right] \tag{4}$$

When $S \le x \le L$,

$$\omega_{F_1}(x) = \frac{F_1L_2^2\left(x - S\right)}{4EI}\left(\frac{3S - L_2}{S^3} - 2\right) \tag{5}$$

When $0 \le x < S$,

$$\omega_{F_2}(x) = \frac{F_2x^2}{12EI}\left[\frac{\left(3L_4 - S\right)\left(3S - x\right)}{S} - 6L_4 + 2x\right] \tag{6}$$

When $S \le x < L_4$,

$$\omega_{F_2}(x) = \frac{F_2}{12EI}\left[S\left(3L_4 - S\right)\left(3S - x\right) - 2x^2\left(3L_4 - x\right)\right] \tag{7}$$

When $L_4 \le x \le L$,

$$\omega_{F_2}(x) = \frac{F_2}{12EI}\left[S\left(3L_4 - S\right)\left(3S - x\right) + 2L_4^2 - 6L_4^2x\right] \tag{8}$$

When $0 \le x < L_1$,

$$\omega_{M_1}(x) = \frac{M_1 x^2}{4EI}\left[\frac{L_1}{S^3}(2S - L_1)(3S - x) - 2\right] \tag{9}$$

When $L_1 \le x < S$,

$$\omega_{M_1}(x) = \frac{M_1 L_1}{4EI}\left|\frac{x^2}{S^3}(2S - L_1)(3S - x) - 4x + 2L_1\right| \tag{10}$$

When $S \le x \le L$,

$$\omega_{M_1}(x) = \frac{M_1 L_1}{4EI}\left[\frac{3}{S^3}(2S - L_1)(x - S) + 4(S - x)\right] \tag{11}$$

When $0 \le x < S$,

$$\omega_{M_2}(x) = \frac{M_2 x^2 (S - x)}{4EIS} \tag{12}$$

When $S \le x < L_3$,

$$\omega_{M_2}(x) = \frac{M_2 (S - 2x)(x - S)}{4EI} \tag{13}$$

When $L_3 \le x \le L$,

$$\omega_{M_2}(x) = \frac{M_2}{4EI}\left[S(3x - S) + 2L_3(L_3 - 2x)\right] \tag{14}$$

The calculation of the deflection of the beam with a distance of $x$ from the fixed end is as the follows.

$$\omega(x) = \omega_q(x) + \omega_{F_1}(x) + \omega_{F_2}(x) + \omega_{M_1}(x) + \omega_{M_2}(x) \tag{15}$$

As shown in figure. 3, the beam is supported by two fixed ends, the beam is under one uniform loading due to self-weight q, at the same time, loaded by the radial force F1, F2, and the concentrated couple M1 and M2. The concentrated couple M1 has the end distance of L1, M2 has the end distance of L3. The radial force F1 has the distance L2, the radial force F2 has the fixed distance L4. The calculation process of the deflection from the left distance for X cross section values on the beam are as follows.The length of the beam is L, the elastic modulus of the beam is E, and the moment of inertia of the cross section to the Z axis is I, and the deflection value of the uniformly distributed load Q. The deflection of the beam caused by radial force: The radial force causes the deflection of the beam to be similar, The deflection of beams caused by the concentrated couple. Concentrated couple caused by beam deflection and beam deflection is similar. From the left distance for X section is as the follows. Other parameters, such as shear force, bending moment, rotation angle, bending normal stress, bending shear stress, tensile normal stress and torsional shear stress, are similar to those calculated by deflection method, superposition addition are utilized.

*2.1 Software Development.*

Development platform and tools: The beam design software module is developed, which are embedded in the domestic CAD software SINOVATION in the form of DLLs (Dynamic Link Libraries) . The programs are based on the SINOVATION ADKs and Visual Studio 2005, C++ platform [6]. SINOVATION is a three-dimensional CAD application software system, with advanced hybrid modeling, parametric design, rich feature modeling function, is the Alex Hua Tian software in the country's support and cooperation with foreign companies [7]. API(Application Program Interface). ADK is the development kit provided by SHANDDA Hua Tian software co, in the form of API(Application Program Interface).

*2.2 Software Interface Design.*

Tab Control, Picture Control, Group Box, List Control, Button, Edit Control, Combo Box, Radio Button and Static Text in the MFC(Microsoft Foundation Classes) are utilized in this software. The design parameters are put in the design parameter interfaces of the beam, the supporting parameters and the load parameters are shown in fig. 4.

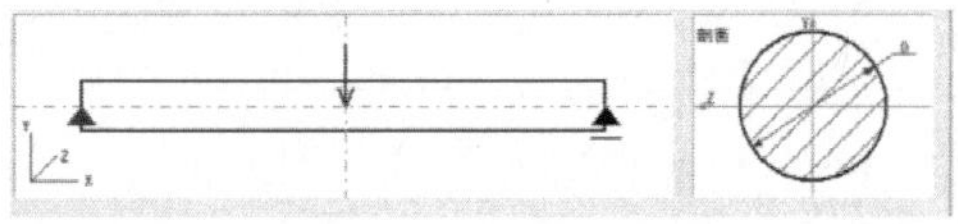

**Figure 4.** "parameter" interface

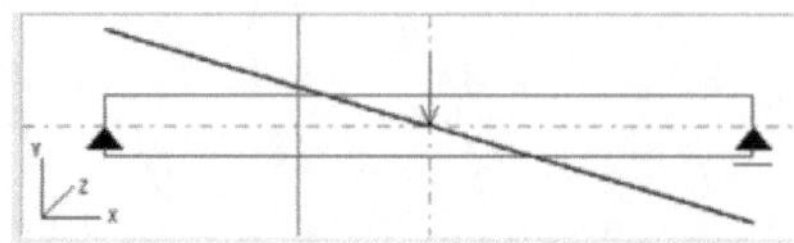

**Figure 5.** The calculation results

The type of beam section are categorized as circle, ring, square, hollow square, rectangle, hollow rectangle, hexagon, triangle, trapezoid, work shape, L shape and C shapes etc. After Modification of the section size of the beam sections, the corresponding area, moment of inertia and the coefficient of cross section will automatically be changed; After th modification of the support type and support distance, the top icon of the supporting types are also automatically changed; Adding or removing of the load/load from, and changing of the load distances, etc, the corresponding icon will also be changed. In the design process of the software interfaces, one functionality need to be developed in the List Control, i.e. When the left mouse button clicked, a drop-down box appeared. For example, the section types are displayed and could be selected in the design parameter interface, when the left mouse button is clicked. Because List Control itself does not have such this function, so, NM_CLICK is utilized in the List Control messages. The "calculation results" page is shown in Fig. 5.

There is a red solid line in upper part of the software "calculation results" page, and the location of the vertical coordinate and the parameter value is shown in the bottom part of the page below Edit Control. When the mouse slides on the parameter graph, a solid red vertical line will move with the mouse, the vertical coordinate parameter in the Edit Control and the corresponding parameters will be changed simultaneously. Select a different button or select different parameters, such as the XY plane, XZ plane, the results of the calculation page will show different parameters. The elastic modulus and density of the beam are changed by selecting different materials, the material in the drop-down box including plain carbon steel, cast steel, nickel steel, stainless steel, gray cast iron, cast iron can be broken, copper, aluminum, magnesium alloy and custom materials.

*2.3 Parameter graph drawing.*

In the "results" page of the software, the algorithm flow chart of parameter drawing is shown in Fig. 6. The maximum parameter values of the parameters are firstly calculated, each pixel value corresponding to the logical unit is then calculated, lastly a line is drawn from the left most beam to the current point, (d_x, * the parameter value line), With the increase of d_x, the line segments are drawn in turn until the right end of the beam, finally, all the segments are combined to form the parameter graph. By using the OnMouseMove message dialog box, the red line moves with the mouse function in the "calculation results" in the parameter map.

The flow chart of this function is shown in fig. 7, The RECT in the client area of the Picture control is obtained and is saved in the variable rc, rc is then mapped to the dialog box coordinates, Then determine whether the current mouse coordinates are in the variable rc, if they are in in the variable rc, then a red line is drawn in the parameter map in the mouse location, if not, the procedure ended. drawing in the calculation result page

*2.4 Parametric design of beam.*

The type of the beam and the size of the profile are selected in the "design parameter" page of the software. Click on the "create" button to create a three-dimensional model of the beam directly, The parametric design of beam is realized, the modeling process of beam is to draw the beam profile, the three-dimensional model of the beam can be obtained by pulling the model section.

IOP Conf. Series: Materials Science and Engineering **224** (2017) 012024 doi:10.1088/1757-899X/224/1/012024

*2.5 The output report.*

There is a "output report" button in the lower part of the "calculation result" page of the software. Click this button, a hybrid report with text, tables, pictures and other forms of HTML etc, will be automatically created and is opened. The report contains the creation time of the report, the creator, the criteria used, the current design parameters, and the results of the calculations etc. Algorithm flow of the output report is as the follows, firstly a folder is created, a HTML file is then created, and is saved in the same fold. And then the design of the beam and the results of the calculation of all the parameters are stored in the folder, All design parameters and calculation results are saved in a CString variable ***str***, Finally, the contents in the ***str*** into a good HTML file, after the report was created, use the Web browser to open the report. The HTML report created by the software is shown in Fig. 10.

## 3. Application cases

The size of a lathe spindle and the force diagram is shown in fig. 8, The spindle is supported by A, B, C three supports, under normal circumstances, B support does not work, Only when the bending deformation of the main shaft B is greater than the clearance between the main shaft and the B support, the B support can play a supporting role.

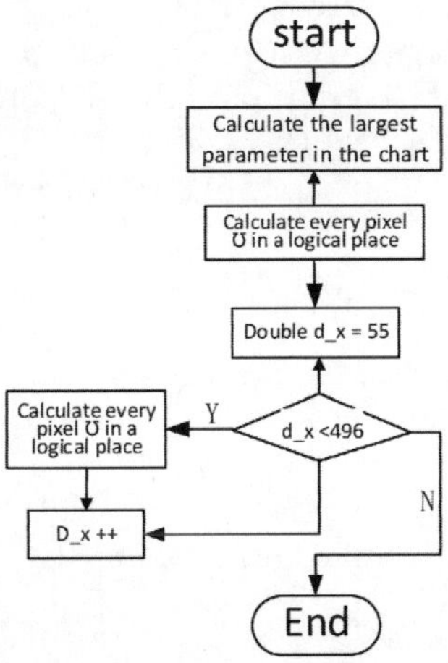


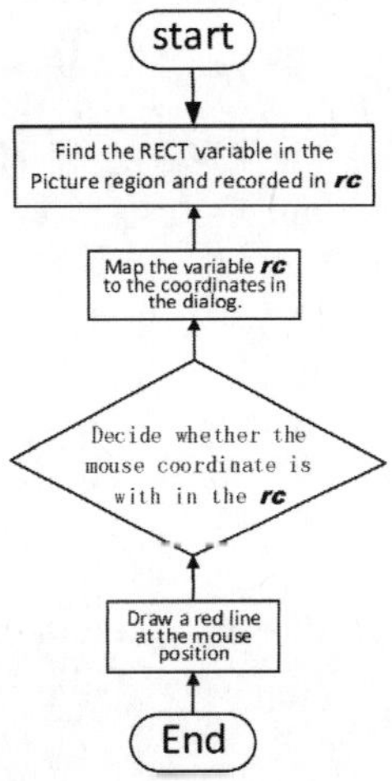


**Figure 6.** Flow chart of parameter **Figure 7.** Red line moving with mouse.

The helical gear driven by the main shaft of E is subjected to external force. F=396N, =860N.

Cutting force at the cutting edge of position H, indirectly make the lathe spindle under a concentrated force couple at D position. Where, $M_{Dy} = 432N \cdot m$, $M_{Dz} = 810N \cdot m$, and, $l_1 = 0.18m$, $l_2 = 0.55m$, $l_3 = 0.17m$, $a = 0.15m$, $b = 0.18m$.It is known that the cross section of lathe spindle is hollow, the outer radius of the cross section is 70mm, inner radius is 50 mm, the shaft material is 45 # cast steel.

The following calculation is to find the dangerous section of spindle, and calculate the bending deformation of the B section of the spindle. The above design parameters are input to the software design parameters page, Click the "calculate" button, The shear diagram, bending moment diagram, deflection diagram, deflection diagram, bending normal stress diagram and bending shear stress diagram are obtained. As shown in fig. 8, fig. 9, fig. 10. The shear diagram, bending moment diagram and deflection diagram of the lathe spindle are displayed.

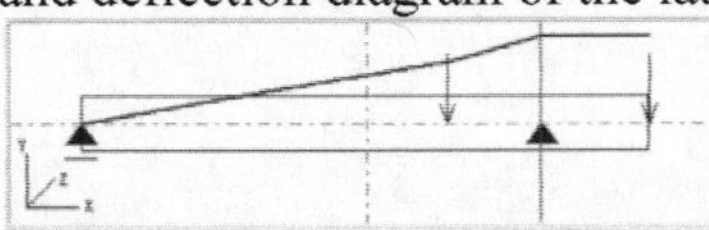

**Figure 8.** Shear diagram

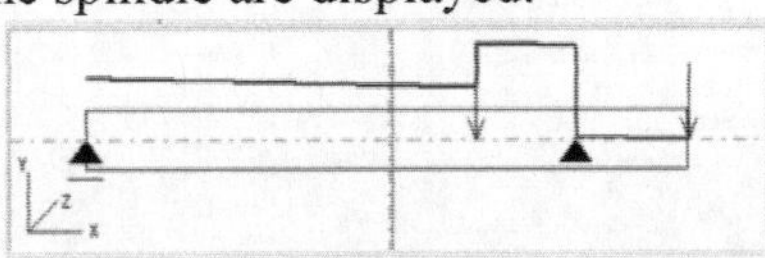

**Figure9.** Bending moment

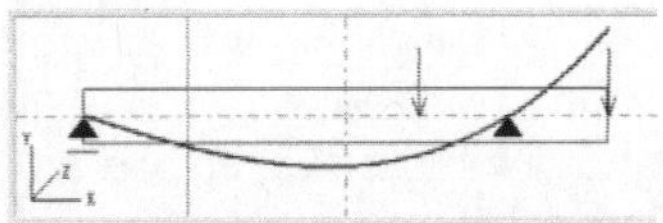

**Figure 10.** Deflection diagram

## 4. Conclusions

In this paper, the beam supported by simply supported beam, cantilever beam, fixed end and movable hinge support and two fixed end supports are analyzed beam, are analyzed, under multiple external loads, to get the shear force, bending moment, deflection, deflection, bending normal stress, shear stress, torsional shear stress etc. The parameter analysis and calculation schemes and software development are put forward.Based on SINOVATION software, Using VS2005 and ADK to develop the beam design software, The mechanical analysis and parametric design of the beam are realized, the design efficiency and reliability of the beam design are improved. The development method and development process used in this paper can provide reference for the other development of CAD systems.

## Acknowledgment

The above-mentioned research work is supported by the Science, Technology and Innovation Commission of Shenzhen Municipality, JCYJ20160510165328965 and the Chinese NSFC, 61272017, P.R. China.

## References

[1] Xu Yan, Li Shubin, Zhou Bing. Parametric design of main girder of crane based on two development technology of [J]. UG/NX coal mining machinery, 2011 (6): 241-243.
[2] Lv Junjiang, Cheng Rong, Chen Shuirong, et al. Research and development of CAD software for crane beam design YJ-CRANE [J]. BEAM steel structure, 2012, 27 (): 73-75.
[3 Parametric design and analysis of bridge girder based on Pro/E., Taiyuan University of Science and Technology, 2012
[4] Han Ling. A new type of bridge crane design system based on AutoCAD, Dalian University of Technology, 2003
[5] Feng Weiming. Mechanics of materials [M]. National Defence Industry Press, 2010
[6] Wang J J, Yuan X X, Lin L I, et al. Development of CAD Module for Riser of Iron Castings Based on SINOVATION[J]. Foundry, 2013, 62(6):519-84.
[7] Sun H, Zhang J. Study on the Parametric Design of Offshore Platform Outfittings Based on Sinovation Software[C]. International Conference on Intelligent Transportation, Big Data and Smart City. IEEE, 2015:797-800.

# Wire in the Cable-Driven System of Surgical Robot

**X F Wang[1,2], N Lv[3], H Z Mu[1,2] and L J Xue[1,2]**

[1]Tianjin Key Laboratory of the Design and Intelligent Control of the Advanced Mechatronical System, Tianjin University of Technology
[2]School of Mechanical Engineering, Tianjin University of Technology
[3]School of Mechanical Engineering, Zhonghuan Information College Tianjin University of Technology

wangxiaofeitjut@163.com

**Abstract.** During the evolution of the surgical robot, cable plays an important role. It translates motion and force precisely from surgeon's hand to the tool's tips. In the paper, the vertical wires, the composition of cable, are mathematically modeled from a geometric point of view. The cable structure and tension are analyzed according to the characteristics of wire screw twist. The structural equations of the wires in different positions are derived for both non-bent cable and bent cable, respectively. The bending moment formula of bent cable is also obtained. This will help researchers find suitable cable and design more matched pulley.

## 1. Introduction

Because of its high strength, good flexibility, small diameter and other characteristics, cable has been widely used in the general industry[1]. With the continuous development of the machinery, materials and other disciplines, a variety of specifications of cables are applied to many areas. In medical applications, tungsten cable is generally selected and is highly applied in surgical robot. When cable is used, cable-driven mechanism can make the whole mechanism has the advantages of compact structure, high transmission precision. It is conducive to the realization of the lightweight design of the robot. Meanwhile, the joint driver can be arranged in the same fixed base, thus will reduce the movement of the recoil.

In the existing institutions, the use of cable as the transmission system, mostly based on practice or experience, there is no corresponding theoretical analysis as a support. Sometimes, cable-driven system is considered as a rigid member when the motion of mechanism is analysed[2, 3], sometimes, cable is just a transmitter[4]. But for the structural design including the guide pulley, cable cannot be simply equated with the spring or the rod[5]. Although the application of cable has a long history, it is not long time to analyse its dynamics. As the special structural, mechanical theories based on rigid rods cannot be applied directly to the cable-driven system.

There are many kinds of cables, but their basic structures are the same. Based on the same character of cable, the paper focuses on the dynamic analysis of wires, which twist together to form cable.

## 2. Cable-driven system

Just the same as chain-driven and belt-driven system, cable-driven system needs to transmit motion and force through pulleys. To make sure that cable can transmit them precisely, some important parameters are need to be analysed.

Different from belt and spring, the structure of cable is a little complicated. First, a wire strand is twisted by a number of wires, and then, cable is formed by the same method of certain number of wire strand.

To analyse the dynamic of cable, the structure of cable should be analysed firstly.

### *2.1. Wire of cable*

Cable is twisted by wires, as shown in Figure 1. Certain number of wires comprises a wire strand. When the cable bent around the pulley, bearing tension, wires at different location will suffer different tension. To ensure that the motion and force can be transmitted precisely, every wire should be in good condition.

Figure 1(b) is the cross-section of the cable, where I is the centre wire of outer wire strand (OWS), II is the outer wire of OWS and III is the centre wire strand (CWS). According to the wire twisting method, the outer wire of III and I are named single spinning wire (SSW), the II is named dual spinning wire (DSW).

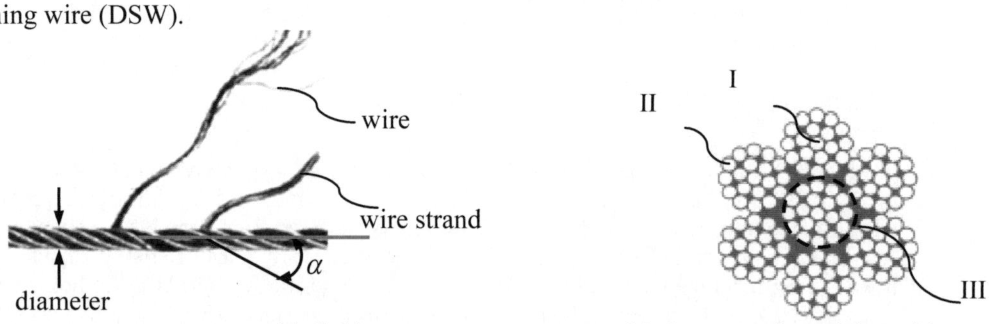


(a) Cable (b) Cross-section of the cable

**Figure 1.** Cable and its cross-section

To know the weak position of cable when cable is in tension, electronic universal testing machine (Model: CSS-44007) is used to do tensile failure experiment. The result shown that no matter what diameter of cable is, no matter what kind of force cable is suffered, the weakest position is the bent part, as shown in Figure 2. So it is necessary to analyse the weakest part bent cable.

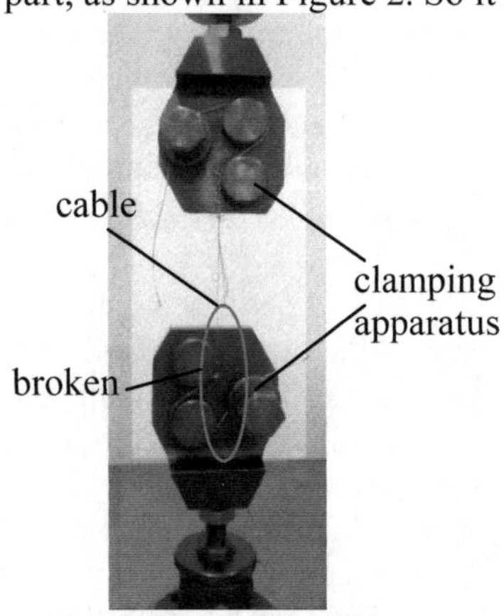


**Figure 2.** Tensile failure experiment

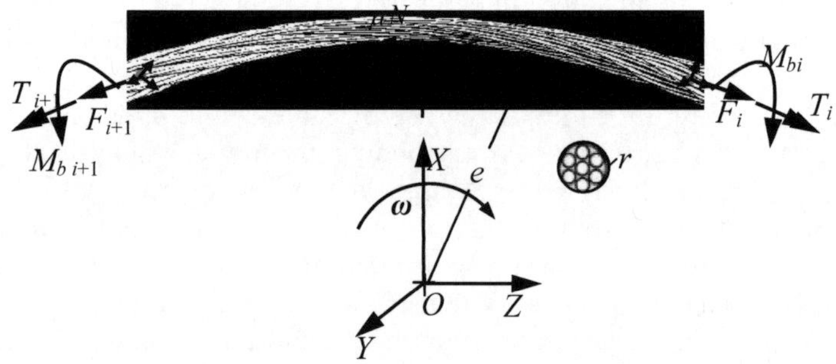


**Figure 3.** Bent cable

When the cable is bent around a pulley, it can be modelled as shown in Figure 3. The forces applied to the cable are pressure, friction, axial force, torsion and bending moment, which are represented by $N$, $\mu N$, $F$, $T$, $M_b$, respectively. According to the formation of cable, the mathematical equation of wire can be written as

$$\mathbf{s}(\varphi)=\left(-\left(e+r\cos n\varphi\right)\cos\varphi \quad r\sin n\varphi \quad \left(e+r\cos n\varphi\right)\sin\varphi\right) \tag{1}$$

where $e$ is the bending radius, $n$ is the twisting number and $r$ is the radius of wire.

IOP Conf. Series: Materials Science and Engineering **224** (2017) 012025 doi:10.1088/1757-899X/224/1/012025

### *2.2. Bending moment of wire*

Set a Frenet coordinate system[6] at the *i*th infinite element, as shown in Figure 4. It is easy to know that the base point of Frenet coordinate system can be written as

$$\left(\tau_0,\ \upsilon_0,\ \beta_0\right)=\left(s\alpha c\alpha/r_0,\ c^2\alpha/r_0,\ 0\right) \tag{2}$$

where $c=\cos$, $s=\sin$, $r_0$ is the distance from the base point of Frenet coordinate to that of the fixed coordinate.

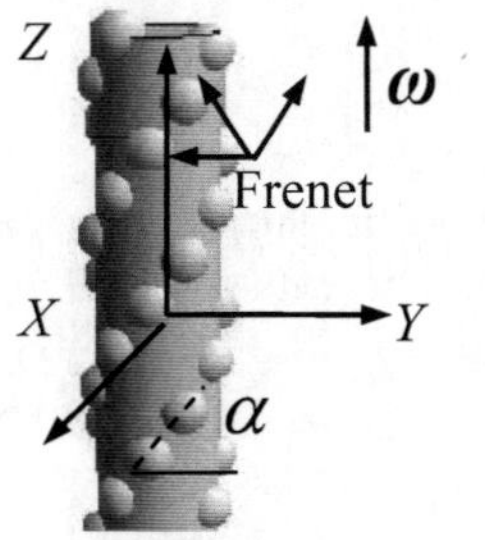


(a) Straight wire

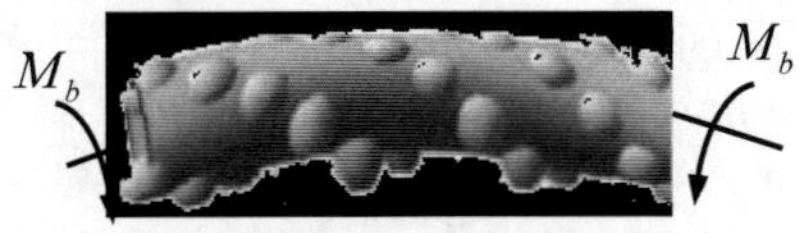


(b) Bent wire

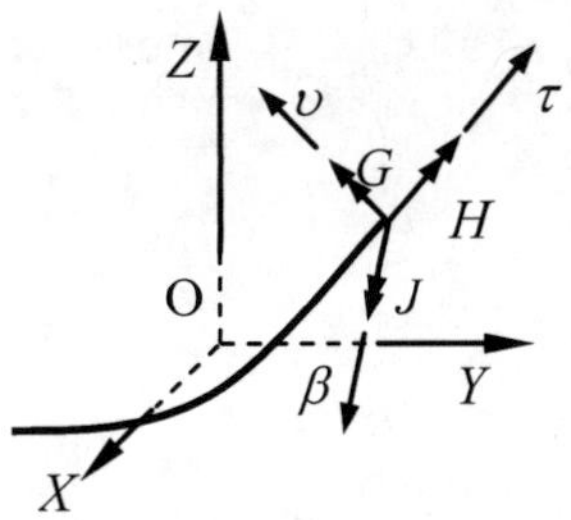


(c) Wire in Frenet coordinate

**Figure 4.** Infinitesimal element of wire

Decomposing the torsion and bending moment of the infinitesimal element wire into the Frenet coordinate system. $G$ and $J$ are the torsion at the direction of $\upsilon$ and $\beta$, respectively, while $H$ is the bending moment at $\tau$ direction. Then, according to the torsion and bending moment formula, the following equations can be obtained:

$$G=EI_\upsilon\left(\upsilon'-\upsilon_0\right) \tag{3}$$

$$J=EI_\beta\left(\beta'-\beta_0\right) \tag{4}$$

$$H=C_pI_p\left(\tau'-\tau_0\right) \tag{5}$$

where $C_pI_p$ is torsional rigidity,

$E$ is the elastic modulus of cable,

$I_\upsilon$ and $I_\beta$ are the moments of inertia of the two axes, which are $I_\upsilon=I_\beta=\left(\pi r^4/4\right)+\pi r^2 r_0^2=\pi r^2\left(r^2+4r_0^2\right)/4$,

$I_p=I_\upsilon+I_\beta$ is polar inertia,

$C_p=E/2\left(1+\rho\right)$ is the shear modulus,

$\rho$ is Poisson ratio.[7]

Let the fixed coordinate at the centre of the pulley, so the bending moment equation of the wire can be obtained as:

$$m_b = \frac{\pi r^2 \left(r^2 + 4r_p^{\ 2} + 4e^2\right) E\phi}{4l} \tag{6}$$

where $\phi$ is bending angle,

$l$ is the arc length of cable curved part,

$r$ is the radius of cable,

$r_p$ is the distance from the centre of cable to the pulley centre,

$e$ is the radius of pulley.

*2.3. Result*

Using the equation (6), the weakest wire is known. According to the surgical instrument, a pulley with 5mm radius, 15mm arc length of cable curved part is chosen. The cable with 0.6mm radius produced by SAVA Company is taken as an example. Its elastic modulus is 460N/mm$^2$. The result of different wires is shown in Figure 5. From the result, it is easy to know that the centre wire of CWS is the most likely broken wire.

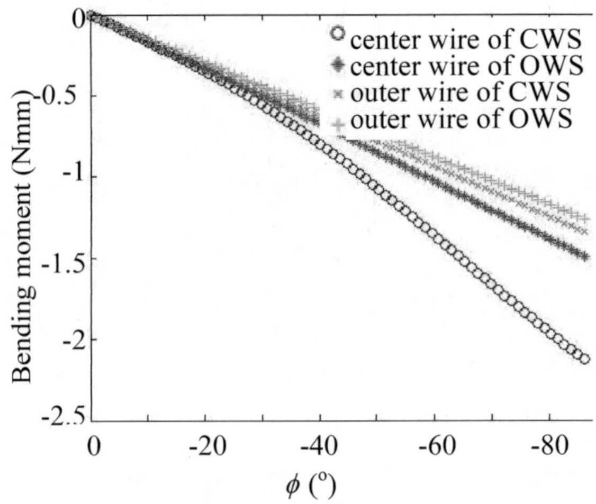


**Figure 5.** Bending moment of different wires

## 3. Conclusion

The structure of cable is very complicated, which is formed by twisting wires. A plurality of wires twist into strands, and a plurality of strands twist into cable. For better analysis of cable, according to the twisting method, the structural equation of wire is obtained, including the unbent centre wire and the outer wire, and bent centre wire.

In surgical operation, safety is of primary importance. Therefore, the choice of the appropriate cable, to ensure the safety of wire transmission system is the primary problem. According to the characteristics of cable structure, this paper put forward the bending moment formula, which provides a theoretical support for the design of cable-driven system in surgical robot.

## Acknowledgments

This paper is supported by Tianjin Municipal University Science and Technology Development Fund Project (No. 20140402), the Third Level of the '131' Training Funds for Innovative Talents (401008003, 401009002) of Tianjin.

## References

[1] Ehtesham N and Suhaib M 2012 *Int. J. Mech. Eng. & Rob. Res.* **1** 350-357
[2] Tsai L 1999 *Robot analysis: the mechanics of serial and parallel manipulators* (New York: Wiley)
[3] Lee Y and Lee J 2003 *Mech. Mach. Theory* **38** 1431-47

[4] Cesare R, Sergio S, Vincenzo Niola and Stefano T 2015 *Robotica*, **33** 1034–1048
[5] Velinsky S A, Anderson G L, Costello G A 1987 *J Eng Mech.* **3** 380-91
[6] Love A E 1944 *A treatise on the mathematical theory elasticity* (New York: Dover Publications)
[7] Shan H Z 2016 *Stress Material* (Beijing: Higher Education Press)

IOP Conf. Series: Materials Science and Engineering **224** (2017) 012026 doi:10.1088/1757-899X/224/1/012026

# Study on Control of Two Axis Linkage of Live-working Robot for Power Transmission Lines

**Yan Yu[1,2] , Zou Dehua[1,2] , Liu Xiaqing[1,2],Li Jinliang[1,2]and Zhou Zhanfan[1,2]**

1.State Grid Hunan Electric Power Company Live Working Center,Changsha 410100,China

2.Hunan Province Key Laboratory of Intelligent live working Technology and Equipment (Robot) ,Changsha 410100,China

whuyanyu@foxmail.com

**Abstract:** A two axis linkage dynamic control method for joint space of a live-working robot for power transmission lines is presented in order to solve the problem of that the working efficiency of the live-working robot is low in single joint serial connection working. This paper established a global dynamical equation during obstacle-crossing through the Newton-Euler approach, providing the basis for establishing any two-axis linkage dynamic equation; analyzed and determined joints of the robot capable of two axis linkage, presenting a group of common dynamical equations for two axis linkage joints; and designed a PD control method with a gravity model. Simulation and experimental results show that the two axis linkage method is feasible and capable of increasing working efficiency of the robot.

## 1. Introduction

The live-working robot for power transmission lines developed by this paper can fulfill working tasks of live replacement of suspension insulators and live current plate bolt tightening by replacing working ends. The working object (current plate bolt) is small and the working space is compact so that many working steps of the robot are required, and therefore two axis linkage can effectively increase the working efficiency of the robot.

At present, linkage control strategy is widely applied to machine tools[1-2], and some researches have been also done on inspection robot for power transmission lines. In essence, the studies on linage of line inspection robot for power transmission lines according to references [3-5] were intended to analyze and solve the problem of motion coordination of two arms without establishing a dynamical model thereof, thus accuracy control can not be achieved. References [6-9] only established a two axis linkage dynamical equation for the gravity adjusting joint and the arm lifting joint of the line inspection robot, without taking the common two axis linkage dynamical modeling approach. References [10-11] introduced a two axis linkage approach which mainly focus on linkage study of the gravity adjusting joint and the arm lifting joint of the line inspection robot, but the gravity adjusting joint is a relatively independent joint which is not coupled to arm staggering joint motion, therefore this method can not be popularized to other two axis linkage control of the robot. In addition, references [12] studied linkage of the gravity adjusting joint and other joints of the line inspection robot.

In conclusion, there are only few researches on two axis linkage control of live-working robot for power transmission lines at present. A two axis linkage approach applicable to the live-working robot is in urgent need so as to save the working time of the robot and increase the working efficiency of the robot

This paper established a global dynamical equation during obstacle-crossing through the Newton-Euler approach [13], providing the basis for establishing any two-axis linkage dynamic equation; then designed a corresponding PD control law for two axis linkage and proved stability of the control system after the designed control law is involved; and then carried out simulated analysis on the control law with Matlab, and determined the optimal PD parameters; and finally verified validity of the approach presented in this paper through experiments.

## 2. Kinematic and dynamic model establishment of double working arm of robot

The robot strokes along a line, stops moving after detecting collision detection sensor signals, then fixes the mechanical arms with the line through a clamping mechanism, and then live working can be completed through the two working arms.

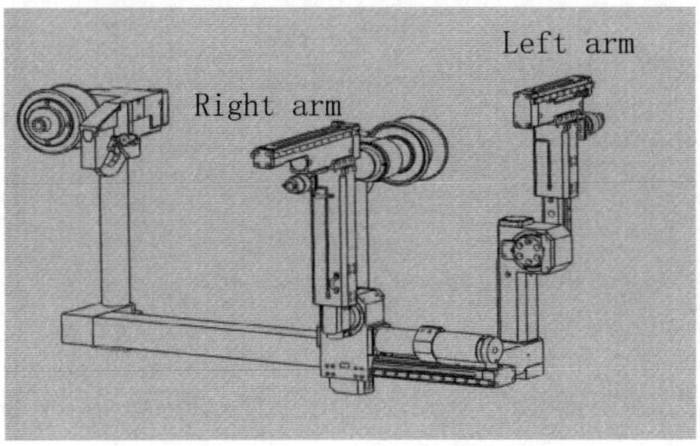


**Figure 1.** Robot structure diagram

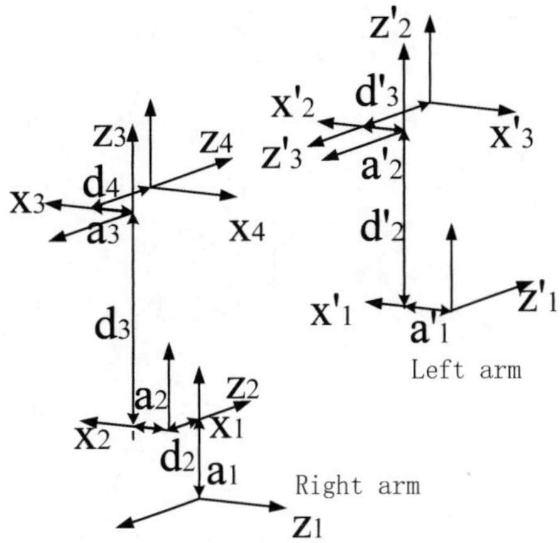


**Figure 2.** D-H coordinate system of double working arm of robot.

The robot studied in this paper, as shown in Figure 1, is equipped with two working arms and capable of fulfilling working tasks of live replacement of suspension insulators and live current plate bolt tightening by replacing ends of the two working arms. The left working arm has 3 degrees of freedom while the right working arm has 4. D-H coordinate systems are established for the two working arms, the coordinate system of the left working arm is X'Y'Z' and the coordinate system of the right working arm is XYZ, as shown in Figure 2. Corresponding link parameters are listed in Table 1 and Table 2.

**Table 1.** The D-H parameters of the right working arm of the live-working robot.

| $i$ | $\theta_i$ | $\alpha_{i-1}$ | $a_{i-1}$ | $d_i$ | Variable range of joint |
|---|---|---|---|---|---|
| 1 | $0^0$ | $0^0$ | 0 | $d_1(0)$ | $0\sim 260(mm)$ |
| 2 | $\theta_2(0^0)$ | $90^0$ | $a_1$ | $d_2$ | $-180^0\sim 180^0$ |
| 3 | $0^0$ | $-90^0$ | $a_2$ | $d_3(445)$ | $445\sim 645(mm)$ |
| 4 | $0^0$ | $90^0$ | $a_3$ | $d_4(-120)$ | $-120\sim 0(mm)$ |

**Table 2.** The D-H parameters of the left working arm of the live-working robot.

| $i$ | $\theta'_i$ | $\alpha'_{i-1}$ | $a'_{i-1}$ | $d'_i$ | Variable range of joint |
|---|---|---|---|---|---|
| 1 | $\theta'_1(0^0)$ | $0^0$ | 0 | 0 | $-180^0\sim 180^0$ |
| 2 | $0^0$ | $-90^0$ | $a'_1$ | $d'_2(425)$ | $425\sim 625(mm)$ |
| 3 | $0^0$ | $-90^0$ | $a'_2$ | $d'_3(120)$ | $0\sim 120(mm)$ |

By means of the kinematic model of the robot, the dynamic model of the robot's working can be established conveniently. In addition, the Newton-Euler approach has the advantages of facilitating programming with the computer language and having an algorithm pattern itself, thus this paper established the dynamic mode of the robot with the Newton-Euler approach.

The dynamic issues refer to calculating required joint torque or force $\tau$ of the joints according to displacement, velocity and accelerated velocity ($\theta$,$\dot{\theta}$ and $\theta$). The entire algorithm consists of step 1 performing outward recursion to calculate velocity and accelerated velocity of each link and working out inertia force and torque of each link with the Newton-Euler approach; and step 2 performing inward recursion on interaction force and torque of each link as well as joint driving force or torque.

1) The whole dynamic equation is shown in formula (1): ($i:0\rightarrow n-1$)

$$
\begin{aligned}
&{}^{i+1}\omega_{i+1}=\begin{cases}{}^{i+1}_{i}R\,{}^{i}\omega_i+\theta_{i+1}\,{}^{i+1}z_{i+1}\\ {}^{i+1}_{i}R\,{}^{i}\omega_i\end{cases}\\
&{}^{i+1}\dot{\omega}_{i+1}=\begin{cases}{}^{i+1}_{i}R\,{}^{i}\dot{\omega}_i+{}^{i+1}_{i}R\,{}^{i}\omega_i\times\dot{\theta}_{i+1}\,{}^{i+1}z_{i+1}+\ddot{\theta}_{i+1}\,{}^{i+1}z_{i+1}\\ {}^{i+1}_{i}R\,{}^{i}\dot{\omega}_i\end{cases}\\
&{}^{i+1}\dot{v}_{i+1}=\begin{cases}{}^{i+1}_{i}R\left[{}^{i}\dot{v}_i+{}^{i}\dot{\omega}_i\times{}^{i}p_{i+1}+{}^{i}\omega_i\times\left({}^{i}\omega_i\times{}^{i}p_{i+1}\right)\right]\\ {}^{i+1}_{i}R\left[{}^{i}\dot{v}_i+{}^{i}\dot{\omega}_i\times{}^{i}p_{i+1}+{}^{i}\omega_i\times\left({}^{i}\omega_i\times{}^{i}p_{i+1}\right)\right]+\\ 2\,{}^{i+1}\omega_{i+1}\times\dot{d}_{i+1}\,{}^{i+1}z_{i+1}+\ddot{d}_{i+1}\,{}^{i+1}z_{i+1}\end{cases}\\
&{}^{i+1}\dot{v}_{c(i+1)}={}^{i+1}\dot{v}_{i+1}+{}^{i+1}\dot{\omega}_{i+1}\times{}^{i+1}r_{c(i+1)}+{}^{i+1}\omega_{i+1}\times({}^{i+1}\omega_{i+1}\times{}^{i+1}r_{c(i+1)})\\
&{}^{i+1}f_{c(i+1)}=m_{i+1}\,{}^{i+1}\dot{v}_{c(i+1)}\\
&{}^{i+1}n_{c(i+1)}={}^{c(i+1)}I_{i+1}\,{}^{i+1}\dot{\omega}_{i+1}+{}^{i+1}\omega_{i+1}\times({}^{c(i+1)}I_{i+1}\,{}^{i+1}\omega_{i+1})
\end{aligned}
\tag{1}
$$

In formula(1), ${}^{i+1}\omega_{i+1}$ is the angular velocity of of the link $i+1$ in the coordination system $i+1$, $rad/s$; ${}^{i+1}_{i}R$ is the transformation matrix from the link i to the link i+1; ${}^{i}\omega_i$ is the angular velocity of of the link i in the coordination system i, $rad/s$; $\dot{\theta}_{i+1}$ is the joint angular velocity of link i+1, $rad/s$; ${}^{i+1}z_{i+1}$ is matrix $[0,0,1]^T$; ${}^{i+1}\dot{\omega}_{i+1}$ is the angular velocity of of the link i+1 in the coordination system i+1, $rad/s^2$; ${}^{i}\dot{\omega}_i$ is the angular acceleration of the link i in the in the coordination system i, $rad/s^2$; $\ddot{\theta}_{i+1}$ angular acceleration of the link $i+1$, $rad/s^2$; ${}^{i+1}\dot{v}_{i+1}$ is the displacement acceleration of the link i+1 in the coordination system i+1, $m/s^2$; ${}^{i}\dot{v}_i$ is the displacement acceleration of the link i in the coordination system i, $m/s^2$; ${}^{i}p_{i+1}$ is the radius vector of link i+1 in the coordination system i+1, $m$; $\dot{d}_{i+1}$ is the velocity of link $i+1$, $m/s$; $\ddot{d}_{i+1}$ is displacement acceleration of the link i, $m/s^2$; ${}^{i+1}\dot{v}_{c(i+1)}$ is the displacement acceleration of the link i+1 in the coordination system i+1, ${}^{i+1}r_{c(i+1)}$, $m/s^2$; ${}^{i+1}f_{c(i+1)}$ is force of the barycenter of the link i+1 in the coordination system i+1, $N$; $m_{i+1}$ is the mass of the link i+1, $kg$; ${}^{i+1}n_{c(i+1)}$ is the torque of the link i+1 in the coordination system i+1, $Nm$; ${}^{c(i+1)}I_{i+1}$ is the inertia tensor of the barycenter of the link i+1.

2) The joint dynamic equation is ($i:n-1\rightarrow 0$)

$$
\begin{aligned}
{}^{i}\mathrm{f}_i &= {}^{i}_{i+1}R^{i+1}\mathrm{f}_{i+1} + {}^{i}\mathrm{f}_{c(i)} \\
{}^{i}\mathrm{n}_i &= {}^{i}_{i+1}R^{i+1}\mathrm{n}_{i+1} + {}^{i}\mathrm{n}_{c(i)} + {}^{i}\mathrm{r}_{c(i)} \times {}^{i}\mathrm{f}_{c(i)} \\
&\quad + {}^{i}\mathrm{p}_{i+1} \times {}^{i}_{i+1}R^{i+1}\mathrm{f}_{i+1} \\
\tau_i &= \begin{cases} {}^{i}\mathrm{n}_i^{Ti} z_i \\ {}^{i}\mathrm{f}_i^{Ti} z_i \end{cases}
\end{aligned}
\tag{2}
$$

In formula (2), ${}^{i}\mathrm{f}_i$ is force of link i in the coordination system i, $N$ ; ${}^{i+1}\mathrm{f}_{i+1}$ is force of link i+1 in the coordination system i+1, $N$ ; ${}^{i}_{i+1}R$ is transformation matrix from link i+1 to link i; ${}^{i}\mathrm{f}_{c(i)}$ force of the barycenter of the link i in the coordination system i, $N$ ; ${}^{i}\mathrm{n}_i$ is torque of link i in coordination system i, $Nm$ ; ${}^{i+1}\mathrm{n}_{i+1}$ is torque of link i+1 in coordination system i+1, $Nm$ ; ${}^{i}\mathrm{n}_{c(i)}$ is torque of barycenter of link i in coordination system i, $Nm$ ; ${}^{i}\mathrm{r}_{c(i)}$ is sector radius of barycenter of link i in coordination system i, $m$ ; ${}^{i}\mathrm{p}_{i+1}$ is sector radius of link i+1 in coordination system i, $m$ ; $\tau_i$ is torque of link i+1 in coordination system i+1, $Nm$ ; ${}^{Ti}z_i$ is torque $[0,0,1]^T$ 。

Transformation matrix between links is shown in the formula as follows.

$$
{}^{i}_{i+1}R = \begin{bmatrix} c_{i+1} & -s_{i+1} & 0 \\ s_{i+1} & c_{i+1} & 0 \\ 0 & 0 & 1 \end{bmatrix} \quad {}^{i+1}_{i}R = \begin{bmatrix} c_{i+1} & s_{i+1} & 0 \\ -s_{i+1} & c_{i+1} & 0 \\ 0 & 0 & 1 \end{bmatrix}
\tag{3}
$$

In this formula, ${}^{i}_{i+1}R$ is transformation matrix from link i+1 to link i; ${}^{i+1}_{i}R$ is transformation matrix from link i to link i+1; suppose mass of links of the robot concentrate on ends of the links ${}^{i}\mathbf{p}_{i+1} = {}^{i+1}\mathbf{r}_{c(i+1)}$ . Kinematic parameters, dynamic parameters and initial conditions when the robot starts working are shown in the following formula.

$$
\begin{cases}
{}^{i}p_{c(i)} = a_i x_i \\
{}^{c(i)}I_i = 0 \\
f_j = 0,\ n_j = 0 (\mathrm{j} = 3,4) \\
w_0 = 0,\ \dot{w}_0 = 0 \\
{}^{0}\dot{v}_0 = g y_0
\end{cases}
\tag{4}
$$

By means of the above preparation, velocity, accelerated velocity and force of each link can be calculated according to the adopted algorithm with the tools of Matlab.

### 3. Design of two axis linkage control law

After the overall dynamic model of the robot is established in a Matlab environment with the Newton-Euler approach, the two axis linkage dynamic equation of the robot can be calculated conveniently.

During two axis linkage, the dynamic equation of the robot can be expressed as follows.

$$
\tau = D(q)\ddot{q} + h(q,\dot{q}) + G(q) \tag{5}
$$

In this formula, $\tau$ is joint force, $N$ ; $D(q)$ is mass matrix; $\ddot{q}$ is joint acceleration, $m/s^2$ ; $h(q,\dot{q})$ is joint force matrix; $G(q)$ is gravity compensation $N$ .

The two axis linkage control law is designed as follows.

$$\tau = k_p e - k_d \dot{q} + G(q) \tag{6}$$

In this formula, $e$ is position error, $m$ ; $\dot{q}$ is joint velocity, $m/s$ ; $k_p$ 、 $k_d$ are position and velocity diagonal gain matrixes respectively. A loop system can be obtained through formula (5) and formula (6):

$$D(q)\ddot{q} + h(q,\dot{q}) + k_d\dot{q} + k_p q = k_p q_d \tag{7}$$

In this formula, $q_d$ is expected displacement, $m$ . Global asymptotic stability of this system can be proved with the Lyapunov theory, and the constructed Lyapunov function is shown as below.

$$v = \frac{1}{2}\dot{q}^T D(q)\dot{q} + \frac{1}{2}e^T k_p e \tag{8}$$

Function (8) is non-negative because the mass matrix D(q) and the position gain matrix $k_p$ are both positive definite. According to derivation of formula (8), it can be calculated that

$$\dot{v} = \frac{1}{2}\dot{q}^T \dot{D}(q)\dot{q} + \dot{q}^T D(q)\ddot{q} - e^T k_p \dot{q} = -\dot{q}^T k_d \dot{q} \tag{9}$$

Provided $k_d$ is positive definite, $\dot{}$ is non-negative, by which the global stability of the control law (9) can be proved. The following is a plurality of controllers of two axis linkage design of the robot with a gravity model.

Each elmo driver adopted by the live-working robot can at most drive 4 motors in a part-time drive mode. The total number of of joints of the two working arms is no more than 4, thus each elmo driver drives one motor of a working arm, as shown in Table 3. In theory, any joint positioned on either working arm can realize two axis linkage.

**Table 3.** Driver and motor distribution

| **Working arm** | **Motor No.** | | | |
|---|---|---|---|---|
| **Left working arm** | 1 Left rotary joint | 2 Left telescopic joint | 3 Left traversing joint | |
| **Right working arm** | 4 Right vertical moving joint | 5 Right rotary joint | 6 Right telescopic joint | 7 Right traversing joint |

## 4. Simulation and program implementation of two axis linkage

### 4.1. *Two axis linkage simulation*

Joint variables of the right vertical moving joint of the right working arm $d_1 = 0$ 、 $\dot{d}_1 = 0$ and $\ddot{d}_1 = 0$ , joint variables of the right rotary joint of the right working arm $\theta_2 = 90^0$ 、 $\dot{\theta}_2 = 0$ and $\ddot{\theta}_2 = 0$ , joint variables of the right traversing joint of the right working arm $d_4 = -120$ , $\dot{d}_4 = 0$ and $\ddot{d}_4 = 0$ ; and joint variables of the joint of the left working arm $d'_2 = 425$ 、 $\dot{d}'_2 = 0$ and $\ddot{d}'_2 = 0$ are plugged in the global dynamic equation of the robot, and a dynamic equation of during linkage of the rotary joint of the left working arm ( $\theta'_1$ ) and the telescopic joint of the right working arm ( $d_3$ ) can be achieved.

$$14\ddot{\theta}'_1-5.7d_3+0.21\dot{\theta}'^2_1-d_3(1.6\ddot{\theta}'_1+5.7\dot{\theta}'^2_1-10d_3\ddot{\theta}'_1-20d_3\dot{\theta}'_1+10g\sin\theta'_1-5.7\ddot{\theta}'_1+10d_3+ 1.6\theta'^2_1-10d_3\theta'_1+10g\cos\theta'_1)+4.1d_3\theta'_1-3.2\dot{d}_3\theta'_1-17g\cos\theta'_1-3.8g\sin\theta'_1+7.3d_3\theta'^2_1- 0.21\dot{\theta}'_1\ddot{\theta}'_1=\tau'_1 \quad (10)$$

$$(-5.7-10d_3)\ddot{\theta}'_1+10\ddot{d}_3+(1.6-10d_3)\dot{\theta}'^2_1+10g\cos\theta'_1=\tau_3 \quad (11)$$

By adopting the control law in formula (9), it can be calculated that

$$\tau'_1=k_{p1}(\theta'_{1d}-\theta'_1)-k_{d1}\dot{\theta}'_1-10d_3g(\sin\theta'_1+\cos\theta'_1)-17g\cos\theta'_1-3.8g\sin\theta'_1 \quad (12)$$

$$\tau_3=k_{p3}(d_{3d}-d_3)-k_{d3}\dot{d}_3+10g\cos\theta'_1 \quad (13)$$

By plugging formula (11) in formula (3), the loop control law of the telescopic joint of the right working arm can be obtained that

$$(-5.7-10d_3)\ddot{\theta}'_1+10\ddot{d}_3+(1.6-10d_3)\dot{\theta}'^2_1=k_{p3}(d_{3d}-d_3)-k_{d3}\dot{d}_3 \quad (14)$$

By plugging formula (12) in formula (14), the loop control law of the rotary joint of the left working arm can be obtained that

$$14\theta'_1-5.7\ddot{d}_3+0.21\theta'^2_1-d_3(-8.2\theta'_1-20d_3\theta'_1-20\dot{d}_3\theta'_1+10\ddot{d}_3)-3.2\dot{d}_3\theta'_1-0.21\theta'_1\theta'_1 =k_{p1}(\theta'_{1d}-\theta'_1)-k_{d1}\dot{\theta}'_1 \quad (15)$$

Suppose the state variable is that $x=[\theta'_1\ \dot{\theta}'_1\ d_3\ \dot{d}_3]^T$, $\dot{x}=[\dot{\theta}'_1\ \ddot{\theta}'_1\ \dot{d}_3\ \ddot{d}_3]^T$ can be obtained by performing differential on it; suppose the expected angle of the rotary joint of the left working arm is $\theta'_{1d}=2$ rad, the expected length of the telescopic joint of the right working arm is $d_{3d}=10$ mm, and the given initial value is $x_0=[1\ \ 5\ \ 1\ \ 5]^T$, $\dot{x}_0=[5\ \ 0\ \ 5\ \ 0]^T$, and then simulation experiments are carried out.

Torques of two joints can be obtained with formula (13) and formula (14), and time response curve of state variable $x=[\theta'_1\ \dot{\theta}'_1\ d_3\ \dot{d}_3]^T$ can be obtained from Matlab. With the adjustment time, the average torque and velocity of the two joints are obtained respectively, providing two more references for selection of PD parameters.

In Table 4, the system vibrate violently when $k_p$>400, the adjustment time is in this table refers to the time that required by the response curve to reach and permanently keep within an error range 5% of the steady-state value. When $k_p$=500, the adjustment time of the system is only 13% of that when $k_p$=100, and other performance parameters have no obvious defect, hence $K_p$=500 is used.

**Table 4.** Proportional gain adjustment of the right arm telescopic joint motor

| $k_p$ | $k_d$ | Average torque ($N\cdot m$) | Average velocity ($m/s$) | Adjustment time ($s$) | Overshoot (1) |
|---|---|---|---|---|---|
| 4000 | -50 | 86.4987 | 0.5623 | 0.325 | 21.05% |
| 3000 | -50 | 84.2741 | 0.4989 | 0.265 | 19.28% |
| 2000 | -50 | 81.7470 | 0.4266 | 0.305 | 16.69% |
| 1000 | -50 | 76.7580 | 0.3446 | 0.305 | 12.06% |
| 500 | -50 | 74.7120 | 0.3176 | 0.435 | 7.71% |
| 100 | -50 | 69.5132 | 0.3658 | 3.465 | 0.36% |
| 50 | -50 | 61.4962 | 0.3913 | 3.990 | 0 |

In Table 5, when $k_d$<-150, overshoots of the system are all zero, indicating that control variables converge in a very smooth manner to the target value, but the adjustment time is long, which is 1.73s. Based on overall consideration, differential gain $k_d$=-150 is selected, although the overshoot is 0.23%, the adjustment is only 0.595s.

**Table 5.** Differential gain adjustment of the right arm telescopic joint motor

| $k_p$ | $k_d$ | Average torque $(N \cdot m)$ | Average velocity $(m/s)$ | Adjustment time $(s)$ | Overshoot (1) |
|---|---|---|---|---|---|
| 500 | -550 | 52.5114 | 0.2050 | 2.825 | 0 |
| 500 | -450 | 56.1581 | 0.2068 | 2.565 | 0 |
| 500 | -350 | 59.8850 | 0.2083 | 2.250 | 0 |
| 500 | -250 | 63.7338 | 0.2095 | 1.735 | 0 |
| 500 | -150 | 69.1871 | 0.2240 | 0.595 | 0.23% |
| 500 | -50 | 74.7120 | 0.3176 | 0.435 | 7.71% |
| 500 | -20 | 81.2558 | 0.5466 | 1.940 | 22.98% |

With the same principle, expected proportional gain and expected differential gain of the rotary joint of the left working arm can be selected, suppose the expected proportional gain is $K_{pm}$ and the expected differential gain is $K_{dm}$, both of which are $2 \times 2$ matrix.

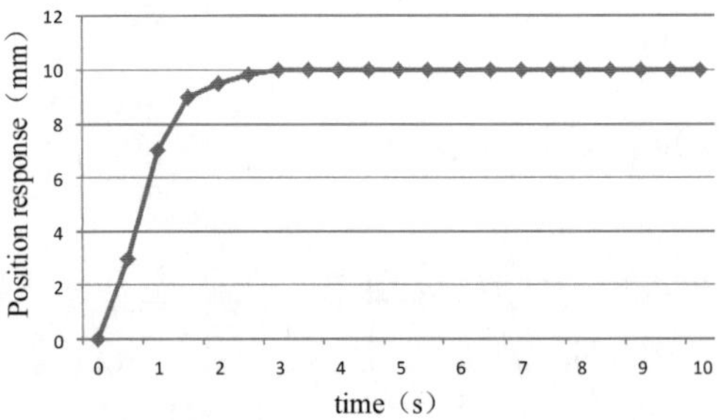


**Figure 3.** Right arm telescopic joint position response

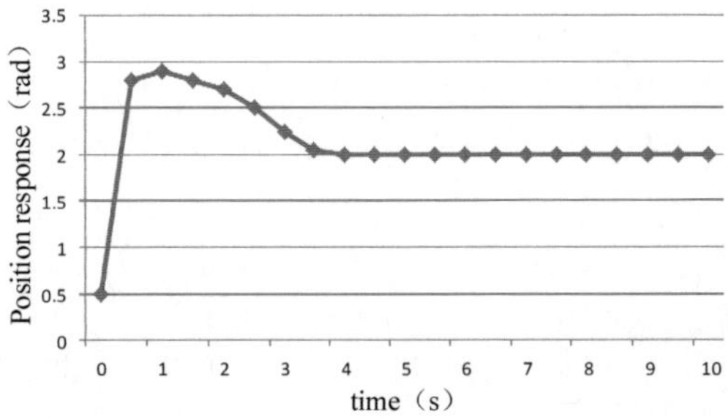


**Figure 4.** Left arm rotation joint position response

After PD parameters of $K_{pm}$ and $K_{dm}$ of the two joints are selected with the tool of Matlab, it can be seen in Figure 3 and Figure 4 that within 4s, positions of the two joints of the working arms quickly respond to expected positions and come into stable state, indicating good control effect. In these two figures, $K_{pm} = \begin{bmatrix} 1000 & 0 \\ 0 & 500 \end{bmatrix}$ and $K_{dm} = \begin{bmatrix} -150 & 0 \\ 0 & -150 \end{bmatrix}$.

*4.2. Program implementation of two axis linkage*

In the process of two axis linkage of the robot, firstly, it must ensure that each shaft in linkage operates in its safe stroke; secondly, from the perspective of control, ensure that each shaft in linkage immediately stops after running to the expected position; and finally, ensure that the robot body can respond to instructions of a base station to stop moving when the base station gives emergency braking instructions.

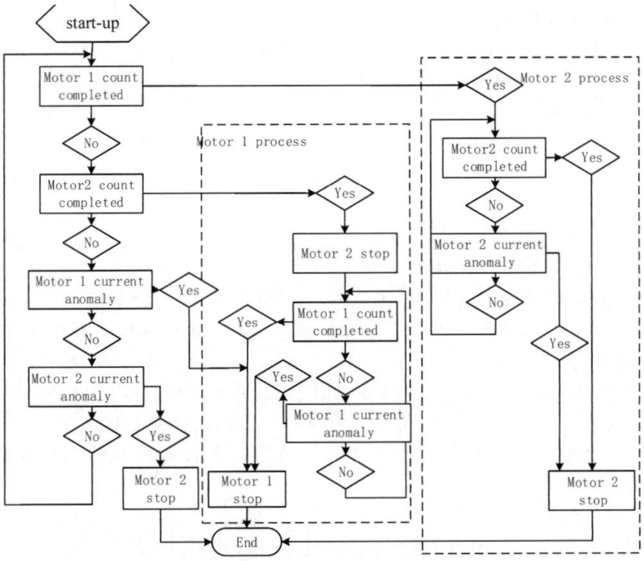


**Figure 5.** Two axis linkage algorithm flow chart

To meet the three requirements mentioned above, each motor shaft in operation is subject to constrain conditions of stroke limit, counting, current abnormality and emergency brake, among which only stroke limit and counting are effective for single shaft while current abnormality and emergency brake require double shafts in motion to respond simultaneously as to guarantee safety of equipment.

In the case that two axis linkage is available, the robot calls a two axis linkage program, as shown in Figure 5. At first two linkage motors are selected and parameters of the motors are set. After the motors are enabled, velocity and stroke of the motors are set so as to drive the motors. The two axis linkage program adopts do-while statement, and takes full counting and current abnormality as termination judging conditions. In the loop of do-while of the main program, two motor motor operating threads are created so that when one motor meets termination conditions, it stops running immediately while the other motor is not affected.

## 5. Experiment and conclusion

The velocity of the motors of the robot is limited by a reducer. Under load condition, the motors generally runs at a velocity of 8000rpm. In order to verify validity of the approach presented in this paper, operation time of each single joint stroke of the left working arm and the right working arm of the robot, as well as period of time that the robot fulfills working tasks of live replacement of suspension insulators and live current plate bolt tightening with a single joint are counted; and then the working time under the two axis linkage control approach is calculated.

Robot two axis linkage experiments were carried out on Phase A (right), #006 Tower, 220kV Wangpei I line, State Grid Hunan Live-working Center. Live working experiment of replacement of single insulator string is shown in Figure 6 (a), and live tightening of tension clamp current plate implemented on Phase A (right) of #003 Tower is shown in Figure 6 (b).

(a) (b)

**Figure 6.** Two axis linkage experiment

The result is shown in Table 6.

**Table 6.** Comparison of experimental results

| **Working type** | | **Single joint motion** | **Two axis linkage** |
|---|---|---|---|
| Suspension insulator | Number of step | 180 | 139 |
| | Time (min) | 30 | 20 |
| Current plate bolt | Number of step | 90 | 70 |
| | Time (min) | 14 | 9 |

Conclusions drawn from the above experiments are as follows:

(1) The global dynamic equation of the robot established in this paper is correct, and the proposed two axis linkage approach is feasible, which can be spread to other live-working robots for power transmission lines.

(2) According to the two axis linkage approach presented in this paper, working steps are reduced by 20% and working time is saved by 30% compared with single joint motion.

(3) Two axis linkage can effectively increase working efficiency of the robot.

## References

[1] Yue, Xi, Wang, Hong-guang.and Jiang, Yi. Design of a novel locomotion mechanism of power transmission line inspection robot [J]. 2017 *Applied Mechanics & Materials* **10** 863

[2] Dong Xin,Gao Guo-qin and Lv Yun-qi. Study on Smooth Sliding Model Control for a 6-DOF Parallel Robot on Numerical Control [J]. 2009 *Machinetool&Hydraulics* **05** 111

[3] Fang Li-jin,Wei Yong-le and Tao Guang-hong. Design of a Novel Dual-arm Inspection Robot with Flexible Cable[J]. 2013 *Robot* **03** 319

[4] Xu Xian-jin,Wu Gong-ping and He Yuan .Mechanism Design of an Inspection Robot for Reformed Ground Wire[J]. 2011 *Machinery Design & Manufacture* **11** 93

[5] Mao Xing-tao ,Bao Gang and Yang Qing-jun. Joint Trque Control for the Pneumatically Robotic ManiPulator with 3 Degrees of Freedom[J]. 2008 *Chinese Journal of Mechanical Engineering* **12** 254

[6] Luo Jan-guo,He Mao-yan and Sun Xue-cheng. Design on the Control System of 5 DOF Series-parallel Robots[J]. 2009 *Machanery Design* **07** 19

[7] Sun Cui-lian,Wang Hong-guang,Zhao Ming-yang. Comparisons in Localization Methods during Obstacle Navigation of Inspection Robot for Power Transmission Lines[J]. 2007 *China Mechanical Engineering* **01** 4

[8] Cai Li,Li En and Liang Zi-ze.Fuzzy Control of the Inspection Robot for Obstacle-negotiation [J]. 2008 *Control & Automation* **34** 24

[9] Zhu Xing-long,Wang Hong-guang and Fang Li-jin.Centroid Adjustment Control on Autonomous Obstacle Negotiating Inspection Robot[J]. 2006 *Robot* **04** 385

[10] Fang Li-jin,Wang Hong-guang. Rrsearch on the Characteristics of the Movement and Obstacle-Clearing Processes of a Wire-suspended Mobile Robot[J]. 2010 *Transactions on Intelligent Systems* **06** 492

[11] Wu Hua,Meng Ling-zhi and Liu Chang-an. Hybrid Anti-swing Control Scheme for Cable Tunnel Inspecting Robot[J]. 2013 *Huazhong Univ.of Sci.&Tech.Natural Science Edition* **s1** 440

[12] Wen Xue-feng,Dian Song-yi and Dong Hang. Study on Climbing Mechanism of Doal- arm Robot for Power Line [J]. 2013 *Machinery Design & Manufacture* **07** 168

[13] Zhou Jun,Leng Hong-bing and Zhu Hong-jun. Analysis of Influence on Natural Characteristics of Gear by Thermal Effect[J]. 2012 *Journal of Machine Design* **12** 25

IOP Conf. Series: Materials Science and Engineering **224** (2017) 012027 doi:10.1088/1757-899X/224/1/012027

# Development of Live-working Robot for Power Transmission Lines

**Yan Yu[1,2] , Liu Xiaqing [1,2] , Ren Chengxian [1,2],Li Jinliang[1,2]and Li Hui[1,2]**

[1]State Grid Hunan Electric Power Company Live Working Center,Changsha 410100,China

[2]Hunan Province Key Laboratory of Intelligent live working Technology and Equipment (Robot) ,Changsha 410100,China

380147536@qq.com

**Abstract.** Dream-I, the first reconfigurable live-working robot for power transmission lines successfully developed in China, has the functions of autonomous walking on lines and accurately positioning. This paper firstly described operation task and object of the robot; then designed a general platform, an insulator replacement end and a drainage plate bolt fastening end of the robot, presented a control system of the robot, and performed simulation analysis on operation plan of the robot; and finally completed electrical field withstand voltage tests in a high voltage hall as well as online test and trial on actual lines. Experimental results show that by replacing ends of manipulators, the robot can fulfill operation tasks of live replacement of suspension insulators and live drainage plate bolt fastening.

## 1. Introduction

Live working refers to a working method of overhauling and testing on high voltage electrical equipment [1] without power outage. Live working is an effective measure to avoid repair with power off so as to guarantee normal power supply. At present, the live working manner for power transmission lines in China is still manual live working, and operator work in an adverse environment with high altitude and high voltage. Thus, developing a live-working robot suitable for power transmission lines is of great significance.

Since 1980s, many countries have done researches on automated mechanical devices successively, such as Japan, Spain, the USA and France [2] in order improve automation level and safety of live working and reduce labor intensity of operators and physical threats of strong electrical field to operators. However, live-working automated mechanical devices at home and abroad are developed focusing on maintenance of distribution network lines, and matching with an aerial life device with an insulating boom to carry an intelligent automatic tool to carry our line breakage and connection and replacement of insulators and lightning arresters.Kyushu Electric Power Company successfully developed Phase-I, Phase-II and Phase-III robots [3-5] in 1990, 1996 and 2001 respectively, among which the intelligence level gradually rises. Robot Robtet [6-7] is successfully developed in Spain in 1998, which is at the same level with Phase-II. China's Shandong Electric Power Research Institute developed a robot for distribution network lines in 2003 [8-9]. The above robots are all based on the technology of aerial life device with an insulating boom with an working arm mounted at the tail end of the aerial life device with an insulating boom, and can only applied to the distribution field with low line construction height

Although power transmission lines are deemed as "arteries" of power transmission, few researches on live-working robots for 110kV and above high voltage transmission lines have done at home and abroad. Representatives are Hydro-Quebec Research Institute of Canada, and Wuhan University and Shandong Electric Power Research Institute [10] of China.

China is vast in territory, its energy distribution does not match with economic development, and power transmission in China has formed a pattern of west-east power transmission and north-to-south power transmission. With implementation of UHV grid construction strategy of State Grid, the coverage area of transmission lines will be more wide and live-working operations for power transmission lines will increase sharply, therefore study on live-working robots for transmission lines, which can replace operators to carry out dangerous live working, is in urgent need.

## 2. Working operation and object

Live replacement of suspension insulators and drainage plate bolt fastening are the most routine live-working items of 110-220kV transmission lines, and for the transmission lines, insulators (as shown in figure 1) are mainly used for supporting lines and keeping enough distance between the lines and the ground. During manual replacement, an operator needs to replace insulators with a wire grip to tension lines at first, and then loosen the wire grip until new insulators are mounted. The difficulty of this replacement process lies in pushing out W pins in the old insulators. The working space is small, thus the safe ground potential working manner can not be used, and the replacement process can be only carried out in a manner that the operators wear shielding clothes to enter an equipotential condition for operation. To liberate the operators from dangerous high electrical field, robots are used to enter an equipotential condition to replace operators to push W pins in old insulators out, and then push the W pins in after new insulators are installed. The robot can fully replace electricians, and then coordinate with a ground potential electrician to realize ground potential live replacement of suspension insulators.Formatting the title

In the past, drainage plate (as shown in figure 2)bolt fastening was carried out only in a manner than an electrician wearing shielding clothes enters the high electrical field to conduct equipotential working. Now, robots can be used to replace equipotential electrician to perform drainage plate bolt fastening working in an automatic mode.

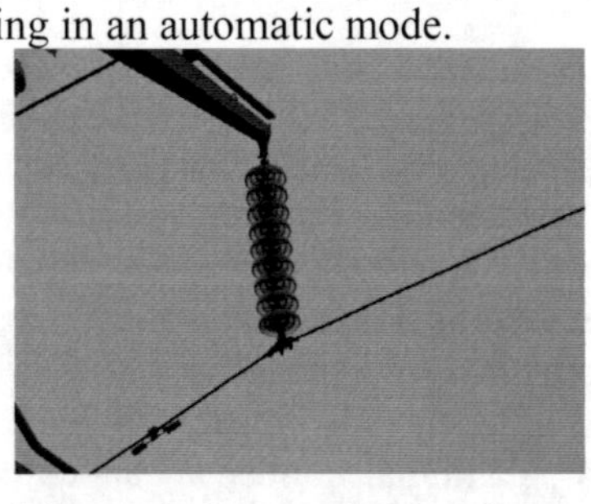

**Figure 1.** Suspension insulator.

**Figure 2.** Drainage plate.

## 3. Mechanical structure part

The Dream-I live-working robot for power transmission lines is shown in figure 11. The prototype main consists of a robot general platform, an insulator replacement end, a bolt fastening end, a control system, an image system and a control base station, wherein the general platform is combined with the insulator replacement end to complete insulator replacement, and the general platform is combined with bolt fastening end to realize bolt fastening. An operator performs information interaction with the robot body through a wireless network on the ground base station, and controls the robot to carry out live working remotely.

### *3.1. Robot general platform*

The general platform includes two moving manipulators, two operation manipulators and a control body, wherein the two moving manipulators are used for connecting the body with lines, the end of each moving manipulator is of a wheel and jaw composite structure, the robot moves relative to lines through wheels, and two clamping jaws of each moving manipulator are used for clamping lines and fix the position of the robot relative to the lines so that ends of the two operation manipulators of the robot can be positioned relative to the working object.

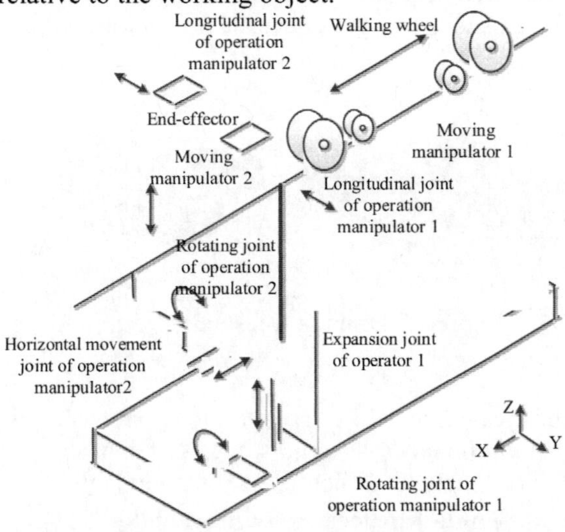


**Figure 3.** Robot general platform structure sketch map.

The two operation manipulators match with each other to enable operating ends to reach designated position. The operation manipulator 1 has three degrees of freedom, which is illustrated by an xyz coordination system established at the lower right corner of Fig. 2, wherein the x axis is parallel to lines, with the moving manipulator 2 pointing to manipulator 1; the y axis is parallel to the ground and perpendicular to the lines to extend outwards; and the z axis is perpendicular to the ground, pointing into the sky. Suppose the position that the end needs to arrive at is (x,y,z), the x-axis coordination of the operation arm 1 is provided by traveling wheels at the end of the moving manipulators, the y coordination is provided by a longitudinal moving joint of manipulator 1 in Fig2, and z coordination is provided by the telescopic joint of manipulator 1. In order to deal with the problem that lines may be not totally perpendicular to insulators, a rotary joint capable of rotating around y axis is added to manipulator 1 correspondingly. Joints of manipulator 2 are similar to those of manipulator 1. The x axis can not adopt the traveling wheel mechanism, thus a traversing joint is additionally designed for manipulator 2.

### 3.2. *Insulator replacing end*

In order to replace insulators, the end of manipulator 1, as shown in figure 4(a) is used for clamping a socket clevis and pushing W pins out or in, and it is provided with three joints. The end of manipulator 2, as shown in figure 4(b) is used for clamping porcelain insulators,and is equipped with only one joint. Porcelain insulators are driven by the traversing joint of manipulator 2 to separate from or connect with the socket clevis.

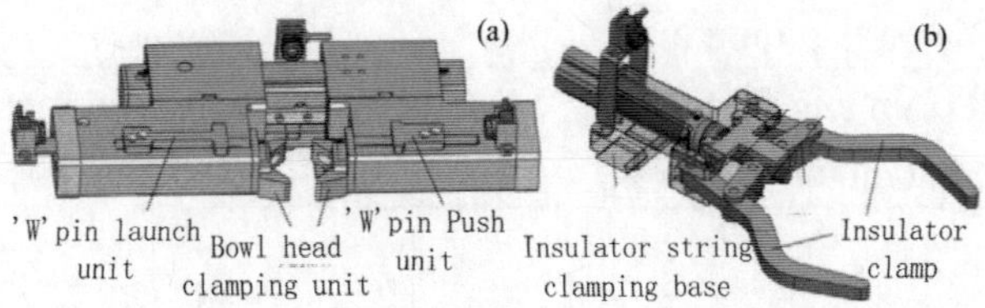


**Figure 4.** Insulator replacement end.

### 3.3. *Bolt fastening end*

Regarding the bolt fastening working, end devices are a nutting device and a bolt fixing device respectively, wherein the nutting device is, as shown in figure 5(a), is used for fixing nuts through a revolute, a rotary shaft is connected with a sleeve in a cross hinge mode to adapt to the drainage plate with different inclination angles, further to sleeve the nut and fasten the bolt. The bolt fixing device, as shown in figure 5(b), reaches the head of the bolt with motion of manipulator 1, the sleeve shaft of the bolt fixing device is divided into two sections which are connected with each other in a cross hinge mode so that the bolt head can be fastened by the sleeve shaft, and further the bolt head can be locked.

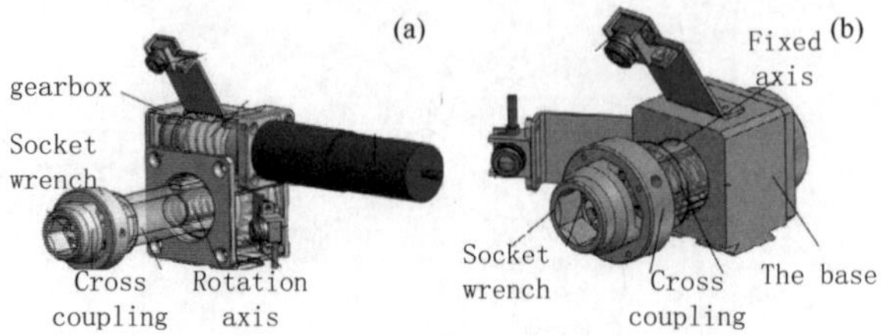


**Figure 5.** Drain bolt fastening end.

## 4. Electrical control part

The control object of the electrical part of the live-working robot comprises motors, sensors and cameras.The robot general platform includes 11 motors in total, the insulator replacing end consists of 4 motors, and the bolt fastening end includes 2 motors, as shown in table 1.

**Table 1.** Allocation of robot motor

| No. | Motor name | Remarks |
|---|---|---|
| 1 | Traveling wheel motor of manipulator 1 | General platform motor |
| 2 | Traveling wheel motor of manipulator 2 | |
| 3 | Clamping motor of manipulator 1 | |
| 4 | Clamping motor of manipulator 2 | |
| 5 | Rotary joint motor of manipulator 1 | |
| 6 | Rotary joint motor of manipulator 2 | |
| 7 | Telescopic joint motor of manipulator 1 | |
| 8 | Telescopic joint motor of manipulator 2 | |
| 9 | Longitudinal moving motor of manipulator 1 | |
| 10 | Longitudinal moving motor of manipulator 2 | |
| 11 | Traversing motor of manipulator 2 | |
| 12 | Socket clevis clamping motor | Special motor for insulator loading and unloading |
| 13 | W pin push-out operation motor | |
| 14 | W pin push-in operation motor | |
| 15 | Insulator clamping motor | |
| 16 | Bolt clamping motor | Special motor for strain clamp bolt fastening |
| 17 | Nutting motor | |

### *4.1. Camera system*

4 micro-cameras and a video server arranged on the live-working robot are used for acquiring video and image information of working environments and transmitting them to the ground base station through Ethernet. Meanwhile, the cameras receive control signals sent by the ground base station through Ethernet interfaces.

### *4.2. Sensor system*

IOP Conf. Series: Materials Science and Engineering **224** (2017) 012027 doi:10.1088/1757-899X/224/1/012027

The live-working robot need to monitor various sensors in real time. Sensors of the robot are divided into two types, one type is internal sensors, such as tilt sensor, photoelectrical limit sensor, Hall location counting sensor, power source monitoring sensor and temperature and humidity monitoring sensor, and the other is external sensors mainly including 4 micro-cameras, wherein the micro-cameras acquire image and video information through the video server, and sensors of the other type acquire sensor data through AD conversion circuits and digital interface circuits.

### *4.3. Power monitoring system*

The live-working robot is powered by a single power source. The power source monitoring management system adopts an STM32F103VB ARM chip as the control core, is mainly used for monitoring voltage of power source and current variation during working of the robot, evaluating remaining capacity, and at the meantime managing energy of devices arranged on the robot.

### *4.4. Wireless communication system*

The live-working robot is in wireless communication with the ground base station with a communication range within 1km, the robot body can transmit video or image information to the ground base station in real time. The proposed robot is a inter-section non-obstacle-surmounting robot, and wireless wifi communication can meet design requirement of the system.

### *4.5. Overall schematic design of electrical control system of robot*

To realize above functions, the control system of the robot adopts the implementation scheme shown in figure 6. The control system consists of a motion control system, an image and video acquiring system, a power source control system and a communication system. The core of the motion control system is a PC104 industrial personal computer which acquires motion status of the mechanism arms through information of the limit sensor and the inclination angle sensor, and then controls action of each joint motor,including 17 motors and 6 elmo motor drivers, among which each of the two traveling wheels is independently driven by one driver respectively, the other 15 motors share 4 drivers, and control is realized by close loop transform. The image and video acquiring system mainly includes 4 channel micro-cameras and video servers arranged on the robot body, which acquire videos and images of the working site and send them to the ground base station through the Ethernet so that users can make corresponding control decisions when controlling the robot. The power source management system is mainly used for effectively managing power source distribution of the robot in a unified mode so as to ensure power supply controllability and reset operation. The power source control model utilizes ARM chips are the main controller, and the ARM chips are also responsible for acquiring information of voltage detection, main circuit current detection, the inclination angle sensor and the temperature and humidity sensor. The communication system is mainly used for intercommunication between each electrical system part and the ground base station. Cable communication is used inside the robot while wireless local network communication is applied outside the robot.

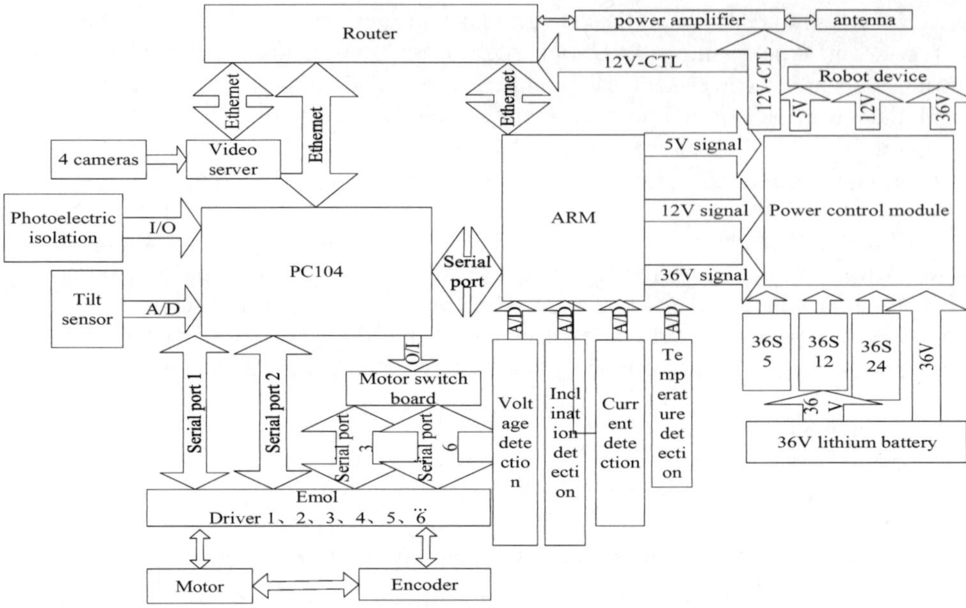


**Figure 6.** Control system block diagram.

## 5. Operation plan and simulation of robot

*5.1. Live suspension insulator replacement with robot*

The robot is adaptive to line environment of 110-220kV single-bundle single-string suspension insulator lines. As required working of robot, a wire tightener, an insulator string hoisting pulley and a robot hoisting pulley should be installed on transmission lines before the robot is applied to the lines. A virtual line environment is shown in figure 7. The operation plan of live suspension insulator string replacement with the robot is described as follows.

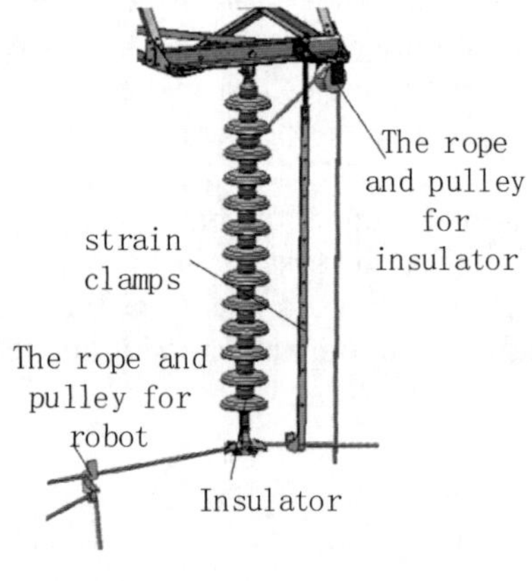


**Figure 7.** Insulator replacement robot operation preparatory environment.

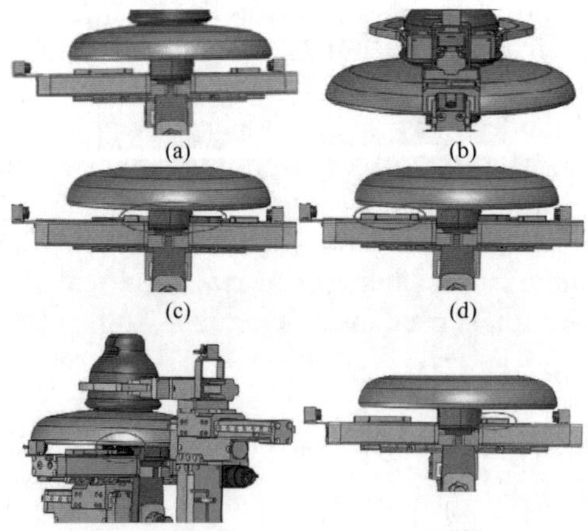


**Figure 8.** Robot replacement for insulator operation plan.

1) Collision detection suspension clamp of the robot

Carrying out rough positioning on collision detection suspension clamp of the front wheels of the robot .

2) W pin pushing device positions and grips the socket clevis

IOP Conf. Series: Materials Science and Engineering **224** (2017) 012027 doi:10.1088/1757-899X/224/1/012027

Enabling the traveling wheels to move backwards, observing pose of the socket clevis clamping jaws of the W pin pushing device relative to the socket clevis through camera 1 and 2, performing fine adjustment on joints of the manipulator 1 so that the socket clevis is positioned in the middle of the clamping jaws, and then enabling the clamping jaws to move oppositely to clamp the socket clevis, as shown in figure 8(a).

3) The insulator string pushing device protects insulator caps

Carrying out fine adjustment on two joint motors of the manipulator 2, observing pose of jaws of the insulator string pushing device relative to caps through camera 4, driving the insulator string pushing device to the position of caps of the insulator string, adjusting flare angle of the jaws to enable the insulator string pushing device to protect insulator caps, as shown in figure 8 (b).

4) Pushing W pins out

The W pin push-out block of the W pin pushing device pushes W pins from the opening of the socket clevis to a bayonet and then returns, as shown in figure 8 (c) and (d).

5)The wire tightener tensions lines.

6) Pushing insulator string out

The traversing joint of manipulator 2 moves backwards, driving the insulator string pushing device to move the insulator string ball out of the opening of the socket clevis.

7) The wire tightener loosens lines

The wire tightener loosens lines to enable the insulator ball to rise above the socket clevis.

8) Insulator string replacement

Releasing jaws of the insulator string pushing device, enabling manipulator 2 to move to a non-working position, lifting a new insulator string through the insulator string hoisting pulley, and connecting the last insulator with the cross beam of a tower by working staff.

9) The insulator string pushing device positions and protects caps

Enabling manipulator 2 to move to the working position again, observing pose of the insulator string pushing device relative to the insulator, carrying fine adjustment on the insulator string pushing device to reposition and protect the cap of the insulator through camera 4, and enabling the robot to return to the status specified in figure 8 (e) relative to the insulator string.

10) The wire tighteners tensions lines

11) Pushing the insulator string in

The traversing joint of manipulator 2 moves forwards to drive the insulator string pushing device to move the insulator string ball into the opening of the socket clevis.

12) The wire tightener releases lines

The wire tightener releases the lines so that the insulator string ball is positioned on the top of a groove of the socket clevis.

13) Pushing the W pins in

The W pin push-in block of the W pin pushing device pushes the W pins from the bayonet to the opening of the socket clevis, as shown in figure 8 (f).

### *5.2. Working simulation of bolt fastening of robot*

The robot runs to the drainage plate, the original bolting position according to images observed by the cameras, as shown in figure 9 (a), the operation plan is described as follows.

IOP Conf. Series: Materials Science and Engineering **224** (2017) 012027 doi:10.1088/1757-899X/224/1/012027

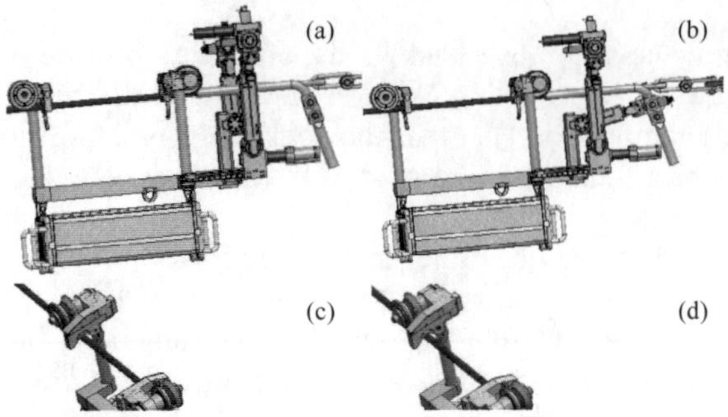


**Figure 9.** Robot fastening bolt operation plan.

1) The bolt fixing device fixes the bolt head

Enabling the rotary joint of manipulator 1 to rotate forwards, observing images of the cameras, carrying out fine adjustment on the rotary joint and the telescopic joint of manipulator 1 to enable the hexagonal socket of the bolt fixing device to coaxial with the bolt head, and then enabling the longitudinal moving joint of manipulator 1 to approach the bolt head until holding the bolt head down so as to limit rotation of the bolt head, as shown in figure 9 (b).

2) The nutting device tightens the nut

Enabling the rotary joint of manipulator 2 to rotate forwards, observing images of the cameras, carrying out fine adjustment on the traversing joint and the telescopic joint of manipulator 1 to enable the hexagonal socket of the bolt fixing device to coaxial with the bolt head,placing the longitudinal moving join to manipulator 2 close to the nut until sully sleeving it, enabling the motor of the nutting device to rotate to tighten the nut, as shown in figure 9 (c).

3) End return

The nutting device and the bolt fixing device are carried to the external limit position through manipulator 1 and manipulator 2 respectively, as shown in figure 9 (d).

## 6. Tests and conclusion

### *6.1. Withstand voltage test*

In order to test the voltage endurance capability of the robot and obtain the working voltage grade of the robot, power-frequency voltage-withstand tests were carried out on the robot in the High Voltage and Insulation Laboratory of Wuhan University, as shown in figure 10,and the testing equipment is YDJC-2000KVA/1000KV power frequency testing transformer/ultraviolet imager. Test results are recorded in table 2.

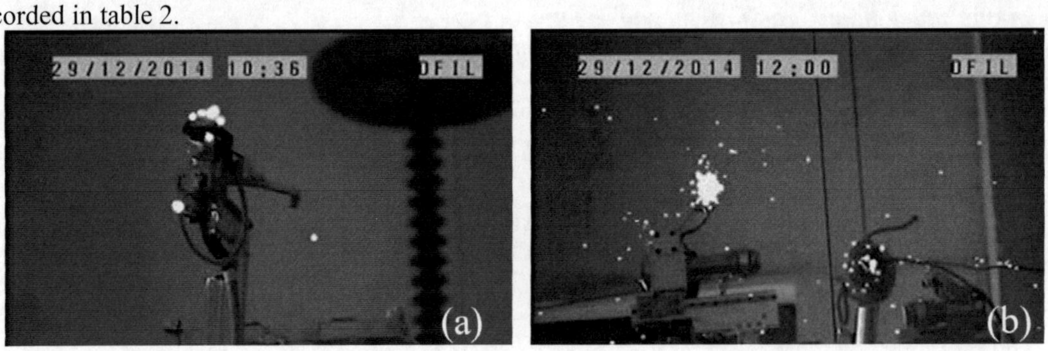


**Figure 10.** Robot high voltage hall test width as the graphic.

According to the test results, under high voltage of 220-330kV, each index of the robot is normal, the robot can accurately receive and respond to control signals sent by the ground base station in real time, the ground base station can accurately receive the status feedback information of the robot in real time, the four cameras operate normally, and transferred video and image information is clearly and smoothly.

**Table 2.** Test case of robot high voltage hall.

| Robot end tool | Insulator replacing device | | Bolt fastening device | |
|---|---|---|---|---|
| Voltage grade | 220kV | 330kV | 220kV | 330kV |
| Test time | 7:10-7:40 | 7:40-8:10 | 8:10-8:40 | 8:40-9:10 |
| Robot responds to base station instructions | Normal | Normal | Normal | Normal |
| Ground base station receives feedback information of robot | Normal | Normal | Normal | Normal |
| Working condition of camera | Normal | Normal | Normal | Normal |

*6.2. In-site experiments*

**Table 3.** Robot performance parameters.

| Item | Parameter |
|---|---|
| Applicable line | 110kV/220kV single-wire transmission lines |
| Conductor type | LGJ240～LGJ400 |
| Power frequency withstand voltage | No less than 330kV |
| Insulator type | XP—7 |
| Maximum push-out/push-in force of W pins | 1609N |
| Maximum clamping force of socket clevis | 1072N |
| Maximum clamping force of caps | 300N |
| Push-pull force of insulator (string)ball | 1015.6N |
| Maximum torque for bolt fastening | 120Nm |
| Power supply for robot body | Power by lithium battery, capacity 36V40Ah |
| Power supply for base station | Power by lithium battery, capacity 36V40Ah |
| Operation mode | Image and video based human-machine interaction operation of ground base station |
| Communication mode | Wireless Ethernet |
| Communication distance | Approx. 2km |
| Image resolution | 628 x 582 |
| Robot weight | Approx. 52kg |
| Overall dimensions of robot body | L*W*H:1000*350*650mm |

By means of withstand voltage tests and simulated line experiments, key performances of the robot are obtained, as shown in Table 3. Finally, live test of the robot is carried out on actual operating transmission lines.

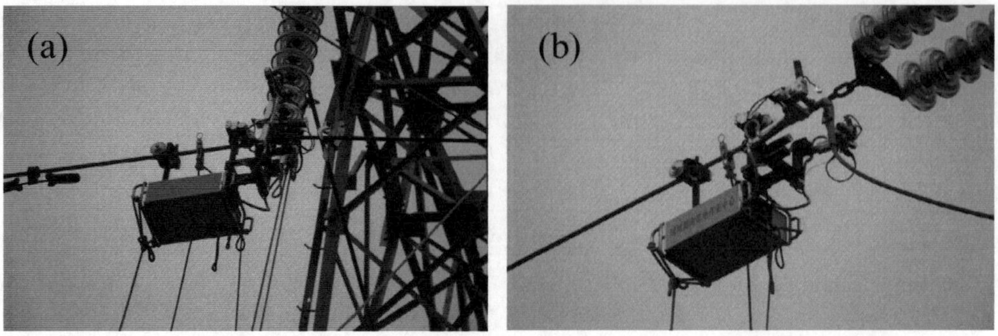


**Figure 11.** Field test of robot replacement for insulators and fastening bolts.

Experiments of single suspension insulator string replacement with the robot were carried out on Phase A(right), #006 Tower, 220kV Wangpei Ⅰ line in State Grid Hunan Live Working Center. Tower type ZB1-24, line type LGJ-400/50, insulator type XP-7, and number of insulators 13. Experiment of single suspension insulator string replacement with the robot is shown in figure 11(a).

Experiments of live strain clamp drainage plate bolt fastening with the robot were carried out on Phase A(right), #003 Tower, 220kV Wangpei Ⅰ line in State Grid Hunan Live Working Center. Tower type is SDN31-18 and line type is LGB20A-95/55. Experiment of live strain clamp drainage plate bolt fastening with the robot is shown in figure 11(b).

According to above experiments, conclusions are drawn as follows.

(1) The robot can realize live working of insulator replacement.

The function of insulator replacement can be achieved through human-machine interactive rough positioning, accurate positioning of socket clevis, W pin push-in/push-out, cap clamping, ball push-in/push-out and other operations in working process. In the working process, the two manipulators and their ends do not interfere with each other in respective working space, and the key action positing can meet requirement.

(2) The robot can achieve live working of strain clamp bolt fastening and unscrewing.

The function of live working of strain clamp bolt fastening and unscrewing can be realized through human-machine interactive operation, using the bolt fixing device to accurately position the bolt head, using the bolt fastening device to align to and accurately position the not, and other control operations in the working process. In the working process, the two manipulators and their ends do not interfere with each other in respective working space, the key action positing can meet requirement, and the nut and bolt head can be connected to and separated from the end operating mechanism steadily.

(3) The end tool of the robot can be replaced.

The end tool of the robot adopts the design of combination of dovetail grooves with pins, the dovetail grooves are connected with the longitudinal moving mechanism of manipulator 1 and manipulator 2, and the pins are used for preventing displacement thereof. Each of the ends can be connected with robot body through a pluggable connector so as to maintain line connection during end replacement. The ends with the mechanical structure and line connectors can be replaced quickly when operations change, and have sufficient stability and reliability, realizing the function of end tool replacement.

## References

[1] Zhao Y,Tsiotras P 2013 *Analysis of energy-optimal aircraft landing operation trajectories. Journal of Guidance, Control, and Dynamics* **36** 833

[2] Zhu Xinglong, Zhou Jiping and Wang Hongguang 2009 *Transmission line inspection and test of robot mecha-nism. Journal of Mechanical Engineering* **02** 119

[3] Lu Shou Yin, Fu Mengchao and Li Bingqiang 2005 *DWR-I teleoperation live working robot. Journal of Shanghai Jiao Tong University* **06** 910

IOP Conf. Series: Materials Science and Engineering **224** (2017) 012027 doi:10.1088/1757-899X/224/1/012027

[4] Lu Shou Yin, Ma Peisun and Qi Hui 2003 *Development of live working robot. Automation of Electric Power System* **17** 56

[5] Zhang F, Liu G and Fang L 2012 *Estimation of battery state of charge with H observer: Applied to a robot for inspecting power transmission lines. IEEE Transactions on Indus-trial Electronics* **59** 1086

[6] Wu Hua,Meng Lingzhi and Liu Changan 2013 *Anti sway control of cable tunnel inspection robot for hybrid dy-namical systems. Journal of Huazhong University of Science and Technology (Natural Science Edition)* **s1** 440

[7] Wen Xuefeng, Dian Songyi 2013 *Study on Mechanism of double arm type climbing robot for overhead lines. Mechanical Design and Manufacturing* **07** 168

[8] Song Bo, Xiao Shide and Hua Shirong 2007 *Automatic inspection robot attitude adjustment method research based on MSP430. Robot* **02** 117

[9] Codino, A., Wang, Z., Razzaghi, R., Paolone, M., and Rachidi, F. 2017 *An alternative method for locating faults in transmission line networks based on time reversal. IEEE Transactions on Electromagnetic Compatibility* **99** 1

[10] Vijay, M., & Jena, D. 2016 *Intelligent adaptive observer-based optimal control of overhead transmission line de-icing robot manipulator. Advanced Robotics* **30** 1

IOP Conf. Series: Materials Science and Engineering **224** (2017) 012028 doi:10.1088/1757-899X/224/1/012028

# Research on Walking Wheel Slippage Control of Live Inspection Robot

**Yan Yu[1,2] , Liu Xiaqing [1,2] , Guo Hao [1,2],Li Jinliang[1,2]and Liu Lanlan[1,2]**

[1]State Grid Hunan Electric Power Company Live Working Center,Changsha 410100,China

[2]Hunan Province Key Laboratory of Intelligent live working Technology and Equipment (Robot) ,Changsha 410100,China

380147536@qq.com

**Abstract.** To solve the problem of walking wheel slippage of a live inspection robot during walking or climbing, this paper analyzes the climbing capacity of the robot with a statics method, designs a pressing wheel mechanism, and presents a method of indirectly identifying walking wheel slippage by reading speed of the pressing wheel due to the fact that the linear speed of the pressing wheel and the walking wheel at the contract point is the same; and finds that the slippage state can not be controlled through accurate mathematical models after identifying the slippage state, whereas slippage can be controlled with fuzzy control. The experiment results indicate that due to design of the pressing wheel mechanism, friction force of the walking wheel is increased, and the climbing capability of the robot is improved. Within the range of climbing capability of the robot, gradient is the key factor that has influence on slippage of robot, and slippage can be effectively eliminated through the fuzzy control method proposed in this paper.

## 1. Introduction

The track path of the live inspection robot is 110-220kV overhead high voltage lines [1-3]. The robot passed through 330kV power frequency withstand test in high voltage hall of Wuhan University, each index response and loaded devices of the robot operate normally, and the control system is immune to interference of high electric field. In practice of actual line, walking wheel slippage accelerates abrasion of walking wheel, reduces repairing efficiency, and causes energy load to the live inspection robot with limited power capacity [4-6]. In the meanwhile, slippage may cause positioning error for robots using encoder or milemeter for positioning. Thus, slippage identifying and controlling of walking wheels will be focus of research on control of live inspection robot walking along transmission lines.

## 2. Research on slippage control of live inspection robot

The robot is prone to slippage due to the line structure characteristics and erection environment such as small cross section of lines, sag caused by gravity of the line itself and height difference between adjacent towers, thus reducing the working efficiency of the robot and increasing energy consumption of the robot. With deepening of research and application of live inspection robot, it turns to be an important part of live inspection robot research to increase its self-adaption climbing capacity and repair working efficiency.

References [7-8] analyzed slippage of spherical robots, established a dynamic mode during robot climbing, and presented corresponding control methods which however, are difficult to apply to suspension moving robots. References [9-10] studied slippage of lunar rovers, established a slippage mode that wheels of the lunar rover interact with the ground. In the meanwhile, a slippage rate formula was present with the gradient of the ground taken into account. In addition, an anti-slip model is established and simulation of the model is carried out, but analysis and simulation in this method were only in theory.

The reference [11] analyzed the climbing capacity of the live inspection robot, divided the walking states of the robot on overhead ground wires into constant speed, deceleration, acceleration and stop, and then adjusted structure of the robot according to different states so as to increase adaptability of the live inspection robot to gradient.

In conclusion, so far, only a few researches on slippage of wheel type moving robots have been done at home and abroad, and there is hardly any research on slippage of live inspection robots at present. The robot works in an environment with rigid-flexible coupling characteristics and force loading conditions of the walking wheel are very complex, thus it is very difficult to identify slippage with a dynamic method. Therefore, this paper is intended to present a new method for identifying and controlling slippage of walking wheel of the live inspection robot so as to increase working efficiency of the live inspection robot and avoid severe abrasion of the walking wheel.

## 3. Analysis of climbing capacity and slippage of robot

### *3.1. Analysis of climbing capacity of robot*

When the robot slips, the common cause lie in large gradient, thus it is necessary to study the maximum climbing capacity of the robot at first.

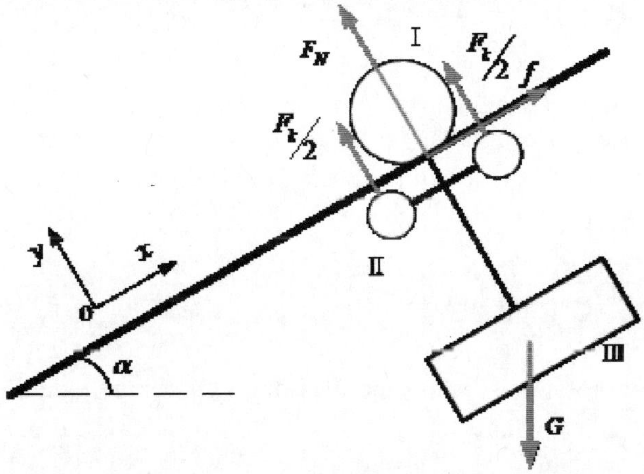


**Figure 1.** Static model of robot walking.

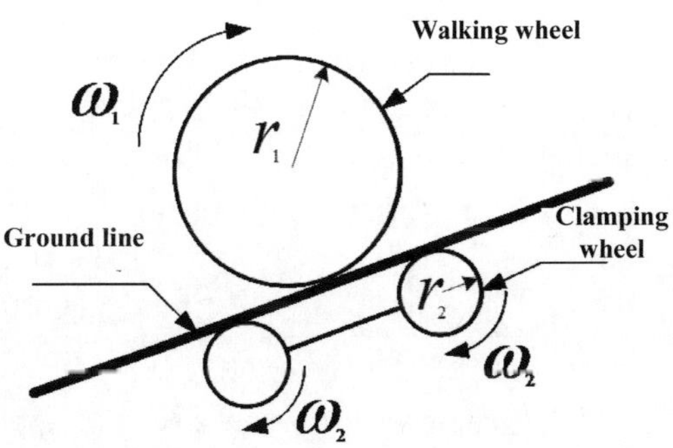


**Figure 2.** Schematic diagram of robot skid identification.

The prototype of the robot is shown in figure 9. The robot is of a double-arm antisymmetric suspension structure, thus force bearing state of only one arm is analyzed herein. The statics model of the robot walking along the ground wire at certain gradient is shown in figure 1, where in I is the walking wheel of the robot, II is the pressing wheel, III is the control cabinet of the robot, α is gradient, and G is gravity of the robot. In order to improve the climbing capability of the robot, a pressing wheel mechanism is designed on the arm to increase positive pressure loaded on the walking wheel to increase friction force.

The statics equation when the robot climbs a gradient is shown in formula (1).

$$f - G_x \geq 0 \tag{1}$$

$G_x$ is component is the robot's gravity along the x direction shown in figure 1, which is described in the following formula.

$$G_x = -\frac{1}{2} G\cos(\alpha) \tag{2}$$

In figure 1, support force by lines along the y direction consists of two parts: reaction support force $F_N$ of lines to the robot after the robot's own gravity is applied to the lines and pressing force $F_k$ that the pressing wheel transmits to the walking wheel.

$$f = -\mu(F_N + F_k) \tag{3}$$

$$F_k = kx \tag{4}$$

In formula (3), $\mu$ is the friction coefficient between the walking wheel of the robot and the ground wire. In formula (4), $k$ is the elastic coefficient of a pressing spring on the pressing wheel. Only one arm is in analysis, thus the support force is only set to be half of the total support force.

$$F_N = \frac{1}{2} G\cos\alpha \tag{5}$$

The maximum transformation of the spring is $x_m$, and in the formula, $x \in (0, x_m)$. By integrating formula (1), (2), (3), (4) and (5) simultaneously, a formula can be obtained as follows.

$$\sin(\alpha - \arctan(\mu)) \le \frac{2\mu kx}{G\sqrt{1+\mu^2}} \tag{6}$$

According to formula (6), without taking the driving capacity of the walking wheel into consideration, positive pressure applied by the pressing wheel is the most important factor influencing climbing capacity of the robot. With the maximum transformation of the pressing spring $x_m$ taken into account, the maximum climbing angle of the robot is that

$$\alpha_{\max} = \arcsin\left(\frac{2\mu kx_m}{G\sqrt{1+\mu^2}}\right) + \arctan(\mu) \tag{7}$$

When the pressing wheel does not apply any pressure, suppose the maximum climbing angle of the robot is $\alpha_{\min}$, then it can be obtained from formula (7) that

$$\alpha_{\min} = \arctan(\mu) \tag{8}$$

In accordance with formula (7), in theory, when the robot rolls on a line with the gradient smaller than $\alpha_{\max}$, slippage of the robot can be totally prevented. Thus, it is necessary to find a slippage identifying method first to judge the state of the robot and then control the slippage.

*3.2. Analysis of robot slippage*

According to analysis of the climbing capacity of the robot, the pressing force $F_k$ is very important for increasing the gradient adaptability of the robot. It can be seen from formula (7) that when $F_k$=0, namely the pressing wheel does not apply any pressure, the maximum climbing angle of the robot is $\arctan(\mu)$, and the pressing pressure can bring out extra $\arcsin\left(\frac{2\mu kx}{G\sqrt{1+\mu^2}}\right)$ climbing angle increment for the robot.

When slippage of the robot is controlled, as in formula (7), it only requires to measure the friction coefficient $\mu$ between the walking wheel of the robot and the ground wire, the elastic coefficient $k$

of the pressing spring, and the gravity $G$ of the robot itself. Then the stroke of the pressing spring can be controlled according to gradient of the lines $\alpha$ .

## 4. Slippage identifying

### *4.1. Slippage identifying model*

When slippage of the robot is controlled, the walking state of the robot is as shown in figure 2. The pressing wheel is not provided with a drive unit, belonging to a driven mechanism. When the walking wheel of the robot does not slip, the walking wheel and the pressing wheel rotate synchronously, and their walking distance in a certain period of time is the same, thus the liner speed $v_1$ of the walking wheel is the same with the linear speed $v_2$ of the pressing wheel, namely $v_1=v_2$. $v_1=\omega_1 r_1$ and $v_2=\omega_2 r_2$, wherein $\omega_1$ is the angular speed of the walking wheel, $r_1$ is the radius of the walking wheel, $\omega_2$ is the actual angular speed of the pressing wheel, and $r_2$ is the radius of the pressing wheel. When the robot does not slip, suppose the non-slip angular speed of the pressing wheel is $\omega_{2m}$, the following formula needs to be met.

$$\omega_1 r_1 = \omega_{2m} r_2 \tag{9}$$

### *4.2. Slippage identifying method*

The angular speed of the walking wheel, $\omega_1$, can be directly read through the driver, and the actual angular speed of the pressing wheel, $\omega_2$ can be acquired by a slippage sensor installed in the pressing wheel. Within a period of time $\Delta t$ when the pressing wheel rotates, the slippage sensor acquires a certain number of pulse signals, and the pulse signals are counted by a counter arranged on a multi-functional data acquisition card. Thus, actual angular speed of the pressing wheel, $\omega_2$, is described in the following formula.

$$\omega_2 = \frac{2\pi N}{n\Delta t} \tag{10}$$

In this formula, $\Delta t$ is acquisition time interval, $N$ is the number of pulse acquired by the counter within the time interval $\Delta t$, $n$ is the number of pulse per revolution of the pressing wheel.

Therefore, the slippage state of the robot can be defined through $\omega_2$, the actual angular speed of the pressing wheel.

1) $\omega_2 = 0$, the robot slips totally.

2) $0 < \omega_2 < \omega_{2m}$, the robot does not slip at all.

3) $\omega_2 = \omega_{2m}$, the robot does not slip.

## 5. Slippage control method

When the walking wheel slips, according to formula (7), the slippage can be eliminated by controlling the pressing force Fk on the pressing wheel. The degree of slippage of the walking wheel directly defines magnitude of pressing force applied to the pressing wheel, and the concept of slippage extent is introduced to quantize the degree of slippage of the robot.

**Slippage degree:** the ratio of $\omega_2$, the actual angular speed of the pressing wheel to $\omega_{2m}$, the non-slip angular speed of the pressing wheel is defined as slippage degree. The slippage degree is represented by $\varepsilon$, and its mathematical expression is as follows.

$$\varepsilon = \frac{\omega_2}{\omega_{2m}} = \frac{2\pi N r_2}{n\Delta t r_1 \omega_1} \tag{11}$$

Slippage degree reflects the degree of slippage of the walking wheel, and its domain is $\varepsilon \in [0,1]$, the smaller $\varepsilon$ is , the higher the degree of slippage is. $\varepsilon_m = 1$ is set as the target slippage degree, representing that the robot does not slip.

By means of the analysis of slippage degree hereinbefore, it can be known that when the slippage degree of the robot returns to $\varepsilon = \varepsilon_m$ from different slippage degrees, the required pressing force applied by the pressing wheel will surely be different. In order to better describe the robot's returning to the slippage $\varepsilon = \varepsilon_m$ from any original state, the concept of slippage increment is introduced.

**Slippage increment:** suppose the slippage degree of the walking wheel of the robot at the time $t_1$ is $\varepsilon$, the bearing pressing force is $F_1$, after the pressing wheel applies or releases certain pressure $\Delta F$ , the slippage degree of the robot reaches $\varepsilon_m$ at the time of $t_2$, and the beating pressing force is $F_2$, and then the slippage increment within the period of time $\Delta t$ is denoted by $\Delta x$ , and its mathematical expression is as follows.

$$\Delta F = F_2 - F_1 = k\Delta x \tag{12}$$

In this formula, $k$ stands for the elastic coefficient of the pressing spring, and $\Delta x$ stands for slippage increment. When $\Delta F$ is a negative value, it represents that the pressing force of the pressing wheel at previous time is too large and the pressing wheel needs to release a pressing force with the magnitude of $\Delta F$ . Correspondingly, when $\Delta F$ is a positive value, it represents that pressing force of the pressing wheel at previous time is too small and the pressing wheel needs to apply a pressing force with the magnitude of $\Delta F$ . The domain is $\Delta x \in [-x_m, x_m]$, when $\Delta x = 0$, it presents that the robot does not slip at the previous time.

*5.1. Relation between slippage increment and slippage degree*

According to the analysis above, in case of the fixed line gradient, the smaller the slippage degree is, the larger the required slippage increment is. The direct control means of slippage of the robot, as shown in formula (12), slippage can be controlled by using the pressing wheel to compress the pressing spring upwards or releasing the pressing spring downwards.

When the line gradient is 12 degrees and the spring is of different original deformations, different slippage degrees of the robot can be controlled by applying different slippage increments. Variations are as shown in figure 3, when $y_1$, $y_2$ and $y_3$ are the required slippage increments for adjusting the robot under different slippage degrees into the non-slip state when the original deformations of the spring are $\Delta x_b = 20mm$ , $\Delta x_b = 0mm$ and $\Delta x_b = 40mm$ respectively. As shown in figure 3, the lower the slippage degree is, the larger the slippage increment is.

*5.2. Relation between slippage increment and gradient*

Under the same slippage degree, the larger the gradient is, the larger the slippage increment is. This kind of qualitative relationship lies between the pressing force increment required for controlling slippage and gradient. Carry out experiments with the robot under different gradients, adjust the original pressing force to enable the walking wheel to keep the same slippage degree $\varepsilon$ , and then supplement pressing force to control the robot not to slip. According to recorded experimental data, the relation between pressing force increment and gradient is as shown in figure 4, $g_1$ stands for variation of the slippage increment $\Delta x$ of the robot with gradient when the slippage degree of the robot meets $\varepsilon = 0$; $g_2$ stands for variation of the slippage increment $\Delta x$ of the robot with gradient when the slippage degree of the robot meets $\varepsilon = 0.3$; and $g_3$ stands for variation of the slippage increment $\Delta x$ of the robot with gradient when the slippage degree of the robot meets $\varepsilon = 0.8$.

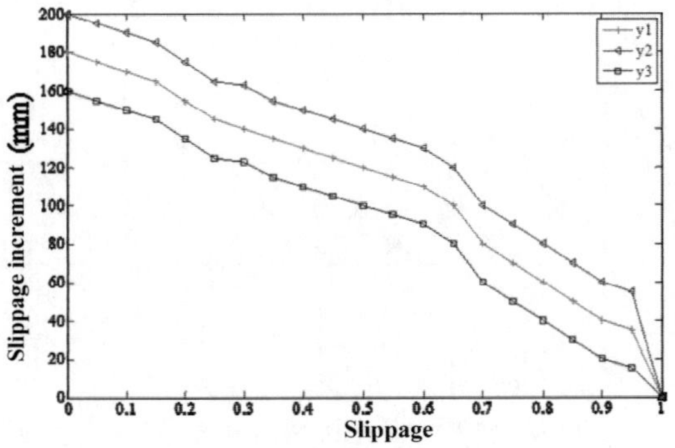


**Figure 3.** Relationship between skid increment and degree of slip.

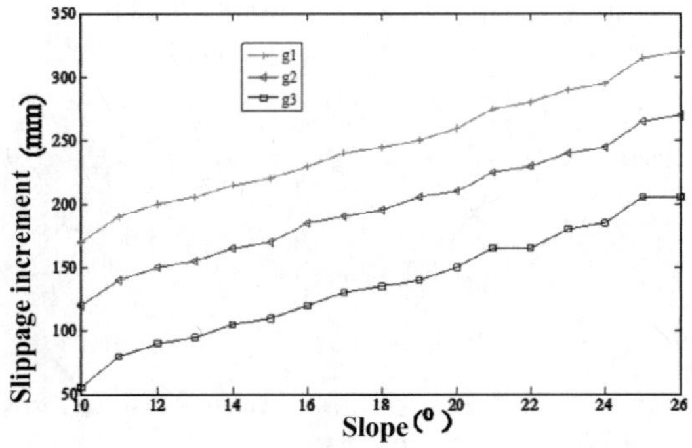


**Figure 4.** Relationship between skid increment and gradient.

*5.3. Design of slippage controller*

According to the analysis of causes of slippage of the robot described above, slippage of the robot is related to line gradient α, the friction coefficient between the walking wheel and the ground wire as well as the slippage degree $\varepsilon$. During actual operation of the robot, reasons of slippage of the robot are more complicated, material of the ground wire, rust degree of the surface, rain, snow or dust covered on the surface and the speed state of the robot can all affect the slippage state of the robot. Even when the robot walks on line of the same type, the slippage conditions are different when the robot is on different line sections. In addition, the surface of the walking wheel is in continuous abrasion so that roughness of the surface of the wheel groove of the walking wheel changes all the time, thus it is very difficult to determine the friction coefficient between the walking wheel and the surface of the line. Slippage can not be effectively controlled by adjusting the pressing force with accurate mathematical models, thus the fuzzy control method can be used to solve this problem effectively.

Figure 5 shows the slippage fuzzy controller designed by this paper, and the controller comprises two inputs and one output. The line gradient α and the slippage degree $\varepsilon$ are used as the inputs of the system respectively, and then close loop of the line gradient is realized through the inclination angle sensor, and the close loop of slippage degree is achieved through the slippage sensor. The output is pressing force increment $\Delta x$, and $T_d$ is external disturbance.

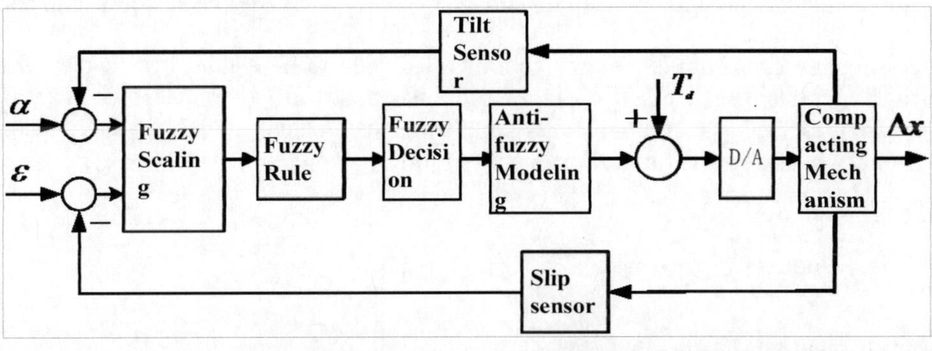


**Figure 5.** Sliding fuzzy controller.

IOP Conf. Series: Materials Science and Engineering **224** (2017) 012028 doi:10.1088/1757-899X/224/1/012028

*5.3.1. Determination of domain of discourse and membership function.* The input variable slippage degree is acquired by the slippage sensor. As shown in formula (11), the range of slippage degree of the robot is $\varepsilon \in [0,1]$. Thus, the domain of discourse of $\varepsilon$ is [0,1] in 6 grades, with the gas between each two adjacent grades of 0.17. The linguistic variable is divided into 7 levels, which comprises, from small to large based on slippage degree, very small (NK3), small (NK2), relatively small (NK1), medium (K0), relatively large (PK1), large (PK2) and very large (PK3). Each fuzzy partition membership function μ is expressed in a triangular function as shown in figure 6.

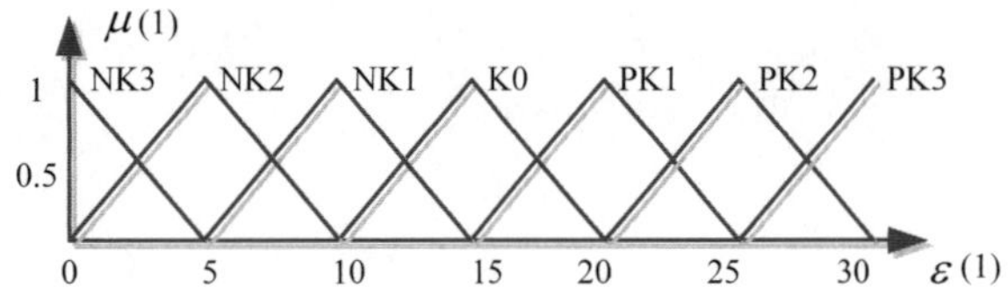


**Figure 6.** Membership function of slipping degree.

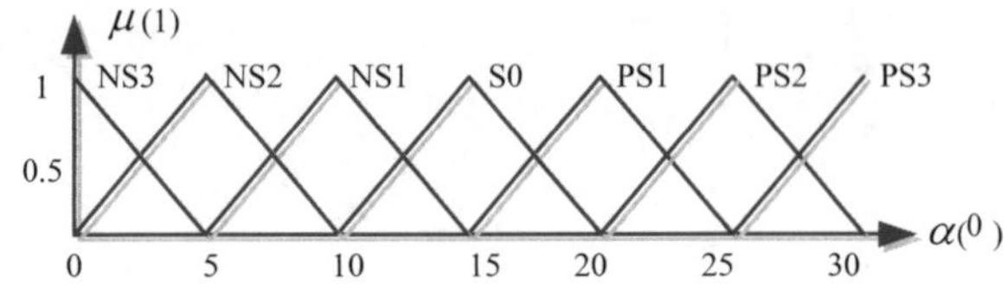


**Figure 7.** Gradient membership function.

The input variable line gradient is acquired by the inclination angle sensor. The climbing capacity of the live inspection robot adopting a roll running manner is 30 degrees. Thus, the domain of discourse of $\alpha$ is [0,30] in 6 grades, with the gas between each two adjacent grades of 5 degrees. The linguistic variable is divided into 7 levels, which comprises, from small to large in accordance with gradient, very small (NS3), small (NS2), relatively small (NS1), medium (S0), relatively large (PS1), large (PS2) and very large (PS3). Each fuzzy partition membership function μ is expressed in a triangular function as shown in figure 7.

The output of the fuzzy controller is the stroke of a pressing motor of the pressing wheel and slippage increment. The stroke reflects magnitude of pressing force increment, its domain of discourse is [0,360] in mm, and can be read through a driver. The domain of discourse of slippage increment is divided into 6 grades with the gas between each two adjacent grades of 60mm, and is divided into into 7 levels in a fuzzy partition mode, with corresponding linguistic variables being extremely small (DN3), very small (DN2), relative small (DN1), medium (D0), relatively large (PD1), very large (PD2) and extremely large (PD3). Its fuzzy partition membership function μ is expressed in a triangular function as shown in figure 8.

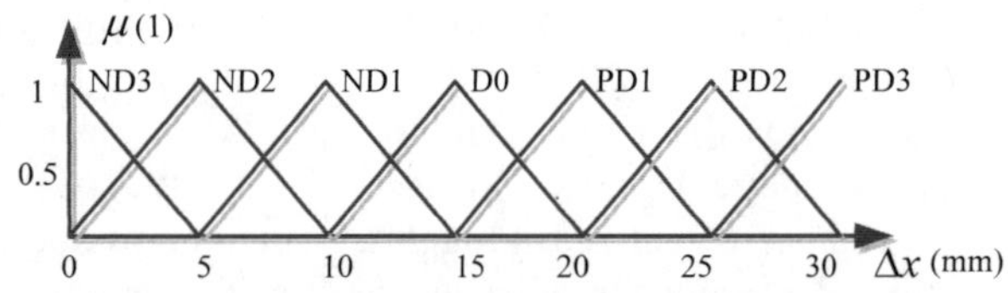


**Figure 8.** Slipping increment membership function.

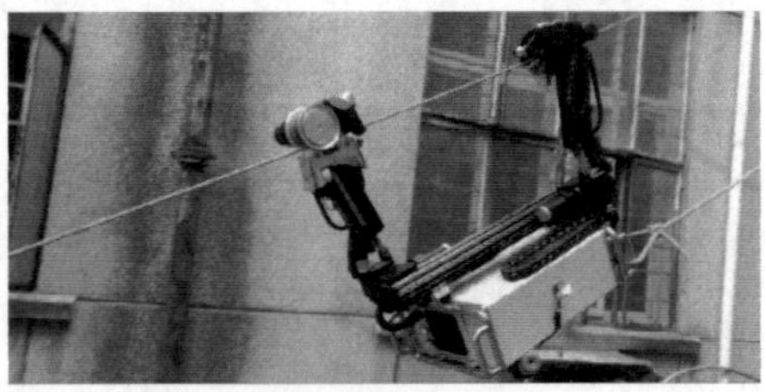

**Figure 9.** Robot skid test.

*5.3.2. Fuzzy slippage control rules.* Fuzzy control rules need to be established in the fuzzy control process, which are established based on knowledge and practical experience of human. The fuzzy control rules are presented as a group of fuzzy conditional statements, such as "if-then"statements. For the slippage fuzzy controller, its form of rules are as follows.

$R_1$ :If $\varepsilon$ is $A_1$ and $\alpha$ is $B_1$,then I is $C_1$

$R_2$ :If $\varepsilon$ is $A_2$ and $\alpha$ is $B_2$ ,then I is $C_2$

……

$R_n$ :If $\varepsilon$ is $A_n$ and $\alpha$ is $B_n$ ,then I is $C_n$

The fuzzy relation corresponding to the general control rule of the fuzzy controller is R, and

IOP Conf. Series: Materials Science and Engineering **224** (2017) 012028 doi:10.1088/1757-899X/224/1/012028

$$R = R_1 \cup \quad \cup \quad \cup \quad \cup \tag{13}$$

$$R_i = (A_i and B_i) \rightarrow C_i \tag{14}$$

Suppose the input fuzzy quantity of the fuzzy controller is that $\varepsilon$ is $A'$ and $\alpha$ is $B'$, then through Mamdani approximate reasoning with the fuzzy control rules, it can be obtained that the output fuzzy quantity I (represented by fuzzy geometry C')is that

$$C' = (A' and B')R \tag{15}$$

The judgment rule is as shown in table 1. In the process of slippage control, the fuzzy control plan is adjusted according to experimental experience, when the gradient α is small, the slippage degree will not bring about medium (K0), relatively large (PK1), large (PK2) and very large (PK3).

**Table 1.** Rule for judging skid control.

| Rank | NK3 | NK2 | NK1 | K0 | PK1 | PK2 | PK3 |
|---|---|---|---|---|---|---|---|
| NS3 | PD1 | D0 | ND1 | ND2 | ND2 | ND2 | ND3 |
| NS2 | PD1 | PD1 | ND1 | ND1 | ND2 | ND2 | ND3 |
| NS1 | PD2 | PD1 | D0 | ND1 | ND2 | ND2 | ND3 |
| S0 | PD2 | PD1 | D0 | D0 | ND1 | ND2 | ND3 |
| PS1 | PD3 | PD2 | PD1 | D0 | ND1 | ND2 | ND3 |
| PS2 | PD3 | PD3 | PD1 | D0 | ND1 | ND1 | ND3 |
| PS3 | PD3 | PD3 | PD2 | PD1 | D0 | ND1 | ND3 |

## 6. Test research and analysis

The experimental line is as shown in figure 9. The line gradient range is [10°,30°], and the gradient can be adjusted according to different test objectives. The rope in figure 9 is used for protect safety of the robot, applying no towing force to the robot.

On a simulated line, the speed is set as 3km/h, and the fuzzy control method is used to carry out climbing slippage tests when the robot is on the line with gradients of 10 degrees, 20 degrees and 30 degrees respectively. The time response curves of slippage control are shown in figure 10, figure 11 and figure 12 respectively. According to figure 10, the robot only slips at the startup stage, and then quickly returns to the normal walking state because a gap is reserved between the pressing wheel of the robot and the ground wire, and the linear speed of the walking wheel is not consistent with that of the pressing wheel from time to time. After the gradient comes up to 20°, figure 11 shows that slippage of the robot aggravates, resulting from the fact that component of the gravity of the robot itself along the direction perpendicular to the line decreases, and support force of the robot is reduced, consequently the pressing wheel needs to supplement support force so as to increase friction force in the walking direction. Compared with figure 10, the robot spends 5 seconds to adjust itself to the normal walking state. Figure 12 shows that the slippage degree of the robot in the adjustment process repeats, this is mainly due to the fact that at large gradient, the two walking wheels are not in the same straight line because of gaps between the two arms of the robot and the guide rail. Thus in the adjustment process, when gaps are caused between the pressing wheel and the ground wire, slippage will be aggravated. However, after the pressing wheel continuously applies pressing force to the ground wire, influence of the gaps is eliminated, and slippage will disappear.

As known from the analysis of test results described above, gradient is the dominant factor having influence on slippage of the robot, while the fuzzy control method presented in this paper can effectively eliminate slippage, and validity of control method is verified.

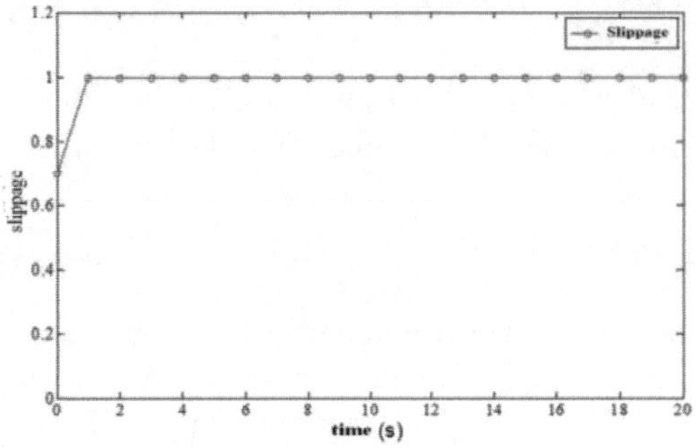


**Figure 10.** Time response curve of gradient degree of 10 degree slip.

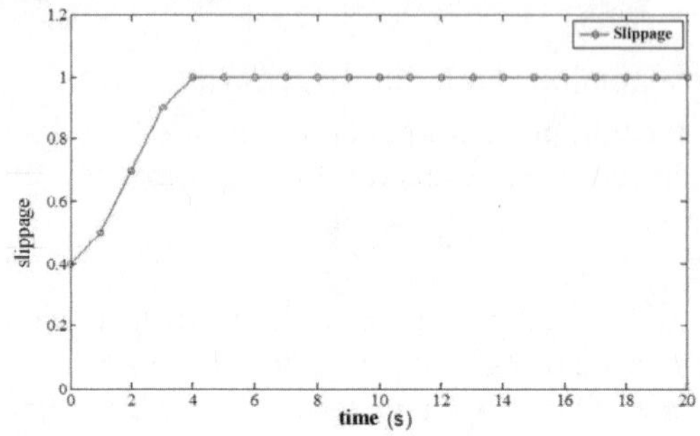


**Figure 11.** Time response curve of gradient degree of 20 degree slip.

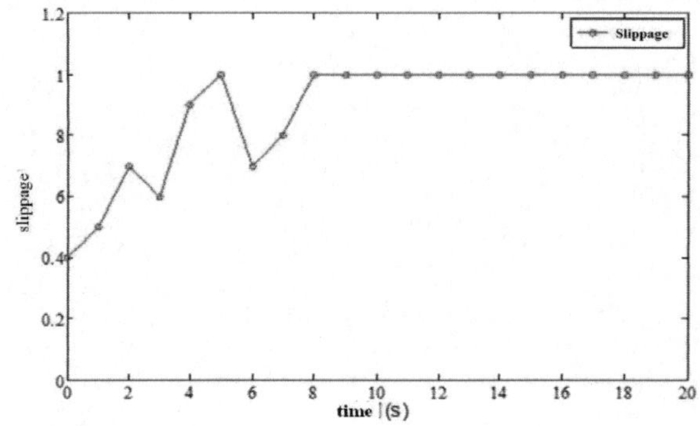


**Figure 12.** Time response curve of gradient degree of 30 degree slip.

## 7. Conclusion

This paper analyzed the material cause of slippage of the robot at first, pointed out that gradient is the dominant factor having influence on normal walking of the robot, then analyzed the climbing capacity of the robot and presented the maximum climbing angle of the robot, and obtained the theoretical basis for why the pressing wheel can effectively increase the climbing capacity of the robot. On this basis, the paper developed the slippage control method, used the method of controlling the stroke of the pressing wheel to control slippage of the robot, and analyzed the relation between slippage increment and slippage degree and relation between slippage increment and gradient. To solve the problem that accurate mathematical models can not be established for slippage, this paper designed a fuzzy controller using slippage degree and gradient as input and slippage increment as output. Finally, tests are carried out for the proposed slippage control method. The test results show that the presented fuzzy slippage control method is correct, and can effectively eliminate slippage of the robot and meet application requirements.

## References

[1] Pouliot N, Richard P and Montambault S 2012 *LineScout power line robot: Characterization of a UTM-30LX LIDAR system for obstacle detection . Intelligent Robots and Systems*, 4327

[2] Pouliot N, Mussard D and Montambault S 2012 *Localization and archiving of inspection data collected on power lines using LineScout Technology. Applied Robotics for the Power Industry (CARPI)*, 197

[3] Montambault S, Pouliot N and Lepage M 2012 *On the latest field deployments of LineScout Technology on live transmission networks . Applied Robotics for the Power Industry (CARPI)*, 126

[4] Toth J , Pouliot N and Montambault S 2010 *Field experiences using LineScout Technology on large BC transmission crossings. Applied Robotics for the Power Industry (CARPI),* 1

[5] Sun Zhiwei, Sun Hanxu and Jia Qingxuan 2012 *One kind of robot motion analysis of spherical robot with climbing ability. Robot* **02** 152

[6] Buehringer M, Berchtold J and Buchel M 2010 *Cable-crawler - robot for the inspection of high-voltage power lines that can passively roll over mast tops. Industrial Robot-an International Journal* **37** 256

[7] Alutei A, Tatar M and Cirebea C 2010 *Model and test of a modular inspection robotic system. Mechanika* **84** 58

[8] Yue Ming ,Deng Zongquan 2009 *Dynamic modeling and optimal controller design of spherical robot climbing state. Journal of Mechanical Engineering* **11** 46

[9] Jiao Zhen, Gao Haibo and Deng Zongquan 2010 *Ground interaction model of lunar rover climbing wheel based on ground mechanics . Robot,* **01** 70

[10] Farinelli, A, Iocchi, L and Nardi, D 2016 *Distributed on-line dynamic task assignment for multi-robot patrolling. Autonomous Robots*, 1

[11] Huang, Q, Shao, L and Li, N 2016 *Dynamic detection of transmission line outages using hidden markov models. IEEE Transactions on Power Systems* **31** 2026

IOP Conf. Series: Materials Science and Engineering **224** (2017) 012029 doi:10.1088/1757-899X/224/1/012029

# Control of autonomous ground vehicles: a brief technical review

**Babak Shahian-Jahromi**[1], **Syed A Hussain**[2], **Burak Karakas**[1], **and Sabri Cetin**[1]

[1] Mechanical and Industrial Eng, University of Illinois at Chicago, Chicago, Illinois, USA, 60607

[2] Electrical and Computer Sciences Eng, University of Illinois at Chicago, Chicago, Illinois, USA, 60607

bshahi2@uic.edu

**Abstract.** This paper presents a brief review of the developments achieved in autonomous vehicle systems technology. A concise history of autonomous driver assistance systems is presented, followed by a review of current state of the art sensor technology used in autonomous vehicles. Standard sensor fusion method that has been recently explored is discussed. Finally, advances in embedded software methodologies that define the logic between sensory information and actuation decisions are reviewed.

## 1. Introduction

Autonomous vehicle technology has been a major research and development topic in the automotive industry during the last decade. The technologies developed in the automotive industry also has direct applications in construction, mining, agricultural equipment, seaborne shipping vessels, and unmanned aerial vehicles (UAVs). Significant R&D activities in this area date back three decades. Despite the heavy intensity of investment in technology development, it will still take a few decades before an entirely autonomous self-driving vehicle navigates itself through national highways and congested urban cities [1]. People have been trying to make self-driving cars, since the invention of the car. The history of automated driving from early stages to automated highways and automated vehicles, are reviewed in [2]. A brief history of major autonomous cars development is shown in Fig. 1.

**Figure 1.** History of Autonomy in Ground Vehicles ([3, 4, 5, 6, 7])

The past, present and potential future of driver assistance systems are reviewed by Benglar et al [8] and summarized in Fig. 2. Early driver assistance systems were based on sensors that measure the internal status of the vehicles. These sensors enable the control of vehicle dynamics so that the trajectory requested by the driver is followed in the best way possible. In 1995, additional dynamic driving control systems such as electronic stability control (ESC) was introduced. The Second generation of driver

IOP Conf. Series: Materials Science and Engineering **224** (2017) 012029 doi:10.1088/1757-899X/224/1/012029

assistance systems was introduced around the 1990s based on sensors that measure the external state of the vehicle with the focus of providing information and warnings to the driver.

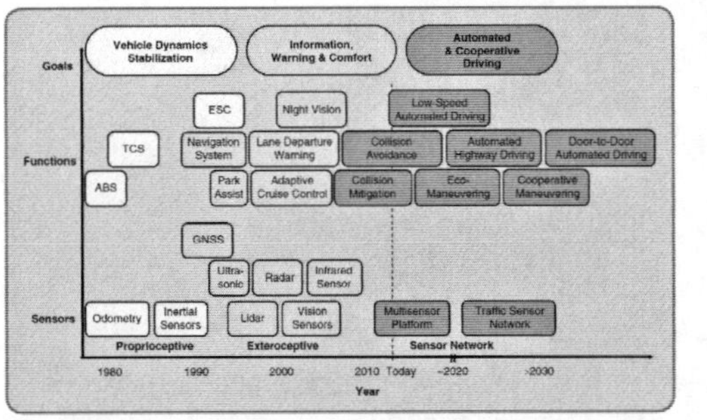


**Figure 2.** Evolution of Advanced Driver Assist Systems

| Level | Name | Dynamic Driving Task (DDT): Sustained lateral and longitudinal vehicle motion control | Dynamic Driving Task (DDT): Object and Event Detection and Response (OEDR) | DDT Fallback | Operational Design Domain (ODD) |
|---|---|---|---|---|---|
| *Driver performs part or all of the DDT* | | | | | |
| 0 | No Driving Automation | Driver | Driver | Driver | N/A |
| 1 | Driver Assistance | Driver and System | Driver | Driver | Limited |
| 2 | Partial Driving Automation | **System** | Driver | Driver | Limited |
| *Automated Driving System (ADS "System") performs the entire DDT (while engaged)* | | | | | |
| 3 | Conditional Driving Automation | System | **System** | Fallback-ready user | Limited |
| 4 | High Driving Automation | System | System | **System** | Limited |
| 5 | Full Driving Automation | System | System | System | **Unlimited** |

**Figure 3.** Level of Automation according to SAE J3016

The following sensors that measure conditions outside the vehicle, and vehicle position relative to its environment are essential in driving assist systems and autonomous vehicle technologies: Vision, lidar, radar, ultrasonic range, GPS, and inter-vehicle communication. The latest generation of Driver Assist Systems (DAS), also called Advanced Driver Assist Systems (ADAS), defines and controls trajectories beyond the current request of the driver, i.e. overriding driver commands to avoid a collision. All these sensors have overlapping and complementary capabilities. Data fusion strategies that combine real-time information from these multiple sensors is an important part of the embedded control software. Actuation technologies, i.e. computer control steering, throttle, transmission, and braking, are mature and do not present any R&D challenges. The embedded control software development (that includes data fusion from sensors, inter-vehicular communication, real-time cloud computing support) is the key technical challenge.

The on-road vehicle automation requires a standard set of regulations and terminology with a taxonomy and definitions. Some regulation and standard have been released [9]. The new standard J3016 from SAE International simplifies communication and facilitates collaboration within technical and policy domains. According to the standard as shown in Fig. 3, the levels of driving automation can be divided into Conditional, High, and Full Automation. The standard does not provide complete definitions applicable to lower levels of automation (No Automation, Assisted, or Partial Automation). Active safety and driver assistance system that intervene to avoid and/or mitigate an emergency situation and then immediately disengage are also not included for the various levels of automation.

The short-term goal is to automate driving in select well-defined situations as implemented by some of the technology companies, for example testing of self-driving taxi cabs in well-defined suburbs. The long-term goal is to achieve door-to-door automated driving in any situation. Some current ADAS examples include traffic jam assist, collision avoidance assist, pedestrian and oncoming traffic detection systems.

## 2. Sensors: Principles & Limitations

Different sensors and systems are used for navigation and control of the autonomous vehicle as shown in Fig. 4. Prominent sensors such as LIDAR, GPS, radar, vision, ultrasonic and inertial measurement unit, their working principle and usage are discussed in the following sections.

### *2.1. LIDAR*

LIDAR (Light Detection and Ranging) transmits a beam of light pulse from a rotating mirror, part of that light will reflect back to the sensor to detect any non-absorbing object or surface (Fig. 5). The target distance is calculated as speed of light times the measured time period at multiple angles. It scans the field of view in 3D with a finite spatial resolution and frequency. Using the calculated distance, the sensor

IOP Conf. Series: Materials Science and Engineering **224** (2017) 012029 doi:10.1088/1757-899X/224/1/012029

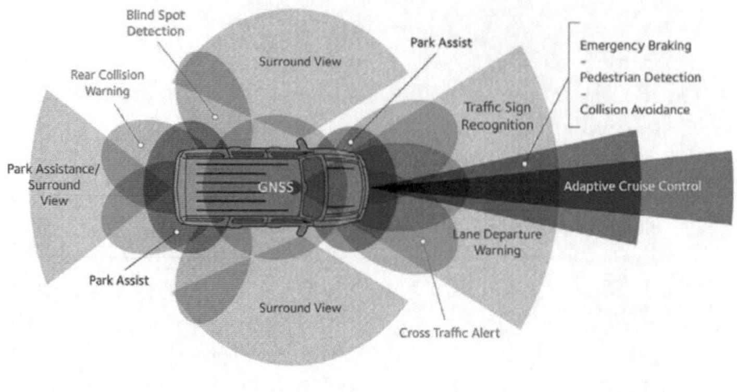


**Figure 4.** Sensors and their functionality in autonomous vehicle (source: Novatel)

constructs a 3D map of the world including the objects around it. LIDAR uses infrared (IR), visible or ultraviolet (UV) waves of the electromagnetic spectrum for different applications and comes in various 1D, 2D and 3D configurations. They are the second most valued sensors, after vision for object detection

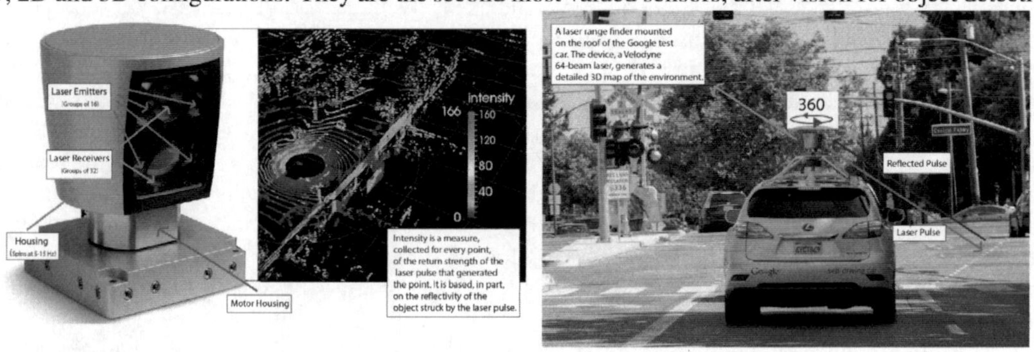


**Figure 5.** 3D LIDAR components, data and generated map by one of Google's self-driving cars

like a vehicle, pedestrian and obstacles, as shown in a typical application of a self-driving cars (Fig. 5). Their performance is poor in rain, snow, dust and foggy environments. A high-end LIDAR sensor can measure multiple distances per laser pulse which is helpful to see through dust, rain and mostly transparent surfaces such as glass windows and porous object like wire fences. To reduce the signal to noise ratio, a higher power laser generation is desired but in order to prevent damage to the human eye, a laser power of 905nm is used to achieve desired range with low duty cycle [10]. Current cost of LIDAR sensor is relatively high [11, 12] and there are some issues with the long-term reliability of their mechanical scanning mechanisms. They have been used heavily in research applications, but not widely used in automotive OEM safety systems until recently.

### 2.2. *Radar*

Radar (Radio Detection and Ranging) transmits electromagnetic pulses and senses the echoes to detect and track objects. Echoes sensed vary in frequency depending on the speed of the object. Radar can measure the relative distance, velocity, and orientation of the object [13, 14]. In case of monostatic radars in which the transmitter and receiver are located at the same location, range of the target is measured by using the round trip travel time of a pulse, times the speed of light divided by two. They are typically available for short ($\approx$ 30), medium ($\approx$ 60) and long-range ($\approx$ 200 m) distances and range from 3 MHz (Very long range) to 100+ GHz (Short range) frequencies (Fig.4). Radar requires less computing resources compared to vision or LIDAR. Typical application include lane keeping, advance cruise control, object detection, etc [15].

### 2.3. *GPS*

Global positioning system (GPS) is used to locate the position of a GPS receiver (x,y,z coordinates) using four or more geostationary satellites. These satellites maintain their position relative to earth and broadcast reference signals. A GPS receiver on earth deciphers the actual location within meters accuracy. However, the so called "differential GPS" can pinpoint a location within centimeter accuracy which is necessary for navigation of autonomous vehicle. GPS consists of three segments: space, control, and user segments. The space segment includes the satellites, control segment manages and controls them and user segment is related to development of user equipment for both military and civil purposes [16]. GPS based navigation application are greatly used to accurately predict the vehicle location with respect to map which is known as localization. However GPS based navigation are inaccurate and can lead to ghosting phenomena.

### *2.4. Vision*

Vision system is composed of a camera and image processing unit (Fig 6). A typical camera is a combination of focusing lens and array of photo-detectors for each pixel in the field of view (FOV). The array of photo-detectors send pixel information to image processing unit. This unit processes the information based on certain algorithms to detect desired objects. Vision sensors capture more visual information, hence tracking the surrounding environment more effectively than other sensors. They are categorized into mono and stereo types. Mono camera systems are often used for lane marking, lane edge detection, basic object detection, road sign detection and localization. Multiple or Stereo camera systems provide depth for objects detection. Their primary advantage are their low-cost, off-the-shelf components and their software implementation. Their primary disadvantage is handling a full range of ambient and uncontrolled conditions such as lighting, shadowing, reflection, weather, dust, smoke. Also processing data from these sensors in real-time require a large amount of computational resources. Even with these limitations they are extensively used in autonomous vehicles.

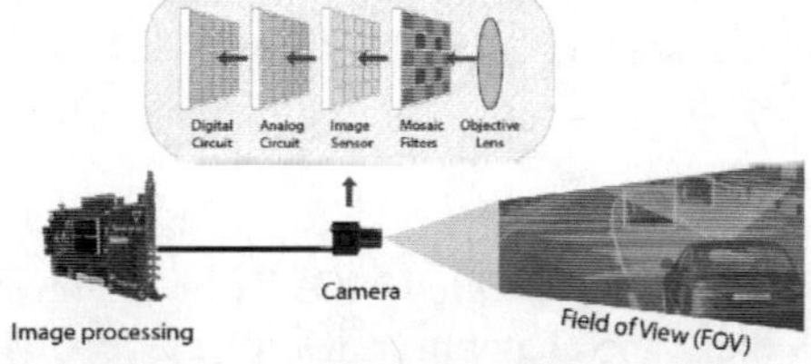


**Figure 6.** Principle of Vision System

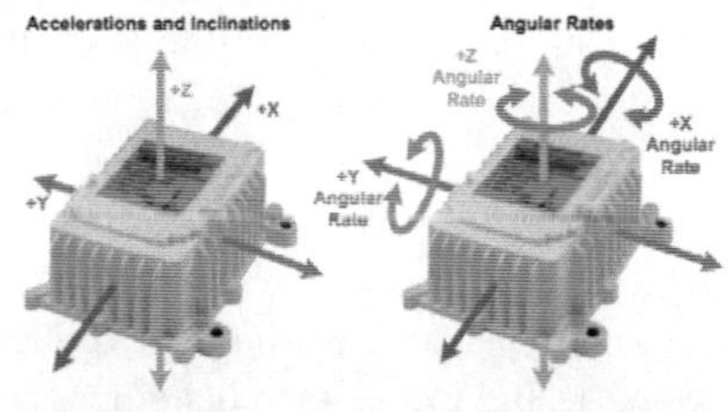


**Figure 7.** Inertial Measurement Unit

### *2.5. Inertial Measurement Unit (IMU)*

A moving vehicle can experience linear and rotational motions along x, y, and z-axes: lateral, longitudinal, and vertical which can be measured by inertial measurement unit including linear and angular accelerations. This information is used to improve GPS measurements. IMU includes accelerometers and gyroscopes as shown in Fig. 7.

### *2.6. Ultrasonic Range*

Ultrasonic sensors are short range sensors (typical $\approx$ 2 m) which send an ultrasonic pulse wave and detect echoes returned from the obstacles using transmitter-receiver pair. These are mainly used in relatively low speed ADAS modules like parking space detection and assistance [17] and obstacle detection during congested traffic conditions [18]. They are reliably detect under any weather conditions like rain, snow and winds. However they have range limitation which restricts them to be used as OEM vehicle sensors.

## 3. Sensor Fusion

In sensor fusion techniques, raw sensor data is received. After receiving the data, feature extraction, clustering and object detection hypotheses are conducted. These hypotheses are then associated with tracks showing state estimates of objects detected. The associated information is used by state estimators like Bayesian filters, to predict the current states [19]. The order of sensor measurements availability can be different from the order of raw data acquisition by the sensors. Buffering of information until all the data is available results in an unneeded dead-time which can degrade the performance of control system. To avoid performance degradation pseudo-measurements that are aligned in time or asynchronous tracking systems that employ every measurement upon availability can be used.

## 4. Embedded Software: Current Research and Development issues

Embedded software for control of the vehicle are codes written mainly to control both longitudinal and lateral maneuvers. This control algorithm in both includes a higher level of strategy control and a lower level of vehicle control. The strategy control involves decisions based on information from all vehicles affected by the maneuver and the infrastructure. The lower level vehicle control involves control of vehicle steering, throttle and brake systems.

### *4.1. Longitudinal Control*

Four types of information are necessary for longitudinal control: Speed and acceleration of host vehicle, the distance to preceding vehicle, the speed and acceleration of the preceding vehicle and in case of platoon speed and acceleration of the first vehicle. The speed and acceleration of the host vehicle can be measured by speed sensors and accelerometers (On-board OEM vehicle sensors). The distance to preceding vehicle can be measured using range sensors like LIDAR, vision, radar, and ultrasonic. Radar has been the most common range sensor for this case [20]. There are two ways to measure speed and acceleration of the preceding vehicle. One way is to derive it from the host vehicle and the measurement from the range sensors. Another way to obtain the speed and acceleration of the preceding vehicle is by communicating this information between the vehicles [21]. The same method can be used in platooning i.e. the speed and acceleration of the lead (first) vehicle of the platoon is transmitted to vehicles in the platoon. It should be noted that the communication reliability can not be completely trusted [22].

### *4.2. Lateral Control*

The strategic level evaluates the environment for lane change maneuver, like the presence of vehicles in the current and adjacent lane and their dynamics. A strategic level model called MOBIL - Minimizing Overall Braking Induced Lane changes, was proposed to deduct lane changing rules for any optional and mandatory lane changes for different car following models [23]. Different lane change trajectories (circular, the cosine approximation to the circular, the polynomial, and the trapezoidal acceleration) were studied, among them trapezoidal acceleration trajectory was the most desirable for best transition time and passenger's comfort [24]. Two different approaches are presented at the vehicle control level [25]. One approach is to treat the maneuvers as a tracking control problem, another approach uses the unified lateral guidance algorithm. In tracking control, a virtual desired trajectory is generated considering the lateral acceleration and jerk using a sliding mode controller. As for unified lateral guidance approach, a yaw rate generator generates the desired yaw rate for a desired maneuvers, either lane change or lane following maneuvers. Commands for steering angle are generated using a reference yaw rate signal and a yaw rate controller for the lane change [26].

## 5. Conclusions and Future Research

This paper presents a review of sensors and current advances in the control methodologies for autonomous vehicle applications. First, a short history of the ADAS was presented and major contributions were summarized. It was followed by highlighting the principles of major sensors used in the autonomous vehicles including LIDAR, GPS, radar, vision, ultrasonic and inertial measurement unit. Sensor fusion was reviewed and its importance in level-3 self-driving cars was highlighted. Finally, advances in the embedded control of autonomous cars were presented and classified as longitudinal and lateral control schemes. As advances in sensor technology, computing and communication takes place, more complex embedded control system will make its way into the autonomous vehicles. More integration of developmental tools, processing, sensing, connectivity, mapping, algorithms and security will be seen in the future driverless cars. Further research from this literature involves but not limited to testing the lane change maneuver in driverless cars using different control algorithms and simulation of ADAS features in various driving scenarios.

## References

[1] Litman T 2014 Autonomous Vehicle Implementation Predictions: Implications for Transport Planning Tech. Rep. 2014

[2] Beiker S 2014 *Road Vehicle Automation*

[3] Rosen D A, Mammano F J and Favout R 1970 *IEEE Transactions on Vehicular Technology* **19** 143–152

[4] Board T R 1998 *National Automated Highway System Research Program: A Review–Special Report 253* vol 253 (Transportation Research Board)

[5] Pomerleau D A 1989 Alvinn: An autonomous land vehicle in a neural network Tech. rep. DTIC Document

[6] Wolf P M 1974 *The future of the city: new directions in urban planning* (Watson-Guptill Publications; For the American Federation of Arts)

[7] Schmidhuber J 2012 *Available fro m: http://www. idsia. ch/˜ juergen/robotcars. html. Accessed February*

[8] Bengler K, Dietmayer K, Maurer M and Winner H 2014 6–22

[9] US Department of Transportation and Safety I R 2016

[10] Thakur R 2016 *IEEE Consumer Electronics Magazine* **5** 48–54 ISSN 21622256

[11] Nashashibi F and Bargeton A 2008 *Intelligent Vehicles Symposium, 2008 IEEE* (IEEE) pp 847–852

[12] Natale D J, Tutwiler R L, Baran M S and Durkin J R 2010 *Image Analysis & Interpretation (SSIAI), 2010 IEEE Southwest Symposium on* (IEEE) pp 21–24

[13] Cho H J and Tseng M T 2013 *Mathematical and Computer Modelling* **58** 438–448

[14] Blanc C, Aufrere R, Malaterre L, Gallice J and Alizon J 2004 *IAV 2004*

[15] Widmann G R, Daniels M K, Hamilton L, Humm L, Riley B, Schiffmann J K, Schnelker D E and Wishon W H 2000 ISSN 0148-7191 URL `http://papers.sae.org/2000-01-0345/`

[16] Misra P and Enge P 2006 *Global Positioning System: Signals, Measurements and Performance Second Edition* (Lincoln, MA: Ganga-Jamuna Press)

[17] Wan-Joo Park, Byung-Sung Kim, Dong-Eun Seo, Dong-Suk Kim, Kwae-Hi Lee, Park W J, Kim B S, Seo D E, Kim D S and Lee K H 2008 *IEEE Intelligent Vehicles Symposium, Proceedings* 1039–1044 ISSN 1931-0587 URL `http://ieeexplore.ieee.org/document/4621296/`

[18] Jiménez F, Naranjo J E, Gómez O and Anaya J J 2014 *Sensors (Switzerland)* **14** 22689–22705 ISSN 14248220

[19] Stiller C, Puente León F and Kruse M 2011 *Information Fusion* **12** 244–252 ISSN 15662535 URL `http://dx.doi.org/10.1016/j.inffus.2011.03.005`

[20] Meinel H H 1995 *IEEE Transactions on Microwave Theory and Techniques* **43** 1639–1653 ISSN 15579670

[21] Rajamani R and Shladover S 2001 *Transportation Research Part C: Emerging Technologies* **9** 15–31 ISSN 0968090X

[22] Rajamani R, Tan H S, Law B K and Zhang W B 2000 *IEEE Transactions on Control Systems Technology* **8** 695–708 ISSN 10636536

[23] Kesting A, Treiber M and Helbing D 2007 *Transportation Research Record: Journal of Transportation Research Board* **1999** 86–94 ISSN 0361-1981

[24] Chee W and Tomizuka M 1994 *California Partners for Advanced Transit and Highways (PATH)*

[25] Chee W 2000 *SAE TECHNICAL PAPER SERIES*

[26] Hatipoglu C, Ozguner U and Redmill K A 2003 *IEEE Transactions on Intelligent Transportation Systems* **4** 13–22 ISSN 15249050

IOP Conf. Series: Materials Science and Engineering **224** (2017) 012030 doi:10.1088/1757-899X/224/1/012030

# A Hierarchical Learning Control Framework for an Aerial Manipulation System

**Ma Le[1,a], Chi yanxun[1], Li Jiapeng[1], Li Zhongsheng[1], Ding Yalei[1] and Liu Lixing[1]**

[1] Robotics Institution, School of Automation Engineering, Northeast Electric Power University, China

[a] male_robot_nedu@sina.com

**Abstract.** A hierarchical learning control framework for an aerial manipulation system is proposed. Firstly, the mechanical design of aerial manipulation system is introduced and analyzed, and the kinematics and the dynamics based on Newton-Euler equation are modeled. Secondly, the framework of hierarchical learning for this system is presented, in which flight platform and manipulator are controlled by different controller respectively. The RBF (Radial Basis Function) neural networks are employed to estimate parameters and control. The Simulation and experiment demonstrate that the methods proposed effective and advanced.

## 1. Introduction

Recently, unmanned aerial vehicles (UAV) of multi-rotors (such as quadrotor, hexrotor) have been deeply investigated and applied in many fields due to their advantages with flexibility and efficiency[1]. The traditional UAVs were equipped with actuators, however, which can be applied in some passive tasks, such as aerial photo and supervise etc[2]. Aerial manipulators (AM), which consist of aircrafts of multi-rotors and manipulators, can make up this problem and can be used in many active tasks, including aerial assembling, rescue etc. Hence, there is the tendency from traditional UAVs to aerial manipulation system.

The design, modeling and control for AM are a series of challenging problems because of their complexities. These include the following profiles: 1) structural complexity. Aerial manipulator belongs to a kind of combined system, and there exist important impact to the whole system from the interactions between subsystems. Simultaneously, it is sensitive for aerial platforms to load, and hence the structer and parameters of manipulator equipped affect the loading capability and working flexibility of the system. 2) Non-holonomy and underactuation, the typical aerial platforms, such as quadrotor and octarotor etc, belong to non-holonomic system and are underactuated, which lead to the difficulty to control. 3) Dynamical interaction, the forces and torques derived from manipulator motion and interaction with environment make great impact to system stablity, some parameters values of which are hard to estimate, which is difficult to design of controllers. And the perturbances from interior and exterior impact the precision of manipulator motion. From the above, design and control of AM are challenging problems and hot research topics.

Recently, there have been various types of AMs, which cover different aerial platform including helicopter[3], quadrotor[4], hexrotor[5] etc, and different manipulators including single joint, multi-joint[6], delta[7] and two-arm[8] etc. In comparison to delta, cascade manipulators have larger work space and better flexibility. However, there exist problems of balance among flexibility, complexity and loading capability for the choosing degree of freedom (DoF)[9]. Now, the non-holonomic and underactuated system are used for aerial platform, which restricts the precisions and flexibilities of manipulators. In [10], a redundant arm of 7 DoF is used to boost system flexibility, enhancing the difficulty of cooperation between subsystems and reducing load simultaneously. In [11] and [12], two types of holonomic aerial platforms are presented. The core idea of these is that the thrust forces are converted from parallel to non-parallel.

For method to modeling for kinematics and dynamics of AMs are classical. In comparison to quaternion, Euler angle is more natural and easier to express in form of rotation matrix used regularly in robotics, which are used in AM kinematics more frequently. And New-Euler is more fit than Lagrange to dynamics modelling of cascade system.

In the last years, the main research work focus on trajectory of end-effector tracking[10] and interaction force control[13]. There have been many control methods for the problems as mentioned above, among which PID controller is used in wide range[14], but parameter adjustment of PID remains still to restrict adaptivity. To address to this problem backstepping controller have been proposed in aerial platform, however, which can arises "Differential Blast" in multi-dimensional situation, and are not fit to multivariable system as a consequence.

In this paper, a hierarchical learning control framework for AM is proposed. The system researched consists of a traditional quadrotor and a manipulator with 4 joints, which are controlled by different methods respectively. An estimation and learning control methods based on RBF neural networks are proposed.

## 2. Description of Mechanical Structure

To combine the power to volume ratio, the terminal load and the operational flexibility available, in this article, we choose the AM which use quadrotor and 4-DOF manipulation as our research object, see figure 1.

The research system in this paper consist of quadrotor's platform and 4-DOF manipulation, using the carbon fiber as body material of quadrotor, and using the technology of 3D printing to support design of manipulator. We use Dynamixel AX-12A digital steering servo motor. The core of controller of the quadrotor and manipulator are STM32-M4 types.

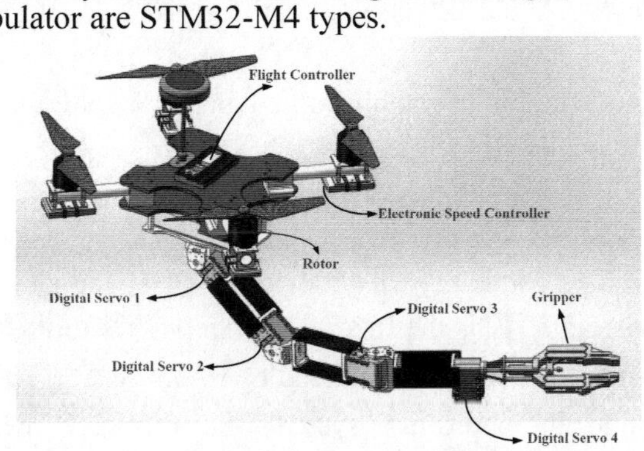


**Figure1.** The mechanical structure of AM researched

## 3 System Modeling

### *3.1 Kinematic Modeling*

The coordinate system of inertia for reference ,the inherent coordinate system of flight platform and the coordinate system of actuator which at the end of the mechanical arm are respectively expressed as $\{W\}$ , $\{B\}$ and $\{E\}$. If Euler angles are used to represent the posture, the positional matrix of $\{B\}$ with respect to that of $\{W\}$ could be expressed as (1):

$$ {}^{\mathrm{W}}_{\mathrm{B}}\mathbf{T} = \begin{pmatrix} \mathbf{R}_{xyz}(\psi,\varphi,\gamma) & {}^{\mathrm{W}}\mathbf{P}_{\mathrm{BORG}} \\ \mathbf{0} & 1 \end{pmatrix} \tag{1} $$

In this matrix, ${}^{\mathrm{W}}\mathbf{P}_{\mathrm{BORG}}$ is the position vector of the coordinate system origin of $\{B\}$ with respect to $\{W\}$. $\mathbf{R}_{xyz}$ is the corresponding Rotation matrix of x-y-z Euler angles .

The transformation matrix of $\{B\}$ relative to $\{W\}$ can be obtained by the standard D-H method:

$$ {}^{\mathrm{B}}_{\mathrm{E}}\mathbf{T} = \prod_{i=1}^{N} {}^{i-1}_{i}\mathbf{A}(\theta_i) \tag{2} $$

in which, $N$ is the joint number of the manipulator, ${}^{i-1}_{i}\mathbf{A}(\theta_i)$ is the transformation matrix of the coordinate system of the $i$ th joint relative to the coordinate system of the $i-1$ th joint (the 0th joint defined as the coordinate system of $\{B\}$ ), $\theta_i$ is the angle variables of the $i$ th joint. We can use the D-H parameter method to determine the transformational matrix of joint's coordinate system.

${}^{B}\mathbf{V}_{E}$, the linear velocity of the coordinate system of $\{E\}$ relative to $\{B\}$, has a kind of relationship with angular velocity $\dot{\mathbf{q}}$ of joints. It can be expressed as follows:

$$ {}^{B}\mathbf{V}_{E}={}^{\mathrm{B}}_{\mathrm{E}}\mathbf{J}_{\mathrm{V}}(\mathbf{q})\dot{\mathbf{q}} \tag{3}$$

where ${}^{\mathrm{B}}_{\mathrm{E}}\mathbf{J}_{\mathrm{V}}(\mathbf{q})$ is Jacobian matrix.

$\mathbf{Q}$ has a kind of velocity relationship with different coordinate systems ,it can be expressed as follows :

$$ {}^{\mathrm{A}}\mathbf{V}_{\mathbf{Q}}={}^{\mathrm{A}}\mathbf{V}_{\mathrm{B}}+{}^{\mathrm{A}}_{\mathrm{B}}\mathbf{R}\,{}^{\mathrm{B}}\mathbf{V}_{\mathrm{Q}}+{}^{\mathrm{A}}\Omega_{\mathbf{B}}\times{}^{\mathrm{A}}_{\mathrm{B}}\mathbf{R}\,{}^{\mathrm{B}}\mathbf{Q} \tag{4}$$

In this formula, ${}^{\mathrm{A}}\mathbf{V}_{\mathrm{B}}$ is the velocity of the origin of the $\{B\}$coordinate system under the $\{A\}$coordinate, and ${}^{\mathrm{A}}_{\mathrm{B}}\mathbf{R}$ is the Rotation matrix of coordinate systems which between $\{A\}$ and $\{B\}$, ${}^{\mathrm{B}}\mathbf{V}_{\mathbf{Q}}$ is the relative velocity of $\mathbf{Q}$ which under the coordinate system of $\{B\}$, ${}^{\mathrm{A}}\mathbf{\Omega}_{\mathrm{B}}$ is the angular velocity of $\{B\}$ which relative to the coordinate system of $\{A\}$, ${}^{\mathrm{B}}\mathbf{Q}$ is the position vector of $\mathbf{Q}$ which under the coordinate system of $\{B\}$ .

The postures and velocities of the end-effector relative to the aerial platform frame are determined according to the equation above.

*3.2 Dynamics Modeling*

We can use the Newton-Euler Equation to establish the dynamic equation model of the multi-propeller platform:

$$\begin{cases} m_{\mathrm{R}}\dot{\mathbf{V}}^{\mathrm{B}}+\mathbf{\Omega}^{\mathrm{B}}\times m\mathbf{V}^{\mathrm{B}}=\sum_{i=0}^{N}\mathbf{F}_i^{\mathrm{R}}+\mathbf{G}_{\mathrm{R}}^{\mathrm{B}}+\mathbf{F}_{\mathbf{A}}\left(\dot{\mathbf{V}}^{\mathrm{B}},\mathbf{V}^{\mathrm{B}},\boldsymbol{\theta},\dot{\boldsymbol{\theta}},\ddot{\boldsymbol{\theta}}\right) \\ \mathbf{I}_{\mathrm{R}}\dot{\mathbf{\Omega}}^{\mathrm{B}}+\mathbf{\Omega}^{\mathrm{B}}\times\mathbf{I}_{\mathrm{R}}\mathbf{\Omega}^{\mathrm{B}}=\sum_{i=0}^{N}\mathbf{r}_i\times\mathbf{F}_i^{\mathrm{R}}+\sum_{i=0}^{N}\boldsymbol{\tau}_i^{\mathrm{R}}+\mathbf{M}\left(\dot{\mathbf{\Omega}}^{\mathrm{B}},\mathbf{\Omega}^{\mathrm{B}},\boldsymbol{\theta},\dot{\boldsymbol{\theta}},\ddot{\boldsymbol{\theta}}\right) \end{cases} \tag{5}$$

where $m_{\mathrm{R}}$ is the quality of airframe. $\mathbf{I}_{\mathrm{R}}$ is the Inertial Tensor Matrix of airframe. $\mathbf{F}^{\mathrm{R}}$ is the propulsion, relative to the coordinate system of airframe, produced by rotors. $\mathbf{G}_{\mathrm{R}}^{\mathrm{B}}$ is used to express the gravity of the aerial platform which under the coordinate system of airframe. $\mathbf{F}_{\mathbf{A}}$ is used to express the force, from the manipulation to the aerial platform ,under the coordinate system of airframe. Meanwhile, $\mathbf{F}_{\mathbf{A}}$ is the functional vector based on the variables of joint angle, angular velocities and angular accelerations. $\mathbf{r}$ is the vector belong to the effect point of lift which relative to the origin of coordinate system. $\tau^{\mathrm{R}}$ is the reverse torque under the coordinate system of airframe which produced by rotors. $\mathbf{M}$ is the torque under the coordinate system of airframe which manipulation send to aerial platform .

Let $\omega$ be the velocity of rotor. $\mathbf{F}^{\mathrm{R}}$ and $\tau^{\mathrm{R}}$ are all proportional to the square of $\omega$ [15] :

$$\begin{cases} f^{\mathrm{R}}=k_{\mathrm{tf}}\omega^2 \\ \tau^{\mathrm{R}}=k_{\mathrm{ct}}\omega^2 \end{cases} \tag{6}$$

where, $k_{\mathrm{tf}}$ is the coefficient of lift and $k_{\mathrm{ct}}$ is the coefficient of torque.

We can use the recursive Newton-Euler method to establish the dynamic model of manipulation [16].

$$\begin{cases} {}^{i+1}\boldsymbol{\omega}_{i+1}={}^{i+1}_{i}\mathbf{R}\,{}^{i}\boldsymbol{\omega}_i+\dot{\theta}_{i+1}\,{}^{i+1}\hat{\mathbf{Z}}_{i+1} \\ {}^{i+1}\dot{\boldsymbol{\omega}}_{i+1}={}^{i+1}_{i}\mathbf{R}\,{}^{i}\dot{\boldsymbol{\omega}}_i+{}^{i+1}_{i}\mathbf{R}\,{}^{i}\boldsymbol{\omega}_i\times\dot{\theta}_{i+1}\,{}^{i+1}\hat{\mathbf{Z}}_{i+1}+\ddot{\theta}_{i+1}\,{}^{i+1}\hat{\mathbf{Z}}_{i+1} \\ {}^{i+1}\dot{\mathbf{v}}_{i+1}={}^{i+1}_{i}\mathbf{R}\left({}^{i}\dot{\boldsymbol{\omega}}_i\times{}^{i}\mathbf{P}_{i+1}+{}^{i}\boldsymbol{\omega}_i\times({}^{i}\boldsymbol{\omega}_i\times{}^{i}\mathbf{P}_{i+1})+{}^{i}\dot{\mathbf{v}}_i\right) \\ {}^{i+1}\dot{\mathbf{v}}^{\mathrm{c}}_{i+1}={}^{i+1}\dot{\boldsymbol{\omega}}_{i+1}\times{}^{i}\mathbf{P}^{\mathrm{c}}_{i+1}+{}^{i+1}\boldsymbol{\omega}_{i+1}\times\left({}^{i+1}\boldsymbol{\omega}_{i+1}\times{}^{i}\mathbf{P}_{i+1}\right)+{}^{i+1}\dot{\mathbf{v}}_{i+1} \\ {}^{i+1}\mathbf{F}_{i+1}=m_{i+1}\,{}^{i+1}\dot{\mathbf{v}}^{\mathrm{c}}_{i+1} \\ {}^{i+1}\boldsymbol{\tau}_{i+1}={}^{i+1}\mathbf{I}^{\mathrm{c}}_{i+1}\,{}^{i+1}\dot{\boldsymbol{\omega}}_{i+1}+{}^{i+1}\boldsymbol{\omega}_{i+1}\times{}^{i+1}\mathbf{I}^{\mathrm{c}}_{i+1}\,{}^{i+1}\boldsymbol{\omega}_{i+1} \end{cases} \tag{7}$$

IOP Conf. Series: Materials Science and Engineering **224** (2017) 012030 doi:10.1088/1757-899X/224/1/012030

$$\begin{cases} {}^{i}f_i = {}^{i}_{i+1}\mathbf{R}^{i+1} f_{i+1} + {}^{i}\mathbf{F}_i \\ {}^{i}\mathbf{n}_i = {}^{i}\boldsymbol{\tau}_i + {}^{i}_{i+1}\mathbf{R}^{i+1}\mathbf{n}_{i+1} + {}^{i}\mathbf{P}^{C}_{i} \times {}^{i}\mathbf{F}_i + {}^{i}\mathbf{P}_{i+1} \times {}^{i}_{i+1}\mathbf{R}^{i+1} f_{i+1} \end{cases} \tag{8}$$

where $\boldsymbol{\omega}$ and $\dot{\boldsymbol{\omega}}$ are angular velocity and acceleration in body frame. $\dot{v}^c$ is the acceleration of center of mass in body frame. $\mathbf{F}$ and $\boldsymbol{\tau}$ are the resultant force and torque in body frame.

## 4. Hierarchical Learning Control

Aerial platform and manipulator are controlled by different controllers respectively. The learning algorithms are used to estimate parameters and build controllers.

Dynamic characteristics of aerial platform can be affected by manipulation operation except for the rotor rotation. Due to the specified tasks of manipulator and its limited degrees of freedom, it's difficult for manipulator to offer control output under the condition of executing task trajectory. Thus the torque of the manipulator to the aerial platform is regarded as disturbance and the platform position is controlled by the rotor output in the paper.

The $N$ rotors generate propulsion force and the unit direction vector of the propulsion force relative to the body coordinate system constitutes a matrix and the matrix is denoted by $\mathbf{e}^{\mathrm{F}} \in \Re^{3\times N}$. The magnitude of the thrust force is approximately proportional to the rotor speed $\omega$ and the relationship is shown as:

$$f = k_{\mathrm{tf}}\omega^2 \tag{9}$$

The propulsion force generated by the $i$ th rotor relative to the body coordinate system can be calculated:

$$\mathbf{F}_i^{\mathrm{b}} = \mathbf{e}_i^{\mathrm{F}} k_{\mathrm{tf}} \boldsymbol{\omega}_i^2 \tag{10}$$

Relative to the body coordinate system, the resultant of the propulsive forces can be obtained:

$$\mathbf{F}^{\mathrm{b}} = \sum_{i=1}^{N} \mathbf{F}_i = \sum_{i=1}^{N} \mathbf{e}_i^{\mathrm{F}} k_{\mathrm{tf}} \boldsymbol{\omega}_i^2 = k_{\mathrm{tf}} \mathbf{e}^{\mathrm{F}} \boldsymbol{\omega} \tag{11}$$

where $\boldsymbol{\omega} = \left[\omega_1^2, \omega_2^2, \ldots\ldots, \omega_N^2\right]^{\mathrm{T}}$.

Let the force point vector of the $i$ th rotor relative to the body coordinate system is represented by $\mathbf{r}_i$, the torques of rotors $\tau$ relative to the body coordinate system are divided into two parts. The first parts are derived from thrust forces. The second parts are the counter torques produced by rotation of the rotor. Thus their relational expression can be represented:

$$\boldsymbol{\tau}_i = \mathbf{r}_i \times \mathbf{F}_i + (-1)^{i-1} k_{\mathrm{ct}} \mathbf{e}_i^{\mathrm{F}} \omega_i^2 \tag{12}$$

On the basis of the equation (2), (12) can be rewritten as:

$$\boldsymbol{\tau}_i = \mathbf{r}_i \times \mathbf{F}_i + k_{\mathrm{ct}} \mathbf{e}_i^{\mathrm{F}} \omega_i^2 = \mathbf{S}_\tau^i(\mathbf{r}_i, k_{\mathrm{tf}}, k_{\mathrm{ct}}) \mathbf{e}_i^{\mathrm{F}} \omega_i^2 \tag{13}$$

where

$$\mathbf{S}_\tau^i(\mathbf{r}, k_{\mathrm{tf}}, k_{\mathrm{ct}}) = \begin{bmatrix} -1^{i-1}k_{\mathrm{ct}} & -k_{\mathrm{tf}}\mathbf{r}_{\mathrm{z}} & k_{\mathrm{tf}}\mathbf{r}_{\mathrm{y}} \\ k_{\mathrm{tf}}\mathbf{r}_{\mathrm{z}} & -1^{i-1}k_{\mathrm{ct}} & -k_{\mathrm{tf}}\mathbf{r}_{\mathrm{x}} \\ -k_{\mathrm{tf}}\mathbf{r}_{\mathrm{y}} & k_{\mathrm{tf}}\mathbf{r}_{\mathrm{x}} & -1^{i-1}k_{\mathrm{ct}} \end{bmatrix} \tag{14}$$

The torque generated by the rotor in the body coordinate system can be expressed:

$$\boldsymbol{\tau} = \mathbf{M}_\tau \boldsymbol{\omega} \tag{15}$$

in which $m_\tau^i = \mathbf{S}_\tau(\mathbf{r}_i, k_{\mathrm{tf}}, k_{\mathrm{tc}}) \mathbf{e}_i^{\mathrm{F}}$.

According to the equation (2) and the formula (4), the generalized force generated by the input speed of the rotor can be obtained:

$$\begin{bmatrix} \mathbf{F}^{\mathrm{b}} \\ \tau^{\mathrm{b}} \end{bmatrix} = \begin{bmatrix} k_{\mathrm{tf}}\mathbf{e}^{\mathrm{F}} \\ \mathbf{M}_\tau \end{bmatrix} \boldsymbol{\omega} \tag{16}$$

Seen as (16), the transform matrix of the rotor speed $\boldsymbol{\omega}$ to the generalized force is the singular matrix, and therefore this system is not decoupled completely.

In response to the coupling of system and the limit of rotors' lift ,We can use behavioral scheduling technique to dispose the captive sub-goal by specific behaviors which handled in a single control cycle .

We hypothesize that the current Euler angle is $\xi^{C}$ and the targeted Euler angle is $\xi^{T}$, so the corresponding rotational matrix are called ${}^{C}\mathbf{R}$ and ${}^{T}\mathbf{R}$.

According to the transformation of coordinates system, we can get the transformational matrix which used to express how the current coordinate system $\{C\}$ transforms to the desired coordinate system $\{T\}$:

$$ {}^{C}_{T}\mathbf{R} = {}^{C}\mathbf{R}^{-1}\,{}^{T}\mathbf{R} \tag{17} $$

The rotational matrix could be expressed in exponential form as follows:

$$ \mathbf{R} = \mathbf{I}_{3\times3} + \hat{\mathbf{w}}\sin\theta + \hat{\mathbf{w}}^{2}(1-\cos\theta) \tag{18} $$

where $\hat{\mathbf{w}} = \mathbf{S}_{w}(\mathbf{w})$ is used to express the matrix which attached by taking the unit vector $\mathbf{w}$ as the axis and rotating the angle of $\theta$. So we can express the deviation of posture control by the rotation angle ${}^{C}_{T}\theta$ and the equivalent spin axis ${}^{C}_{T}\mathbf{w}$ which is corresponding to ${}^{C}_{T}\mathbf{R}$. And we can the following:

$$ \begin{cases} {}^{C}_{T}\theta = \cos^{-1}\left(\dfrac{tr\left({}^{C}_{T}\mathbf{R}\right)-1}{2}\right) \\ {}^{C}_{T}\mathbf{w} = \dfrac{1}{2\sin {}^{C}_{T}\theta}\begin{pmatrix} {}^{C}_{T}r_{32} - {}^{C}_{T}r_{23} \\ {}^{C}_{T}r_{13} - {}^{C}_{T}r_{31} \\ {}^{C}_{T}r_{21} - {}^{C}_{T}r_{12} \end{pmatrix} \end{cases} \tag{19} $$

According to the equation (5), we can use $\tau^{b}$ as the input of posture control. From the equation (19), we can know that the velocity of rotor can be controlled by $\tau_{x}^{b}$ and $\tau_{y}^{b}$ independently. The other control tasks of aerial platforms are all at the basis of the pitch-roll control task.

The RBF neural networks with Gaussian kernel functions are used to estimate parameters and to control.

## 5. Simulations and Experiments

A pitch-roll control task, with stochastic perturbation whose magnitudes are distributed in 0-1.8Nm, is simulated to compare the performances of three methods, including a good parameters of PID, PID with states estimation (PID-Est), and learning control with estimation (LC-Est). The simulation results are shown in figure 2, which show that learning control with estimation has advantages in dynamic and static state performances over other methods.

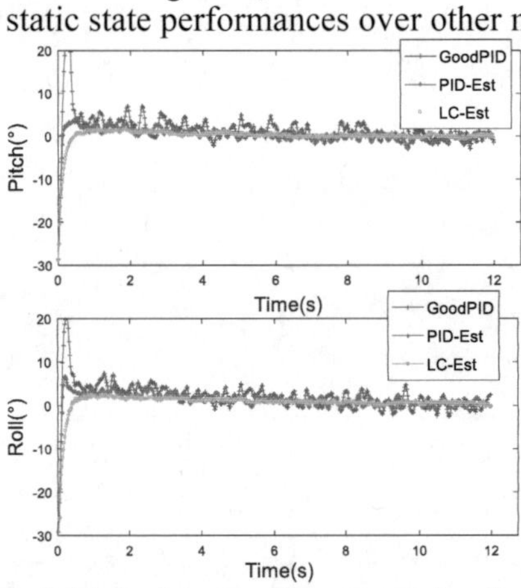


**Figure2.** Methods comparison for pitch-roll control with perturbation

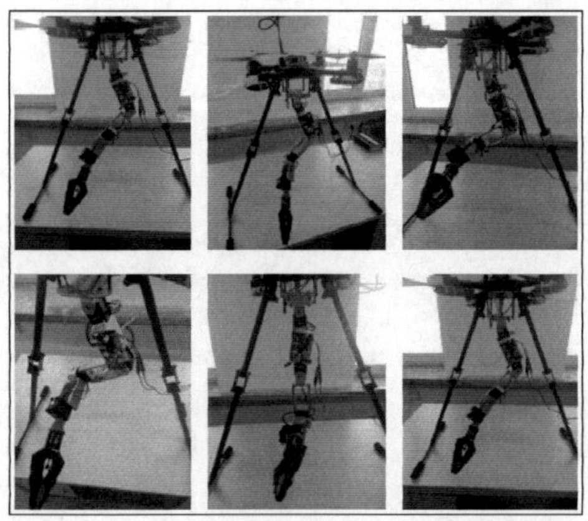

**Figure3.** Testing of manipulator

We make indoor experiment to test the performance of manipulator, shown as figure 3, in which it demonstrated that manipulator can run in stable when switch states of configuration.

## 6. Conclusions

In this paper, a control framework of the aerial manipulation system is proposed based on learning. On account of the analysis for the controlled system, we propose a set of parameter estimation and learning control methods based on RBF network. Simulation and experimental results demonstrate

IOP Conf. Series: Materials Science and Engineering **224** (2017) 012030 doi:10.1088/1757-899X/224/1/012030

that the control framework proposed can boost the performance of the AM effectively, in comparison to PID control.

## References

[1] Mahony R, Kumar V, Corke P. Multirotor Aerial Vehicles Modeling, Estimation, and Control of Quadrotor. IEEE Robotics and Automation Magazine. 2012, 19(3): 20-32.

[2] Zhang X, Li X, Kangwang, et al. A Survey of Modelling and Identification of Quadrotor Robot. Abstract and Applied Analysis. 2014: 1-16.

[3] Song Dalei, Meng Xiangdong,QI Juntong. Strategy of Dynamic Modeling and Predictive Control on 3-DoF Rotorcraft Aerial Manipulator System. Jiqiren. 2015, 37(2): 152-160.

[4] Operating an Unknown Drawer Using an Aerial Manipulator. 2015.

[5] Attitude Stabilization Control of an Aerial Manipulator using a Quaternion-Based Backstepping Approach. 2015.

[6] Design, Modeling and Control of a 5-DoF Light-Weight Robot Arm for Aerial Manipulation. 2015.

[7] A Parallel Manipulator for Mobile Manipulating UAVs. 2015.

[8] Towards Valve Turning using a Dual-Arm Aerial Manipulator. 2014.

[9] Evaluation of Visual Servoing Control of Aerial Manipulators Using Test Gantry Emulation. 2014.

[10] Integral Action in First-Order Closed-Loop Inverse Kinematics Application to Aerial Manipulators. 2015.

[11] Hexrotor UAV Platform Enabling Dextrous Interaction with Structures-Flight Test. 2013.

[12] Modeling, Control and Design Optimization for a Fully-actuated Hexarotor Aerial Vehicle with Tilted Propellers. 2015.

[13] Giglio G, Pierri F. Selective compliance control for an unmanned aerial vehicle with a robotic arm. Palermo, Italy: 2014.

[14] Stability of Helicopters in Compliant Contact Under PD-PID Control. IEEE Transactions On Robotics. 2014, 30(6): 1472-1486.

[15] Nikou A, Gavridis G C, Kyriakopoulos K J. Mechanical Design, Modelling and Control of a Novel Aerial Manipulator. 2015.

[16] Siciliano B, Khatib O. Handbook of Robotics. 2ed. Springer, 2007

IOP Conf. Series: Materials Science and Engineering **224** (2017) 012031 doi:10.1088/1757-899X/224/1/012031

# Research on Fault Rate Prediction Method of T/R Component

**Xiaodong Hou [1,a], Jiangping Yang [1], Zengjun Bi [1] and Yu Zhang [2]**

[1]Dept. of Land-based Early Warning Equipment, Air Force Early Warning Academy, 430019 Wuhan, China

[2]College of Electrical and Electronic Engineering, Hubei University of Technology, 430068 Wuhan, China

[a] hxdscylla@163.com

**Abstract.** T/R component is an important part of the large phased array radar antenna array, because of its large numbers, high fault rate, it has important significance for fault prediction. Aiming at the problems of traditional grey model GM(1,1) in practical operation, the discrete grey model is established based on the original model in this paper, and the optimization factor is introduced to optimize the background value, and the linear form of the prediction model is added, the improved discrete grey model of linear regression is proposed, finally, an example is simulated and compared with other models. The results show that the method proposed in this paper has higher accuracy and the solution is simple and the application scope is more extensive.

## 1. Introduction

Large phased array radar is the main equipment of china's strategic early warning system, which plays an important role in strategic air defense and antimissile warning. Compared with conventional radars, Large phased array radar has strong adaptability of the targets, more achieved functions, short reaction time and high system reliability, more advantage in the target search, tracking and measurement. Due to the use of phased array technology system, and the antenna array is very large, multicomponent and maintenance timeliness requirements, equipped with a small number, high fault rate, its fault diagnosis and isolation are very difficult, the spare parts cannot satisfy the actual demand of the troops. Therefore, the fault prediction for large phased array radar antenna array can predict the time and quantity of the fault in advantage, based on fault rate prediction, the troops can prepare required manpower and material resources in advantage, improve its operational readiness and combat effectiveness, fault prediction has important practical significance for equipment support[1-2].

At present, many scholars have done a lot of researches on fault prediction[3-5], the main methods of fault prediction are:(1) Neural network prediction, the drawbacks are that neural network training requires a lot of data, the convergence speed is very slow, network structure is difficult to determine. (2) Gray prediction, the disadvantages are the prediction accuracy and stability are not high. (3) Particle filter, the disadvantages are particle degradation and the traceability of mutation status is poor. (4) Regression prediction, the disadvantages are requiring a large number of samples and having a good regularities of distribution.

Due to the highly unstable fault rate of T/R component, this paper analyzes and researches the original monitoring data, establishes a discrete grey model, and improves GM(1,1) model by ameliorating the background value and introducing linear regression. The new model overcomes the defects of forecast sequence singleness in the grey model and improves the accuracy of the model, good prediction results

IOP Conf. Series: Materials Science and Engineering **224** (2017) 012031 doi:10.1088/1757-899X/224/1/012031

are obtained in practical application.

## 2. Traditional GM(1,1) model

The grey theory was putted forward by Professor Deng Julong of Huazhong University of Science and Technology at the International Economics Conference in 1982, and was widely used and studied. GM(1,1) model is focuses on the study of establishing differential equations through sequential accumulation and differential fitting to predicting the future trend[6-7].

The main steps of traditional GM(1,1) model prediction are:

(1) Incremental operation of original data.

Set $X^{(0)}$ is non-negative original sequence,

$$X^{(0)} = \left\{x^{(0)}(1), x^{(0)}(2), \cdots, x^{(0)}(n)\right\} \quad (1)$$

A cumulative generated sequence is :

$$X^{(1)} = \left\{x^{(1)}(1), x^{(1)}(2), \cdots, x^{(1)}(n)\right\} \quad (2)$$

Which, $x^{(1)}(t) = \sum_{i=1}^{k} x^{(0)}(i) \quad (t = 1, 2, \cdots, n)$.

(2)The establishment of grey differential equation.

About $x^{(1)}(t)$,the differential equations are :

$$\frac{dx^{(1)}}{dt} + ax^{(1)}(t) = b \quad (3)$$

In the unit time :

$$\frac{dx^{(1)}}{dt} = x^{(1)}(t+1) - x^{(1)}(t) = x^{(0)}(t) \quad (4)$$

Obtain:

$$x^{(0)}(t) + ax^{(1)}(t) = b \quad (5)$$

(3)Generate background values using closely averaging method.

$$Z^{(1)} = (z^{(1)}(2), z^{(1)}(3), \cdots, z^{(1)}(n)) \quad (6)$$

In formula above,

$$z^{(1)}(t) = 0.5(x^{(1)}(t) + x^{(1)}(t-1)) \quad (7)$$

Obtain:

$$x^{(0)}(t) + az^{(1)}(t) = b \quad (8)$$

In the formula,$a$ is development coefficient,$b$ is ash content. $Z^{(1)}$ is the generating sequence of $X^{(1)}$ using closely averaging method..

(4)Calculate the model.

Estimated by least square method:

$$\begin{bmatrix} a \\ b \end{bmatrix} = (B^T B)^{-1} B^T Y \quad (9)$$

In the formula,

$$B = \begin{bmatrix} -z^{(1)}(2) & 1 \\ -z^{(1)}(3) & 1 \\ \vdots & \vdots \\ -z^{(1)}(n) & 1 \end{bmatrix}; \quad Y = \begin{bmatrix} x^{(0)}(2) \\ x^{(0)}(3) \\ \vdots \\ x^{(0)}(n) \end{bmatrix}$$

Obtain the value of $a$ and $b$,into the formula(3):

$$\hat{x}^{(1)}(t+1) = \left[x^{(0)}(1) - \frac{b}{a}\right] e^{-at} + \frac{b}{a}, \quad (t = 1, 2, 3, \cdots, n-1) \quad (10)$$

(5)Obtain prediction model.

$$\hat{x}^{(0)}(t+1)=\hat{x}^{(1)}(t+1)-\hat{x}^{(1)}(t) \tag{11}$$

Residual error is(absolute error) $\varepsilon^{(0)}=x^{(0)}-\hat{x}^{(0)}$ ,relative error is $E^{(0)}=\varepsilon^{(0)}/x^{(0)}\times 1000$ .

## 3. The problems of traditional GM(1,1) model in prediction

The time distribution curve of the T/R component fault rate of a phased array radar is obtained by state monitoring as shown in Fig. 1.

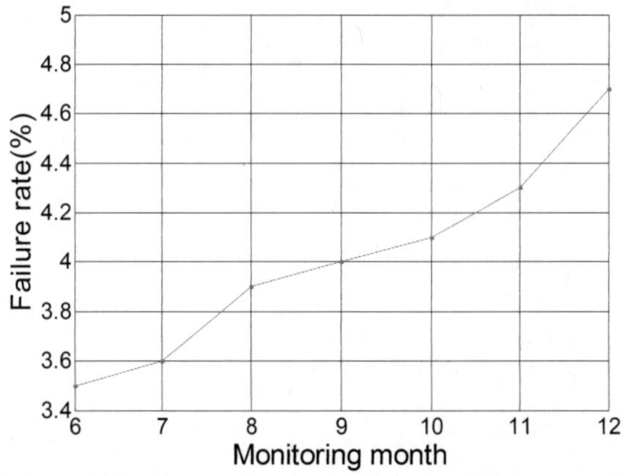


**Figure 1.** fault rate curve in the second half of 2015

Through the analysis of Fig.1, the data basically obey exponential distribution, but there is obvious volatility, the fault rate increases with time. Problems using traditional GM(1,1) model to predict are:

(1) Traditional GM(1,1) model in prediction of functions obey exponential distribution has certain advantages, while in practical application, through analyzing the function curve of fault rate, the most of data collected are undulatory, do not accord with the strict exponential function, there is a large error in prediction of exponential growth sequence using GM(1,1) model.

(2) In the traditional GM(1,1) model, the method of constructing of background value is closely mean value, such as formula (7), and the prediction accuracy of the model lies in the construction of $z^{(1)}(t)$, the different data selected in the same sequence makes the values of $z^{(1)}(t)$ different. For the accuracy of the model, the method of selecting closely mean value in the structure of $z^{(1)}(t)$ is inappropriate in some cases.

(3) In different stages, the fault rate curve of equipment has different forms, which basically obey exponential distribution and linear distribution[8], according to less fault rate of the original data, use GM(1,1) model to predict, but GM(1,1) model is used to simulate the data of exponential function, generally applies only to the exponential change, it is difficult to describe the change of other forms of sequence data. For the data with linear variation, the GM(1,1) model not only cannot obtain better prediction results, but also has a large prediction error.

## 4. Improved GM(1,1) model

Through the above analysis combined with the actual situation of equipment fault prediction, in order to obtain better prediction results, the traditional GM(1,1) model is improved from the following aspects[9-13].

### *4.1 Dispersed GM(1,1) model*

The dispersed GM(1,1) model is :

$$x^{(1)}(t+1)=\beta_1 x^{(1)}(t)+\beta_2 \tag{12}$$

In the formula, the estimation formula of $\hat{\beta}=[\beta_1, \beta_2]^T$ and formula(9) are similar, that is $\hat{\beta}=[\beta_1, \beta_2]^T=(B^T B)^{-1}B^T Y$ .Which,

$$B=\begin{bmatrix}-z^{(1)}(2) & 1\\ -z^{(1)}(3) & 1\\ \vdots & \vdots \\ -z^{(1)}(n) & 1\end{bmatrix};\quad Y=\begin{bmatrix}x^{(0)}(2)\\ x^{(0)}(3)\\ \vdots \\ x^{(0)}(n)\end{bmatrix}$$

According to the formula of $x^{(1)}(1)=x^{(0)}(1)$, the recursive function of the discrete GM(1,1) model is :

$$\hat{x}^{(1)}(t+1)=[x^{(0)}(1)-\frac{\beta_2}{1-\beta_1}]\beta_1^t+\frac{\beta_2}{1-\beta_1},\quad t=1,2,\cdots,n-1 \tag{13}$$

Obtain the estimated value of $\hat{x}^{(1)}(t+1)$ through putting the value of $\beta_1$ and $\beta_2$ into the formula (13).

Obtain predicted value of the improved GM(1,1) model by subtracting consecutively and deoxidizing :

$$\hat{x}^{(0)}(t+1)=\hat{x}^{(1)}(t+1)-\hat{x}^{(1)}(t),\quad t=1,2,\cdots,n-1 \tag{14}$$

*4.2 Background value optimization*

Restructure the formula (7)by introducing optimization factor $w$ ,that is

$$z^{(1)}(t)=wx^{(1)}(t)+(1-w)x^{(1)}(t-1),\quad t=2,3,\cdots,n \tag{15}$$

The $w$ is coefficient of background value optimization $(0<w<1)$,formula (7) is the special case when $w=0.5$.The value of $w$ is when the average relative error of the original value subtracts the predicted value reaches the minimum,here the average relative error is :

$$\overline{\Delta}=\frac{1}{n}\sum_{k=1}^{n}\frac{\left|x^{(0)}(t)-\hat{x}^{(0)}(t)\right|}{x^{(0)}(t)}=\min \tag{16}$$

*4.3 Grey linear regression forecasting model*

Fit formula (10)with the linear regression equation,obtain:

$$\hat{x}^{(1)}(t)=C_1e^{vt}+C_2t+C_3 \tag{17}$$

Here,$v$ and $C_1, C_2, C_3$ are parameters to be determined.

The model can be used to various forms of generate sequences, when $C_1=0$ ,the model can be used for linear growth sequences.When $C_2=0$ ,the model is the ordinary GM(1,1) model.In following steps, the parameters in formula (17) are calculated :

Order $z(t)=\hat{x}^{(1)}(t+1)-\hat{x}^{(1)}(t)$,obtain by formula (17) :

$$z(t)=C_1\left(e^{v}-1\right)e^{vt}+C_2,\quad t=1,2,\cdots,n-1 \tag{18}$$

Order $y_m(t)=z(t+m)-z(t)$,obtain by formula (18) :

$$y_m(t)=C_1e^{vt}\left(e^{v}-1\right)\left(e^{vm}-1\right),\quad m=1,2,\cdots,n-3;n=1,2,\cdots,n-m-2 \tag{19}$$

Obtain by formula(19): $y_m(t+1)/y_m(t)=e^v$ ,so the value of $v$ is:

$$v=\ln\left[y_m(t+1)/y_m(t)\right] \tag{20}$$

In formula(17),change $x^{(1)}(t)$ to $\hat{x}^{(1)}(t)$,the approximate value $v_m$ of $v$ can be obtained ,different $m$ values will obtain different $v_m$,in order to minimize the error,thus using their average $\overline{v}_m$ as the value of the last $v$.

$$\overline{v}_m=\ln\left[y_m(t+1)/y_m(t)\right] \tag{21}$$

Among it, $m=\frac{(n-2)(n-3)}{2}$ ,obtain

$$\overline{v}_m = \frac{\sum_{m=1}^{n-3}\sum_{t=1}^{n-m-2} v_m(t)}{(n-2)(n-3)/2} \tag{22}$$

Order $l(t)=e^{vt}$ ,then formula(17)is expressed as:

$$\hat{x}^{(1)}(t) = C_1 l(t) + C_2 t + C_3 \tag{23}$$

The estimated value of parameter $C_1, C_2, C_3$ can be obtained by least square method.Order :

$$X^{(1)} = \begin{bmatrix} x^{(1)}(1) \\ x^{(1)}(2) \\ \vdots \\ x^{(1)}(n) \end{bmatrix}, A = \begin{bmatrix} l(1) & 1 & 1 \\ l(2) & 2 & 1 \\ \vdots & \vdots & \vdots \\ l(n) & n & 1 \end{bmatrix}, C = \begin{bmatrix} C_1 \\ C_2 \\ C_3 \end{bmatrix}$$

Obtain :

$$C = \left(A^T A\right)^{-1} A^T X^{(1)} \tag{24}$$

Calculate the value of $v, C_1, C_2, C_3$,into formula(17),obtain the predicted value of the primary data order by subtracting consecutively and reduction:

$$\hat{x}^{(0)}(t) = \hat{x}^{(1)}(t) - \hat{x}^{(1)}(t-1) \tag{25}$$

The relative average error $\Delta$ of the model is :

$$\Delta = \frac{1}{n}\sum_{t=1}^{n}\Delta(t) = \frac{1}{n}\sum_{t=1}^{n}\left(\hat{x}^{(0)}(t) - x^{(0)}(t)\right) \tag{26}$$

*4.4 Computational procedures*

On the basis of previous analysis,the application steps of improved GM(1,1) model are:

**Step 1** Select $n$ data of the state monitoring data to establish initial data $X^{(0)}$,obtain order $X^{(1)}$ by accumulating procession.

**Step 2** Construct background value function $z^{(1)}(t)$ of discrete grey model by formula(15),obtain the optimal factor $w$ to satisfying the minimum value of $\overline{\Delta}$.

**Step 3** By using the obtained value of $B$ ,obtain the value of $\hat{\beta}=[\beta_1, \beta_2]^T$ using least square method,put into formula(13), obtain the predicted value of $\hat{x}^{(1)}(t+1)$.

**Step 4** Structure grey linear regression model,by formula (22)(23)(24), calculate the parameters $v, C_1, C_2, C_3$,obtain $\hat{x}^{(1)}(t)$.

**Step 5** Obtain the predicted value $\hat{x}^{(0)}(t)$ of the model through subtract consecutively reduction finally.

## 5. Example analysis

T/R component is an important part of large phased array radar antenna array, which mainly completes the functions of transmitting power amplification, transmitting receives isolation and receiving low noise amplification. Account for more than 80% of the total equipment , due to the huge numbers and high fault rate, its BITE coverage reached 100%,the state monitoring of the phased array antenna array T/R fault rate[14-15] in the second half of 2015, as show in Tab.1.

**Table 1.** T/R component fault rate statistics

| Month($t$) | 6 | 7 | 8 | 9 | 10 | 11 | 12 |
|---|---|---|---|---|---|---|---|
| Fault rate $\lambda(t)$ | 3.5 | 3.6 | 3.9 | 4.0 | 4.1 | 4.3 | 4.7 |

As can be seen from the data in Tab.1, the fault rate of T/R components shows an increasing trend with the increase of time, but it exists a greater volatility, according to the calculation steps of the improved GM(1,1) model provided in this paper, combined with MATLAB simulation, the fault rate of T/R components is predicted, and the prediction results are obtained. In order to show the superiority of the improved GM(1,1) model, respectively, compared with linear regression model and GM(1,1) model, the predicted and relative error results are shown in Tab.2.

**Table 2.** Comparisons of prediction results of different models

| Month | Actual value | Linear regression model | | GM(1,1) model | | Proposed method | |
|---|---|---|---|---|---|---|---|
| | | Predicted value | Average error% | Predicted value | Average error % | Predicted value | Average error % |
| 7 | 3.6 | 3.51 | 2.5 | 3.41 | 5.28 | 3.59 | 0.28 |
| 8 | 3.9 | 3.80 | 2.56 | 3.69 | 5.38 | 3.85 | 1.28 |
| 9 | 4.0 | 4.08 | 2.0 | 3.99 | 0.25 | 3.97 | 0.75 |
| 10 | 4.1 | 4.37 | 6.6 | 4.31 | 5.1 | 4.14 | 1.0 |
| 11 | 4.3 | 4.65 | 8.1 | 4.67 | 8.6 | 4.43 | 3.0 |
| 12 | 4.7 | 4.94 | 5.1 | 5.04 | 7.2 | 4.74 | 0.9 |
| Average relative error % | | 4.58 | | 4.11 | | 1.03 | |

The analysis of Tab.2 can be concluded that although three methods can predict the fault rate of the T/R component, but through comparisons and analysis of the results, the average relative error from the point of view, the improved GM(1,1) model of the minimum error, the prediction accuracy compared with linear regression model, increased by 77.5%, and GM(1,1) model, the prediction accuracy is improved by 74.9%.in order to show the accuracy of three methods more clearly, the comparison curves of predicted and actual values are given respectively, and the relative error curves of the three methods are shown in Fig.2 and Fig.3.

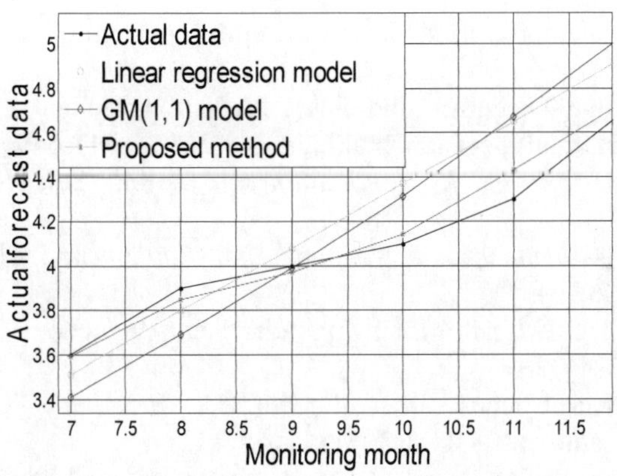


**Figure 2.** Comparison curve of predicted and actual values

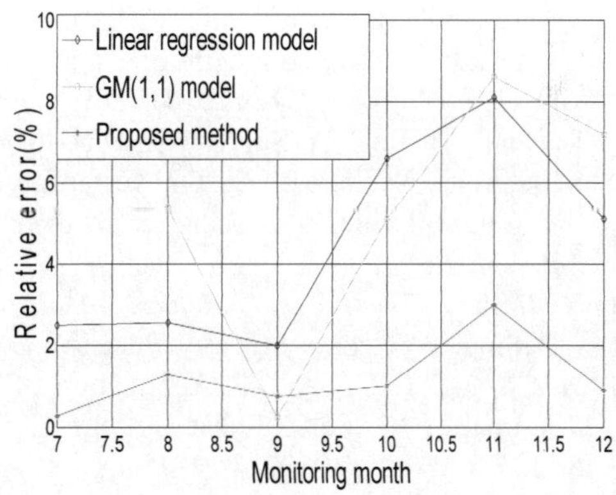


**Figure 3.** Relative error curves of the three methods

From Fig.2, compared with the actual data curve, the predicted data of proposed method and the actual data are more closer, more in line with the actual situation, showing strong volatility, and linear regression and GM(1,1) model and the actual data difference is large, cannot well reflect the fluctuations of the data.

From Fig.3, the error of proposed method in this paper is minimal in the errors of three methods, the errors of linear regression model and GM(1,1) model are larger, the prediction error of GM(1,1) model is maximum, it is said that the exponential growth of the sequence, proposed method in this paper is better than linear regression model , linear regression model is better than GM(1,1) model.

Through the analyses above, the improved GM(1,1) model is used to fit the fault rate curve by linear regression equation and exponential equation, gives full play to the advantage of less data modeling of grey system and regression model factors correlation, complex utilizes many information such as linearity and exponential, and does not require data which must accord with a typical regularities of distribution, it is more coincident with the factual situation. Therefore, the improved GM(1,1) model is better than the linear regression model and the traditional GM(1,1) model in the accuracy prediction of T/R component fault rate, which has some practicability in the prediction of radar equipment fault rate.

## 6. Summary

In the prediction of T/R component fault rate, according to the problems arisen in the prediction of traditional GM(1,1) model, in this paper, the model is improved, discrete GM(1,1) model is established to solve the problems of large error in predicting exponential growth sequence, introduce optimization factor to reconstruct model background value, which can adjust the value of the background according to the change of the original data and improve the accuracy of the model, based on that, linear regression model is added to solve the situation of different distribution of fault rate in different prediction stages , greatly expand the prediction range and accuracy, the model has universal applicability in practical fault prediction. At last, the prediction results of linear regression model 、 GM(1,1) and this method are compared and analyzed with examples. Results show that the improved GM(1,1) model proposed in this paper not only improves the accuracy of the model, but also broadens the application scope of the model and it has wider application value.

## References

[1] Y P Wang. Research On Maintenance Planning Method For Large Phased Array Radar[D]. Wuhan: Air Force Early Warning Academy(2016).

[2] M C Hu, Z P Zhou, W Yan. Phased Array Radar Transceiver Module Technology[M]. Beijing: National Defence Industry Press, (2010).

[3] L Z Cui. Grey Prediction Technique and Its Application[D]. Nanjing: Nanjing University of Aeronautics & Astronautics,(2010).

[4] Q Sun, J G Yue. Summary of Fault Prediction Method Based On Uncertainty[J]. Control and Decision Making,,**29**,5(2014): 769-778.

[5] S Deng, B Jing. A Survey On The Comprehensive Diagnosis and Fault Prediction Method of Electronic System Based On Testability[J] Control and Decision Making, **38**,5(2013): 641-649.

[6] S F Liu, T B Guo, Y G Dang, et al. *Grey System Theory and Its Application*[M]. Beijing: Science Press,(1999).

[7] Y J Shao. *Theory and Method of Preventive Maintenance of Weapons and Equipment* [M]. Beijing: Electronic Industry Publishing House, (2015).

[8] H Z Wang, J H Zou, L F. Application of Improved UGM(1,1) Model In Radar Fault Prediction[J]. Journal of Air Force Radar College,**23**,5(2009):321-323.

[9] S J Xu, H Z Wang, Y J Pang, et al. Improved Unequal Interval GM(1,1) Model and Its Application[J]. Practice and Understanding of Mathematics,**41**,8(2011): 108-114.

[10] S J Xu, X S Tan, H Z Wang, et al. Radar Fault Prediction Based On Multi Factor Unequal Time Interval Grey Model[J]: Modern Radar,**33**,8(2011):25-28.

[11] H Z Wang, J P Yang, M R Huang, et al. An Improve Unequal Interval Grey Prediction Model[J]. Journal of Air Force Engineering University (Natural Science Edition),**11**,6(2010):75-79.

[12] S F Liu, Z Beng, J F Liu, et al. Research On Several Basic Forms of GM(1,1) Model and Its Application[J]. System Engineering and Electronic Technology,**36**,3(2014):501-508.

[13] N M Xie, S F Liu. Development of Discrete Grey Model and Its Optimization[J]. System Engineering Theory and Practice,**6**,6(2006):108-112.

[14] L Zhou, J Zhao, G F Feng. *Equipment Fault Prediction and Health Management Technology*[M]. Beijing: National Defence Industry Press,(2015).

[15] X N Mie, C Y Zhu, J Zheng. Expansion modeling of discrete grey model based on multi-factor information aggregation [J]. Journal of Systems Engineering and Electronics, **25**,5(2014):833-839.

IOP Conf. Series: Materials Science and Engineering **224** (2017) 012032 doi:10.1088/1757-899X/224/1/012032

# DC Bias Simulation of 1000 kV UHV Transformer Based on MATLAB / Simulink

**Hu Peng[1] Yu Yongjun[1] Wu Weili[2] Sun Fan[1] Qi Xiaoxiao[1] Liu Yong[3]**

1. Xinjiang Electric power Research institute of State Grid. Urumqi 830000
2. YiLi Normal University Xinjiang, YiNing 835000
3. Anhui Electric Power Technology Co., Ltd. Hefei 230000

wwllxm2015@163.com

**Abstract:** In order to study the effect of Direct Current bias on the excitation current of 1000kV UHV transformer caused by the parallel operation of DC transmission and AC transmission, this paper builds a simulation model of 1000kV three-winding autotransformer based on MATLAB/Simulink This paper introduces the simulation model of simulation model parameters, and builds a simulation model of 1000kV UHV AC transmission system. Based on the analysis of the change trend of excitation current when the DC voltage source is applied at the neutral point of the transformer, 1000kV transformer, the conclusion is that the DC bias suppression measures and the transmission line design work to provide a theoretical reference.

## 1. Introduction

China's major load centers and energy centers of the geographical characteristics of the distribution of energy-intensive areas led to energy over, while the central and eastern regions in short supply, so large-scale, long-distance, low loss of Ultra High Voltage （UHV）Direct Current （DC） transmission technology increasingly important[1], China's first ± 1100kV UHV DC project Zhundong - Wannan line will be completed and put into operation in 2018, when the line can be through the Wannan converter station 1000kV UHV transformer group and Huainan - Wannan - Shanghai 1000kV UHV Alternating Current （AC） transmission Network connection to further improve China's UHV transmission network structure, improve the overall economic efficiency of the power grid [2]. UHVDC transmission system into the earth's DC current is relatively large, when the DC current through the 1000kV transformer neutral point into the transformer winding, seriously affect the transformer performance, so the impact of the ground current on the 1000kV transformer is very Necessary [3-4].

According to the J-A hysteresis theory[5], the influence of CGI on the excitation current of 750 kV single-phase autotransformer is studied in this paper. In [6], the influence of DC bias on the excitation current of 750kV transformer is studied based on the finite element method. In [7], a 500 kV autotransformer model was established based on J-A theory, and the DC bias resistance of the transformer was simulated. Because of the large difference in the structure of the transformer, the results of the literature [5-7] can not correctly describe the mechanism of 1000kV UHV transformer affected by DC bias. Because the structure of 1000kV transformer is complex, this paper is based on Matlab / Simulink simulation. The change of 1000 kV autotransformer main transformer and the variable voltage excitation current under the action of DC bias, and provide the theoretical support for the optimization design and protection work of 1000kV transformer.

## 2. Modeling of 1 1000kV UHV Transformer

*2.1 1000kV UHV transformer structure*

1000kV UHV transmission line transmission capacity is very large, calculated by the transformer single-phase design capacity to reach 1000MVA, if the use of ordinary transformer to achieve such a large capacity, then had to design a large volume of the transformer, which will bring the transformer loss increases, The operating efficiency is reduced, the system economy is bad and so on many problems. In order to reduce the cost of the transformer to choose a relatively low insulation requirements of the neutral point without excitation voltage regulator, but the voltage regulator will change and result in over-excitation and low-voltage winding voltage offset phenomenon, in order to eliminate these problems 1000kV transformer selection of the regulator transformer and autotransformer split operation design principles [8]. Voltage regulator by the regulator compensation transformer and low-voltage compensation transformer, high, medium and low voltage side windings were used Yn, Yn, dll connection, as shown in Figure 1. In order to minimize the fluctuation of the low-side winding voltage of the main transformer, it is necessary to adjust the core magnetic flux density by adjusting the voltage regulator coil in series with the medium-voltage side coil during the pressure regulation process. In order to further stabilize the level of the transformer core excitation, the low-voltage compensation winding is designed and its compensation winding is connected in series with the main transformer low-voltage side winding, and the excitation winding is connected in parallel with the voltage regulating coil.

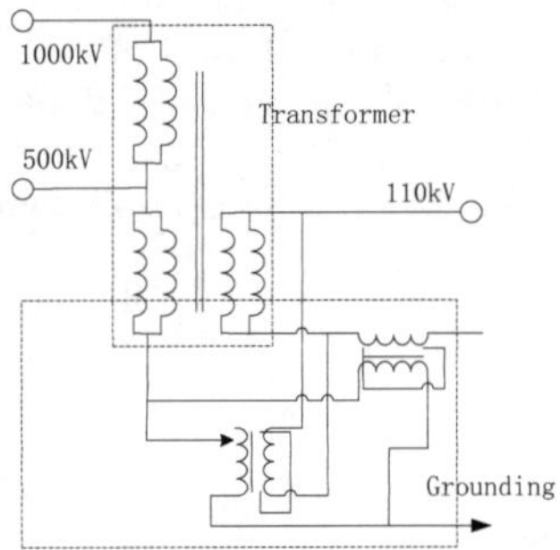


**Figure 1.** Structure diagram of 1000kV single phase Auto transformer

*2.2 UHV transformer modeling method*

The simulation system is changed in general three-winding transformer actual wiring to replace the three-winding autotransformer, as shown in Figure 2, the ordinary three-winding transformer second winding connected to the first end of the first winding to form an autotransformer series winding and common winding,The third winding simulator autotransformer low-voltage winding, voltage regulation and compensation variable excitation coil and the main transformer low-voltage winding in parallel, voltage regulator coil and compensation variable compensation coil and the main transformer winding in series, the literature [9-10]. The feasibility of the above modeling method is verified.

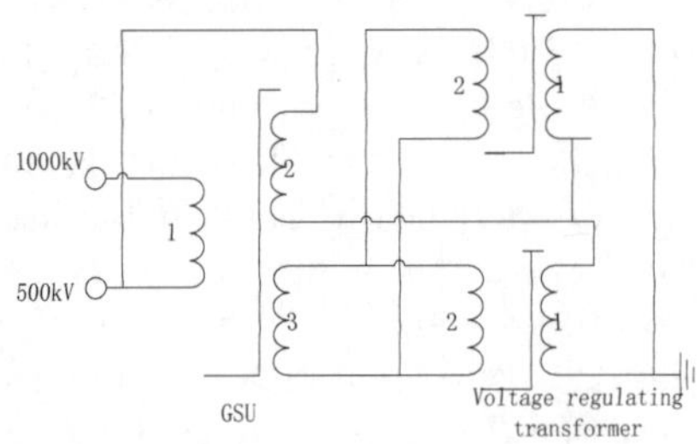


**Figure 2.** Equivalent simulation model of 1000kV single phase Auto transformer

*2.3 Model parameter correction calculation*

The use of matlab / simulink to establish the transformer model need to enter the three winding rated voltage effective value (V), resistance per unit value (p.u.) and reactance per unit value (p.u.). Transformer nameplate will provide the transformer rated voltage of the winding voltage (rated voltage), the three windings between the two winding short circuit test short circuit loss and short circuit voltage percentage (short circuit voltage), where the short circuit loss is used to calculate the winding resistance. The short-circuit voltage is used to calculate the reactance [11]. In view of the three-phase autotransformer and ordinary three-winding transformer electrical characteristics are different, the actual transformer winding parameters and simulation transformer input parameters are not equal, the need for auto-couple transformer winding relationship between the actual transformer parameters converted into analog transformer parameters [12].

*2.3.1 Rated voltage conversion.* Assuming that the rated voltage of the high, medium and low voltage windings on the transformer nameplate are UH, UM and UL respectively, the series winding voltage UC, the common winding voltage UG and the low voltage winding voltage UD obtained as $U_C = U_H - U_L, U_G = U_M$ and $U_D = U_L$.

*2.3.2 Reactance per unit value conversion.* Assuming that the percentage of the short-circuit voltage between the high-middle and medium-low, high-low voltage windings on the transformer nameplate is $U'_{H-M}$ %、 $U'_{M-L}$ %、 $U'_{H-L}$ %,so $U_{H-M}\% = U'_{H-M}\%$ , $U_{M-L}\% = U'_{M-L}\%(\frac{S_N}{S_{N3}})$ and $U_{H-L}\% = U'_{H-L}\%(\frac{S_N}{S_{N3}})$. Where, $U_H$%、 $U_M$%、 $U_L$ %are the short-circuit voltage (/%) for groups, common windings and low-voltage windings, respectively. Then the corresponding kink per unit value of each winding is $X_H = \frac{U_H\%}{100} \times \frac{S_j}{S_N}$, $X_M = \frac{U_M\%}{100} \times \frac{S_j}{S_N}$ and $X_L = \frac{U_L\%}{100} \times \frac{S_j}{S_N}$. And,

$$\begin{cases} U_H\% = \frac{1}{2}[U_{H-M}\% + U_{H-L}\% - U_{M-L}\%] \\ U_M\% = \frac{1}{2}[U_{H-M}\% + U_{M-L}\% - U_{H-L}\%] \\ U_L\% = \frac{1}{2}[U_{H-L}\% + U_{M-L}\% - U_{H-M}\%] \end{cases} \quad (1)$$

where, SN is the transformer capacity and Sj is the reference capacity. The above reactance is the autotransformer parameter, so it should be converted into the normal three-winding transformer winding reactance per unit value. Assuming the autotransformer high - medium, high - low, medium - low voltage winding reactance per unitary values are $X'_{HM} = X_H + X_M$ 、 $X'_{HL} = X_H + X_L$ , and $X'_{ML} = X_M + X_L$, then the conversion formula is as follows:

$$\begin{cases} X_{HM} = X'_{HM} \cdot (\frac{U_H}{U_H - U_M})^2 \\ X_{ML} = X'_{ML} \\ X_{HL} = X'_{HM} \cdot \frac{U_H U_M}{(U_H - U_M)^2} + X'_{HL} \cdot \frac{U_H}{U_H - U_M} - X'_{ML} \cdot \frac{U_M}{U_H - U_M} \end{cases} \quad (2)$$

Then, the series winding, the common winding and the low voltage winding per unit value of the simulated autotransformer after calculation are calculated according to the relationship between the perimeter value of the winding and the per unit value of each winding.

2 Simulation of DC Bias Magnetic Field of 1000kV Self - coupling Transformer Transformer

After the operation of Zhundong-Wannan ± 1100kV UHVDC transmission project, there will be ± 1100kV, ± 800kV, ± 500kV UHVDC transmission lines and 1000kV, 750kV UHV AC transmission lines in the central region of China. The following will be based on China's independent research and development of ODFPS-1000MVA / 1000kV single-phase self-coupled UHV transformer parameters and a special high-voltage AC system line parameters in MATLAB / Simulink in the establishment of UHV AC system simulation model, AC transformer grounding point access DC voltage Source to simulate the distortion law and degree of excitation current of 1000 kV UHV transformer.

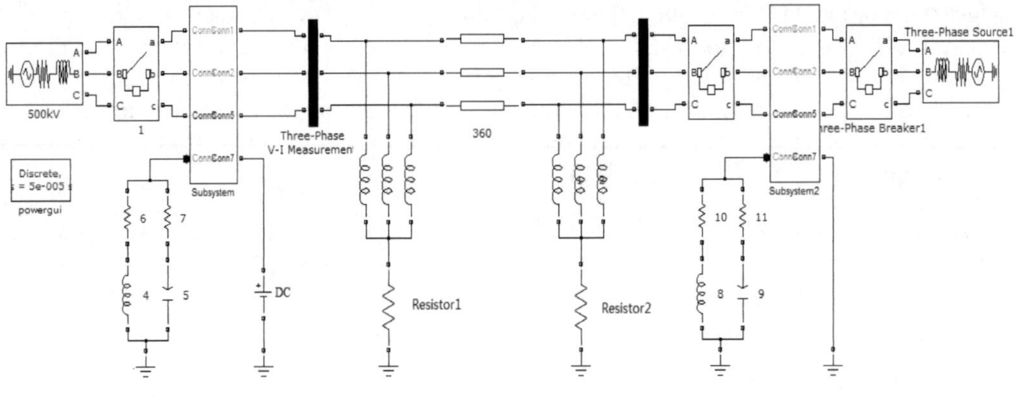


**Figure 3.** Simulation model of real AC system contained 1000kV tree phase Auto transformer

Ultra-high voltage AC system simulation model parameters are as follows, ODFPS-1000MVA / 1000kV transformer parameters: the main transformer high, medium and low three-rated capacity and rated voltage were 1000: 1000: 334MVA and 1050 /: 525 /: 110kV, The percentage of voltage is%,%,% is 18%, 62%, 40%, respectively, and the rated capacity is 59 / 18MVA, the rated voltage is 110: 5% 1050 / and 5% 1050 /: 5.4kV And the ratio is 3.63 and 5.16 respectively. The three single-phase autotransformers are integrated into a single three-phase autotransformer sub-system. The transmission line parameters are: 360km long, [0.263,0.8306] Ω / km, The resistance is [7.36e-5,2.5786e-3] Ω / km and the capacitance is [1.397e-2,0.3e-2] μF / km. The capacity of the starting side capacitive reactive power compensation reactor is 9600Mvar, the main reactance and grounding resistance are 1300Ω and 280Ω respectively. The capacity of the end capacitive reactive power compensation reactor, the main reactance and the ground reactance are 720Mvar, 1680Ω and 370Ω respectively. UHV transformer 110kV bus capacitance, reactance compensation capacity are 240Mvar, Figure 3 for the system matlab simulation diagram.

In the UHV transformer neutral point when the application of different DC voltage waveform distortion start time comparison chart and the main transformer excitation current waveform is as follows.

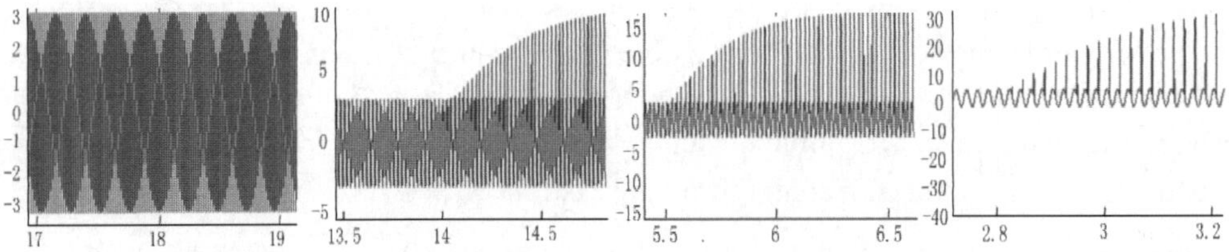

$U_{dc}$=10v $U_{dc}$=50v $U_{dc}$=100v $U_{dc}$=200v

**Figure 4.** The start time contrast of waveform distortion

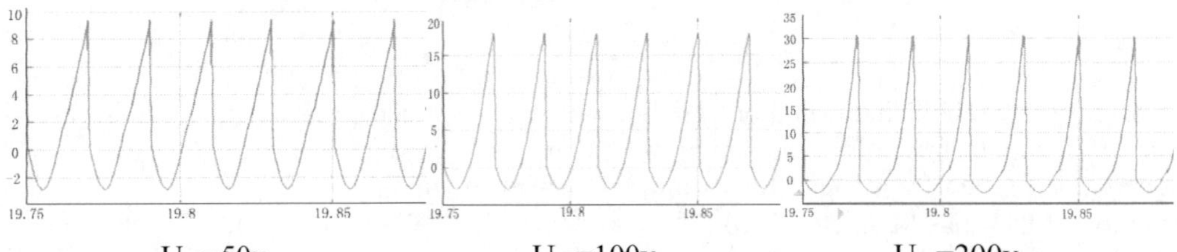

$U_{dc}$=50v $U_{dc}$=100v $U_{dc}$=200v

**Figure 5.** Waveform of excitation current of main transformer under different DC bias

It can be seen from Figure 4, with the increase in the voltage applied to the main excitation current began to appear more and more front-time harmonics, indicating the greater the neutral point DC current magnetic flux saturation faster, when Udc <10v When the excitation current is almost no distortion, which shows that if the use of reasonable measures to reduce the center of the transformer electromotive force to a certain value can effectively avoid the impact of DC bias.

It can be seen from Fig. 5 that when the Udc = 50v, the positive half-wave of the main transformer excitation current has a severe spike and the negative half-axis waveform is not affected, Udc is 100v, 200v, the excitation current phase exhibits a nonlinear growth trend. Half-wave distortion is more serious, the amplitude of inrush even more than 10 times the rated excitation wave.

## 3. Conclusion

Firstly, the basic structure of 1000kV three-winding autotransformer is introduced, and a simulation model is put forward. Based on the reasonable calculation model of the transformer, the rated voltage and reactance per unit value of the autotransformer simulation model are established. The real-time UHV AC transmission system simulation model is used to analyze the excitation current waveform of the 1000kV autotransformer under different DCs. It is found that the negative half-wave of the excitation current is not affected by the DC bias. The positive half-wave appears in the spire, the greater the DC voltage, The greater the value, the more severe the excitation current waveform distortion.

## Acknowledgments

This paper is supported by the research and application of the national power grid Xinjiang Electric Power Company and the Yili Normal University doctoral Fund (YLSFBS01).

## References

[1] GUO Huan, CAO Junzheng, TANG Guangfu, et al. Optimization Design of Main Circuit of Converter System for 1100kV UHVDC Transmission System[J] *Power System Technology*, 2013,09:2383-2389.(in Chinese).

[2] LIU Haojun, YAN Guozeng, WANG Shaohua.et al. Analysis of Electrostatic Induced Voltage on 1 000 k V Wannan-Zhebei UHV AC Transmission Line[J]. *High Voltage Engineering* 2015,11:3687-3693.(in Chinese).

[3] ZHU An-ming, WANG Sen, HU Pan-feng.et al. Influences of the Deyang-Baoji DC Transmission Line on the 750 kV Power Transformer in Baoji Power Substation[J].*Power System and Clean Energy*.2010,26（12）： 67-70..(in Chinese).

[4] YUAN Shun, WANG Tian-shi.et al. Summary of the Research on Transformer DC Magnetic Bias. *High Voltage Apparatus*,2010,03:83-87.(in Chinese).

[5] PENG Chen-guang, LIU Lian-guang.etal. tudyon Direct Current Bias of7 50kV Autotransformer Basedon JAT heory.[J]*East China Electric Power* 2010,03:349-353. (in Chinese).

[6] LIU Yugen, LENG Di, TIAN Zi, CHENG Wenjie .etal. DC Bias Simulation of 750KV Autotransformer Based on ANSYS Maxwell[J].*high voltage Engineering* 2013,01:218-225.(in Chinese).

[7] YANG Xiao-hui, WANG Feng-hua, et.al. DUAN Ruo-chen. Simulation study of DC bias enduring ability of 500 k V power transformer[J].*Power System Protection and Control*.2014,14:60-66.(in Chinese).

[8] SUN Dou.et.al. The Choose of Voltage Regualtion in Transformer and The Running and Maintance of Transformer[J] *Electric Power*,2010,07:29-33.in Chinese).

[9] ZENG Lin-jun, LIN Xiang-ning, HUANG Jing-guang.et al. Modeling and Electromagnetic Transient Simulation of UHV Autotransformer [J]*Proceedings of the CSEE*.2010,01:91-97.(in Chinese).

[10] ZHANG Jie. A Method for Inrush Detection in Ultra-high Voltage Transformer.[D].*North China Electric Power University*,2012..(in Chinese).

[11] FENG Jian Qin, FENG Gong Ling. The Basis of Electrical Engineering[M] *China Electric Power Press*,2010，29.

[12]QIU Gui Hua .Modeling Simulation and Adaptability Rsearch for 500kV Main Transformer Single Phase Replacement[D].*South China University of Technology*.2012

# Analysis and Countermeasure Study on DC Bias of Main Transformer in a City

**Wang PengChao [1], Wang Hongtao [1], Song Xinpu[1], Gu Jun[2], LIU yong[3] WU weili[3]**

1. Xinjiang Economic and Technology Research Institute of State Grid. Xinjiang 830000
2. Urumqi electric power company of State Grid. Xinjiang 830000
3. Anhui Electric Power Technology Co., Ltd. Anhui 230000

wwllxm2015@163.com

**Abstract.** According to the December 2015 Guohua Beijing thermal power transformer DC magnetic bias phenomenon, the monitoring data of 24 hours of direct current is analyzed. We find that the maximum DC current is up to 25 and is about 30s for the trend cycle, on this basis, then, of the geomagnetic storm HVDC and subway operation causes comparison of the mechanism, and make a comprehensive analysis of the thermal power plant's geographical location, surrounding environment and electrical contact etc.. The results show that the main reason for the DC bias of Guohua thermal power transformer is the operation of the subway, and the change of the DC bias current is periodic. Finally, of Guohua thermal power transformer DC magnetic bias control method is studied, the simulation results show that the method of using neutral point with small resistance or capacitance can effectively inhibit the main transformer neutral point current.

## 1. Introduction

In December 2015, Guohua Beijing thermal power plant (referred to as Guohua thermoelectric) 1# main transformer noise increased significantly, after the measurement, the main transformer neutral Direct Current(DC) current reached 16A. Changes in the main substation grounding mode, that is, 2# main transformer neutral grounding, 1, 3# main transformer neutral grounding mode of operation, found that 2# main transformer neutral point DC current reached 25A. The real reason for revealing the cause of DC transformer neutral point, the main transformer neutral point were monitored for 24 hours, found that the neutral point DC current is about 60s cycle changes, in a certain period of time gap interrupt situation.

Previous research suggested that the cause of the DC bias of transformer has two main reasons [1-4], one is the geomagnetic storm, two is the reason of DC transmission at home and abroad; DC transmission completion time is not long, people first discovered the transformer DC magnetic bias is caused by geomagnetic storms. The storm event broke out in 1940, high latitude countries first discovered the transformer DC magnetic biasing [1] reference [2] is proposed for the first time will make the transformer neutral point DC half wave saturation voltage offset will cause power system point of view, initially pointed out the impact of transformer DC bias of power system. With the development of power system, the intensity of storms in 1940 1989 storm the Quebec blackout of [3] Kappaenman, this system was analyzed by [4], and the measured data show that transformer DC bias is the cause of the accident in power grid. Since 2000, with the development of power industry in China, the transformer DC magnetic biasing appeared in the actual operation of the power system of our country: Guangdong power grid, Jiangsu River, Guohua Yuedian Taishan and Xinjiang Hami power grid and other regions have reported a lot due to the DC transmission DC bias of [5-11] events; in addition, although China is located in the low latitude, but still in the transformer substation in Guangdong Lingao, Jiangsu Shuanglong River, Zhejiang and Fujian East China grid system occurs and transformer noise and strong vibration due to abnormal geomagnetic storm events [12]. Based on the above analysis, it is found that the characteristics of magnetic bias current caused by geomagnetic storm and DC transmission are not consistent with the DC current of the event.

The neutral point DC current monitoring and its characteristics as the basis, the location of the main transformer, the surrounding environment and electrical contact and other factors are analyzed, and finally determine the operation of the subway is the main cause of transformer DC bias, and put forward the corresponding control methods, the effectiveness of the method is validated by the simulation of governance.

## 2. Analysis of DC bias and monitoring data of main transformer neutral point

### *2.1Analysis of DC bias of main transformer in thermal power plant*

The typical phenomena of DC magnetic bias include the increase of transformer noise, the increase of temperature, the increase of reactive power loss and harmonic wave. At the beginning of December 2015, China 1# main transformer noise power increases obviously, preliminary determination of transformer DC current exceed the standard, the main transformer neutral DC current reaches 16A. Change in the main substation grounding mode, 2# main transformer neutral DC current up to 25A. 2# thermal power plant main transformer is a three-phase five column structure, rated current is 1002A, according to the standard DL/T 437-2012 "HVDC technology guide" the ends of 6.3, allowing the winding DC limit of the rated current is 5 per thousand 2# main transformer, 7.55 per thousand and the actual flow through the DC transformer capacity, beyond 50%, with an increase in noise the obvious phenomenon, can determine the thermal power plant transformer DC magnetic bias has occurred.

In order to find out the main causes of the main transformer DC bias, the continuous DC current monitoring should be carried out.

### *2.2 main transformer neutral point DC monitoring data*

The monitoring of the neutral point of Guohua thermal power main transformer was carried out for 24 hours, as shown in figure 1.

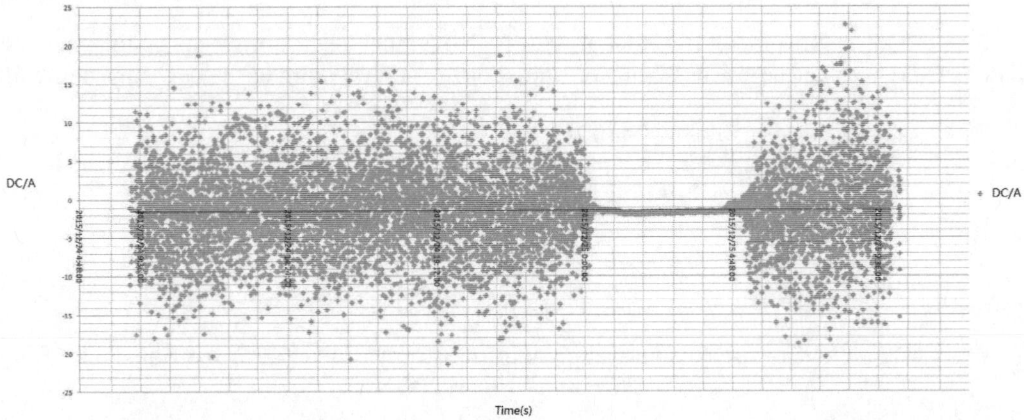


**Figure 1.** Distribution of neutral point DC current distribution of Guohua 2# main transformer

As can be seen from figure 1:

①The DC current is positive and negative, the variation period is about 60s, far less than the AC frequency 50HZ;

②If the inflow of the main transformer neutral point in the direction is positive, the outflow of main transformer neutral point in the direction is negative, then the maximum DC current value is 22.7A,

appeared in 25 8:32:53; the negative maximum value of -21.4A, appeared in 24 19:32:49;
③DC current concentration occurs at 24 9:15 - 25 0:00 and 25 5:00 - 25 10:19. In 25 0:00 - 25 5:00 this time period, the absolute value of the DC current is less than 1A, belonging to the device zero drift, can be considered in this period of time the main transformer neutral DC current;
④In 24 hours, the 2# main transformer appeared serious DC magnetic bias, the DC magnetic bias time in 5:00 - 24:00.

## 3. Analysis on the causes of DC bias of Guohua thermal power transformer

### *3.1. DC magnetic bias analysis*

Geomagnetic storms and High Voltage Direct Current (HVDC) project is regarded as the main cause of the DC bias: the complex interaction with the earth's magnetic field solar wind to the earth, change the earth's magnetic field intensity, the formation of a geomagnetic storm. The variation of the geomagnetic field induced electric field on the earth surface, the formation of potential gradient, when the low frequency and has a certain induction duration of the electric field in the grounding of power transformer, the equivalent of an external voltage source. HVDC monopole operation, the electromagnetic field distribution near the surface potential is changed, the earth surface will also present different potential distribution, when the grounding is located at different potentials, the grounding potential difference exists between the location, the equivalent of an external voltage source function. It can be seen that the DC bias phenomenon is caused by the existence of DC potential difference between the neutral point of transformer grounding in the power system under the two causes. However, the DC bias caused by the two different characteristics of the problem. Figure 2 is the monitoring of the neutral point current curve of the main transformer in Guangdong Ao Ao nuclear power plant in September 16, 2005, when the three - DC - DC operation.

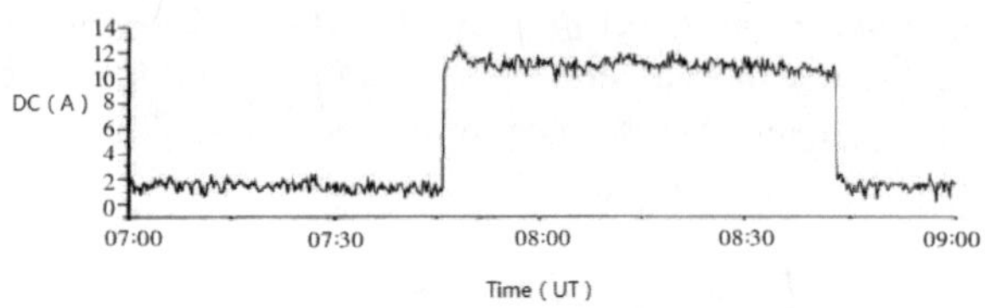


**Figure 2.** Results of the DC current in transformer neutral point in Ling Ao Nuclear Power Station, September 16, 2005

Figure 3 is at 2:30 on November 10, 2004 at 6:30 on November 11th, during the geomagnetic storm, Guangdong Ling'Ao Nuclear Power Plant No. 1 transformer neutral point of the GIC current monitoring results.

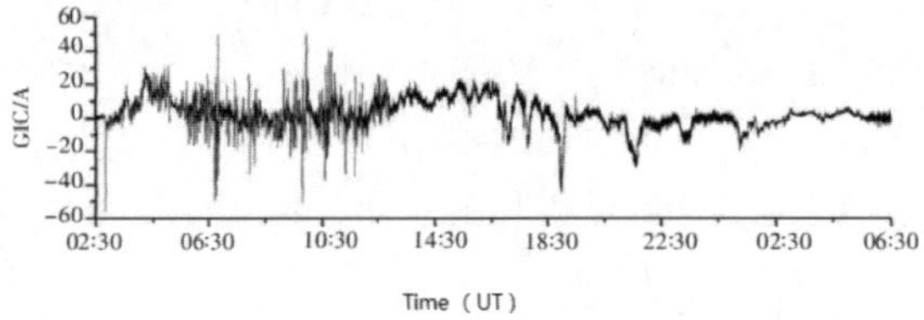


**Figure 3.** GIC monitoring curve of neutral point of transformer in Ling Ao Nuclear Power Station, November 10-11, 2004

As compared with figure 2 and figure 3, the rate of change of GIC is greater than that of HVDC. This is mainly due to geomagnetic storms, the variation of the geomagnetic field is violent, the change of GIC but also more intense; High Voltage Direct Current(HVDC) transmission power change is very small, so the changes in the size of DC current in transformer neutral HVDC monopole operation point is very small, basically stable.

### *3.2.Cause analysis of transformer DC bias in thermal power plant*

IOP Conf. Series: Materials Science and Engineering **224** (2017) 012033 doi:10.1088/1757-899X/224/1/012033

Comparison of 1, 2 and 3 showed that the basic characteristics of thermal power plant DC current neutral DC current trends and geomagnetic storms and HVDC causes under the changing trend, period and frequency difference: solar geomagnetic storms produce geomagnetic current (GIC) AC frequency 1Hz, which can be approximated by DC. The inflow of the AC line can also trigger the main transformer DC magnetic bias. The geomagnetic storm caused by solar activities to make the grid in high latitude regions appear bias this phenomenon in middle and low latitudes, will result in high resistivity area, long distance EHV transmission system of DC bias, but DC bias caused by the geomagnetic storm in time should be a few ten years of time. For a few minutes to a few days in the region should be great geographical scope; grounding position, grounding current of HVDC is certain, but in exchange for a considerable period of time the grid architecture, the soil resistivity will not change. Therefore, the main transformer bias current should be a fixed value.

The main cause of transformer DC bias is judged as follows:

Firstly, according to the release of Key Laboratory of space weather information shows that geomagnetic storms during December 2015 24-25, the level of geomagnetic activity is low, Guohua thermal power DC magnetic bias duration more than 7 days, and 3 transformer voltage level is not high, it is not connected to the long-distance transmission line, so it can be initially ruled out causes of geomagnetic storms.

Secondly, thermal power plant position analysis. There is no HVDC transmission project in Beijing, Hebei, there is no DC transmission project. The monitoring data is different from the DC current caused by the DC transmission, so the origin of HVDC transmission can be excluded.

The third step, thermal power plant surrounding environment analysis. Guohua Beijing gas power plant is located in Beijing City, Chaoyang District Jin Yu Lu, belonging to the Beijing northeast gas power center. Distance from the subway line 6, the airport line straight distance of 3 km, 10 km, schematic see figure 4.

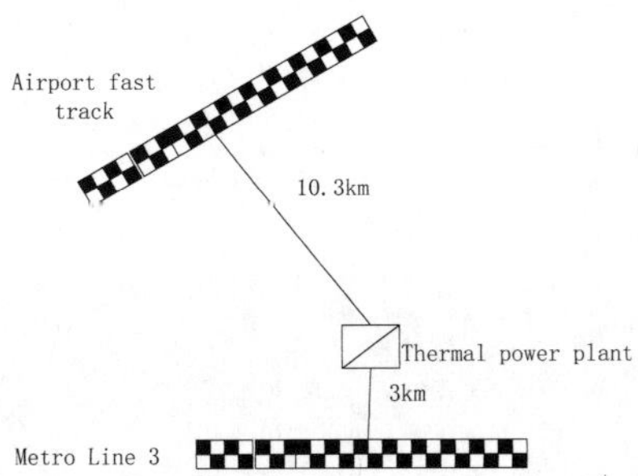


**Figure 4.** Schematic diagram of Guohua thermal power and Metro Line

In addition, figure 1 shows that the 0:00 - 5:00 time, 2# main transformer neutral point DC component, in the meantime, Beijing subway stop running, monitoring the current positive and negative frequency change basically accord with the characteristics of power supply system of Metro DC bias, can determine the initial cause of thermal power plant, the DC magnetic bias is around the subway.

### *3.3. Analysis of transformer DC bias mechanism caused by metro operation*

Beijing metro power supply system using DC power supply mode, mostly DC750V, due to the short distance of 750V power supply voltage, stray current, now use 1500V. Divided into unilateral power supply system and bilateral power supply system. Power supply principle see figure 5:

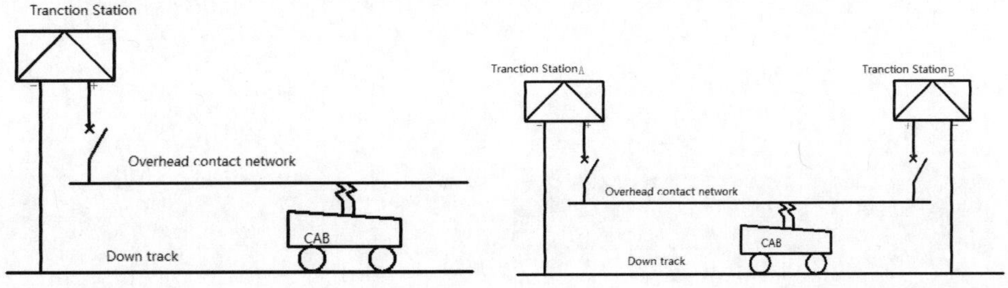

**Figure 5.** single / double power supply system

Due to the locomotive running track and ground is not completely insulated, part of the DC component will flow into the earth. This part of the current is called stray current. As shown in Figure 6 (single side power supply has such problems, as shown in the unilateral power supply system):

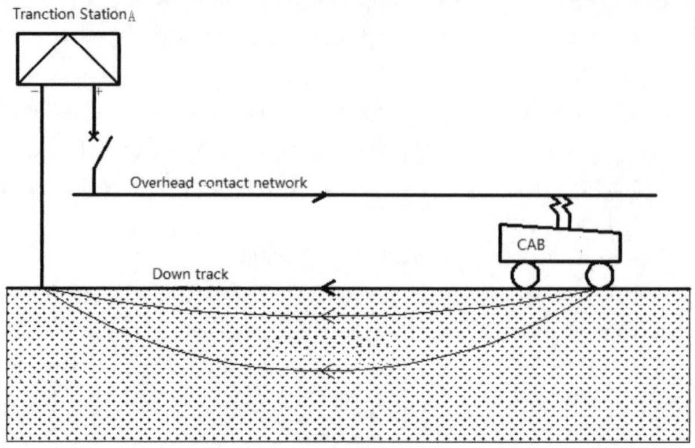


**Figure 6.** sketch of stray current in subway power supply system

Due to the effect of stray current in the ground, the distribution of ground potential. Similar to the HVDC Ground electrode. If at this time, the two connecting terminals with AC connection are located at different potential, which will produce a certain bias current. As shown in figure 7:

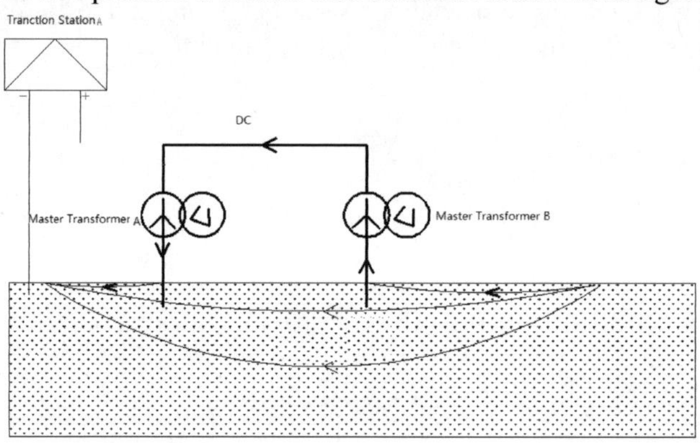


**Figure 7.** DC magnetic bias of main transformer caused by stray current in subway power supply system

In the actual operation of the subway, the same line on a number of locomotives two-way traffic. In fact, the running position of the locomotive is the location of the stray current. Therefore, the influence of the DC component of the neutral point of the main transformer is the combined action of the stray current generated by multiple locomotives. At different times, the location of the locomotive is different, the location of the stray current is different, the specific impact of the way and the direction of the DC current generated see Figure 8 and figure 9.

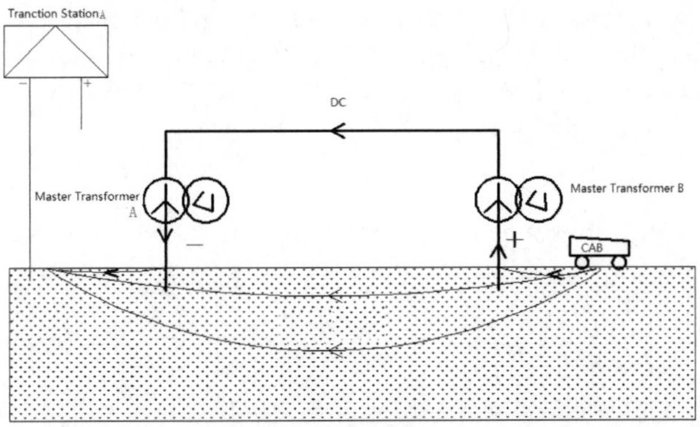


**Figure 8.** schematic diagram of DC current in neutral point of main transformer in locomotive position 1

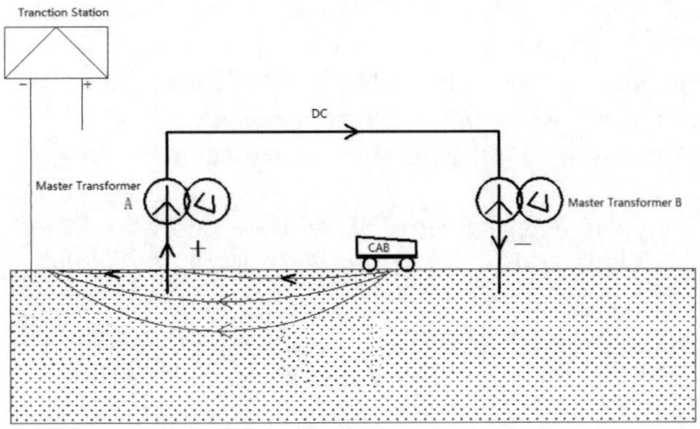


**Figure 9.** schematic diagram of DC current in neutral point of main transformer in locomotive position 2

All locomotives are equivalent to one, when the locomotive is located at position 1, the neutral current of the main transformer A is negative, and when the locomotive is located at position 2, the neutral current of the main transformer A is positive. The actual situation, subway locomotive number itself, speed, will cause a change in positive and negative DC current transformer ten seconds, and flights a day subway locomotive is generally fixed, so the main variable changes the frequency of positive and negative DC current is relatively stable, which also explains the trend of monitoring data in the period ca. 60s in the visible, the subway is the main reason for the DC bias of transformer thermal power plant.

## 4. Main transformer DC bias control measures

According to the above analysis, the main transformer substation will be affected by the operation of the subway and the phenomenon of DC bias. Considering the importance of the main transformer in main transformer neutral point series capacitor blocking device, this method is more practical in the actual project, with its simple principle, obvious effect, after the installation does not need to modify the protection and automatic safety device settings to get the industry recognized technical characteristics.

## 5. Conclusion

Previous studies suggest that geomagnetic storms and HVDC power system is the main reason causing the DC bias [1-4],However, Guohua thermal power transformer DC magnetic bias events show that the subway can also cause the surrounding transformer DC bias. Firstly the paper introduces the process of determining the city power plant DC bias causes, and then analyzes the cause mechanism and characteristics of the DC bias of the subway operation methods are given in urban heat and power plant transformer DC magnetic bias control.

The subway operation caused by the DC bias of transformer is a new phenomenon, as China's power plant modernization and modern city construction and development of electric rail transportation, is established in the vicinity of the main transformer in power plant and substation will be more and more influence power supply system for urban rail transit and the DC bias phenomenon. This paper through the analysis of the incident, the system was put forward, the method to determine the causes of DC bias and countermeasures and some problems that need further study, in order to similar DC bias events and provide theoretical reference for governance.

**Acknowledge**
This paper is supported by the project of the State Grid Xinjiang Electric Power Company Economic and Technical Research Project (the influence of the subway power supply system on the DC bias of the power grid,3016B5-QT-04.

**Reference**

[1] Davidson, W. F., The magnetic storm of March 24, 1940 - Effects in the power system[J], *Edison Electric Institute Bulletin*, July 1940: 365-366
[2] Albertson, V.D. Thorson, J.M., Clayton, R.E.,Tripathy, S.C.Solar-Induced-Currents in Power Systems: Cause and Effects *.Power Apparatus and Systems*[J], *IEEE Transactions* .1973,92(2):Page(s): 471 - 477
[3] NERC report,March 13,1989 Geomagentic Disturbances[R/OL], *http://www.nerc.com/files/1989-Quebec-Disturbance.pdf*
[4] Kappaenman J G, Albertson V D. Bracing for the geomagnetic storms[J]. *IEEE Spectrum,* 1990, 27(3): 27–33
[5] Yu Yongjun,Hou Zhiyuan, Yang Qi, et al. The influence of the DC ground current on the AC power network in Xinjiang Hami [J], *Power System Technology*, 2014,38 (8):31-17 (in Chinese)
[6] Wang Zhenhao, Tong Xin, Qi an. The direct current control of the transformer by the method of capacitance isolation and direct control [J]. *Journal of electrical engineering*, 2014,28 (10): 120-126.
[7] Zhang Lu, Ruan Ling, Pan Zhuohong, et al. Application analysis of DC bias suppression equipment for transformer [J]. *Electric Power Automation Equipment*, 2014,33 (9): 151-156.
[8] Kovan B, De Leon F. Mitigation of Geomagnetically Induced Currents by Neutral Switching[J]. *IEEE Transactions on Power Delivery*, 2015, 30(4):1-1.
[9] Jiang Tao, Tong Xin, Wen Xi Shan, and other complex operating conditions. The inhibition of [J]. *High Voltage Technology, transformer DC bias of 2015*, 41 (7): 2464-2471.
[10] Du Zhongdong, Dong Xiaohui, et al. Study on DC bias suppression by DC potential compensation [J]. *High Voltage Technology*, 2006, 32 (): 69~72 (in Chinese)
[11] Zeng R, Yu Z, He J, et al. Study on restraining DC neutral current of transformer during HVDC monopolar operation[J]. *IEEE Transactions on Power Delivery*, 2011, 24(4): 2785-2791
[12] LIU Lian Guang LIU Chun Ming ZHANG Bing, et al. Strong magnetic storm's influence on China's Guangdong power grid[J]. *Chinese Journal of Geophysic*s, 2008, 51(4):694-699.

IOP Conf. Series: Materials Science and Engineering **224** (2017) 012034 doi:10.1088/1757-899X/224/1/012034

# The Anomaly Power Transmission and Analysis at a 500kV GIS Substation

**Qi Geqing[1,2], Wu Gang[1], Qi Zhuang[1], Meng Fanlong[1], Jin Daxin[1], Zhang Wei[1], Zhang Nan[1], Wang Lulu[1,3]**

[1]State Gird Dalian Electric Power Supply Company, Dalian, Liaoning, China

[2] qgqing777@163.com, [3] wluluu@126.com

**Abstract.** In view of the extension which steps up from 220kV to 500kV at the switching station, induced the flashover of insulation spacer in gas chamber from 500kV line side, the voltage fault of 500kV bus, the discharge of junction which between oil-gas casing of 500kV main transformer from 220kV bus side and GIS branch bus, due to improper installation. The paper analyzes connection diagram, graph of fault recorder and repair with disassembly point for point. It also suggests the recommendations for improvement, to avoid the appearance.

## 1. Introduction

The newly-built or expanded substation and overhead line, have finished overall acceptance of correlation specialty. The operating crews will execute power transmission according to the scheme plan beforehand. The power transmission order is: checking the new line→ checking the line interval devices→ checking the bus and the devices belong to bus( such as potential transformer, arrester, etc.) → checking the main transformer, etc. After the assessment of charged and the correct of phase position, the system would be on-load operation. The paper has analyzed anomaly power transmission that extension steps up from 220kV to 500kV at the switching station. It also has suggested the recommendations for improvement, how to handle such commissioning test.

## 2. The power transmission situation of substation

The set-up transform of 220kV Yanshui substation would be showed by one-line graph under 500kV.

### *2.1. The first power transmission situation*

The first power transmission was on December 6th, 2010. At 11:43, the dispatcher gave an order that charging 500kV Jinyan line, 500kV Ⅰ and Ⅱ bus in Yanshui substation. All of their voltages were normal. The phasing of PT secondary side in 500kV Ⅰ and Ⅱ bus was accurate.

At 13:32, the dispatcher gave an order that disconnecting the 5012 switch of 500kV Ⅰ and Ⅱ bus in Yanshui substation. After that, finding the three phase voltages of 500kV Ⅱ bus were unusual. Such as: $U_a$=60kV, $U_b$=60kV, $U_c$=126 kV, but the voltages of Jinyan line and 500kV Ⅰ were regular. Then the dispatcher gave an order that disconnecting the 5051 switch of Jinyan line in Yanshui substation. The line voltage was regular. The three phase voltages of 500kV Ⅰ were 10kV or so. The operator executed the order next, disconnecting the Jinyan line. The background displayed the voltages of 500kV Jinyan line, 500kV Ⅰ and Ⅱ bus PT were 0kV.

In consideration of above conditions, provincial dispatcher commanded to check on site. The result showed no exception. So the operation went on. When charging the 500kV Jinyan line and the

500kVIbus, current differential protection operated, both sides of the switch tripped. The fault differential current was 7100A, the fault phase was C, fault distance was 0km from the Yanshui substation. The fault distance of opposite terminal was 56.2~57.7 km (The line was 65.2 km). The fault distances matched each other.

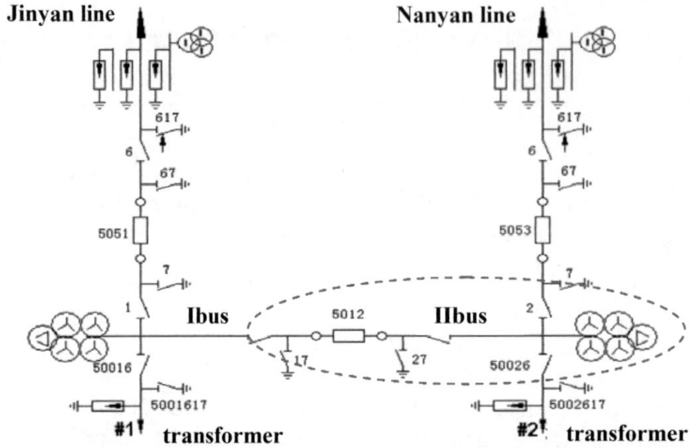


**Figure 1.** The portion of one-line graph under 500kV in Yanshui substation

After the failure, power supplier and operator checked the 500kV Jinyan line and related devices. The visual position indicated normally. So it said that the GIS had a fault in a preliminary estimate. To aim at finding the fault position, tester tested all analysis on $SF_6$ of the Jinyan line GIS of C phase in Yanshui substation. The analysis showed the abnormality in Jinyan C phase outlet chamber ($SO_2$+$SOF_2$:87.2 μ L/L, $H_2S$:10.1 μ L/L, CO:90.4 μ L/L). Hereby, the insulation of Jinyan C phase outlet chamber had broken down.

At 19:20, both sides of Yanshui and Jinjia substation switches were disconnected, also the Jinyan line.

*2.2. The second power transmission situation*

On December 6th, after replacing the disc insulator of C phase. On December 18th, power supplied. The same order induced the same fault like before. When #2 transform got power, finding that A and B phase voltage of 220kVⅣ bus were unusual (The 220kV side connected to the 220kVⅣ bus). Hearing 'flat-flat' following the electric spark at the connection of GIS case flange, that A and B phase casing connected to the GIS. The voltage waveform record of 220kV side bus is shown below.

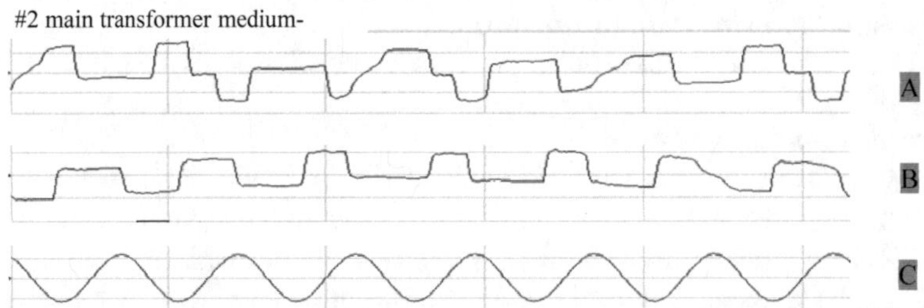


**Figure 2.** The voltage waveform record of 500kV transform by 220kV side bus

## 3. The checking on site after twice power transmissions

*3.1. The checking the chamber C phase of GIS line on 500kV Jinyan line*

The installer checked the C phase chamber, which had an insulation breakdown on December 6th. The disc insulator of outlet casing bottom discharged breakdown. See figure 3.

The figure 3 shows a penetrating discharge trail of the 500kV disc insulator face. The trail was due to the system short circuit current after discharging.

**Figure 3.** The disc insulator ablation of C phase line casing bottom on 500kV Jinyan line

*3.2. The checking of 500kV #2 transformer abnormal transmission*

On December 18th, found 500kV #2 transformer abnormal transmission. The installer found nothing from the checking, which the equipotential bonding screws of electric spark between the connection of GIS case flange and the conduction were normal. When the tester tested the conduction of ⅣBus 67# grounding switch, which the three phase were connected to the neutral point of 500kV main transformer. The open circuit was found between A and B phase. By using the 1000V megohmmeter, the insulation resistance value was 0MΩ. The value of A and B phase was ∞MΩ. The connection between the 220kV outlet casing of A and B phase, 220kV Ⅳ bus was abnormal. Opening the handhole of 220kV outlet casing of A, B phase and 220kV GIS guide, see figure 4 and 5 below.

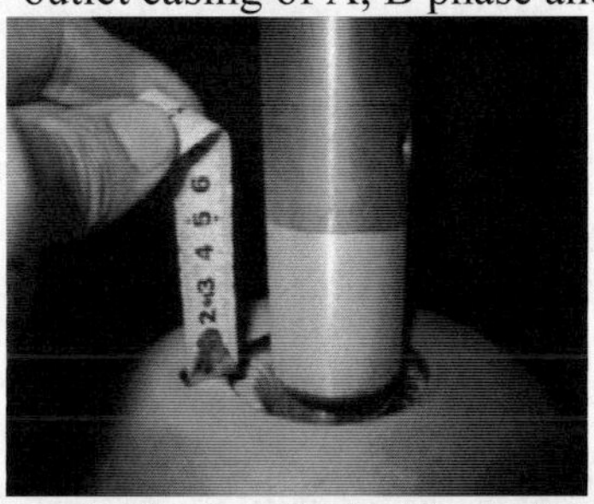

**Figure 4.** The handhole of opening the transformer 220kV side A phase

**Figure 5.** The handhole of opening the transformer 220kV side B phase

From the figures above, there was installed fault between the outlet casing of A and B phase, 220kV GIS guide, which due to clearance. The largest clearance was A phase, it reached 15mm or so. There was a penetrating discharge trail on mei flower button contact of A and B phase casing end, GIS guide. The C phase and other parts of GIS were normal.

## 4. The abnormal phenomenon analysis

*4.1. The abnormal voltage analysis of 500kV transmission*

See figure 1, when discharging 5012 bus connection switch, the equivalent circuit in dashed line frame is showed in figure 6.

In figure 6: (a) is equivalent, (b) and (c) are simplified equivalent circuit.

- $C_1$,$C_2$-5012 parallel capacitor of breaker fracture
- $C_3$-5012 breaker pillar capacitance to earth
- $C_4$-equivalent capacitance to earth of high voltage conductor between 500kVⅡbus PT and 5012 breaker
- C'-$C_1$,$C_2$ series capacitor
- C-the total equivalent capacitance of simplified circuit
- L-inductance of 500kV Ⅱ bus PT
- U-equivalent supply voltage

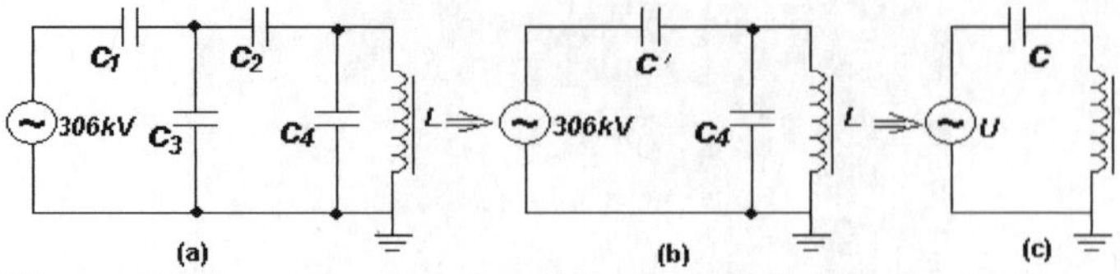


**Figure 6.**The equivalent circuit in dashed line frame in figure1

In figure 6, $C_1$ and $C_2$ are 5012 grading capacitances of breaker fracture. The both of capacity value are 600pF or so. $C_3$ is the capacitance to earth, the value is several tens pF. $C_4$ is the equivalent capacitance to earth of high voltage conductor between 500kV II bus PT and 5012 breaker, the value is 1000—1500 pF. L is the inductance of 500kV II bus PT.

Due to the value of $C_3$ is the least, than $C_1$, $C_2$ and $C_4$. The influence can be ignored. So figure (a) can be simplified to figure (b). In figure (b):

$$C' = C_1C_2/(C_1+C_2)=300\text{pF};$$

Figure (b) can be simplified to figure (c) according to the venin theorem. In figure (c):

$$C= C' // C_4=1500\text{pF (convenient for calculating, } C_4 \text{ takes 1200 pF as value);}$$

$$U= 306\times C'C_4/(C'+C_4)=61.2\text{kV};$$

The PT exciting characteristic curve which manufacturer provided shows the impedance:

$$Z_L=61.2\text{M}\Omega\approx X_L\text{(the coil DC resistance value is 38k}\Omega\text{ or so, can be ignored), so:}$$

$$U_L= X_L\times U/(X_L- X_c)=63.4\text{ kV};$$

By this token, the voltage value of 500kV PT (63.4kV) matches the voltage value of A and B (60kV). The voltage value of C phase transmission is the largest. Maybe the PT resistance and capacitance to earth of C phase have a difference among there phase, same as the abnormity of 500 kVI bus voltage.

*4.2.The breakdown analysis of 500kV disc insulator*

On December 6th, at the first power transmission, it was very strange that 500kV disc insulator broke down. Through anatomizing the disc insulator, there were several long cracks on the discharge channel section, see figure 7.

**Figure 7.** The surface discharge channel section figure of 500kV disc insulator

The disc insulator passed the 592kV in 1min AC voltage test, but didn't pass 306kV operating voltage test. It was explained that the disc insulator had the congenital defect. The AC withstand voltage test projects have the low sensitivity to this kind of disfigurements. When the switches and brakes operating, following the expanded disfigurements of shock, leading the aberrance of local electric field. The disc insulator discharged flashover under low working voltage which was lower than AC withstand voltage[1-4], matching the characteristic of $SF_6$.

*4.3.The abnormal phenomenon analysis of 220kV Ⅳ bus voltage*

When 500kV transform 220kV side powered on 220kVⅣ bus, the oscillograph recorded the abnormal voltage waveform of 220kVⅣ bus A and B phase, which were irregular step wave or square wave. Taking the wave of figure 2 dashed line frame part for example.

According to the tin-opener checking, there was intermittent discharge formed by gap in the transformer gas-oil bushing of A and B phase, GIS conductor. It explained: in figure 8 at $t_1$, the gap broke down. The transient voltages of 220kV Ⅳ bus and transformer 220kV side were same. Then the gap discharge stopped. Then the bus voltage held the discharge value (ignore the bus leakage).

Until $t_2$, both of gap end voltages had reached breakdown value, broke down again. The bus voltage had been the transformer side AC voltage instantaneous value until $t_3$. At $t_3$, the gap broke down by reverse AC voltage. The bus voltage was repeated several times. Because of the dispersion and randomness of discharge, there were differences between A and B phase waveform at different time.

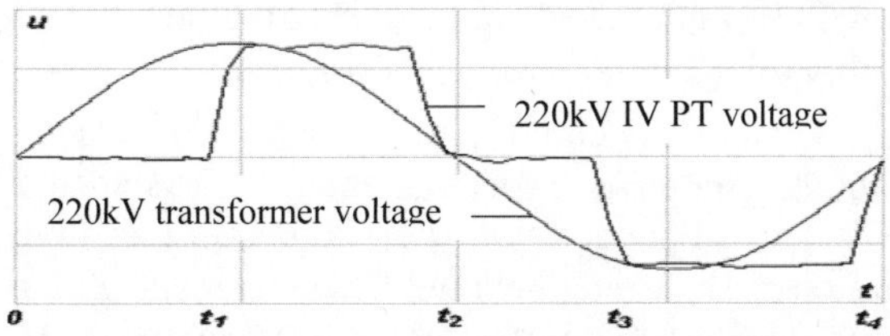


**Figure 8.** The partial enlarged drawing of dashed line frame part in figure 2

In fact, process above, the voltage of 220kVⅣ bus and PT could be denoted:

$$u_s = \sum_{n=1}^{N} U_{sn}\, e^{-\frac{t}{T_n}} + \sum_{m=1}^{M} U_{sm} \sin(\omega_m + \theta_m)\, e^{-\frac{t}{T_m}} \tag{1}$$

$U_{sn}$, $U_{sm}$: the voltage amplitude of 'n' and 'm'; $T_n$, $T_m$: the voltage decay time constant of 'n' and 'm'; $\omega_m$, $\theta_m$: the voltage angular frequency and phase angle.

The voltage of PT has two parts, the attenuation DC and high frequency shock in equation (1). The voltage of PT approximates the first part of equation (1). It's showed the PT response condition of attenuation DC voltage. See the equivalent circuit below for analysis.

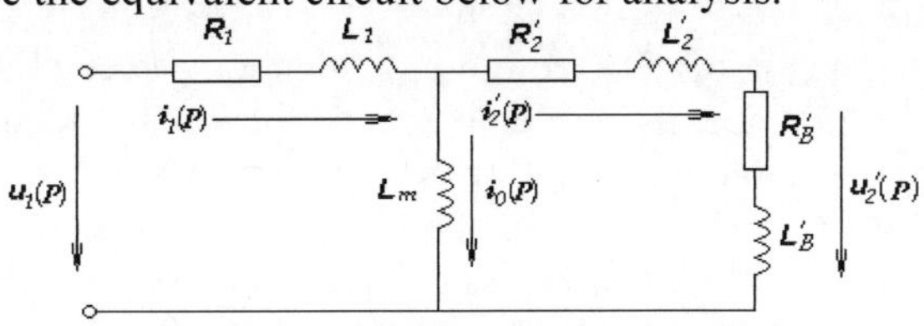


**Figure 9.** The equivalent circuit of PT

$R_1$, $R_2$': the conversion value of primary winding resistance and secondary winding resistance; $L_1$, $L_2$': the conversion value of primary winding leakage inductance and secondary winding leakage inductance; $L_m$: leakage inductance; $R_B$', $L_B$': the conversion value of load resistance and inductance.

According to the equivalent circuit in figure 9, the equation is obtained:

$$\left.\begin{aligned} u_1(p) &= i_1(p)(R_1 + pL_1) + i_0(p)pL_m \\ i_0(p)pL_m &= i_2'(p)\left(R_2' + pL_2'\right) + u_2'(p) \\ i_1(p) &= i_2'(p) + i_0(p) \\ u_2(p) &= i_2'(p)\left(R_B' + pL_B'\right) \end{aligned}\right\} \tag{2}$$

In formula, p is laplacian operator, if there is the attenuation DC voltage waveform of PT primary side:

$$u_1(t) = U_{1s} e^{-\frac{t}{T_n}} \tag{3}$$

In equation (3), $T_n$ is as the time constant, so:

$$u_1(p) = U_{1s}\frac{1}{p+\frac{1}{T_n}} \tag{4}$$

Substituting equation (3) into equation (4), to counter inverse laplace transform, obtain:

$$u_2'(t) = U_{1s}\left[\frac{1}{1-\frac{T_n}{T_m}} \cdot e^{-\frac{t}{T_n}} - \frac{T_n}{T_m} \cdot e^{-\frac{t}{T_m}} + \frac{R_1+R_2'}{R_B'} \cdot \frac{T_n}{T_n - T} \cdot e^{-\frac{t}{T}}\right] \tag{5}$$

In equation (5), $T = \frac{L_1+L_2'+L_B'}{R_1+R_2'+R_B'}$; $T_m = \frac{L_m+L_1}{R_1}$, $T_m$ is as the voltage transformer time constant.

In general, the voltage transformer of 220kVabove, $T_n \ll T_m$, $T_n/T_m \approx 0$, the second part of equation (5) can be ignored. By the time, T is small. The third part decays very quickly, ignore the influence. So the primary decides the secondary voltage. The secondary waveform of voltage transformer reflects the connected bus voltage change, they have a nice consistency[5-8].

Following voltage abnormality, the connection of GIS case flange had a spark phenomenon. Owing to the similar between the intermittent pulse discharge and the wave propagation property, when the pulse discharge travelled in the medium which were GIS conductor and cylinder wall. Then the connected flange wave resistance had changed, the refraction and reflection of wave would discharge.

## 5. The solving strategies

For this case, incorporated with location condition, focusing on the following tasks:

- To improve the quality of electrical equipment installation

  From the abnormal voltage and GIS tin-opener checking, the main reason is the installer's false installing without installation process standards, existed the demotivation, the installer approved. Otherwise, lack of supervision is an important reason. So to improve the employee responsibility consciousness, strengthen field management and site supervision assessment. Redesign installation process, stop the faults from taking place again[9]. To respect for science, reasonable installation, commissioning duration are the basis and precondition of quality.

- To strengthen the professional communication of the site equipment installation

  Objectively speaking, even the installation is improper, it must be found before the transmission. If the installation process has a well design, a good communication of installer and debugging, with a simple circuit conduction test, this problem will be found in time. So to strengthen and improve the work is needed.

- To improvement and increase the detection means

  AC withstand voltage test of GIS complies with requirements[10-13], the breakdown had happen all the time in transmission. Some defectives and hidden troubles can't be tested though the available technology. In GIS, increasing the AC withstand voltage test value (State Grid requires the site AC withstand voltage test value is 90% of delivery voltage test value, 220kV site withstand test value is same as delivery voltage test value). Monitoring ultrasound and UHF PD are effective means. While recent experience shows that lightning impact test, which detects GIS abnormal electrode and dirty has an effective means. It should be applied on site [14-16]. The delivery test of disc insulator should be tested, the impulsive check which are ultrasonic inspection, x-ray inspection, changes to the whole check.

Footnotes should be avoided whenever possible. If required they should be used only for brief notes that don't fit conveniently into the text.

## 6. Discussion

In conclusion:

- The reasonable schedule, reasonable installation process, perfect site supervision are the base of equipment installation quality;
- To strengthen the professional communication of the site equipment installation is the important means of reducing vulnerability.
- To increase the detection means is the premise of mastering the equipment condition.
- To change the system operation mode reasonably, is the effective method of removing 500kV bus voltage abnormality (Such as the 500kV transformer is operating with a prepared, the prepared transformer is in no-load operation).

## References

[1] Li Bin M 2007 SF6 High-voltage electrical appliances design (Beijing: China Machine Press)
[2] Luo Xueshen M 1999 SF6 gas insulative totally enclosed combined electric equipment (Beijing: China Electric Power Press)
[3] Li Xiaofeng Zhen Li Zhou Xinwei J 2010 Analysis of withstanding voltage test on 220 kV gas-insulated metal-enclosure switchgear(GIS) vol 43 (Beijing: Electric Power) p 43
[4] Zhao Erdong J 1997 500kV SF6Withstand voltage test research of composite apparatus vol 58 (Shenyang: Liaoning Journal of Electrical Engineering) p 16

[5] Wang Hongxin He Jingliang M 2003 Electromagnetic compatibility in power system (Wuhan: Wuhan University Press)
[6] Li Jinglu M 2008 Electromagnetic compatibility in power system (Beijing: China Electric Power Press)
[7] He Jingliang M 1994 Electromagnetic compatibility in power system (Wuhan: Wuhan University Press)
[8] Chen Jinfa Yang Lichuan Li Sinan J 2007 Analysis of discharge in GIS caused by disconnector failure no 3 (Hangzhou: Zhejiang Electric Power) p 23
[9] Li Runchun Wang Lugang Diao Zhiqun J 2010 GIS Combined Electrical Apparatus Installation Technology and Key Control Points vol 31 (Beijing: China Electric Power Press) p 37
[10] S 2004 DL/T 555-2004 Guide for withstand voltage and insulated test of gas-insulated metal-enclosure switchgear on site (Beijing: China Electric Power Press)
[11] S 2007 DL/T 402-2007 Specification of high-voltage alternating-current circuit- breakers (Beijing: China Electric Power Press)
[12] S 2010 DL/T 617-2010 Specification for gas-insulated metal-enclosed swithchgear (Beijing: China Electric Power Press)
[13] S 2008 GB 7674-2008 Gas-insulated metal-enclosed switchgear for rated voltages of 72.5kV and above (Beijing: China Electric Power Press)
[14] Wang Feng Zhang Qiaogen Qiu Yuchang J 2003 Research on the flashover behavior between electrode and insulator caused by bad contract in GIS no 1 (Guilin: Insulating Materials) p 45
[15] Li Jisheng Zhao Xuefeng Sun Zhenquan J 2009 The simulation study of oscillating impulse voltage for power equipment on-site testing vol 45 (Xi'an: High Voltage Apparatus) p 54
[16] Tian Kexin Wang Gang J 2003 Analysis of the surface flashover on GIS basin insulator under lightning impulse voltage vol 39 (Xi'an: High Voltage Apparatus) p 14

IOP Conf. Series: Materials Science and Engineering **224** (2017) 012035 doi:10.1088/1757-899X/224/1/012035

# Reliability Evaluation of Distribution System with Distributed Generation

**Guoyan Chen[1], Feng Zhang[2], Dahai You[2], Yong Wang[1], Guojun Lu[1], Qi Zou[2], Hengwei Liu[2], Junjie Qian[2] and Heng Xu[2]**

[1] Guangzhou Power Supply Co., Ltd., Guangzhou, china
[2] Huazhong University of Science and Technology, Wuhan, china

stuzhangfeng@sina.com

**Abstract.** Distribution system reliability assessment is an important part of power system reliability assessment. In recent years, distributed generations (DG) are more and more connected to distribution system because of its flexible and friendly environment features, which imposes a great influence on distribution system reliability. Hence, a reliability evaluation method suitable for distribution system with DG is imperative, which is proposed in this paper. First, a probabilistic model of DG output is established based on the generation characteristics of DG. Second, the island operation mode of distribution system with DG is researched, subsequently, the calculation method of the probability of island successful operation is put forward on the basis of DG model and the load model. Third, a reliability assessment methodology of distribution system with DG is proposed by improving the traditional minimal path algorithm for reliability evaluation of distribution system. Finally, some results are obtained by applying the proposed method to the IEEE-RBTS Bus6 system, which are consistent with the well-known facts. In this way, the proposed method is proved to be reasonable and effective.

## 1. Introduction

Distribution system is directly connected with the customers, which has a great influence on the power supply reliability. Statistics show that most of the power outages are caused by the failure of distribution system, so the reliability evaluation of distribution system has been widely concerned. In recent years, distributed generations are more and more connected to the distribution system for its flexible and friendly-environment features, which changes the structure and operation mode of traditional distribution system and brings significant impact to the reliability evaluation of the distribution system [1].

At present, some researches have been made on the reliability evaluation of distribution system with DG. In reference [2], the island partition model is established according to the importance of load, based on which, the reliability evaluation model of distribution system with constant DG is proposed, but the impact of intermittent DG is not considered. A three-state model of DG output is presented in [3], however, the variation range of intermittent DG output is large and three state can't accurately reflect the DG's actual state. In reference [4], multi-state output power model of wind turbine is researched and the reliability of distribution system is assessed using the minimal cut set method, which does not consider the effect of different types of DG on distribution system. In reference [5], Monte Carlo simulation method is used to calculate the reliability of distribution system with DG, which takes a long time.

In this paper, a probabilistic model of DG output is established through the research on the generation characteristics of DG. Then the calculation method of probability of island successful operation is put forward based on DG model and the load model of the island. Finally, the reliability of distribution system with DG is evaluated by improved minimal path algorithm.

## 2. Reliability model of DG and load

### *2.1. Model of traditional DG*

Traditional DG (e.g., diesel generator, micro gas turbine) is based on fossil fuels, whose output power is stable and continuously adjustable and can be represented by two state model: when DG fails to work, its output power is 0 and the corresponding probability is the unit forced outage rate (FOR); when DG works normally, its output power is the unit rated power ($P_{rated}$) and the corresponding probability is (1-FOR). Output power probability model of traditional DG is shown in Table 1.

**Table 1.** Output power probability model of traditional DG

| Level of output power | Output power | Probability |
|---|---|---|
| 1 | 0 | FOR |
| 2 | $P_{rated}$ | 1-FOR |

### *2.2. Model of renewable DG*

Renewable DG (e.g., wind turbine generator, photovoltaic power generation) is much more complicated than the traditional DG. Both the unit forced outage rate and the uncertainty of output power of renewable DG should be considered. Due to the intermittent nature of renewable energy, the renewable DG output power is uncertain. In this paper, output power probability model of wind turbine generator (WTG) is considered. The output power of WTG is affected by two factors: wind speed distribution and output power characteristic of the selected WTG.

*2.2.1. Model of wind speed distribution.* Weibull distribution model is one of the most commonly used models for simulating wind speed distribution. And the weibull probability density function is expressed as:

$$f(v)=(k/c)\cdot(v/c)^{k-1}\cdot\exp\left[(v/c)^{k}\right] \tag{1}$$

Where $k=(\sigma/v_m)^{-1.086}$ is called the shape parameter, $c=v_m/\Gamma(1+1/k)$ is called the scale parameter, $v_m$ and $\sigma$ represent the mean and standard deviation of the local wind speed, respectively.

*2.2.2. Output power characteristics of WTG.* Output power characteristic curve of WTG describes the relationship between the output power and the wind speed, which can be represented by the rated power ($P_{rated}$), cut-in wind speed ($v_{ci}$), rated wind speed ($v_r$) and cut-out wind speed ($v_{co}$) of the WTG. The relationship between output power of WTG ( $P(v)$ ) and the wind speed ($v$) can be expressed as:

$$P(v)=\begin{cases}0 & 0\le v\le v_{ci}\\ P_{rated}\dfrac{v-v_{ci}}{v_r-v_{ci}} & v_{ci}\le v\le v_r\\ P_{rated} & v_r\le v\le v_{co}\\ 0 & v_{co}\le v\end{cases} \tag{2}$$

*2.2.3. Output power probability model of WTG.* Output power of WTG is uncertain and two-state model can't accurately reflect the actual output state of it. Thus, multi-state model that can reflect the output power level of renewable DG is necessary. Output power of WTG is directly related to wind speed, so it can be calculated using the local statistical wind speed information within one year. As the wind speed range is large, the real-time wind speed is insteaded by average hourly wind speed in this paper. Wind speed is divided into several levels with the interval of 1m/s. Then the output power state and the corresponding probability of WTG can be obtained according to the output power characteristics and the local wind speed distribution. Output power probability model of a type of WTG is shown in Table 2.

Some relevant parameters are as follows:

Parameters about the WTG: $P_{rated}$ is 1MW, $v_{ci}$ is 4m/s, $v_r$ is 14m/s, $v_{co}$ is 25m/s.

Information about the local wind speed: $v_m$ is 10.0643m/s, $\sigma$ is 4.8426m/s.

**Table 2.** Output power probability model of WTG

| Level of output power | Wind speed ($m \cdot s^{-1}$) | Output power as a percentage of the rated power | Probability |
|---|---|---|---|
| 1 | 0~4 or >25 | 0 | 0.0977 |
| 2 | 4~5 | 5% | 0.0556 |
| 3 | 5~6 | 15% | 0.0660 |
| 4 | 6~7 | 25% | 0.0738 |
| 5 | 7~8 | 35% | 0.0788 |
| 6 | 8~9 | 45% | 0.0808 |
| 7 | 9~10 | 55% | 0.0799 |
| 8 | 10~11 | 65% | 0.0763 |
| 9 | 11~12 | 75% | 0.0707 |
| 10 | 12~13 | 85% | 0.0636 |
| 11 | 13~14 | 95% | 0.0555 |
| 12 | 14~25 | 100% | 0.2013 |

### *2.3. Load model*

The annual time series load model [6] is used in this paper, based on which, the load level is divided into several grades using the clustering technique. It is verified that in order to balance the calculation precision and time, the load level can be divided into 10 grades [7].

## 3. Probability of island successful operation

Distribution system has become a multi-power network with DG connected to it and islanding becomes a new operation mode of distribution system. Under normal circumstances, the main power and DG supply for the load. In the case of a fault, the circuit breaker and associated control components start to operate, leading to the separation of the partial distribution system containing DG from the main distribution system. Then DG can still supply for its independent distribution system, thus forming an island. However, the island can't operate successfully in any case after it is formed. Only when the output power of DG is greater than or equal to the load demand in the island and the DG works normally, can the island operate successfully, otherwise the island will fail to operate.

In this paper, the probability of island successful operation ( $\rho_{so}$ ) is calculated based on DG model and the load model mentioned above, and some provisions are made as follows: 1) only when the output power of DG is greater than or equal to the load demand, can the island operation be carried out; 2) any DG in the island fails to work , the island fails to operate; 3) during the normal operation of distribution system, the load in the original planned island can be supplied by both the main power and the DG at the same time, and when the distribution system gets fault, the DG can supply the load in the planned island continuously. Thus, $\rho_{so}$ can be calculated as:

$$\rho_{so} = \prod_{g=1}^{N_g}\left(1-\rho_g\right)\cdot\sum_{s=1}^{N_s}\frac{\max\left\{P_s^g - P_s^d, 0\right\}}{P_s^g - P_s^d}\rho_s \tag{3}$$

where $N_s = \prod_{i=1}^{N_d} nl_i^d \prod_{j=1}^{N_g} nl_j^g$ represents the total number of combinations of all possible states of all loads and DGs in the island, $N_d$ is the number of all load points (LPs) in the island, $N_g$ is the number of all DGs in the island, $nl_i^d$ represents the number of load levels of LP $i$, $nl_j^g$ represents the number of output power levels of DG $j$, $P_s^d$ and $P_s^g$ respectively represent the total amount of load and DG output power in the island at the *s-th* combination state, $\rho_g$ is the failure rate of DG *g*, $\rho_s = \rho_{1,l}^d \rho_{2,l}^d \cdots$ $\cdots$ represents the probability of the *s-th* combination state.

## 4. Reliability evaluation of distribution system with DG

### *4.1. Reliability indices*

Reliability indices of distribution system is an important criterion and basis for measuring system reliability [8], which can be usually divided into two levels: load point reliability indices and system reliability indices.

Load point reliability indices reflects the power supply reliability of a single load point, which include failure rate ($\lambda$), average outage time ($\gamma$) and annual outage time (*U*). The reliability indices of LP *i* are calculated as:

$$\lambda_i = \sum \lambda_{i,k} \tag{4}$$

$$U_i = \sum \left(\lambda_{i,k} \cdot \gamma_{i,k}\right) \tag{5}$$

$$\gamma_i = U_i / \lambda_i \tag{6}$$

Where $\lambda_{i,k}$ and $\gamma_{i,k}$ represent the failure rate and repair time of component *k* that has effect on the reliability of LP *i*.

In the distribution system with DG, the island is formed when failure occurs, and DG still supply for the LP within the island, reducing the LP's failure rate and outage time. As for the LP outside the island, DG has no effect on it. Thus, reliability indices of LP *j* within the island are calculated as:

$$\lambda_j = \rho_{so}\lambda_j^{'} + \left(1-\rho_{so}\right)\lambda_j^{"} \tag{7}$$

$$U_j = \rho_{so}U_j^{'} + \left(1-\rho_{so}\right)U_j^{"} \tag{8}$$

$$\gamma_j = U_j / \lambda_j \tag{9}$$

Where $\lambda_j^{'}$ and $U_j^{'}$ respectively represent the failure rate and annual outage time of LP in the island when the island succeed to operate, $\lambda_j^{"}$ and $U_j^{"}$ respectively represent the failure rate and annual outage time of LP in the island when the island fails to operate.

System reliability indices reflects the power supply reliability of the whole distribution system, and the typical indices include system average interruption frequency index (SAIFI), system average interruption duration index (SAIDI), customer average interruption frequency index (CAIFI), customer average interruption duration index (CAIDI), and average service availability index (ASAI). Some indices used in this paper are calculated as:

$$\text{SAIFI} = \frac{\sum_{i=1}^{n_{LP}} \left(N_i \cdot \lambda_i\right)}{\sum_{i=1}^{n_{LP}} N_i} \tag{10}$$

$$\text{SAIDI} = \frac{\sum_{i=1}^{n_{LP}} \left(N_i \cdot U_i\right)}{\sum_{i=1}^{n_{LP}} N_i} \tag{11}$$

$$\text{ASAI} = \frac{8760\sum_{i=1}^{n_{LP}} N_i - \sum_{i=1}^{n_{LP}} \left(U_i \cdot N_i\right)}{8760\sum_{i=1}^{n_{LP}} N_i} \tag{12}$$

Where $n_{LP}$ is the number of LP, $N_i$ is the number of customer connected to LP *i*.

### *4.2. Reliability evaluation of distribution system based on improved minimal path method*

When distribution system with DG gets a fault, the circuit breaker and associated control components start to operate, forming an island. DG in the island still supply for its load, thus the power supply reliability of the load within the island is improved. For the load outside the island, DG has no effect on its reliability. In this paper, traditional minimal path method [9] is improved, which is suitable for reliability calculation of distribution system with DG. The basic idea of the improved minimal path method is as follows:

**Step1:** First of all, get the minimal path of each LP to the main power. All the components in the system are divided into two parts: the minimal path components and the non-minimal path components.

**Step2:** Determine the type of LPs according to the island partition. For LPs outside the island, calculate the reliability indices using the traditional minimal path method. For LPs within the island, calculate the reliability indices using the improved method described as below.

**For the minimal path components:**

**a.** If the component is within the planned island and it gets fault, the DG in the island automatically stop working. The LPs are not supplied due to the failure of component and the outage time is the component's repair time.

**b.** If the component is outside the planned island and it gets fault, the island breaker and associated control components automatically operate and the island will operate with a certain probability. If the island succeeds to operate, the island LPs will be supplied by DG and the component failure has no effect on the reliability of the LPs; if the island fails to operate, the island LPs will be unsupplied and the outage time is the component's repair time.

**For the non-minimal path components:**

**c.** If the component of the load branch to which LP is not connected gets fault, the fuse connected to the load branch automatically work, and the LP reliability is not affected.

**d.** For the non-minimal path component on the main feeder and the branch feeder, search for the upstream breaker and switch nearest to it, then determine the effect of the component failure on the LP reliability according to the position of circuit breaker and switch.

**d1.** If breaker is not on the LP minimal path and the component gets fault, the breaker will be automatically disconnected and the LP reliability is not affected.

**d2.** If breaker is on the LP minimal path and switch is not on the LP minimal path, the effect of the component failure on the LP reliability is similar to case **b** above, and the difference is that when the island fails to operate, the LP outage time is the switch operation time.

**d3.** If breaker is on the LP minimal path and switch is also on the LP minimal path, the effect of the component failure on the LP reliability is the same as case **b** above.

**Step3:** Calculate LP reliability indices based on the above cases, and then calculate the system reliability indices.

## 5. Case study

### *5.1. Case data*

In this paper, taking the main feeder F4 of IEEE-RBTS Bus6 system [10] as an example to evaluate the reliability. There are 23 load points, 1 disconnecting switch, 23 distribution transformers, 23 fuses and 4 circuit breakers in the system. The original data and reliability parameters of each component are consistent with that in [10].

It is considered to combine the WTG with some diesel generators to supply for the island load. That is, DGs connected to the distribution system contain both traditional DG and renewable DG. The failure rate of diesel generator and WTG are 0.02 and 0.04, respectively.

Connecting DG to branch line 19 and 25, respectively. When failure occurs in the distribution system, island breaker and associated control components start to operate, then island 1 and island 2 will be formed, as shown in Figure 1. Assuming that each LP is at the same level at a time, the island load level classification and the corresponding probability are shown in Table 3.

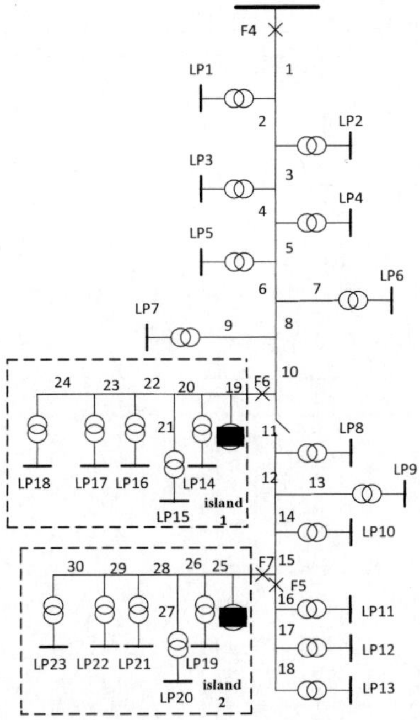


**Figure 1.** Connection diagram of distribution system with DG

**Table 3.** Load level and the corresponding probability

| Island 1 | | Island 2 | |
|---|---|---|---|
| Load (kW) | Probability | Load (kW) | Probability |
| 2400.90 | 0.0001 | 2597.20 | 0.0001 |
| 2140.16 | 0.0429 | 2315.14 | 0.0429 |
| 1968.50 | 0.0992 | 2129.44 | 0.0992 |
| 1816.28 | 0.1079 | 1964.78 | 0.1079 |
| 1653.02 | 0.1411 | 1788.17 | 0.1411 |
| 1510.41 | 0.1598 | 1633.90 | 0.1598 |
| 1346.18 | 0.1223 | 1456.25 | 0.1223 |
| 1197.09 | 0.1477 | 1294.96 | 0.1477 |
| 1039.59 | 0.1165 | 1124.59 | 0.1165 |
| 869.61 | 0.0625 | 940.71 | 0.0625 |

*5.2. Study results*

In this paper, the reliability evaluation of distribution system with DG is researched based on the following several scenarios.

**1) scenario 1:** evaluate the reliability of distribution system without DG.

**2) scenario 2:** connect DG to island 1, where the rated capacity of WTG is 1000kW and the capacity of diesel generator is 50% of the island peak load.

**3) scenario 3:** connect DG to island 2, where the rated capacity of WTG is 1000kW and the capacity of diesel generator is 50% of the island peak load.

**4) scenario 4:** connect DG to island 2, where the rated capacity of WTG is $2\times 1000$kW=2000kW and the capacity of diesel generator is 50% of the island peak load.

**5) scenario 5:** connect DG to island 2, where the rated capacity of WTG is 1000kW and the capacity of diesel generator is 20%, 40%, 60%, 80%,100% of the island peak load , respectively.

According to equation (3), calculating the $\rho_{so}$ of scenario 2~5 , respectively. And the $\rho_{so}$ of scenario 2,3,4 is 0.7123, 0.6844, 0.8205, the $\rho_{so}$ of scenario 5 is shown in Table 4.

IOP Conf. Series: Materials Science and Engineering **224** (2017) 012035 doi:10.1088/1757-899X/224/1/012035

**Table 4.** Probability of island successful operation in scenario 5

| Percentage of diesel generator capacity of island peak load | 20% | 40% | 60% | 80% | 100% |
|---|---|---|---|---|---|
| $\rho_{so}$ | 0.1620 | 0.5062 | 0.8253 | 0.9764 | 1 |

Calculating the reliability indices using the proposed evaluation method. Some load points' reliability indices are shown in Table 5 and Table 6, and the system reliability indices are shown in Table 7. (scenario 5.1~5.5 corresponds to the five cases in scenario 5)

**Table 5.** Failure rate index of some load points

| Scenario | | LP1 | LP6 | LP9 | LP15 | LP20 |
|---|---|---|---|---|---|---|
| 1 | | 1.6725 | 1.7115 | 1.7115 | 2.5890 | 2.5598 |
| 2 | | 1.6725 | 1.7115 | 1.7115 | 1.4083 | 2.5598 |
| 3 | | 1.6725 | 1.7115 | 1.7115 | 2.5890 | 1.4253 |
| 4 | | 1.6725 | 1.7115 | 1.7115 | 2.5890 | 1.1998 |
| 5 | 5.1 | 1.6725 | 1.7115 | 1.7115 | 2.5890 | 2.2913 |
| | 5.2 | 1.6725 | 1.7115 | 1.7115 | 2.5890 | 1.7207 |
| | 5.3 | 1.6725 | 1.7115 | 1.7115 | 2.5890 | 1.1918 |
| | 5.4 | 1.6725 | 1.7115 | 1.7115 | 2.5890 | 0.9413 |
| | 5.5 | 1.6725 | 1.7115 | 1.7115 | 2.5890 | 0.9022 |

**Table 6.** Annual outage time index of some load points

| Scenario | | LP1 | LP6 | LP9 | LP15 | LP20 |
|---|---|---|---|---|---|---|
| 1 | | 8.4015 | 8.5965 | 11.4825 | 12.9840 | 15.7237 |
| 2 | | 8.4015 | 8.5965 | 11.4825 | 9.1364 | 15.7237 |
| 3 | | 8.4015 | 8.5965 | 11.4825 | 12.9840 | 10.0516 |
| 4 | | 8.4015 | 8.5965 | 11.4825 | 12.9840 | 8.9239 |
| 5 | 5.1 | 8.4015 | 8.5965 | 11.4825 | 12.9840 | 14.3815 |
| | 5.2 | 8.4015 | 8.5965 | 11.4825 | 12.9840 | 11.5287 |
| | 5.3 | 8.4015 | 8.5965 | 11.4825 | 12.9840 | 8.8839 |
| | 5.4 | 8.4015 | 8.5965 | 11.4825 | 12.9840 | 7.6316 |
| | 5.5 | 8.4015 | 8.5965 | 11.4825 | 12.9840 | 7.4362 |

**Table 7.** System indices

| Scenario | | SAIFI | SAIDI | ASAI |
|---|---|---|---|---|
| 1 | | 1.9778 | 11.0747 | 0.998736 |
| 2 | | 1.8201 | 10.5608 | 0.998794 |
| 3 | | 1.8263 | 10.3171 | 0.998822 |
| 4 | | 1.7962 | 10.1665 | 0.998839 |
| 5 | 5.1 | 1.9420 | 10.8954 | 0.998756 |
| | 5.2 | 1.8658 | 10.5144 | 0.998800 |
| | 5.3 | 1.7951 | 10.1611 | 0.998840 |
| | 5.4 | 1.7617 | 9.9939 | 0.998859 |
| | 5.5 | 1.7564 | 9.9678 | 0.998862 |

Some results can be obtained as follows by analysing the calculation results of reliability indices.

**1) Effect of DG on the reliability**

It can be obtained by comparing reliability indices of scenario 1 and the other scenarios that: after DG is connected to the distribution system, reliability indices of load points outside the island have no changes; the failure rate and annual outage time of load points within the island are all reduced; and the system indices are also improved.

**2) Effect of DG position on the reliability**

It can be obtained by comparing the calculation results of scenario 2 and scenario 3 that: the improvement of system indices (except for SAIFI) when DG is connected to island 2 is better than that when DG is connected to island 1, which shows that the more the DG is closer to the end of feeder, the more obvious the system reliability is improved.

**3) Effect of DG capacity on the reliability**

By comparing the calculation results of scenario 3 and scenario 4, or the various cases of scenario 5, it can be obtained that: with DG (WTG or diesel generator) capacity increases, probability of island successful operation increases, at the same time, the reliability improves but the degree of improvement is getting smaller and smaller.

**4) Effect of DG intermittency on the reliability**

It can be obtained from the calculation results of scenario 3, scenario 4 and scenario 5.5 that: when WTG capacity and diesel generator capacity increases at the same ratio, the latter one is better than the former one in the degree of improvement for the distribution system reliability. It shows that the contribution of WTG to the system reliability is less than that of diesel generator, which reflects the adverse effects of intermittency of the renewable DG.

## 6. Conclusions

The structure and operation mode of traditional distribution system are changed as DG is connected to it, which presents a new challenge to the reliability evaluation of distribution system. In this paper, a probabilistic model of DG output is established. Then the calculation method of probability of island successful operation is put forward based on DG model and the load model. Finally, the reliability of distribution system with DG is assessed by the improved minimal path algorithm. By applying the proposed method to a typical distribution system, some results are obtained which are consistent with the well-known facts, thus, the proposed method is proved to be reasonable and effective. The results are as below: 1. Accessing DG to distribution system can improve the reliability of load point of the island and whole system; 2. The more the DG is closer to the end of feeder and the greater the DG capacity is, the more obvious the system reliability is improved; 3. Traditional DG has a better improvement on the distribution system reliability compared to renewable DG.

## References

[1] Kang L, Guo H, Wu J and Chen S 2010 Characteristics of distributed generation system and related research issues caused by connecting it to power system *Power System Technology* 34(11):43-47

[2] Liu C and Yan Z 2007 Distribution network reliability considering distribution generation *Automation of Electric Power Systems* 31(22):46-49

[3] Fotuhi-Firuzabad M and Rajabi-Ghahnavie A 2005 An analytical method to consider DG impacts on distribution system reliability *Transmission & Distribution Conf. & Exhibition: Asia and Pacific* pp 1-6

[4] Xu Y and Wu Y 2011 Reliability evaluation for distribution system connected with wind-turbine generators *Power System Technology* 35(4):154-158

[5] Alkuhayli A A, Raghavan S and Chowdhury B H 2012 Reliability evaluation of distribution systems containing renewable distributed generations *North American Power Symp.* pp 1-6

[6] Subcommittee P M 1979 IEEE reliability test system *Power Apparatus & Systems IEEE Transactions on* PAS-98(6):2047-2054

[7] Singh C and Kim Y 1988 An efficient technique for reliability analysis of power systems including time dependent sources *IEEE Transactions on Power Systems* 3(3):1090-1096

[8] Conti S and Rizzo S A 2015 An Algorithm for Reliability Assessment of Distribution Systems in Presence of Distributed Generators *International Journal on Electrical Engineering & Informatics* 7(3):502-516

[9] Bie C and Wang X 1997 Reliability analysis of distribution networks *Electric Power* (5):10-13

[10] Allan R N, Billinton R, Sjarief I and Goel L 1991 A reliability test system for educational purposes-basic distribution system data and results *IEEE Transactions on Power Systems* 6(2):813-820

IOP Conf. Series: Materials Science and Engineering **224** (2017) 012036 doi:10.1088/1757-899X/224/1/012036

# The Dynamic and Optimized Modeling Method of Secondary Winding for Transformer

**Xiudong Xu, Rui Li, Liang Zhao, Zhiyuan Shen, Jie Cheng, Bo Zeng, Yu Zhang and Pengcheng Gao**

Northwest Institute of Nuclear Technology, Xi'an, Shaanxi, 710024, China

xuxiudong@nint.ac.cn

**Abstract.** In order to analyze dynamic characteristics of secondary winding for Transformer, one joint interface dynamic modeling and optimized methodology is provided in this paper. A layer of equivalent material between secondary winding and clamp is established, which can simulate stiffness and damp in the joint interface. The initial elastic modulus, poisson ratio and density of equivalent material are deduced from Hertz contact theory and fractal theory. The final property parameters of equivalent material are obtained by using optimized model. Simulation and experiments results show that the dynamic and optimized model can simulate experimental results exactly. The relative errors between the simulation frequencies of new model and experimental frequencies are controlled within 2%.

**Key words.** Secondary winding, Transformer, dynamic and optimized model, equivalent material, joint interface

## 1. Introduction

There are many influence factors for dynamic characteristics of complicated machine, such as structure, material, stiffness of interface, damp of interface etc. It is said that dynamic characteristics can be improved by designing the styles of structure. Besides, the characteristics of interface are the important factors, which influence on dynamics. The stiffness and damp of interface are the most important properties.

There are about three modeling methods for interface of two mechanical parts: rigid link for two parts, stiffness and damp elements modeling, non-linear modeling method. The stiffness is infinite for rigid link method, which can not simulate truly for two mechanical parts. Therefore, many researchers focus on other two methods. Multi-spring and damp method is given between two metal joint interfaces by Namazi etc, which also study dynamic characteristics for joint interface [1]. It proposes FEM joint interface model for zero thickness by Mayer etc, which analyze the non-linear response [2]. Mottershead identifies dynamic parameters through comparing input accelerations and output response[3]. Fang identifies friction damp between fixed part and movement part. Besides, he analyzes system stabilization for damp [4]. Unfortunately, these researches can not simulate accurately for non-linear dynamic problems. Therefore, one dynamic modeling field can be proposed based on Hertz contact theories [5]. GW[6] model and MB[7] model are classical research methods among Hertz contact theory and fractal theory. Tian Hongliang gives new dynamic modeling method based on above researches [8]. He converts the dynamic modeling problem to parameters identification problem, and gives calculated method for elastic modulus and poisson ratio of virtual material. Thus, this method must obtain relative surface parameters.

Transformers have been applied widely in the field of electric power. In different and complex applications of electrical Transformer, mechanical vibration may destroy the Transformer. Generally, Transformer contains about primary winding, secondary winding, inner cylinder, outer cylinder and insulator etc, which are shown in Figure 1. The secondary winding can be destroyed more easily. Therefore, we focus on dynamic modeling of secondary winding in this paper. The secondary winding contains two parts, which are basic cone and cable. The cable is enlaced around the basic cone. The dynamic characteristics of secondary winding should be analyzed and understood in order to avoid cable loosening from basic cone.

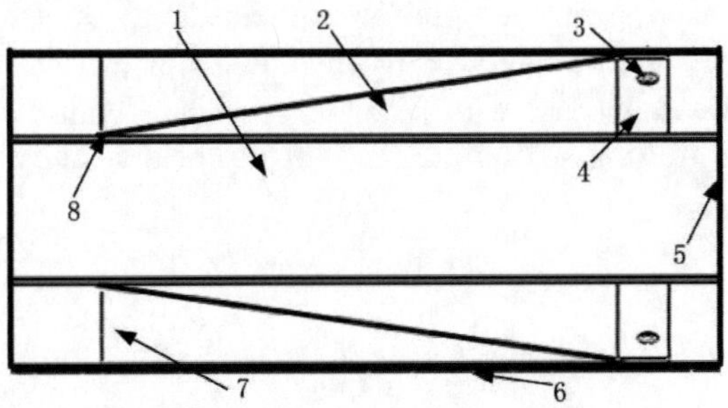


**Figure 1.** Schematic diagram of a Transformer

(1,6- inner and outer cylinder of Transformer; 2,7- secondary and primary windings of Transformer;3- the screw of larger winding side; 4-the larger winding side;5- insulator of Transformer; 8-the smaller winding side)

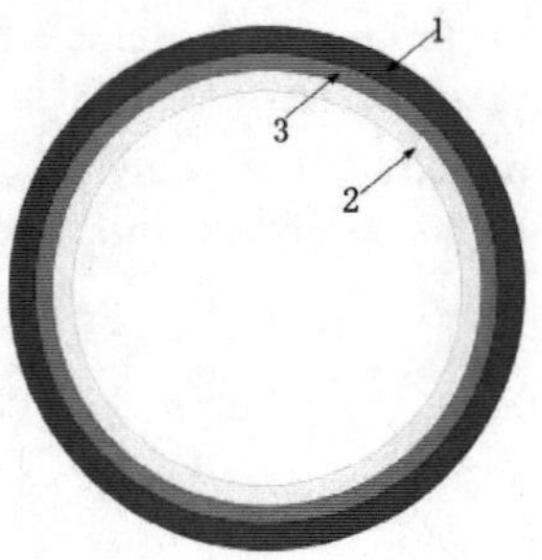


**Figure 2.** The diagram of equivalent material

(1- outer cylinder of Transformer; 2-secondary winding; 3-equivalent material)

Generally, the model of basic cone is regarded as important for secondary winding. Actually, the joint interface between secondary winding and clamp can impact on dynamic characteristic directly according to former researches. Therefore, one dynamic and optimized model of joint interface is presented for secondary winding in this paper. It is assumed that one equivalent material layer is existed between outer magnetic cores and secondary winding for Transformer, which is shown in Figure 2. We should get initial parameters of equivalent material firstly through calculated model. And the final parameters can be obtained by comparing simulation and experimental response of secondary winding. It is not necessary to obtain accurate surface parameters, which is the advantage for this methodology.

## 2. Simplified structure model of secondary winding

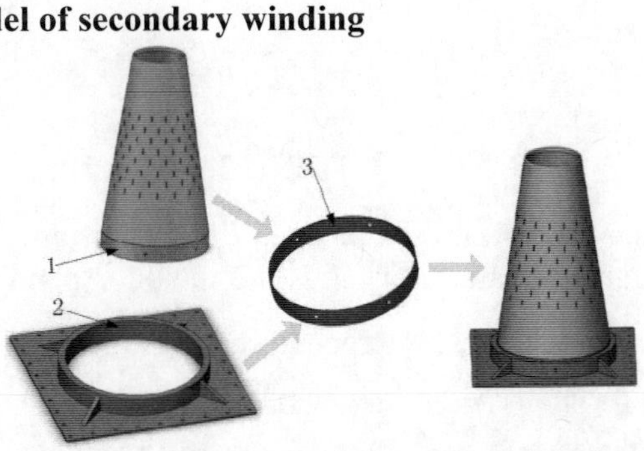


**Figure 3.** The simulation model of the secondary winding (1-outer joint interface of secondary winding; 2-inner joint interface of clamp 3-the layer of equivalent material)

The actual model of secondary winding is shown in Figure 1. The secondary winding of Transformer contains: inner and outer cylinder, secondary and primary windings, two insulators. The inner cyliner are laid between with two insulators. The larger side of secondary winding connects to outer cylinder with six screw hole. The smaller side of secondary winding contacts inner cylinder for circle mode. The actual model can be converted to simplified structure model in Figure 3. There are secondary winding and clamp in the simplified model, which can simulate actual situation for larger side of secondary winding. It is assumed that one layer of equivalent material exists between outer joint interface of secondary winding with inner joint interface of clamp. It is seen that this simplified model is one worse cantilever. Therefore, the dynamic characteristics of secondary winding can be checked fully in this condition.

## 3. Dynamic and optimized methodology for secondary winding

In the simplified model, the joint interface between secondary winding with clamp may be considered equivalent material in the same cross area. A series of analytic solutions of elastic modulus $E$, poisson ratio $\mu$, density $\rho$ and thickness $h$ could be obtained by interface dynamic model, which show in (1)~(4) respectively. The equivalent material is rigidly connected with secondary winding with clamp situated in both side of joint interface. The parameters of $E$, $\mu$ and $\rho$ are the property of the equivalent material. And the $h$ is the physical parameter of the material layer.

$$E = f_1(E_1, E_2, \mu_1, \mu_2, R_{a1}, R_{a2}, P) \tag{1}$$

$$\mu = f_2(E_1, E_2, \mu_1, \mu_2, R_{a1}, R_{a2}, P) \tag{2}$$

$$h = f_3(h_1', h_2', R_{a1}, R_{a2}) \tag{3}$$

$$\rho = f_4(\rho_1, \rho_2) \tag{4}$$

In which $f_1$ is the mathematics model of $E$, $f_2$ is the mathematics model of $\mu$, $f_3$ is the mathematics model of $h$, $f_4$ is the mathematics model of $\rho$. And the $E_1$ is the elastic modulus of secondary winding, $E_2$ is the elastic modulus of clamp, $\mu_1$ is the poisson ratio of secondary winding, $\mu_2$ is the poisson ratio of clamp, $R_{a1}$ is the roughness of outer joint interface for secondary winding, $R_{a2}$ is the roughness of inner joint interface for clamp, $P$ is the pressure between secondary winding with clamp, $\rho_1$ is the density of secondary winding, $\rho_2$ is the density of clamp, $h_1'$ is the largest height of microcontact for secondary winding, $h_2'$ is the largest height of microcontact for clamp.
The mathematics model $f_1$, $f_2$ and $f_4$ are provided directly respectively via Hertz contact theory and fractal theory [5], [8].

$$E = \frac{\sqrt{2}}{6\pi\sqrt{\pi}} D E' G^{1-D} \psi^{1-\frac{D}{2}} a_L'^{\frac{D}{2}} (a_L'^{\frac{1}{2}} - a_C'^{\frac{1}{2}}) \tag{5}$$

$$\mu = \frac{(1+\mu')E}{G_x^* E'} \tag{6}$$

$$\rho = \frac{\rho_1 h_1 + \rho_2 h_2}{h_1 + h_2} \tag{7}$$

Where $D$ is the fractal dimension of surface profile, $E'$ is the equivalent elastic modulus of two surfaces, $G$ is the fractal roughness parameter. $\psi$ is the region expanding coefficient, which can be calculated by $D$. $a_L'$ is the area of largest microcontact, $a_C'$ is the area of critical regimes. $E'$ is the equivalent elastic modulus of material, which can be given by (8). $\mu'$ is the equivalent poisson ratio of material, which can be given by (9), $G_x^*$ is the equivalent shear modulus of material [9].

$$E' = \frac{E_1 E_2}{E_1(1-\mu_2^2) + E_2(1-\mu_1^2)} \tag{8}$$

$$\mu' = \frac{E'G_2(2-\mu_1) + E'G_1(2-\mu_2)}{2G_1G_2} - 1 \tag{9}$$

Where $G_1$ and $G_2$ are the shear modulus of secondary winding and clamp, separately.

Generally, $h$ is chosen as 1mm [8]. And the density $\rho$ is obtained by (7). It is concluded that $\rho$ is less influence on dynamic characteristics in the former researches. Otherwise, the precise $E$ and $\mu$ are decided by interface parameters $D$ and $G$. In this paper, we need only imprecise $E_0$ and $\mu_0$, which can be seen as initial parameters of equivalent material.

It is assumed that $q_i^{end}(i=1,2,3,4,5)$ are experimental frequency of secondary winding, in which $i$ stands for phase index. The simulation frequency of secondary winding are $q_i\ (i=1,2,3,4,5)$. And the optimized model is shown as (10).

$$\min\left[g(\boldsymbol{x})\right]=\sum_{i=1,2,3,4,5}\left|q_i-q_i^{end}\right|\cdot n_i \tag{10}$$

Where, $\boldsymbol{x}=\begin{bmatrix}E & \mu\end{bmatrix}^T$ are optimized variable, $n_i(i=1,2,3,4,5)$ are coefficients. It is seen that PATRAN software is used to optimize relative parameters, and the optimized algorithm is GA.

## 4. Dynamic experiment and simulations result

The material of secondary winding is PI, whose elastic modulus is 3.2Gpa, poisson ratio is 0.36, density is 1300 kg/m$^3$. The material of clamp is Q235, whose elastic modulus is 205Gpa, poisson ratio is 0.29, and density is 7850 kg/m$^3$.

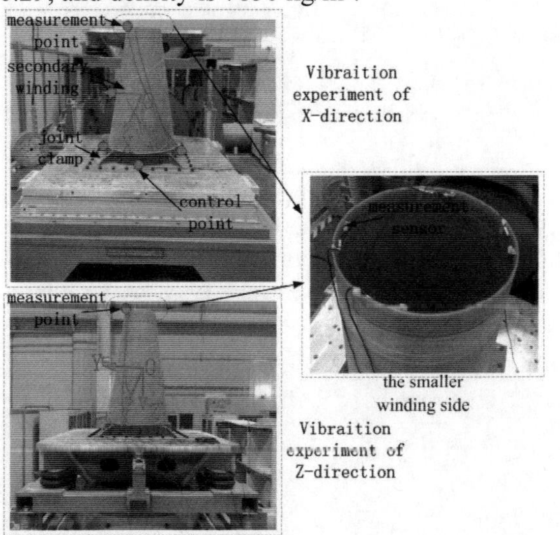


**Figure 4.** The diagram of vibration experimental system.

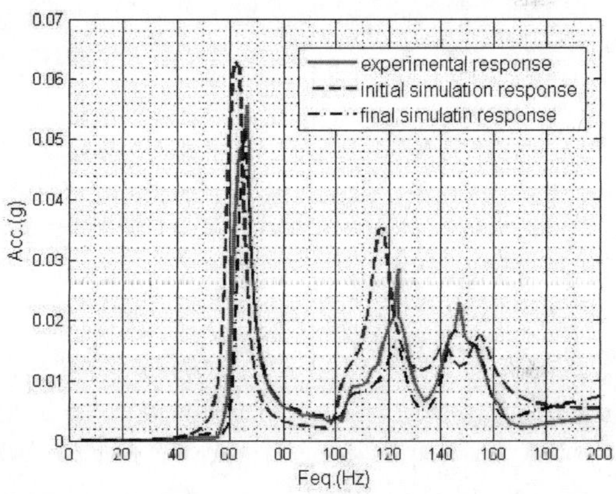


**Figure 5.** Comparative response of x-direction for 1g z-vibration.

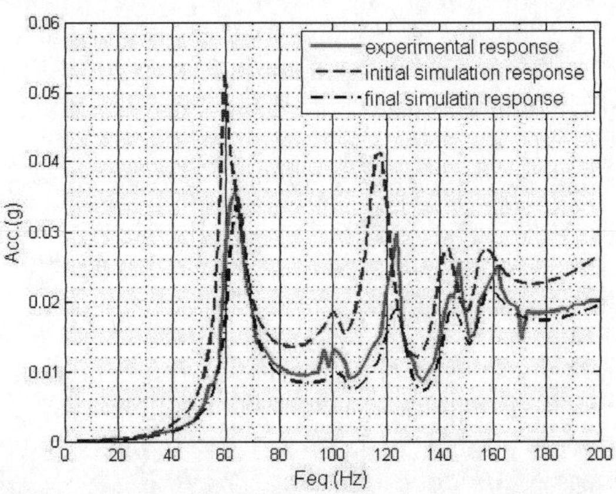


**Figure 6.** Comparative response of y-direction for 1g z-vibration.

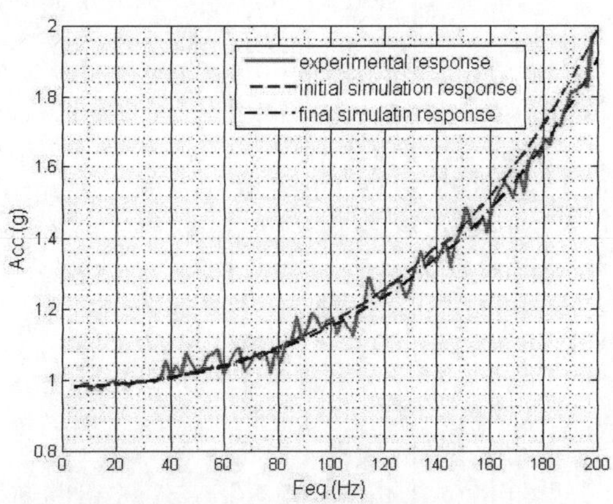


**Figure 7.** Comparative response of z-direction for 1g z-vibration.

The dynamic experiment of secondary winding has been carried on, which is shown in Figure 4. We select three control points, which laid on clamp. The measurement point is fixed on the smaller side of secondary winding. The experiments of x-direction and z-direction 1g sine sweep frequency have been implemented. In order to show typical frequency, we select z-direction 1g sweep experiment, which is seen as Figure 5~Figure 7. And the final comparison between experiment and simulation are detailed by Table 1. The elastic modulus of equivalent material becomes stably from 20.21Mpa to 48.57Mpa, and poission ratio becomes steadily from 0.362 to 0.1811. The errors between simulation and experiment are smaller than the model for initial parameter value.

**Table 1**. Relative results of experiment and simulation

| Item | The first frequency(Hz) | The second frequency(Hz) | The third frequency(Hz) | The fourth frequency(Hz) | The fifth frequency(Hz) |
|---|---|---|---|---|---|
| Vibration experiment | 66.45 | 102.27 | 124.37 | 147.35 | 161.95 |
| Final frequency value | 65.15 | 102.5 | 122.1 | 145.4 | 159.05 |
| errors | 0.45% | 0.22% | 1.83% | 1.32% | 1.79% |
| Parameters | Initial value：elastic modulus（20.21Mpa）poisson ratio（0.362） | | | | |
| | Final value：elastic modulus（48.57Mpa）poisson ratio（0.1811） | | | | |

## 5. Conclusions

The paper interests in the problems of joint interface modeling for secondary winding. Therefore, one joint interface modeling and optimized methodology is provided. This method can simulate the damp and stiffness between secondary winding and other part. The exact acceleration responses are obtained by using this simulation method. The simulation results show that the response curves are similar with experiment curves. The frequency errors can be controller within 2%. It can improve the simulation veracity for secondary winding.

## References

[1] Namazi M, Altintas Y, Abe T, et al. 2009 Modeling and identification of tool holder-spindle interface dynamics *International Journal of Machine tool and Manfacture* **47** pp 1333-1341.

[2] Mayer M H, Gaul L 2007 Segment-to-Segment contact elements for modelling joint interfaces in finite element analysis *Mechanical Systems and Signal Processing* **21** pp 724-734.

[3] Mottershead J E, Stanway R 1986 Identification of structural vibration parameters by using a frequency domain filter *Journal of Sound and Vibration* **109** pp 495-506.

[4] Fang B, Devor R E 2002 Influence of friction damping on workpiece-fixture system dynamics and machining stability *Journal of Manufacturing Science and Engineering* **124** pp 226-233.

[5] Campbell P, Abhyankar S 1978 Fractals, forms, chance and dimension *The Mathematical Intelligencer* **1** pp 35-37.

[6] Fisher R 1953 Dispersion on a sphere *Proceeding of the Royal Society of London. Series A. Mathematical and Physical Sciences* **217** pp 295-305.

[7] Majumdar A, Bhushan B 1991 Fractal model of elastic-plastic contact between rough surfaces *Journal of Tribology* **113** pp 1-11.

[8] Tian Hongliang, Li Bin, Liu Hongqi, Mao Kuanmin, Peng Fangyu, Huang Xiaolei 2011 A New Method of Vitual Material Hypothesis-Based Dynamic Modeling on Fixed Joint Interface in Machine Tools *International Journal of Machine Tools and Manufacture* **51** pp 239-249.

[9] Ray Sudipto, Roy Chowdhury S K 2007 Prediction of Flash Temperature at the Contact Between Sliding Bodies with Nanoscale Surface Roughness *ASME Journal of Tribology* **129** pp 467-480.

IOP Conf. Series: Materials Science and Engineering **224** (2017) 012037 doi:10.1088/1757-899X/224/1/012037

# Research for the Working Method of Live Replacing the L-type Insulator String on ±800kV DC Transmission Lines

**Jinliang Li[1,2], Zhigang Wang[1,2], Dongyun Lei[1,2], Miao Yang[1,2], Zengming Xia[1,2], Yun Xiang[1,2], Yu Yan[1,2], Zhiguo Liu[1,2]**

1.State Grid Hunan Electric Power Company Live Working Center,Changsha 410100,China

2.Hunan Province Key Laboratory of Intelligent live working Technology and Equipment (Robot) ,Changsha 410100,China

530089657@qq.com

**Abstract.** It would be hard to be repaired in the case of a power outage if the ±800kV ultra high voltage DC (UHVDC) transmission line was put into operation, so live working is the guarantee for its security and stability. Insulator strings of duplex L-type are widely used in UHVDC small angle towers. In the meantime, the tower structure and the insulator strings' arrangement type is different than the V-type. In order to ensure the security and stability of the transmission line, this paper presents a method of live replacing the duplex L-type insulators based on the actual parameter of the UHVDC L tower type and develops apparatus as a complementary supporting by mechanics analysis. In order to ensure the safety of working staff, the safe distance and combined gap of the hanging basket method are checked. Safety precautions are also proposed in the paper by the calculation of body surface field strength of the persons doing live working in the finite element method. It is shown by the field application that the live working task of replacing the L-type insulator strings can be accomplished successfully by the method put forward in this paper, which provides reference for the following live working on the UHV transmission line.

## 1. Introduction

The large capacity, long distance, low loss transmission ability can be realized by the UHV transmission system, which can promote energy structure adjustment and optimized layout. With the development of the UHV lines, Insulator strings of duplex L-type are widely used in UHVDC small angle towers. The insulators in service will be damaged because of the influence of the working voltage, switching overvoltage, lightning overvoltages, the lead weight, wind and the external force, which will affect the safe operation of the lines. However, The UHV transmission lines have less maintenance chance. In order to ensure the power supply reliability, the broken insulators need to be replaced timely [1-4].

As the tower structure and the insulator strings' arrangement of the L-type is different than the V-type, the method and matching tools of live replacing V-type insulators are not suitable for the L-type insulators. Take the ±800 kV UHV small angle tower of the binjin line as an example, a new method of live replacing the duplex L-type insulators based on the actual parameter of the UHVDC L-type tower is put forward and the matching tools are developed. Moreover, the insulation pull rods can apply to the insulator string's replacing in the two different orientation by the adjustable structure of the length. In the meantime, the electric field intensity of the equipotential worker are calculated by the finite method and so the shielding efficiency of the shielding clothes is confirmed. The task of live replacing the L-type insulator strings can be accomplished successfully by using the method put forward in this paper, which provides the reference for the following live working on the UHV transmission line.

## 2. Presentation of New Method

The ±800 kV small angle tower is used as the selected model and its field arrangement of duplex L-type insulator string is shown in Fig.1, its total height is 57.8 m and the nominal height is 51 m. The rotation degree is about 12°.The wire is arranged as regular hexagon and its type is ASCR-720/50; its bundle spacing is 450 mm and the diameter of the branch wire is 36.24 mm. The length of the composite insulator strings is 9.6 m, however, the lengths of the connecting fittings are different because of the unique tower structure. And the length of the insulator string and fittings in one direction is 11.925m; the length in the other direction is 10.96m. Considering the practicability and economy, the construction holes and united board are determined as the fulcrum according to the field situation. In order to meet the requirements of generality, the insulation pull rods' length should be adjustable.

**Figure 1.** The field arrangement of L-type insulator string

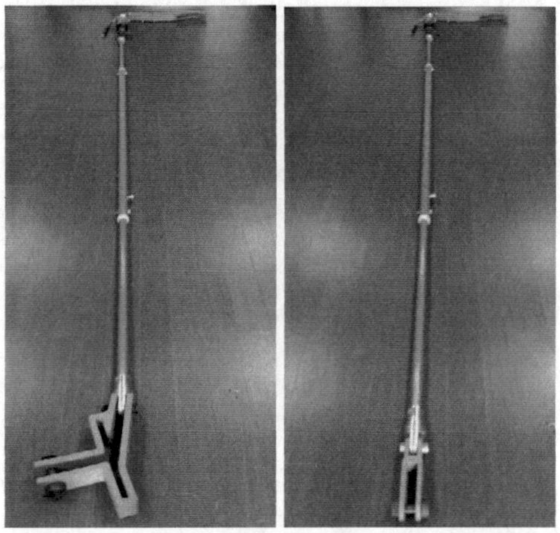

**Figure 2.** The tools of replacing the L-type insulator string

The developed matching tools are shown in Fig.2. In the device, a broadsword card, a hydraulic tightener, a mechanical screw, insulation pull rods and cross-arm fixtures are connected with one another sequentially in turn. The load-transition of the insulator string can be realized when the equipotential worker tighten the screw and then the insulator string can be replaced. In order to meet the actual demand, the cross-arm fixtures need to be changed and the length of the insulation pull rods need to be adjusted when the workers need replacing the different-direction insulator strings.

## 3. The mechanical analysis and verification

The total cross-sectional area of the wire is 973.16 $mm^2$; the stress of the wire is 92.075 Mpa. The technical parameters of the calculated tower and the adjacent ones are shown in table 1. The nominal height of the 1382 tower is 51 m; the elevation is 304.3 m; the span of the front side is 556 m and the one of the rear side is 404 m. In the meantime, the nominal height of the 1381 tower is 69 m and the elevation is 364.9 m; the nominal height of the 1383 tower is 51 m and the elevation is 327.8 m.

**Table 1.** Related line parameters

| Tower number | 1381 | 1382 | 1383 |
|---|---|---|---|
| Tower type | ZC27154 | ZJC2715 | JC27151 |
| Nominal height | 69 m | 51 m | 5 1m |
| Elevation | 264.9 m | 304.3 m | 327.8 m |
| Span of the front side | 404 m | 556 m | 463 m |

| | | | |
|---|---|---|---|
| Span of the rear side | 784 m | 404 m | 556 m |
| Assembly type of the fittings | X3-1A1 | XL2-1B1 | N1-2/N1-2 |
| Type of the insulator strings | duplex V-type composite insulator strings | duplex L-type composite insulator strings | three porcelain insulator strings |

The vertical span can be obtained by the equation (1) and it is 511.5 m through the calculation .

$$l_V = \frac{1}{2}(l_1+l_2)+\frac{\sigma}{g}(\frac{\pm h_1}{l_1}+\frac{\pm h_2}{l_2}) = \frac{1}{2}(404+556)+\frac{92.075}{0.03126}(\frac{21.4}{404}+\frac{-23.5}{556}) = 511.5m \quad (1)$$

The vertical load G can be obtained by the equation (2). In equation (2), *n* presents the amount of the bundle conductor and it is 6; the wire's weight *m* is 3.1047 kg/m; the vertical span of the tower $l_v$ is 511.5 m; *g* presents the gravity acceleration and it is 9.8 N/kg; $G_j$ presents the total weight of the insulator string, the fittings and the operator, which is calculated according to 5000 N. The vertical load *G* is 98.4 kN through the calculation.

$$G=nml_v g+G_j \quad (2)$$

According to the arrangement of the duplex L-type composite insulator strings, the stress analysis diagram was shown in Fig.3. And then the equation (3) can be obtained from the diagram.

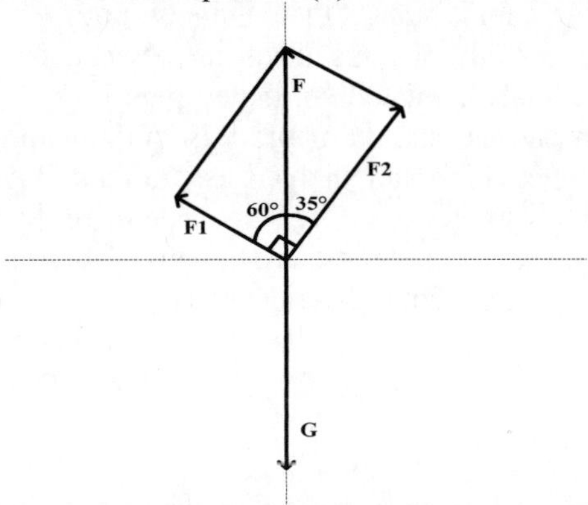


**Figure 3.** The stress analysis diagram of L-type insulator strings

$$\begin{cases} F_1\cos 60° + F_2\cos 35° = G \\ F_1\sin 60° - F_2\sin 35° = 0 \end{cases} \quad (3)$$

The load $F_2$ of the insulator string in the inner side is 85.6 kN and the load $F_1$ in the lateral side is 56.6 kN through the calculation. The load that the developed tools can sustain is 90 kN, therefore, the developed tools meets the actual requirement of live replacing the L-type insulator strings.

## 4. CHECK

In order to ensure the safety of the equipotential worker, the safe distance and the combined gap need to be verified. The minimum safety distance is 6.8 m and the minimum combined gap is 6.6 m for ±800 kV transmission lines according to the relevant standards [5].

### 4.1. The Safe Distance Check

The safe distance refers to the distance between the ground potential worker and the charged body or the distance between the equipotential worker and the grounding body. The particular analysis of the safe distance is shown in Fig.4 when the worker enters the electric field by the hanging basket method. Through the calculation and analysis, the horizontal distance between the cross-arm end and the tower center is 20.826 m; the horizontal distance between the tower edge and the tower center is 2.2 m; the horizontal distance between the tower edge and the divided conductor center is 8.5 m; the length of the insulator string and fittings in the inner direction is 11.925m; and the grading ring's short connected length of the air gap is 1.0m. Considering the influence of the grading ring's short section, the minimum distance $S_1$ is about 10.925 m; the minimum distance $S_P$, *H* is about 7.5 m, 8.36 m respectively after the worker entered the electric field, which accord with the requirement that the safety distance is greater than 6.8 m.

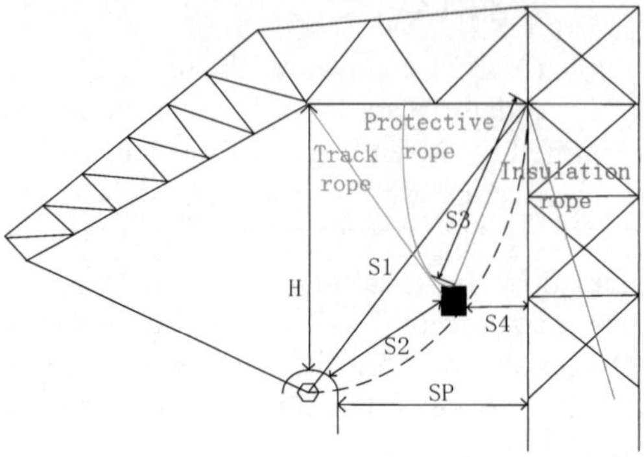


**Figure 4.** The calculation analysis of the safe distance and the combined gap by the hanging basket

### 4.2. The Combined Gap Check

The combined gap refers to the sum distance that includes both the distance between the ground potential worker and the charged body and the distance between the equipotential worker and the grounding body. The particular analysis of the combined gap is shown in Fig.4 when the worker is entering the electric field by the hanging basket method. Both the width *h* and height *b* of the hanging basket are 0.6 m. Through the calculation and analysis as shown in equation (4) and equation (5), the minimum gap distance $S_{g1}$, $S_{g2}$ is greater than 7.764 m, 6.9 m respectively, which accord with the requirement that the combined gap is greater than 6.6 m. Therefore, the equipotential operation method is used and the hanging basket method is selected as the way of entering the electric field.

$$S_{g1}=S_2+S_3>H-h=8364-600=7764 \text{ mm} \tag{4}$$

$$S_{g2}=S_2+S_4>S_P-b=7500-600=6900 \text{ mm} \tag{5}$$

## 5. Electric Field Analysis

The electric field distribution of the live worker is calculated by the finite element method [6-8]. The established model of using the hanging basket is imported into the simulation analysis software. The electric field intensity of body surface at the equipotential location is obtained as shown in Fig.5 by the steps of the subdivision, loading and solving.

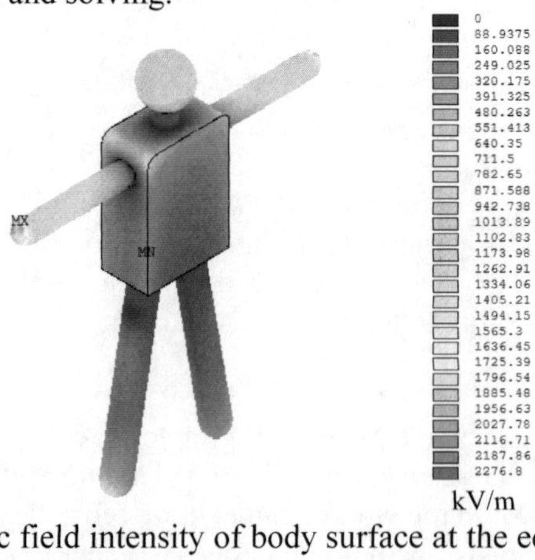


**Figure 5.** The electric field intensity of body surface at the equipotential location

Analyzing the calculated results, we can know the largest electric field intensity at the equipotential location by the hanging basket method is 2276.8 kV/m for the ±800 kV small angle tower. In the process of live working on UHV DC transmission lines, the internal maximum electric field intensity of the screen clothes should be less than or equal to 15 kV/m and the maximum electric field intensity of the human body's exposed areas should be less than or equal to 240 kV/m [9]. Therefore, the 60 dB shielding clothes should be used in order to ensure the equipotential workers' safety.

## 6. Field Application

The live working can be carried out around the premise that the weather, temperature and humidity meet the live working requirements. The hanging basket method is selected as the way of entering the electric field and the equipotential worker wearing the 60 dB screen clothes successfully entered the strong electric field. The specific field application cases are shown in Fig.5. With the cooperation of the tower operation workers and the ground workers, the L-type insulator strings are replaced smoothly by using the new method put forward in this paper and the developed accompanying tools shown in Fig.6, which ensures the safe and stable operation of the ±800 kV UHV DC transmission lines.

**Figure 6.** Field application of replacing the L-type insulator string by the new method

## 7. Conclusion

A new method of live replacing the duplex L-type insulators is put forward and the matching tools are developed based on the actual parameter of the UHVDC L-type tower. And the safe distance and combined gap of the new working method are checked. In addition, the 60 dB shielding clothes should be used.

## References

[1] HU Yi. Analysis on Safety Protective Equipments for Live Working on ±800kV UHV DC Transmission Line[J]. High Voltage Engineering, 2010,36(10)：2357-2361.

[2] Qingfeng LI, Protection for Live Working on ±800 kV DC Transmission Line[J]. Proceedings of the CSEE. vol. 29, no. 34, pp. 96-101, 2009.

[3] Yi HU, Live Working on EHV/UHV Transmission Lines[J]. High Voltage Engineering. vol. 38, no. 8, pp. 1809-1819, 2012.

[4] YANG Xinfa. Experimental Research on Safety Protection for Live-wire Operation on ±660 kV DC Power Transmission Line[J]. Electric Power Construction, 2012, 33(3)：1-5.

[5] "Live Working Technical Guideline of Electricity Transmission and Transformation," in DL/T 973-2005, 2005.

[6] LIU Chao, RUAN Jiangjun, LIAO Caibo. Approaching paths of helicopter live-line work platform to 1000kV UHV AC transmission lines[J]. Electric power automation equipment, 2015, 36(5)：64-70.

[7] LEI Dongyun. Solution for Heat-Failure of Drainage Plate of ±800kV UHV DC Transmission Line[J]. Electric Power, 2015, 48(7)：31-34.

[8] LI Jin-liang，RUAN Jiang-jun，DU Zhi-ye，et al. Analysis of Interference Current for High Voltage Arresters Based on Resistance-Capacitance Network[J]. Insulators and Surge Arresters, 2013，(5)：101-106.

[9] XIAO Yong. Analysis of Live Working on Yun-Guang ± 800 kV DC Transmission Line[J]. High Voltage Engineering, 2010,36(9)：2206-2211.

# Research for the Calculation of the Potential Transfer Energy on ±800kV DC Transmission Lines

**Huisong Ren[1], Jinliang Li[2]**

[1]School of Electrical Engineering, Shandong University
[2]Hunan Province Key Laboratory of Intelligent Live Working Technology and Equipment (robot) of Live Working Center of State Grid Corporation of Hunan

615695537@qq.com

**Abstract.** It would be hard to be repaired in the case of a power outage if the ±800kV ultra high voltage DC (UHVDC) transmission line was put into operation, so live working is the guarantee for its security and stability. It's necessary to calculate potential transfer energy to ensure the safety of the live working personnel. The transfer energy can be calculated by establishing the equivalent model of the process of potential transfer. Potential differences and capacitance parameters can be calculated by the finite element method, thereby transfer energy in different transfer distances can be worked out. The results show that the transfer energy increases with the increase of the transfer distance and the transfer energy reaches to 1.0 J when the distance is 0.5 m. So the potential transfer rod is necessary for the security of the live working personnel on ±800kV DC transmission lines.

## 1. Introduction

Once the ultra high voltage DC (UHVDC) transmission lines were put into operation, it is hard to repair at outage. The live working is the forceful guarantee for the security and stability of operation of the UHVDC transmission lines [1].The potential shift is necessary in live working, it means the worker uses conductive gloves or other special tools to connect himself with the wire, which can make him to get the same potential with the wire [2]. Because of the high voltage and strong field in the live working area, the electric arc, transferring current and transient energy would generate during the moment of the potential shift [3-5]. If the safety prevention measures are not enough or the operations are not proper in the process the safety accident would occur. The current through the worker is the pulse current with high frequency. Because of its complex change, the energy of the spark discharge is used to measure the severity that the pulse current act on the worker [4-6]. Therefore the research and calculation of the transfer energy on UHVDC transmission lines are necessary.
Take the calculation of the potential transfer energy on ±800kV DC transmission lines, and the hanging basket method is used as the way of entering the electric field. The partial capacitances and potential differences can be obtained by the finite element method. Then the transfer energy under different transfer distances can be obtained based on the established equivalent calculation model, which provides reference for the confirm of the potential transfer way and distance on the following live work on the UHV transmission lines.

## 2. The calculation method of the energy transfer

The equivalent circuit diagram of potential shift is shown as Fig.1, it is used to simulate the process of the worker entering the equipotential field of the ±800kV DC transmission lines by the hanging basket.

The $U_1$ presents the voltage of the line needed to be hooked, the $U_2$ presents the voltage of the other line, the $L$ presents the inductance of the screening clothes and potential rod, $C_{12}$ presents the capacitance between the worker and the line needed to be hooked, $C_{13}$ presents the capacitance between the worker and the other line, $C_{11}$ presents the capacitance between the worker and the ground. The worker and the basket is defined as the conductor 1, the line needed to be hooked is defined as the conductor 2, the other line is defined as the conductor 3, the tower, the ground wire and the ground are defined as the conductor 4.

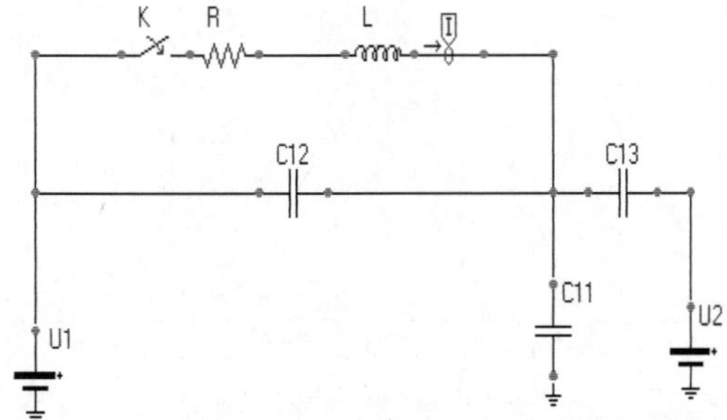


**Figure 1.** Equivalent circuit diagram of potential shift

The moment that the operator contacting the wire is simulated as the switch K closed in Fig.1. The energy stored in $C_{12}$ is released through the operator's body, it is calculated by the equation (1), the $U_{12}$ is the potential between the operator and the wire needed to be hanged and it can be calculated by the finite method. Because of the small value and short discharge time of the three capacitors, the potential of every wire is constant in the process of the operator hanging himself on the certain wire.

$$W = \frac{1}{2} C_{12} U_{12}^2 \tag{1}$$

## 3. The calculation of the capacitances

The capacitance of $C_{ij}$ ($i \neq j$, $i,j$=1,2,3) and $C_{ii}$ ($i$=1,2,3) in the equivalent circuit shown in Fig.1 is calculated according to the total electrostatic energy stored in electrostatic system multiple conductor, which is shown as the equation (2).

$$W = \frac{1}{2}\int_V \mathbf{E} \cdot \mathbf{D} \mathrm{d}V = \sum_{i=1}^{n} \frac{1}{2} u_i q_i \tag{2}$$

The $C_{ij}$ ($i \neq j$, $i,j$=1,2,3) presents the capacitance between the conductor i and the conductor $j$, the $C_{ii}$ ($i$=1,2,3) presents the capacitance between the conductor $i$ and the ground. In the equation (2), the **E** presents the electric field intensity, the **D** presents the electric flux density, the $u_i$ presents the voltage between the conductor $i$ and the ground, the $q_i$ presents the charge stored in the conductor $i$ and it can be expressed as the equation (3).

$$q_i = C_{i1}u_1 + C_{i2}u_2 + \cdots + C_{ii}u_i + \cdots + C_{in}u_n \tag{3}$$

Plug the equation (3) into the equation (2), we can get the equation (4).

$$W = \frac{1}{2}\sum_{i=1}^{n}\left( C_{ii}u_i^2 + \sum_{j=1\ j\neq i}^{n} C_{ij}u_i u_j \right) \tag{4}$$

The processes of capacitance calculation follows:

Step 1) The calculation of the principal diagonal elements $C_{ii}$: the voltage of the $i$th conductor is assigned to be unit voltage and the voltage of the others is set to be zero potential, the three-dimensional electrostatic field can be calculated by the finite element method. The energy $W_{ii}$ can be calculated based on the equation (2), and then the $C_{ii}$ ($i$=1,2,3) can be calculated based on the equation (4).

Step 2) The calculation of the off-diagonal elements $C_{ij}$: the voltage of the ith and the jth conductor is assigned to be unit voltage and the voltage of the others is set to be zero potential, the three-dimensional electrostatic field can be calculated by the finite element method. The energy $W_{ij}$ can be calculated based on the equation (2), and then and then the $C_{ij}$ ($i \neq j$, $i$=1,2,...,n-1, $j$=$i$+1...n) can be calculated based on the equation (4). And the capacitances can be calculated and shown in Tab.1 according to the above formulas when the potential transfer distance is 0.3m.

**Table 1.** Calculated capacitance values

| Type | Capacitance/pF |
|---|---|
| $C_{11}$ | 13.10 |
| $C_{22}$ | 737.20 |
| $C_{33}$ | 722.50 |
| $C_{12}$ | 44.56 |
| $C_{13}$ | 0.53 |
| $C_{23}$ | 64.65 |

The curve of the $C_{12}$ variable against the time is available through the calculation and it is shown in Fig.2. It shows that the capacitance between the live working people and the line needed to be hooked decreases with the increase of the transfer distance.

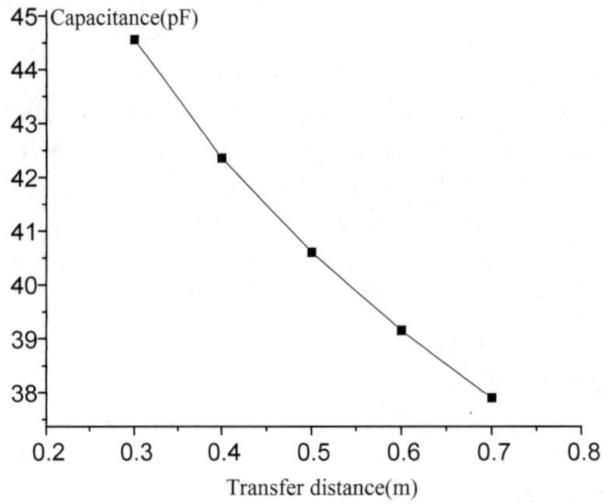


**Figure 2.** The capacitance values at different transfer distances

## 4. Calculation of the potential difference

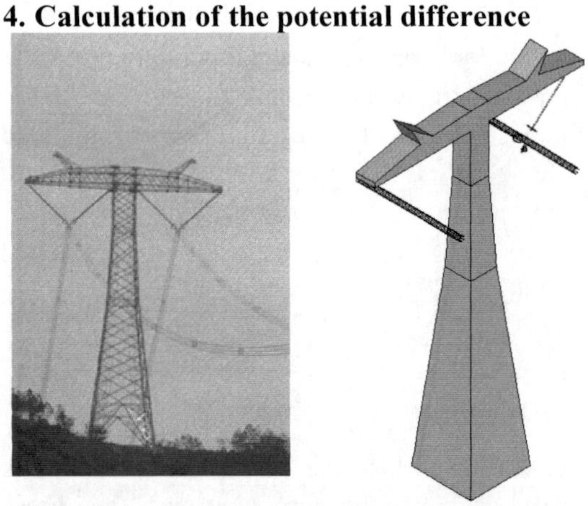

**Figure 3.** The finite element model in the process of potential shift

**Figure 4.** Field application of entering the electric field by the basket

In order to obtain the potential difference between the equipotential worker and the line needed to be hooked, the three-dimensional finite element model is established according to the actual filed situation of ±800kV DC transmission lines[7-8]. The ±800 kV straight-line tower is used as the tower model, its

IOP Conf. Series: Materials Science and Engineering **224** (2017) 012038 doi:10.1088/1757-899X/224/1/012038

total height is 63.5m and the nominal height is 57m. The wire is arranged as regular hexagon and its type is ASCR-720/50; its bundle spacing is 450 mm and the diameter of the branch wire is 36.24 mm. The whole model is shown in Fig.3, the diagram of entering the electric field by the basket is shown in Fig.4, and the partial enlarged view of the finite element model in the process of the potential shift is shown in Fig.5.

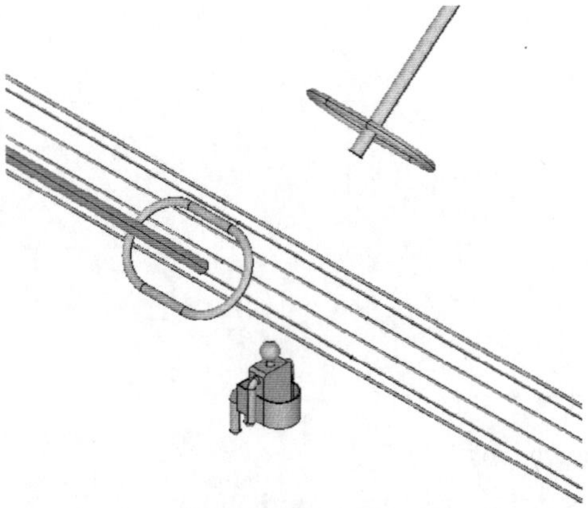

**Figure 5.** The enlarged view of the finite element model in the process of potential shift

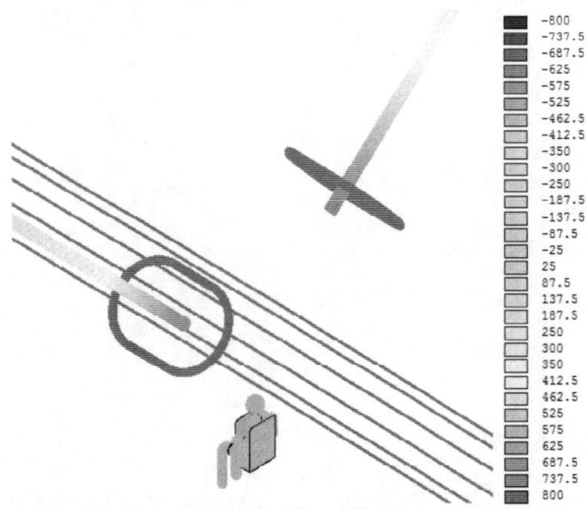


**Figure 6.** Potential distribution of entering the electric field by the basket

In order to ensure the solution accuracy on simulating calculation, the border area is 5 times of the actual model and apply truncated boundary conditions. The ground is defined as the zero boundary, the line length is set as 30 m along the line direction. Impose +800 kV potential on the positive line, impose -800 kV potential on the negative line and impose zero potential on the tower, the ground lines and the ground. Couple the potential for the worker and the hanging basket if the worker does not touch the line. The potential of the worker is equal to the one of the touched line if the worker has touched the line [9]. The calculated potential distribution is shown in Fig.6. As can be seen from the Fig.6， the calculated result consistent with the actual situation.

Potential of the live working personnel and the potential differences between the worker and the wire needed to be hooked can be calculated based on the results of the finite element analysis. The results of the potential difference is shown in Tab.2.

**Table 2.** Potential difference values

| The transfer distance/m | The potential of the live working personnel/kV | The potential difference $U_{12}$/kV |
|---|---|---|
| 0.3 | 605.4 | 194.6 |
| 0.4 | 591.0 | 209.0 |
| 0.5 | 577.6 | 222.4 |
| 0.6 | 565.0 | 235.0 |
| 0.7 | 552.9 | 247.1 |

The Tab.2 shows that the potential of the live working personnel decreases and the $U_{12}$ increases gradually with the increase of the transfer distance.

## 5. The calculation of the potential transfer energy

According to the calculation results of the formula (1), the curve of the transfer energy against the transfer energy can be calculated by the values of the $C_{12}$ and the $U_{12}$ in different transfer distances. The curve is shown in the Fig.7.

The Fig.7 shows that the transfer energy increases with the increase of the transfer distance. The calculated result of the transfer energy in the paper exceeds the pain energy threshold of the worker's physiological reaction to transient electric shock. Besides, the research results in relevant literature show that when the energy of the discharge system is about 1.37J, the root of the arc would act directly on the conductive gloves and the conductive gloves would be ablated[10]. The results of this paper show that, during the live working on ±800kV DC transmission lines, the potential transfer energy is 1.0J when the transfer distance is 0.5m and the potential transfer energy is 1.16J when the transfer distance is 0.7m. Therefore the potential rod should be used to ensure the worker's security during the equal-potential process.

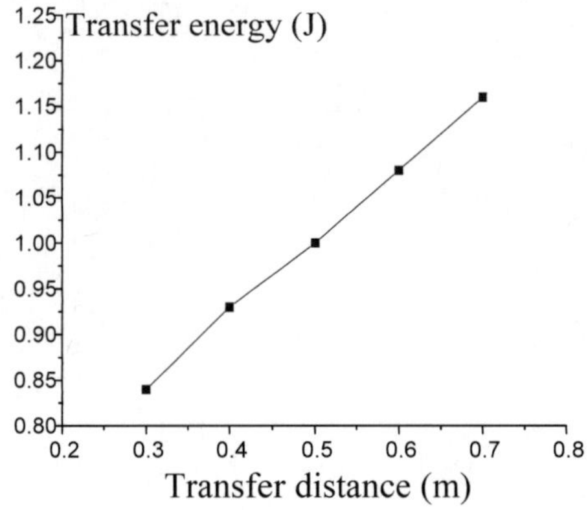


**Figure 7.** The potential transfer energy under different transfer distances

## 6. Conclusion

The potential transfer energy increases with the increase of transfer distance. And the potential transfer energy reaches to 1.0J when the transfer distance is 0.5m, therefore, the potential transfer rod should be used in the process of potential transferring on the ±800kV DC transmission lines.

## References

[1] LEI Dongyun. Solution for Heat-Failure of Drainage Plate of ±800kV UHV DC Transmission Line[J]. Electric Power, 2015, 48(7): 31-34.
[2] State Grid Corporation of China. Method of operation of live working[M]. Beijing, China: China Electric Power Press, 2009: 26-30.
[3] HU Tao, HU Yi, LI Jinglu, et al. Safety protection for live working on transmission line[J]. High Voltage Engineering, 2006, 32(5): 22-25.
[4] HU Yi, LIU Kai, HU Jianxun, et al. Analysis on safety protective equipments for live working on ±800 kV UHV DC transmission line[J]. High Voltage Engineering, 2010, 36(10): 2357-2361.
[5] LIU Kai. Characteristics of potential transfer current of ±800 kV UHVDC transmission line[J]. High Voltage Engineering, 2013,39(3): 568-576.
[6] XIAO Yong. Analysis of Live Working on Yun-Guang ±800 kV DC Transmission Line[J]. High Voltage Engineering, 2010,36(9): 2206-2211.
[7] LIU Chao, RUAN Jiangjun, LIAO Caibo. Approaching paths of helicopter live-line work platform to 1000kV UHV AC transmission lines[J]. Electric power automation equipment, 2015, 36(5): 64-70.
[8] HU Yi, HU Jianxun, LIU Kai. Field Strength of Body Surface During the Live Working on the UHV AC and DC Transmission Lines[J]. High Voltage Engineering, 2010, 36(1): 13-18.
[9] ZHU Jianxiong. Safety Protection for Live Working of AC/DC Parallel Transmission Lines[J]. Electric Power Construction, 2015, 36 (2): 61-66.
[10] CAI Huanqing. Influence of Potential Shift Transient Energy on Personnel Safety During Live Working [J]. High Voltage Engineering, 2015, 41(10): 3479-3483.

IOP Conf. Series: Materials Science and Engineering **224** (2017) 012039 doi:10.1088/1757-899X/224/1/012039

# Analysis on Inductance and Torque of PMSM Considering Magnetic Saturation

**Xiao-Hua CAO**[1], **Xin WANG**[1,2] **and Heng WEI**[1]

[1] Wuhan University of Technology, No.1178.Heping Road, Wu-Chang District, Wuhan City, Hubei Province, China

[2] 497724761@qq.com

**Abstract.** This paper analyses the surface-mounted PMSM which controlled by Id=0 vector control based on Ansoft in which finite element simulation of 2D static magnetic field can be operated on, then calculating and analysing the data with MATLAB, and then operating on the analysis of the change law of torque and inductance under different load conditions, and then paying more attention on the impact of magnetic saturation to torque and inductance. With the analysis of magnetic saturation, this paper puts forward a scheme of control and design used by PMSM.

## 1. Introduction

Permanent magnet motor comparing with the traditional electric field motor, not only more flexible and diverse on shape and size, but also has less loss, higher effectiveness and the strength of high power factor, high Power density on performance [1]. However, Permanent magnet motor is easy to appear the magnetic saturation phenomenon when it is working on low speed but large torque, which lead hard control [2]. So, Permanent magnet motors designed on all walks of life especially crane, are always have a larger maximum torque than they are needed actually. Although the design of having too much magnetic circuit margin can effectively prevent the occurrence of magnetic saturation, it results in waste of materials and energy which is neither economic nor energy saving when contrary to the country's sustainable development strategy. So, the study of permanent magnet synchronous motor's characteristics on magnetic saturation can not only promote the actual application on more occasions, but also can provide a theoretical basis on the design optimization and research on control strategy of modern permanent magnet motors [3].

## 2. Calculation principle

### *2.1. Calculation principle of three-phase current excitation in simulation analysis*

Stator d, q straight axis current change into A,B,C three-phase current [4]:

$$\begin{bmatrix} i_A \\ i_B \\ i_C \end{bmatrix} = \sqrt{\frac{2}{3}} \begin{bmatrix} \cos\theta & \sin\theta \\ \cos(\theta - \frac{2}{3}\pi) & \sin(\theta - \frac{2}{3}\pi) \\ \cos(\theta + \frac{2}{3}\pi) & \sin(\theta + \frac{2}{3}\pi) \end{bmatrix} \begin{bmatrix} i_d \\ i_q \end{bmatrix} \tag{1}$$

Where: $i_d, i_q$ respectively, for the two-phase rotating coordinate system in the vertical axis current and cross-axis current with unit A. $i_A, i_B, i_C$ is three-phase current, unit A. $\theta$ is the angle (electrical angle) between the d axis and the A axis during rotation of the rotor.

### 2.2. The Calculation Principle of Right Angle Inductor

Since the d and q axes are imaginary virtual coordinate systems, the fluxes cannot be directly detected and therefore need to be detected indirectly. In the d-axis calculation, assuming that the d-axis and A-axis coincidence, then the d-axis flux( $\varphi_d$ ) and A-axis flux( $\varphi_A$ ) are equal, so you can get [5]:

$$L_d = \frac{\psi_d}{i_d} = \frac{\varphi_d N_2}{i_d} = \frac{\varphi_A \times \sqrt{\frac{3}{2}} N_1}{\sqrt{\frac{3}{2}} i_A} = \frac{\varphi_A N_1}{i_A} = \frac{\psi_A}{i_A} \tag{2}$$

Similarly, when calculating the q-axis inductance, the q-axis coincides with A and is calculated by detecting the flux value of the A-axis:

$$L_q = \frac{\psi_A}{i_A} \tag{3}$$

Where: $\psi_A$, $\psi_d$ respectively is the flux chain of A-axis, d-axis. $L_d, L_q$ respectively is the d-axis and q-axis inductance. Respectively, $\varphi_A$ , $\varphi_d$ is the A-axis and d-axis flux. $N_1$, $N_2$ is turns of three-phase coordinate and turns of direct-axis and quadrature-axis.

### 2.3. Calculation principle of torque

In this paper, as using surface mount permanent magnet synchronous motor and $i_d = 0$ vector control. Combined with three-phase coordinate system to convert the two-phase rotating coordinate system after the principle of constant power, the stator current space vector( $i_s$ ) is $\sqrt{3}$ times than the effective value of the phase current( $I_e$ ), so:

$$\begin{aligned} T_{em} &= \frac{3}{2} p(\psi_d i_q - \psi_q i_d) = \frac{3}{2} p[\psi_m i_d + (L_d - L_q) i_d i_q] = \frac{3}{2} p\psi_m i_q = \frac{3}{2} p\psi_m \times i_s \cos\theta \\ &= \frac{3}{2} p\psi_m \times \sqrt{3} I_e \cos\theta = \frac{3\sqrt{3}}{2} p\psi_m I_e \end{aligned} \tag{4}$$

Where: $T_{em}$ is the electromagnetic torque; $\psi_m$ is the permanent magnet flux; $p$ is the differential operator.

## 3. Motor parameters and finite element simulation and analysis based on Ansoft

In order to study the variation of torque and direct axis inductance with different loads, in this paper, a three-phase four-pole surface-mount permanent magnet synchronous motor is designed, with the $i_d = 0$ vector control adopted. Then, the permanent magnet motor is modeled and analyzed by finite element simulation based on 2d static magnetic field of Ansoft [6,7]. The specific motor model parameters are shown in Table 1.

**Table 1.** Motor model parameters

| | |
|---|---|
| series | 4 |
| stator outer diameter/ (mm) | 120 |
| stator inner diameter/(mm) | 75 |
| number of stator slots | 24 |
| rotor outer diameter/(mm) | 62 |

| | |
|---|---|
| magnetic pole thickness(mm) | 6 |
| winding turns | 30 |
| rated speed/(rpm) | 1500 |

*3.1. The establishment of simulation model*
Combine table 1 parameters in Ansoft to establish two dimensional simulation model, as shown in figure 1, figure 2.

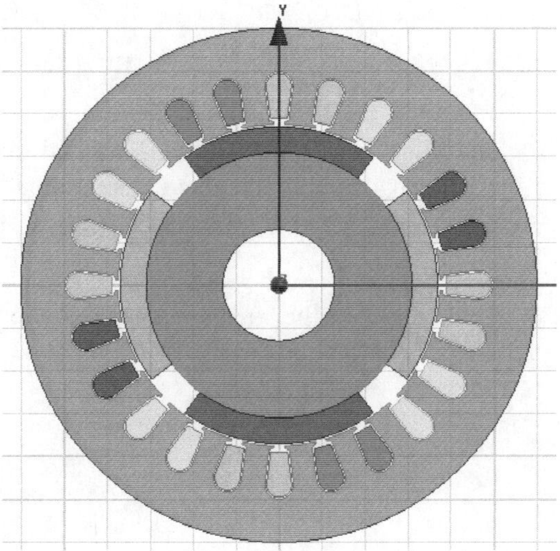


**Figure 1.** 2d Simulation Model of Electric Machine

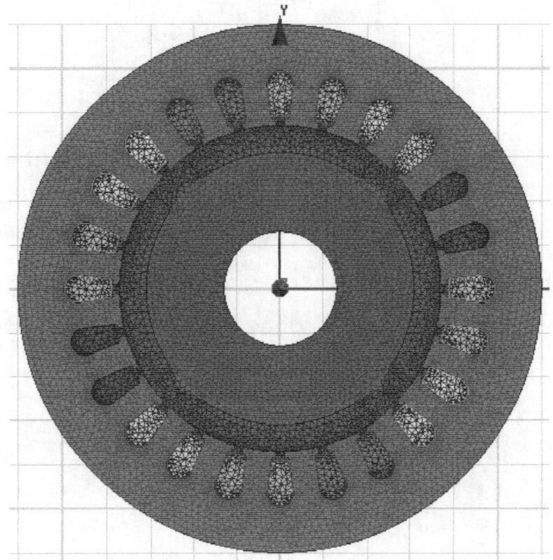


**Figure 2.** Mesh after the model

Figure 1 is a two-dimensional simulation model studied in this paper, using distributed winding. In order to complete the finite element simulation analysis, we need to divide the model into the grid, figure 2 is the division of the grid after the model. In this paper, we use meshing based on the length of the internal unit of the model to divide the grid.

*3.2. Simulation results analysis*
According to the equation (4), this paper studies the inductance and torque variation under different loads by applying different phase current valid valued.

Through the current from 0~500A scan, we can get the inductance matrix curve in figure 3:

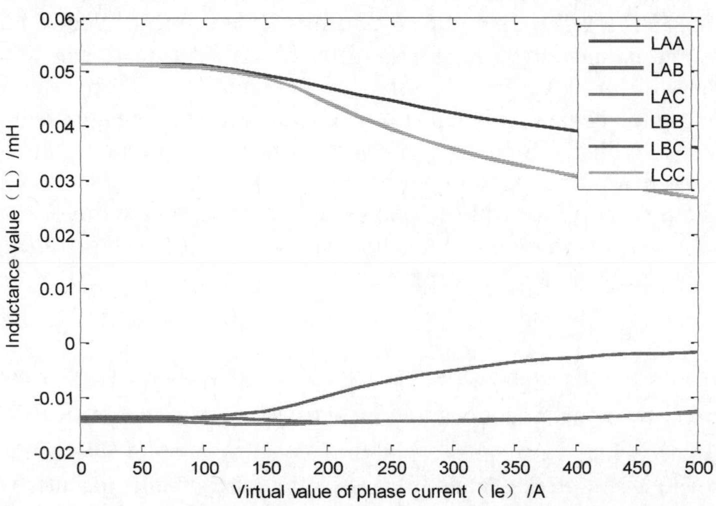


**Figure 3.** Inductance matrix curve

In order to study the inductance of the quadrature axis, the inductance matrix obtained by the full model simulation on the 2d static magnetic field of Ansoft also needs to be transformed:

$$\begin{bmatrix} L_d \\ L_q \end{bmatrix} = \mathbf{C}^{\mathrm{T}} \mathbf{L}_{\mathbf{ABC}} \mathbf{C} \tag{5}$$

Where: $\mathbf{C} = \sqrt{\frac{2}{3}} \begin{bmatrix} \cos\theta & \sin\theta \\ \cos(\theta - \frac{2}{3}\pi) & \sin(\theta - \frac{2}{3}\pi) \\ \cos(\theta + \frac{2}{3}\pi) & \sin(\theta + \frac{2}{3}\pi) \end{bmatrix}$, $\mathbf{L_{ABC}}$ is the inductance matrix.

Through the equation (5), combined with the simulation of the inductance matrix obtained under different load current under the axis of inductance, graph is drawn as shown in figure 4. As can be seen from figure 4: When the load current is applied between 0 and 175 A, $L_d$ and $L_q$ are equal, and basically remain unchanged, about 0.06mH; when the current is applied between 175 and 200A, $L_d$ and $L_q$ are both decline litter, and they are not equal anymore; when the load current is applied to more than 200A, $L_d$ maintain a slow downward trend, and then $L_q$ decline sharply.

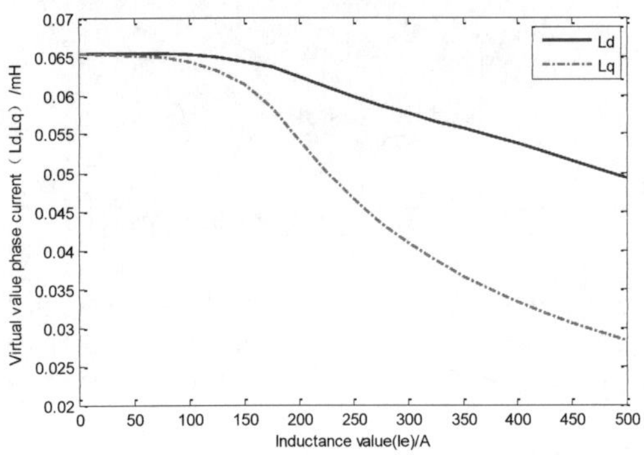


**Figure 4.** Curve of Inductance

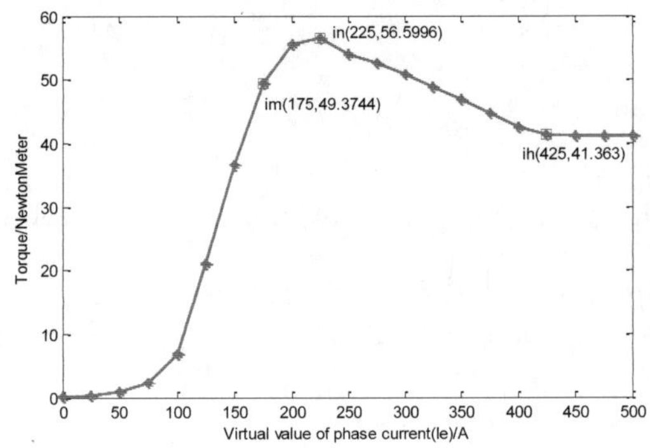


**Figure 5.** Variation Curve of Torque

Figure 5 shows the variation of the torque at different load currents obtained by Ansoft simulation. As shown in Figure 5,when the load current is applied between 0 and 75 A, torque shows a slow growth trend as the unit current increases the load is not high at this time; when the load current increases between 75 and 175A,Torque and phase current have a linear relationship, consistent with the formula (4),as the motor enters the torque high efficiency growth stage; when the current is added to 175 ~ 225A,torque with the load growth is no longer maintain a linear relationship, showing a slowing trend, as the motor begins to reach the magnetic saturation phase, so that the motor has passivation phenomenon with the load growth torque ;when the current reaches 225A, the torque growth reaches the peak, that is, the maximum torque is reached; When the current is applied to 225A ~ 425A, the torque starts to decrease with the increase of current, as the motor torque is decreasing with the deepening of magnetic saturation; When the current is loaded to 425A, the magnetic circuit is fully saturated and the torque is no longer changed as the current increases.

### *3.3. Magnetic saturation analysis*

Combined with the above simulation results, the direct axis inductance ($L_d$) shows a slow decline trend with the saturation of the magnetic saturation; quadrature axis inductance ($L_q$) shows a sharp decline trend with the deepening of magnetic saturation; With the deepening of magnetic saturation, the trend of linear growth is slowed down and the passivation phenomenon occurs. With the deepening of magnetic saturation, the torque reaches the maximum value and begins to descend until the last magnetic circuit is completely saturated and the torque is kept constant value unchanged.

In order to facilitate the proposed control scheme, the magnetic saturation is defined as three stages in conjunction with the change of torque in figure 5: The stage when the electromagnetic torque begins to passivate, and the torque continues to grow to the maximum is magnetic saturation first stage( $i_m \sim i_n$ ); The stage when the electromagnetic torque starts to decrease from the maximum value as the load current increases until the full magnetic saturation is magnetic saturation second stage ( $i_n \sim i_h$ ); The stage when the torque maintains a constant value as full of magnetic saturation is magnetic saturation of the third stage ( $\geq i_h$ ) .

## 4. Conclusion

In order to improve the utilization of the magnetic circuit, it is not necessary to completely discard all the stages of magnetic saturation, we can still implement the appropriate control when permanent magnet synchronous motor gets magnetic saturation until the motor torque reaches the maximum torque combined with the law that the first phase of magnetic saturation torque will gradually increase with the increased load .Magnetic saturation of the second stage torque appears to decline, when the control has been lost meaning, thus, the magnetic saturation of the second and third stages is completely discarded. In short, the control scheme is to make full use of the first phase of magnetic saturation. Combined with the control scheme, when we design permanent magnet synchronous motor, we can design the maximum torque value for the torque limit, as it can greatly reduce the motor size.

## Acknowledgments

This research was financially supported by the Applied fundamental research project of Ministry of Communications(2015329811290)and the Excellent Dissertation Cultivation Funds of Wuhan University of Technology(2016-YS-081）.

## References

[1] Li-Chan TANG, Liang QI. Application Status and Development Trends of Permanent Magnet Synchronous Motor[J]. *Equipment Machinery*, 2011, 7-12(in Chinese)

[2] Li ZHOU. The Application of Rare-earth Permanent Magnet Electric Machine in Crane[J]. *Hoisting and Conveying Machinery*, 2014(8): 70-73(in Chinese)

[3] Meng YANG. Research of Low Speed and High Torque Permanent Magnetic Synchronous Machine for Crane Use[D]. Wuhan: Huazhong University of Science and Technology, 2013(in Chinese)

[4] Ren-Yuan TANG. Modern Permanent Magnet Machines Theory and Machines [M]. Beijing: China Machine Press, 1997(in Chinese)

[5] Yan-Jun AO. Inductance Simulation of Quadrature Axis and Direct Axis Based on Permanent Magnetic Synchronous Motor Used in Electrical Vehicle. *Small & Special Electrical Machines*, 2012,40(8): 14-16(in Chinese)

[6] Li DENG. Finite Element Analysis of Permanent Magnetic Synchronous Machine based on Maxwell14_0 [J]. *Mechanical and electrical engineering technology*, 2011, 40(9): 47-49(in Chinese)

[7] Bo ZHAO, Hong-Liang ZHANG. Apply of Ansoft_12 to Engineering Electromagnetic Field.[M]. Beijing: China Water & Power Press, 1997(in Chinese)

# Study on Determination Method of Fatigue Testing Load for Wind Turbine Blade

**Gaohua Liao**[1,2] and **Jianzhong Wu**[1]

[1] College of Mechanical Engineering, Tongji University, Shanghai 201804, China
[2] Jiangxi Province Key Laboratory of Precision Drive & Control, Nanchang Institute of Technology, Jiangxi 330099, China

ngtjko@163.com

**Abstract.** In this paper, the load calculation method of the fatigue test was studied for the wind turbine blade under uniaxial loading. The characteristics of wind load and blade equivalent load were analyzed. The fatigue property and damage theory of blade material were studied. The fatigue load for 2MW blade was calculated by Bladed, and the stress calculated by ANSYS. Goodman modified exponential function S-N curve and linear cumulative damage rule were used to calculate the fatigue load of wind turbine blades. It lays the foundation for the design and experiment of wind turbine blade fatigue loading system.

## 1. Introduction

As an important component of the wind turbine, the wind turbine blade is subjected to the changing wind load in the actual operation of the wind farm. Because of the bad climate environment and the Long time operation of blade, people pay much attention to wind blade fatigue life [1-2]. The analysis and calculation of the load is the basis for the subsequent design and test, which affects the dynamic characteristics of the blade structure and the analysis and calculation of fatigue. The failure analysis of the blade is divided into two parts: prediction analysis and test detection. The analysis can be used to control the fatigue of the blade, and provide guidance for the test. Fatigue analysis and prediction is still not accurate enough. Fatigue loading test is an important way to study the fatigue failure of blade. Wind turbine blade fatigue evaluation involves many uncertain factors, but more and more attention has been paid to [3]. In order to cause the blade to be subjected to sufficient fatigue damage in a reasonable test period, the test load is usually greater than the design load to speed up the testing process [4]. The finite element analysis and the blade vibration test method are combined to analyze the fatigue of the wind turbine blade [5]. This paper analyzes the characteristics of wind load and the equivalent load of blade, and studies the fatigue property and damage theory. According to the uniaxial constant amplitude loading, the load of wind turbine blade fatigue is calculated.

## 2. Wind turbine blade load analysis

Wind turbine blade is composed of airfoil with different sections. The section of the blade root is round and thick, and the middle part is the aerodynamic airfoil, which gradually thinning to the tip direction. In order to ensure that the blade as light as possible, the general use of hollow interior configuration, according to the structure is divided into: stiffened plate - bearing structure, shell - pneumatic structure and blade root - connection structure. Wind turbine blades are subject to time-varying loads from the outside, including aerodynamic force, centrifugal force and gravity. Usually along the blade is divided into many micro segments, each segment is called a leaf. The leaf as a two-

IOP Conf. Series: Materials Science and Engineering **224** (2017) 012040 doi:10.1088/1757-899X/224/1/012040

dimensional cascade, and assuming that there is no interference between the force of blade's role in blade force on only related to momentum change swept the ring gas. Usually, the blade is taken as the research object to carry on the stress analysis. For a specific load condition, the air load of the blade is simplified as a plurality of concentrated force loads along the spanwise direction of the blade, as shown in Figure 1. The force acting on the blade is obtained by integrating the force acting on the blade and the blade. The bending moment produced by normal force are $M_y$, and $M_x$,

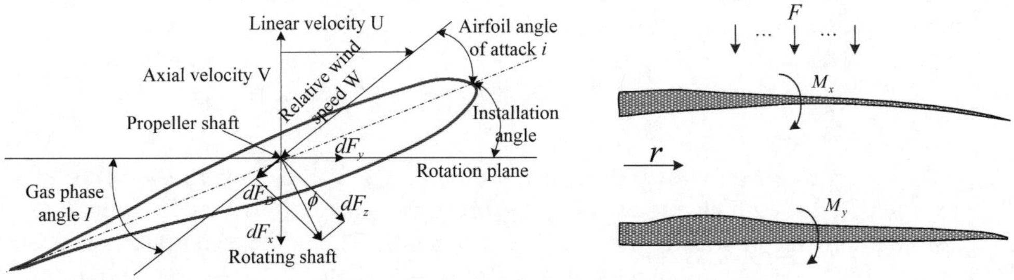


**Figure 1**. Blade load analysis and simplification of blade section.

The flapping direction of shear $F_y$, the shear direction of shear $F_x$, the blade root vibration moment $M_x$ and the flapping moment $M_y$ can be expressed as:

$$\begin{gathered} F_x = \frac{1}{2}\int_{r_0}^{R} \rho v_0^2 BcC_n dr, F_y = \frac{1}{2}\int_{r_0}^{R} \rho v_0^2 BcC_t dr \\ M_x = \int_{r_0}^{R} r dF_x, M_y = \int_{r_0}^{R} r dF_y \end{gathered} \tag{1}$$

Because of the gravity of the blade, the gravity changes periodically with the azimuth of the blade. The gravity load causes the blade to produce a moment $M_{xG}$ in the direction of vibration. In order to facilitate the calculation of the gravity load, it is necessary to calculate the density and area of the blade. The blade rotates, the centrifugal force is produced, and the action point is located at the center of gravity of the blade section, and the component of the centrifugal force moment in *x, y* direction is respectively $M_{xp}$ and $M_{xp}$. The wind wheel rotation caused by centrifugal force on the blade is always outward along the blade, when acting on the blade flapping moment direction make flexible blade from wind wheel rotating plane, the centrifugal force on the blade can reduce the blade deviation in moment of wave direction is generated.

$$\begin{cases} M_{xG} = -\cos\phi \int_{r_0}^{R} (r - r_0)\rho_0 A_0 g dr \\ M_{xp} = \int_{r_0}^{R} (r - r_0)\rho_0 \omega^2 A_0 Y_G(r) dr \\ M_{yp} = \int_{r_0}^{R} [X_G(r) - Y_G(r_0)]\rho_0 \omega^2 A_0 r dr \end{cases} \tag{2}$$

Where $\phi$ is the rotating blade azimuth angle, $\rho_0$ and $A_0$ are the density and area of blade small segment period after conversion.

## 3. Fatigue property and damage theory of blade material

The selection of the correct S-N curve is the precondition for the fatigue life analysis of the blade, and the fatigue loading characteristics are evaluated by defining the maximum stress $\sigma_{max}$, the minimum stress $\sigma_{min}$, the stress amplitude $\sigma_a$ and the mean stress $\sigma_m$. There is no clear limit of fatigue for leaf glass fiber reinforced plastics. It is usually used to estimate the stress value of the cycle number $5\times10^4 - 10^7$ as the fatigue limit. The relationship between the stress and fatigue life of glass fiber reinforced plastics, in the stress ratio $R=\sigma_{min}/\sigma_{max}=-1$, in the S-lg*N* coordinates on a straight line:

$$\sigma_i + B \lg N_i = \sigma_b \tag{3}$$

Where $\sigma_b$ is the static strength of the material, $B/\sigma_b$ is 0.1 approximation, $\sigma_i$ and $N_i$ are the corresponding stress levels and the number of cycles at failure.

The S-N curves of glass fiber reinforced plastic blades were obtained under the condition of cyclic $R$=-1.After the rain flow counting algorithm, the asymmetric cyclic loading spectrum is obtained, and the conditional fatigue limit can be obtained according to the modified Goodman curve equation under the specified cycle life:

$$\sigma_a = \sigma_{-1} \cdot (1 - \sigma_m / \sigma_b) \tag{4}$$

Where $\sigma_a$ is the stress amplitude, $\sigma_{-1}$ is the conditional fatigue limit, $\sigma_m$ is the average stress.
The fatigue performance of the blade is based on the fatigue curve S-N of glass fiber reinforced plastics, and the Goodman stress limit diagram is given according to the GL, and the theoretical limit of fatigue limit is obtained by $R$, which is known as the stress ratio $N$.

$$N = \left[ \frac{R_{\mathrm{k,t}} + |R_{\mathrm{k,c}}| - |2\gamma_{\mathrm{Ma}} S_{\mathrm{k,M}} - R_{\mathrm{k,t}} + |R_{\mathrm{k,c}}||}{2S_{\mathrm{k,A}}(\gamma_{\mathrm{Mb}} / C_{\mathrm{lb}})} \right]^m \tag{5}$$

Where $S_{k,A}$ is the dangerous section of the stress amplitude, $S_{k,M}$ is the dangerous section of the mean stress, $R_{k,t}$ is the tensile strength characteristics of short term structural elements, $R_{k,c}$ is the short-term compressive strength characteristics of structural components, $R_{k,A}$ is the maximum stress amplitude, and $\mathrm{R_{k,A}} = (\mathrm{R_{k,t}} + |\mathrm{R_{k,c}}|)/2$; $\gamma_{\mathrm{Ma}}$ is the local material safety, $\gamma_{\mathrm{Mb}}$ is the local material safety factor; m is the slope parameter of S-N curve.
To quantitatively study the effects of fatigue, this paper defined in a stress level of $R_i$, and the corresponding number of blade load $n_i$. Single load damage degree is $d_i$. A strain of the allowable load cycle number is $N_i$. The fatigue failure is defined as $d_i = 1/N_i$. The fatigue damage value $D$ of the fatigue load design point can be calculated according to the Miner linear damage theory.

$$D = \sum_i^k D_i = \sum_i^k \frac{n_i}{N_i} = 1 \tag{6}$$

## 4. Fatigue damage calculation

In the actual operation of the blade fatigue load spectrum is very complex, must use the actual operation situation of simplified and accelerated method of equivalent simulation of blade in a relatively short period of time, the blade applied equivalent damage fatigue load equivalent damage and service life in the leaves of the same. According to the analysis of fatigue strength, fatigue safety margin selection section as the minimum fatigue test verification section. In order to make the test time can not be too long, but also to avoid too few times because the test can not reflect the true situation, the general test number of $N=2\times10^6 \sim 1\times10^7$ times. Test frequency and blade natural frequency are similar, and the maximum damage location is as fatigue calculation concerns. In order to facilitate the interaction between calculation and test, the leaves of the outer section of the maximum strain damage for fatigue calculation concerns.
According to the fatigue load calculated by Bladed and the stress calculated by ANSYS, Goodman modified exponential function S-N curve and linear cumulative damage rule are selected. The fatigue load calculation method for a certain type of 2MW blade is studied under uniaxial constant amplitude loading. The blade length 56.6m is applied to simulate the blade's working condition at the root of the blade, and the blade finite element model is established. At the tip of the blade, the strain values of each section of the blade finite element model are calculated by using the three directions of the load and the vibration. Strain and fatigue load fatigue loads calculated using Bladed software under the blade value of every section, theoretical calculation of fatigue damage of glass fiber composite materials based on the calculation and drawing the maximum coefficient of each section of the blade fatigue damage by using Matlab software, The maximum coefficient of fatigue damage is shown in Figure 2. According to the evaluation of the maximum coefficient of action, the fatigue test range in the direction of swing is confirmed to be L=13m to L=38m, and the fatigue test range is L=4m to L=27m. According to the calculation results of blade fatigue damage, the blade section girder cap 23m part most prone to fatigue failure of the specific location, determine the location of dangerous points of blade fatigue test of dangerous section (which is the fatigue load design point) at the same time, the direction of fiber strain the outermost layer of the location of the fatigue test load as the design reference value.

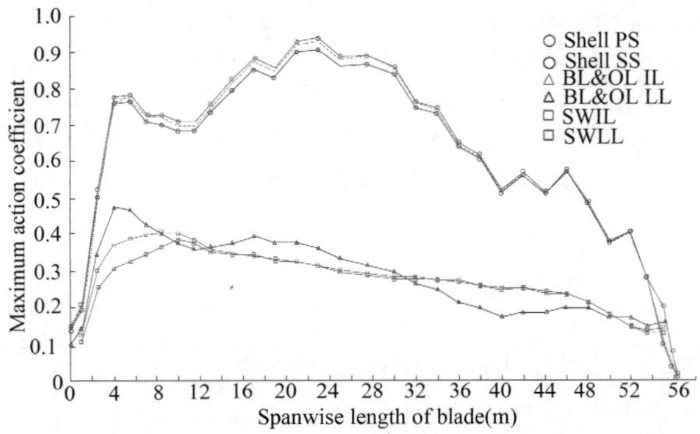


**Figure 2.** Maximum coefficient of fatigue damage.

## 5. Blade fatigue test load calculation

After determining the design points of fatigue load, it is necessary to calculate the fiber orientation strain value of the blade under the action of the unit bending moment in order to design the fatigue load. Finite element software is used to calculate the strain value $\varepsilon$ of the fatigue load design point under the action of unit bending moment. The dynamic strain amplitude $S_{k,A}$ under the cyclic loading is calculated by calculating the strain value under the unit load multiplied by the load amplitude. Then the unit load under the locomotive load are worth to strain under cyclic load strain mean $S_{k,M}$. The blade was reinforced epoxy matrix, $m$ value was 10. By using the formula of fatigue life of glass fiber composites, the fatigue life of fatigue load design points under the action of fatigue load $M_y$ is obtained $N$. The cycle number of the fatigue test is $2\times10^6$ in the direction of the swing and the direction of vibration. Equation (5) transformed into:

$$S_{k,A} = \left[ \frac{R_{k,t} + |R_{k,c}| - |2\gamma_{Ma} S_{k,M} - R_{k,t} + |R_{k,c}||}{2\cdot(\gamma_{Mb})\cdot N^{\frac{1}{10}}} \right] \tag{7}$$

The parameters in the equation (7) are calculated as shown in table1:

**Table 1.** Data of fatigue life calculation

| $R_{k,t}$ | $R_{k,c}$ | $\gamma_{Ma}$ | $2(\gamma_{Mb}/C_{Ib})$ | m |
|---|---|---|---|---|
| 0.02 | -0.016 | 2.205 | 3.594 | 10 |

Due to the actual wind turbine blade in the wind field, the fatigue load is variable amplitude load spectrum, and the actual test process applied to the fatigue load is equivalent to the constant amplitude fatigue load. Therefore, it is necessary to convert the variable amplitude fatigue load into equal amplitude fatigue loading. According to the theory of fatigue damage, it can be known that there is a certain proportion between strain amplitude $S_{k,A}$ and strain mean $S_{k,M}$:

$$\frac{S_{k,A}}{S_{k,M}} = \frac{1-R}{1+R} \tag{8}$$

Where $R$ represents the load ratio.

The value of $R$ is considered in the case of the weight and the weight of the blade in the fatigue test. Then according to equation (8) can be calculated from the mean strain blade corresponding amplitude under fatigue load, at the same time using equation (7) can be calculated from the strain amplitude fatigue loading:

$$S_{k,M} = \frac{|R_{k,c}|}{\frac{1-R}{1+R}(\gamma_{Mb}/C_{Ib})^{1/m} - \gamma_{Ma}} \tag{9}$$

According to the requirements of the test specification and the position of the blade fatigue risk section, the 25m section of the fatigue test load is selected. Considering the dispersion coefficient of the blade and the uncertainty of the fatigue formula, it is necessary to consider the load factor of the fatigue test on the actual calculation of the fatigue test load. $F_{df}$ is the initial load. The load of the fatigue test is $F_{target} = F_{df} \cdot \gamma_{nf} \cdot \gamma_{sf} \cdot \gamma_{ef}$, and the static load amplitude is calculated by using the test load coefficient in listing table 2.

**Table 2.** Load factor of fatigue test

| Test load factor | Numerical value |
|---|---|
| Local coefficient of difference between blades $\gamma_{sf}$ | 1.1 |
| Uncertainty of fatigue formula $\gamma_{nf}$ | 1.04 |
| Local failure factor $\gamma_{ef}$ | 1.15 |

The fatigue life of the blade is $N_{ref}=1\times10^7$, and the damage equivalent load is the value of the cycle number. The cyclic range of each load component is calculated to calculate the damage equivalent load of the target test cycle, and then the cyclic range is multiplied by the load factor of the fatigue test to obtain the target load. When setting the cycle number of $2\times10^6$, the calculated results of the load of the fatigue test target are shown in table 3.The shear strain induced by the shear force accounts for only a small proportion of the whole structural damage. Therefore, the test design will focus on the bending moment, for the same reason, the load of the *FZP* and *MZP* groups are also ignored.

**Table 3.** Fatigue test load

| Radi(m) | FXP (kN) | FYP (kN) | MXP (kN·m) | MYP(kN·m) |
|---|---|---|---|---|
| 0.00 | 213.18 | 507.72 | 8597.29 | 5644.29 |
| 7.00 | 178.77 | 341.85 | 5846.19 | 4416.43 |
| 13.00 | 172.58 | 268.92 | 4068.02 | 3452.52 |
| 25.00 | 146.56 | 158.46 | 1742.18 | 87.47 |
| 34.00 | 109.56 | 93.58 | 841.15 | 41.96 |
| 42.00 | 69.18 | 46.09 | 328.13 | 15.67 |
| 50.00 | 27.14 | 13.71 | 57.73 | 4.01 |

## 6. Conclusion

The characteristics of wind load and blade equivalent load are analyzed. According to the Miner linear fatigue cumulative damage theory and damage equivalent principle, the fatigue performance curve of FRP is calculated. Based on the Bladed simulation, the fatigue load matrix of a 2MW blade is obtained. Based on the theory of fatigue damage analysis and finite element software, the fatigue damage value of the fatigue load design point is determined, and the static load parameters are obtained.

## Acknowledgments

This work is supported by the Jiangxi Province Key Laboratory of Precision Drive & Control project (PLPDC-KFKT-201622).

## References

[1] MalhotraP,HyersRW, ManwellJF, McGowanJG .*A review and design study of blade testing system for utility-scale wind turbines*.RenewSustain Energy Rev, 2012,**16**(1):p.284–92.

[2] Nitin Tenguria ect. *Design and Finite Element Analysis of Horizontal Axis Wind Turbine blade* .International journal of applied engineering research, 2010, 1:p.500-507.

[3] William Michael Haupfear.*The Development of an Adaptive Control System for a Phase Locked Excitation (PhLEX) Method for Advanced Wind Turbine Blade Fatigue Testing*. Daytona Beach:Embry-Riddle Aeronautical University, 2010:p.21- 23.

[4] Jackson K L.*A method of fatigue life assessment of wind turbine structures*.Wind Energy ASME,1993,**14**(2): p.179-180.

[5] Musial W and Allread J. *Test methodology and control of Full-Scale fatigue tests on wind turbine blades*.Wind Energy ASME,1993,**14**(2):p.199-206.

# Design of a Single Motor Based Leg Structure with the Consideration of Inherent Mechanical Stability

**Muhammad Taha Manzoor[1*], Umer Sohail[1], Noor-e-Mustafa[1], Muhammad Hamza Asif Nizami[1] and Yasar Ayaz[1*]**

[1]School of Mechanical and Manufacturing Engineering, National University of Sciences and Technology (NUST), Sector H-12, Islamabad, Pakistan

m.tahamanzoor@gmail.com, yasar@smme.nust.edu.pk

**Abstract.** The fundamental aspect of designing a legged robot is constructing a leg design that is robust and presents a simple control problem. In this paper, we have successfully designed a robotic leg based on a unique four bar mechanism with only one motor per leg. The leg design parameters used in our platform are extracted from design principles used in biological systems, multiple iterations and previous research findings. These principles guide a robotic leg to have minimal mechanical passive impedance, low leg mass and inertia, a suitable foot trajectory utilizing a practical balance between leg kinematics and robot usage, and the resultant inherent mechanical stability. The designed platform also exhibits the key feature of self-locking. Theoretical tools and software iterations were used to derive these practical features and yield an intuitive sense of the required leg design parameters.

## 1. Introduction

Biomimics is an emerging field of robotics which aims at mimicking the motions of naturally existing organism and to utilize these systems in real life applications. Majority of the mammals use legs for locomotion but wheeled robots have been preferred because of enhanced stability and easy control of actuators [1][2]. Undoubtedly, legged robots have certain advantages over wheeled robots in natural environment [3] but their lower stability limits their use in real life applications. In order to minimize the stability issues related to the legged platforms, number of legs can be increased. As a rule, more the number of legs, more is the stability; whereas, increasing the number of legs adds to the difficulty of control problem [4]. Thus, a trade off approach is required. Therefore, researchers are in a quest to develop stable legged robots for the future.

In literature, some researchers have reported their work which can be used as basic criteria for the leg design. Lee has reported a compartmentalized leg structure inspired by nature [4]. A biped named Raptor (made by KAIST), reported to run at high speeds, has made use of a nine bar mechanism. Recently, Oak and Narwane have investigated the use of modified four bar mechanism (two DOF per leg) [5]. Seok et al have developed advanced legged robot mimicking the movement of a Cheetah [6]. A detailed literature review has revealed that in an attempt to ensure the stability of legged robots the researchers have depended heavily on sensory equipment and control algorithms rather than the mechanical design. Moreover, controlling a large number of actuators, simultaneously, is troublesome. Thus, we propose an efficient leg design based on a simple four-bar mechanism with only one motor per leg. This leg design would yield an inherently stable robot with an exciting feature of self-locking.

The design of one DOF based leg requires iterative and complex steps; the first of which is the development of a sound mechanical system.Quadrupeds are a moderate choice with respect to the stability of the platform [4]. Thus, the present study focuses on the mechanical design of the legs of a quadrupedal

IOP Conf. Series: Materials Science and Engineering **224** (2017) 012041 doi:10.1088/1757-899X/224/1/012041

robot which can have a simplified control architecture.

## 2. Choice of one Degree of Freedom

As mentioned before, the aim of this study is to investigate different possibilities of utilizing efficient leg design with one actuator per leg in quadruped platforms. Many possible leg designs were considered and altered in order to realize the idea of one motor per leg. The existing robots have a tendency to use actuators at each joint of the leg; however, this makes the control algorithms extremely complex [6]. Specially, during fast running it is hard to predict the next positions of the legs and a large sensory network is required. Considering the ultimate goal of incorporating least number of actuators per leg and simplified control architecture, leg design based on a simple four bar mechanism with one motor was chosen.

## 3. Design Phase

The major steps required to design this novel four bar mechanism include (a) selection of a suitable foot trajectory; (b) calculation of the link lengths of the four bar mechanism which can follow this trajectory by using only one motor; (c) placing the designed mechanism on the body of the robot so that it functions as a leg and (d) simulating all four legs, attached to the main body (complete robot), in solidworks in order to analyse the stability of the platform. All the above mentioned steps required iterations and modifications at every stage of the design.

### *3.1 Suitable foot Trajectory*

First task was to select a suitable foot trajectory. Foot trajectory plays a central role in the path planning and stability of robot. It also ensures the robot gets enough ground friction during the touch down phase; frictional force actually keeps the platform moving in the desired direction. Foot trajectories of many living species like cheetah, zebra etc. were analysed along with the existing robots (Raptor, Big Dog) [2][4][5][7]. It was realized that the foot trajectory must have two distinct features: a) it should be as flat as possible during the touch down phase, b) touch down phase should be long enough and c) there should be adequate clearance from the ground after touch down. The reason behind (a) is to make the main body of the robot stable during motion. If trajectory is not flat it can cause the main body to oscillate during motion. Then, point (b) is related to the frictional force as stated before but a very long touchdown phase can make the robot's motion very slow [4]. Hence, a suitable solution was required. The last feature also required a trade-off approach. If clearance from the ground is too small, robot will not be able to pass over small obstacles. On the other hand, a large clearance can generate large inertial forces during fast motion and can topple the platform. Thus, all these factors were kept in mind while designing the leg structure. The limitation of using one motor per leg was the main challenge in obtaining a suitable foot trajectory, which replicates the motion like a robot. Figure 1 shows different foot trajectories considered during the design phase and the selected one.

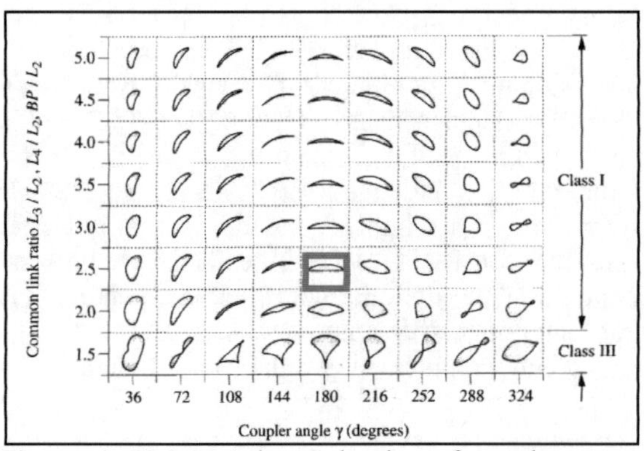


**Figure 1:** H & N Atlas, Selection of coupler curve mechanism and Coupler angle 180 degrees [8]

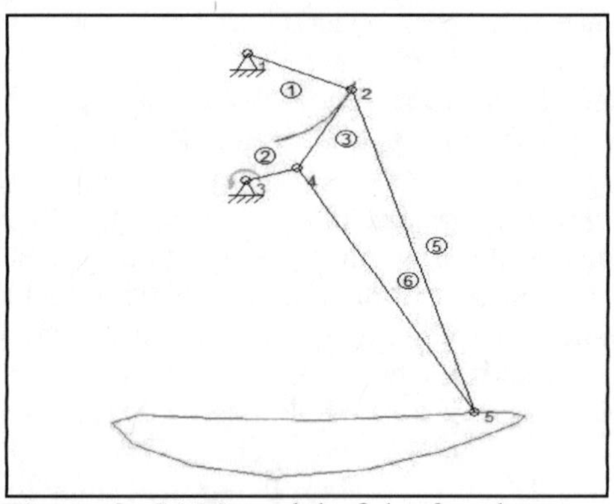


**Figure 2:** SAM model of the four bar to confirm the trajectory

IOP Conf. Series: Materials Science and Engineering **224** (2017) 012041 doi:10.1088/1757-899X/224/1/012041

*3.2 Link length calculations*

After deciding the foot trajectory, next task was to formulate a workable four bar mechanism which would follow this trajectory. The coupler of a four-bar mechanism is theoretically a plane and if any point is selected on this plane, the mechanism is capable of moving that point by just providing a torque to the crank. Therefore, if the point follows a specific trajectory then only one motor, placed at the crank, would be able to achieve the required motion. Hence, a bottom up approach was used by utilizing the information in H&N catalogue [8]. The H&N catalogue serves as the data base of different possible curves that a point on the coupler of a four bar mechanism can follow. The curve which closely matched our pre-defined foot trajectory was chosen. Using the catalogue, link length ratios were calculated; thereby, giving a four bar mechanism. The key factor which played a major role in the calculations was that link lengths should be the one which can be manufactured with ease and don't cause hindrance (collision of links other than coupler with the ground) in the real locomotion. For example, same trajectory could be obtained in which the crank length is 1/100 times smaller than rocker but such a design would be impractical to fabricate. Thus, practicality played a vital role and SAM 1.0 facilitated the iterations and tweaking of four bar mechanism. Table 1 shows the final link lengths used in this study and fig.4 shows the completed mechanism with trajectory.

**Table 1:** Link Lengths

| Point | Length |
|---|---|
| 1 to 3 (Ground Link) | 12.6 cm |
| 1 to 2 (Link # 1) | 9.0 cm |
| 3 to 4 (Link # 2) | 4.3 cm |
| 2 to 4 (Link # 3) | 8.8 cm |
| 4 to 5 (Link # 6) | 28.1 cm |
| 2 to 5 (Link # 5) | 33.5 cm |

*3.3 Attachment of the mechanism on main body*

Next challenge was to attach the designed four bar mechanism to the main body of the robot in such a way that it serves the purpose of the leg. It is obvious that the mechanism is a mere collection of link and the placement of these links on the body required intuitive approach. Moreover, it was ensured that the actuators are placed nearer to the main body so that the leg has low inertia; thus achieving dynamic decoupling [9]. Various different schemes were modelled in solidworks and finally the structure shown in fig.3 was obtained. This structure acts as a leg of the robot, achieves a forward motion, follows the pre-defined trajectory and has a low leg inertia.

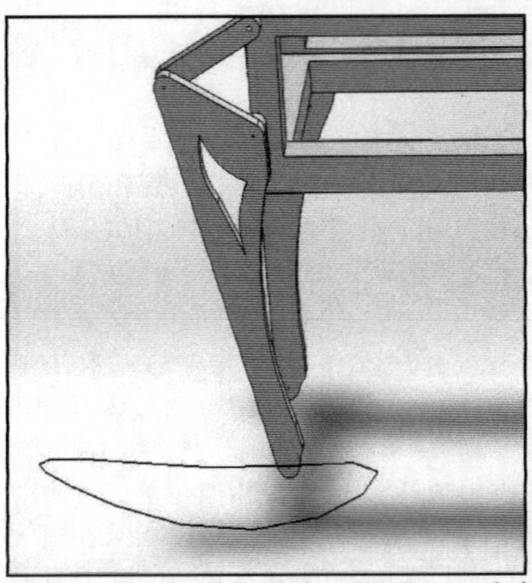

**Figure 3:** Solidworks CAD model of the leg attached to the main body

**Figure 4:** Complete CAD model

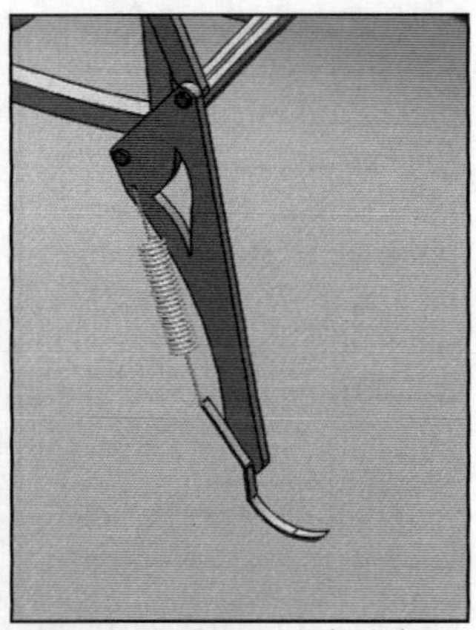

**Figure 5:** Proposed spring mechanism

*3.4 Stability Analysis*

Finally, the main feature related to any robotic platform i.e. the stability of the quadruped robot was analysed. Once again, solidworks was used as it provided a good simulation environment for legged robots. Some other software like pro-e and Adams were also tried but it was found that they work better for wheeled robot simulations only.

A step-wise scheme for testing the stability of the platform was developed. In the current scheme, first of all the joints in all four legs (four-bar mechanisms) of the robot were locked (so that the whole four-

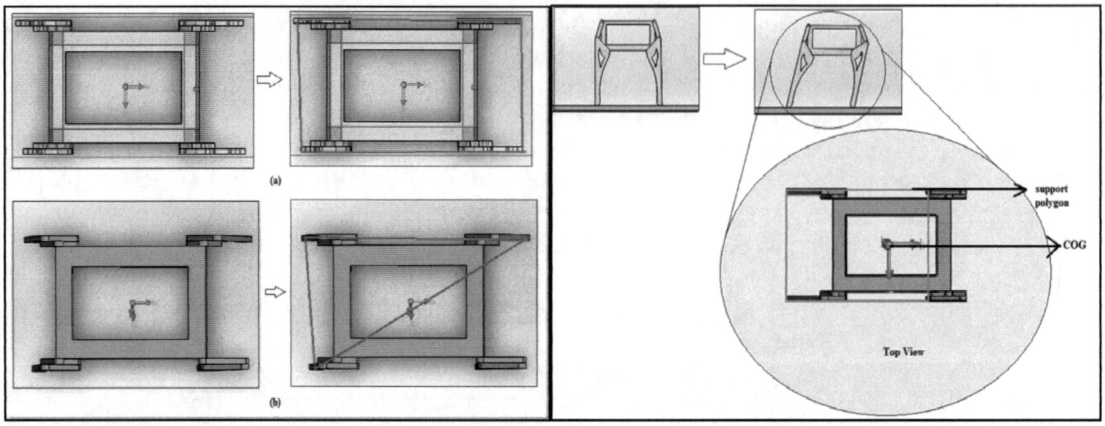


**Figure 6:** Solidworks simulation scheme (a) when all legs on ground (b) one leg lifted in air (c) robot moved forward and one leg lifted

bar acts as one rigid link) and the robot was dropped from a measured height. It was analysed that whether robot maintains stability on the ground or topples away. The height of the fall was increased until robot toppled. This maximum height was recorded for different models and compared. In the next stage, all joints were unlocked and no motor torque was given. The platform was again dropped from a certain height. An interesting feature was noted that the robot moved to and fro for a slight moment and stabilized itself on its own. This feature was simulated several times and is named as self-locking ability. Actually all the four-bar mechanism in the four legs (when arranged in the scheme as shown in fig 4) cancel the net torque produced due to the gravitational and impact force; thus, self-locking the robot. It is a novel aspect of this unique design. In real environment, robots have to be supported on some additional platforms when their motors are turned off (i.e. they cannot stand without motor torques). Whereas, this design requires no additional platforms and can stand on its own; hence, it would be able to perform tasks in real life application more efficiently. Moreover, this self-locking feature imparts inherent mechanical stability to the quadruped, meaning, that it is statically stable (even under slight impacts it is stable) and requires minimum sensory equipment for maintaining its stability during motion.

The basic criteria for performing static stability based gates (Longitudinal Stability Margin) is that robot should remain stable when one of its four legs is not in contact with the ground [10]. Actually, if robot is stable on three legs it means that we can move fourth leg to the next desired point, in this way if all four legs are moved to the next desired point(one at a time) the robot attains a new position. Therefore, the robot was tested for its stability on three legs. One of the legs was held up and locked so that it does not touch the ground and it was tested that whether the quadruped topples or not. It was found that this unique design remains stable on three legs. Later, robot was placed on the ground and motors were given angular displacements to calculate the maximum angle up to which robot can be moved forwards or backwards without toppling. In fact, if the centre of gravity of the platform remains inside the stability polygon, the robot would not fall (see fig. 6(a) (b)). The maximum angle calculated was 17°. In the last stage, the motion of the quadruped was simulated. The robot was moved 17° forwards and was held there for a few second (actually it shifts the centre of gravity of the platform towards the front, see fig. 6) then one of the back legs was lifted upwards, by providing torque to the motor (only one motor in leg at crank), and was placed to the next point. These simulations were repeated multiple times with multiple versions of the leg designs and modified as required. During all these simulations the unique self-locking four-bar mechanism based legged robot successfully cleared all the aforementioned stability tests.

## 4. Passive mechanical impedance

An important aspect in leg design is that the leg should have a low mechanical impedance when it makes an impact with the ground but once the impact is made, the impedance should be high enough to transfer the ground reaction forces (mainly friction) to the main body [11]. Thus, a spring mechanism is arbitrarily proposed (as shown in fig. 5) which imparts the passive mechanical impedance to the leg structure. A comprehensive study on the calculation of the required spring stiffness would be done in future.

## 5. Conclusion

To conclude the issues related to the complex control problems of legged robots inspired us to design an efficient leg design which can be actuated using only one motor per leg. A suitable leg trajectory incorporating the basic required features was selected. Later, link length ratios of a four-bar mechanism capable of replicating this trajectory were calculated. Afterwards, a CAD model was developed and simulated in solidworks. A simulation scheme was formulated to ensure that the robotic platform is stable under different initial conditions. Our unique quadruped exhibited the property of self-locking.

The problems related to passive mechanical impedance were highlighted and solved by proposing a spring mechanism attached to the foot. The next phase of this study aims at the fabrication of the quadruped based on the design reported in this study in order to verify the results of these simulations.

## References

[1] J. E. Bertram and A. Gutmann. Motions of the running horse and cheetah revisited: fundamental mechanics of the transverse and rotary gallop. *Journal of The Royal Society Interface,* 6(35):549-559, October 2008.

[2] A. A. Biewener. Biomechanics of mammalian terrestrial locomotion. *Science,* 250(4984):1097, 1990.

[3] B. Wilcox, A. Nasif, and R. Welch. Implications of martian rock distributions on rover scaling. *Planetary Society International Conference on Mobile Planetary Robots and Rover Roundup, Santa Monica CA,* 1997.

[4] W Lee. *Designing articulated legs for running machines*. PhD thesis, Massachusetts Institute of Technology, (1990).

[5] Oak S, Narwane V. Design, Analysis and Fabrication of Quadruped Robot with Four bar Chain Leg Mechanism. *International Journal of Innovative Science, Engineering & Technology*. 2014;1:340-345.

[6] Seok, S, A Wang, and S Kim. "Design Principles For Highly Efficient Quadrupeds And Implementation On The MIT Cheetah Robot". *IEEE International Conference On Robotics And Automation (ICRA)*. Karlsruhe, Germany: IEEE, 2013. Print.

[7] R. M. Walter and D. R. Carrier. *Ground forces applied by galloping dogs. Journal of Experimental Biology,* (January 2007) 210(2):208-216,.

[8] R.L. Norton, *Design of Machinery,* Third Edition, McGraw-Hill (2004), 116

[9] Z. Vafa and S. Dubowsky. *On the dynamics of manipulators in space using the virtual manipulator approach. Robotics and Automation. Proceedings.(1987) IEEE International Conference,* (1987), 4:579-585,

[10] Garcia, Elena, Joaquin Estremera, and P. Gonzalez De Santos. "A classification of stability margins for walking robots." *Proc. of the 2002 Int. Symp. on Climbing and Walking Robots*. Vol. 20. 2002.

[11] D. A. Lawrence. Actuator limitations on achievable manipulator impedance. In *Robotics and Automation, 1989. Proceedings., 1989 IEEE International Conference on,* pages 560-565, 1989.

IOP Conf. Series: Materials Science and Engineering **224** (2017) 012042 doi:10.1088/1757-899X/224/1/012042

# Design and Simulation of Control Technique for Permanent Magnet Synchronous Motor Using Space Vector Pulse Width Modulation

**Mansoor Khan[1], Wang Yong[2], Ehtasham Mustafa[3]**

[1,2]School of Automation Science Electrical Engineering, Beihang University, Beijing, China
[3]Department of Electrical Engineering, Faculty of Engineering & Technology, Gomal University, Dera Ismail Khan, Pakistan

Khan007_bet@yahoo.com, wangyong@buaa.edu.cn, engr.emustafa@gmail.com

**Abstract.** After the rapid advancement in the field of power electronics devices and drives for last few decades, there are different kinds of Pulse Width Modulation techniques which have been brought to the market. The applications ranging from industrial appliances to military equipment including the home appliances. The vey common application for the PWM is three phase voltage source inverter, which is used to convert DC to AC in the homes to supply the power to the house in case electricity failure, usually named as Un-interrupted Power Supply. In this paper Space Vector Pulse Width Modulation techniques is discussed and analysed under the control technique named as Field Oriented Control. The working and implementation of this technique has been studied by implementing on the three phase bridge inverter. The technique is used to control the Permanente Magnet Synchronous Motor. The drive system is successfully implemented in MATLAB/Simulink using the mathematical equation and algorithm to achieve the satisfactory results. PI type of controller is used to tuned ers of the motothe parametr i.e. torque and current.

## Introduction

These Permanent Magnet Synchronous Motors (PMSM), has been under a lot of research and has been considered as the backbone of modern engineering products [1]. Its applications range from robots, electric automobiles, and heavy industries to domestic products such as washing/drier machines, water pumps, vacuum cleaners, etc. Due to its improved characteristic regarding precise speed and position control, it is considered as one of the best options to be used in the portable devices. To achieve the maximum performance of the PMSM with the use of low cost enhanced essential materials, modern digital processors have made it possible to implement sophisticated control algorithms and techniques for the purpose. A number of control studies have been proposed by the researchers to improve the specific parameters of the motors such as current, speed and position depending upon its usage to achieve maximum efficiency.

A lot of research has been conducted to improve the controllability efficiency and dynamics of drives in the last two decades. Researchers proposed some new designs, in 1986, related to Interior PMSM (IPMSM) for the speed control, contrary to the AC drives present at the moment, resulting robust, efficient systems bearing dynamic speed regulation [2, 3]. Since PMSM is designed to work under

FOC barring damping windings, so the equations of synchronous motor can be used to produced d-q model without considering the effects of the damper winding and the field currents. Two level inverter based detail explanation of SVPWM is described in [4, 5]. While the modelling, analysis, controlling and losses of PMSM have been discussed in different literatures [6-18].
In this research work, the SVPWM technique is simulated using MATLAB/ SIMULINK using the PI control for speed and current control. The technique is adopted using the three phase bridge inverter. The insight of the simulation is gained for the closed loop scenario, speed, and current plots are drawn against dynamic load and reference speed. The plots of the torque, speed and current of SVPWM technique show satisfactory results.

## Analytical Model of PMSM

The PMSM has been extensively studied for the past two decades. In widely held methods used for PMSM, there is a set of equations dependent on rotor position. In this work, the motor equations are represented in rotor reference frame. As a result a set of equations is derived which is independent of the rotor position. The currents in the d-q axis are found using the two transformations. The Park transformation converts the three phase quantitates (*abc*) to the two phase ($\alpha\beta$). The second part transfers the quantities from the stationary frame to rotational frame.

$$\begin{bmatrix} i_\alpha \\ i_\beta \end{bmatrix} = \frac{2}{3}\begin{bmatrix} 1 & -\frac{1}{2} & -\frac{1}{2} \\ 0 & \frac{\sqrt{3}}{2} & -\frac{\sqrt{3}}{2} \end{bmatrix}\begin{bmatrix} i_a \\ i_b \\ i_c \end{bmatrix} \tag{1}$$

$$\begin{bmatrix} i_d \\ i_q \end{bmatrix} = \begin{bmatrix} cos\theta & sin\theta \\ -sin\theta & cos\theta \end{bmatrix}\begin{bmatrix} i_\alpha \\ i_\beta \end{bmatrix} \tag{2}$$

Where $\theta$ represents the rotor position. Keeping the rotor reference frame as a reference, the PMSM voltage equations are:

$$V_q = R_s i_q + \omega_r \lambda_d + \rho\lambda_q \tag{3}$$

$$V_d = R_s i_d - \omega_r \lambda_q + \rho\lambda_d \tag{4}$$

The equations (5) and (6) show the relationships for the flux linkages.

$$\lambda_q = L_q i_q \tag{5}$$

$$\lambda_d = L_d i_d + \lambda_f \tag{6}$$

Using the equations of flux linkages into the voltages' equations result in

$$V_q = R_s i_q + \omega_r(L_d i_d + \lambda_f) + \rho L_q i_q \tag{7}$$

$$V_d = R_s i_d - \omega_r L_q i_q + \rho L_d i_d + \lambda_f \tag{8}$$

Equation (7) and (8) can be written in the matrix form as:

$$\begin{bmatrix} V_q \\ V_d \end{bmatrix} = \begin{bmatrix} R_s + \rho L_q & \omega_r L_d \\ -\omega_r L_q & R_s + \rho L_d \end{bmatrix}\begin{bmatrix} i_q \\ i_d \end{bmatrix} + \begin{bmatrix} \omega_r \lambda_f \\ \lambda_f \end{bmatrix} \tag{9}$$

According to what was mentioned above, electromechanical behaviour of the PMSM is as follows:

$$T_e = \frac{P}{2}\lambda_m \begin{bmatrix} i_a & i_b & i_c \end{bmatrix}\begin{bmatrix} -\sin\theta_e \\ \frac{1}{2}sin\theta_e + \frac{\sqrt{3}}{2}cos\theta_e \\ \frac{1}{2}sin\theta_e - \frac{\sqrt{3}}{2}cos\theta_e \end{bmatrix} \tag{10}$$

Where the $\theta_e$, is the rotor electrical position and $\theta_m$, the rotor mechanical position.

## Space Vector Pulse Width Modulation

In three Phase Bridge VSI, with the help of high frequency pulses given at the gate an output voltage of low frequency is generated. The generated voltages is controllable. In this VSI structure, a total of eight states, also known as vectors, are generated, $\mathbf{V_0}$ to $\mathbf{V_7}$. Out of these states, the non-zero active voltage states are from V1 to V6 while zero voltage vectors are $\mathbf{V_0}$ and $\mathbf{V_7}$. Figure 1 shows a diagram of the vectors. The eight states are described from the equation as below:

$$V_k = \begin{cases} \frac{2}{3} V_{dc} e^{j(k-1)\frac{\pi}{3}} \\ 0 \end{cases}, \quad k = 1, 2, 3, 4, 5, 6 \tag{11}$$

For a preferred value of the fundamental components for the output phase voltages, a reference quantity, $V_{ref}$, is used which is voltage space vector. At equal intervals of the sampling time period, $T_s$, the reference voltage is sampled. Within this time period, the inverter produces different voltage vectors at different time durations. During the time period, the average vector and $V_{ref}$, are equal in magnitude and angle. The space vector pulse width modulation (SVPWM) for three phase VSI is based on transformation of the three phase vector quantities to a two dimensional (αβ) quantitates.
In the SVPWM technique, the inverter will switch, producing switching vectors ($\mathbf{V_0} - \mathbf{V_7}$) with the consideration of having angle (θ) of desirable voltage. The switching frequency for the inverter is variable, which is considered as a major drawback of the SVPWM method.

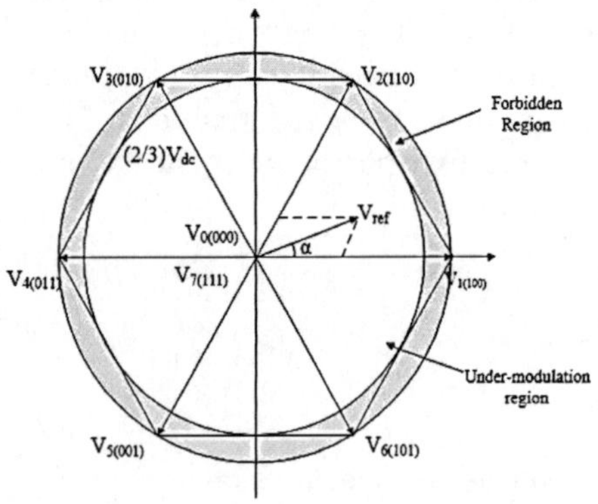


**Figure 1.** SVPWM Voltage Vector.

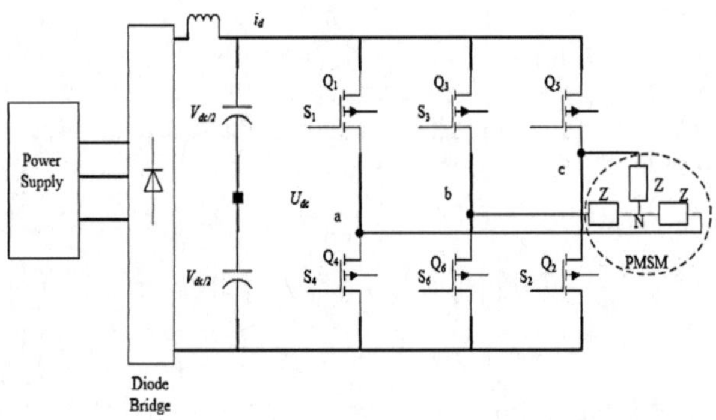


**Figure 2.** Three Phase Bridge Inverter.

## Proposed Work

A classic two-level full bridge inverter is designed with the help of 6 switches that produce 3 phase voltage. A thorough illustration of a 3-phase inverter is presented in Figure 2.Once an upper switch gets turned on, the respective switch is swapped to the opposite state. $S_0$, $S_1$, $S_3$ and $S_5$ switches are govern to produce the voltage at the output by switching the upper switches. The transistors in a pair cannot be switched ON at the same time.
According to the SVPWM principle, binary codes are used to characterize the switching arrangements, the upper switches $S_1$, $S_3$ and $S_5$ as depicted in Figure 2. Every switching circuit produces three sovereign voltages values at pole $V_{ao}$ ,$V_{bo}$ and $V_{co}$ , which represents the output voltages referenced to the midpoint of source. The pole voltages have two values i.e. $V_{dc} / 2$ and $V_{dc} / 2$. The eight inverter positions can be converted into eight equivalent space vectors. For every shape, '0' characterizes the negative voltage level while '1' characterizes the positive voltage level. Association among these vectors and equivalent switching situations is agreed in Table 1. Right-angle axis is used to characterize the under consideration inverter when it comes to phase illustration.
The block diagram of SVPWM control of PMSM drive is shown in Figure 3. The three phase currents are detected from the motor through the closed loop control and are then changed to 2 phase system using Clark transformation. The (αβ) system is then Park transformed into d-q system. The d-q currents are then compared to the reference current and fed to the controller, which is the Proportional Integral (PI) one. The result of which is the d-q referenced voltages which are Park transformed to the (αβ) referenced and fed to the SVPWM system. The three phase inverter then controls the PMSM through the switching pattern formed, which has been discussed above.

**Table 1.** Switching States of the Space Vectors.

| Space Vector | Switching State | On-State Switch | Magnitude and Angle |
|---|---|---|---|
| $\vec{V_0}$ | [000] | $S_4, S_6, S_2$ | $\vec{V_0} = 0$ |
| $\vec{V_1}$ | [100] | $S_1, S_6, S_2$ | $\vec{V_1} = \frac{2}{3}V_{dc}$ |
| $\vec{V_2}$ | [110] | $S_1, S_3, S_2$ | $\vec{V_2} = \frac{2}{3}V_{dc}e^{j\pi/3}$ |
| $\vec{V_3}$ | [010] | $S_4, S_3, S_2$ | $\vec{V_3} = \frac{2}{3}V_{dc}e^{j2\pi/3}$ |
| $\vec{V_4}$ | [011] | $S_4, S_3, S_5$ | $\vec{V_4} = \frac{2}{3}V_{dc}e^{j\pi}$ |
| $\vec{V_5}$ | [001] | $S_4, S_6, S_5$ | $\vec{V_5} = \frac{2}{3}V_{dc}e^{j4\pi/3}$ |
| $\vec{V_6}$ | [101] | $S_1, S_6, S_5$ | $\vec{V_6} = \frac{2}{3}V_{dc}e^{j4\pi/3}$ |
| $\vec{V_7}$ | [111] | $S_1, S_3, S_5$ | $\vec{V_7} = \frac{2}{3}V_{dc}$ |

**Simulation and results**

The paper discusses, the control of PMSM, which is achieved by means of the SVPWM. The controlling is achieved with aid of 3-phase bridge inverter. The d-q currents graph is shown in Figure 4. While the electrical torque and speed of the PMSM is plotted in Figure 5. The Figure 6 shows the three phase currents.

MATLAB/SIMULINK has been used to develop and simulate the whole drive system. The motor equations have been modelled in SIMULINK. The SVPWM signal generator, PI Controller, the bridge inverter and Park transformations models are also developed in SIMULINK. The simulation results depict satisfactory response for the developed model when the command speed was being tracked. The results of the simulation have demonstrated that the adopted SVPWM based control of PMSM drive has a satisfactory performance for the speed, torque and position control with being implemented on three phase bridge inverter. The results obtained also show that the SVPWM control strategy can be adopted effectively with low cost.

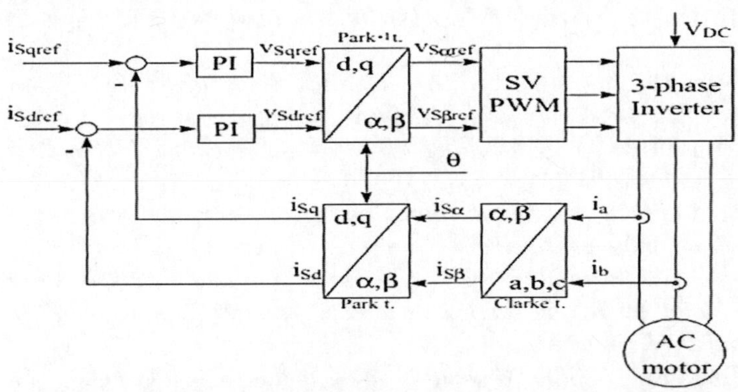


**Figure 3.** Block Diagram of the SVPWM control of PMSM.

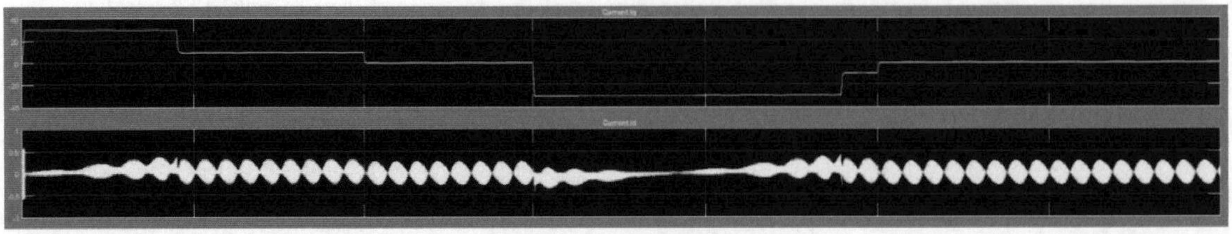

**Figure 4.** *d-q* Currents

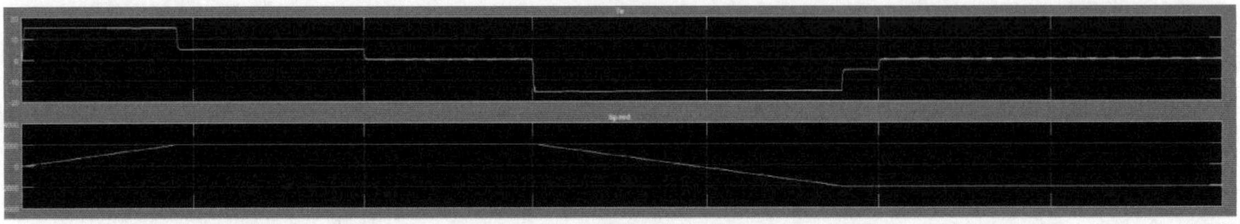

. **Figure 5.** Torque and Speed Curve of PMSM

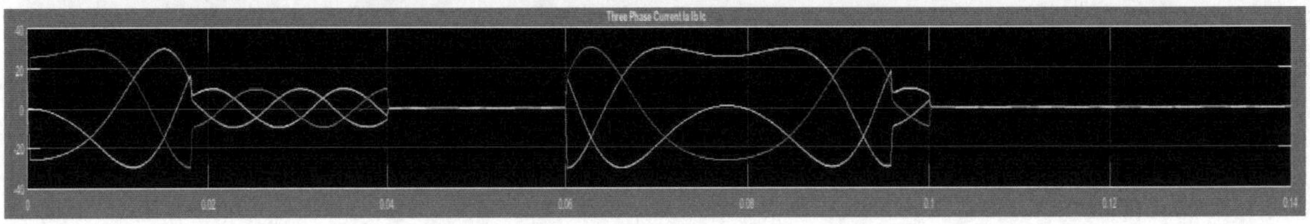

**Figure 6.** Three Phase Currents

## References

[1] Lyshevski S E 2008 *Electromechanical Systems and Devices* (CRC Press)

[2] Jahns T M, Kliman G B and Neumann T W 1986 *Interior Permanent-Magnet Synchronous Motors for Adjustable-Speed Drives IEEE Trans. on Industry Appl.* vol 22 (4) pp 738-747

[3] Salehfar H 2005 *DSP-Based Implementation of Vector Control of Induction Motor Drives Electrical and Computer Engineering* vol 125 p 405

[4] Neacsu D O 2001 *Space vector modulation–An introduction* 27th *Int. Conf. on Industrial Electronics Control and Instrumentation* IECON (2001)

[5] Iqbal A, Lamine A, Ashraf I 2006 *MATLAB/Simulink Model of Space Vector PWM for Three-Phase Voltage Source Inverter IEEE 41st Int. Universities Power Engineering Conf.* UPEC (2006)

[6] Ogbuka C and Agu M 2009 *A Generalized Rectified Sinusoidal PWM Technique for Harmonic Elimination Pacific J. of Science and Tech.* vol 10(2) pp 21-26

[7] Pillay P and Krishnan R 1988 *Modelling of Permanent Magnet Motor Drives IEEE Tran. on Industrial Electronics* vol 35(4) pp 537-541

[8] Morimoto S *et al.* 1994 *Loss Minimization Control of Permanent Magnet Synchronous Motor Drives IEEE Tran. on Industrial Electronics* vol 41(5) pp 511-517

[9] Wijenayake A H and Schmidt P B 1997 *Modelling and Analysis of Permanent Magnet Synchronous Motor by Taking Saturation and Core Loss into Account Int. Conf. on Power Electronics and Drive Systems*

[10] Cui B, Zhou J and Ren Z 2001 *Modelling and Simulation of Permanent Magnet Synchronous Motor Drives 5th International Conference on Electrical Machines and Systems* ICEMS (2001)

IOP Conf. Series: Materials Science and Engineering **224** (2017) 012042 doi:10.1088/1757-899X/224/1/012042

[11] Mademlis C and Margaris N 2002 *Loss Minimization in Vector-Controlled Interior Permanent-Magnet Synchronous Motor Drives IEEE Trans. on Industrial Electronics* vol 49(6) pp 1344-1347
[12] Pongiannan R, Selvabharathi P and Yadaiah N 2011 *FPGA Based Three Phase Sinusoidal PWM VVVF Controller 1st Int. Conf. on Electrical Energy Systems* ICEES (2011)
[13] Halasz S, Csonka G and Hassan A A M 1994 *Sinusoidal PWM Techniques with Additional Zero-Sequence Harmonics 20th Int. Conf. on Industrial Electronics Control and Instrumentation* IECON (1994)
[14] Fitzgerald A E *et al.* 1990 *Electric Machinery* (New York: McGraw-Hill)
[15] Ocen D 2005 *Direct Torque Control of a Permanent Magnet Synchronous Motor*
[16] Holtz J Lotzkat W and Khambadkone A M 1993 *On Continuous Control of PWM Inverters in the Over-Modulation Range Including the Six-Step Mode IEEE Trans. on Power Electronics* vol 8(4) pp 546-553
[17] Bose B 2002 *Modern Power Electronics and AC Drives* (Prentice-Hall. Inc) pp 70-74
[18] Di Gabriele R Parasiliti F and Tursini M 1997 *Digital Field Oriented Controller for Induction Motors: Implementation and Experimental Results The 32nd Universities Power Engineering Conf.* UPEC (1997)

IOP Conf. Series: Materials Science and Engineering **224** (2017) 012043 doi:10.1088/1757-899X/224/1/012043

# A Thermoelastic Damping Model for the Cone Microcantilever Resonator with Circular Cross-section

**Pu Li[a] and Hongyue Zhou**

School of Mechanical Engineering, Southeast University, Jiangsu, China

[a]seulp@seu.edu.cn

**Abstract.** Microbeams with variable cross-section have been applied in Microelectromechanical Systems (MEMS) resonators. Quality factor (*Q*-factor) is an important factor evaluating the performance of MEMS resonators, and high *Q*-factor stands for the excellent performance .Thermoelastic damping (TED), which has been verified as a fundamental energy lost mechanism for microresonators, determines the upper limit of *Q*-factor. TED can be calculated by the Zener's model and Lifshits and Roukes (LR) model. However, for microbeam resonators with variable cross-sections, these two models become invalid in some cases. In this work, we derived the TED model for cone microcantilever with circular cross-section that is a representative non-uniform microbeam. The comparison of results obtained by the present model and Finite Element Method (FEM) model proves that the present model is valid for predicting TED value for cone microcantilever with circular cross-section. The results suggest that the first-order natural frequencies and TED values of cone microcantilever are larger than those of uniform microbeam for large aspect ratios ($l/r_0$). In addition, the Debye peak value of a uniform microcantilever is equal to $0.5\Delta_E$, while that of cone microcantilever is about $0.438\Delta_E$.

## 1. Introduction

Micro-electro-mechanical System (MEMS) beam resonators with discontinuously or continuously variable cross-section are considered seriously due to their enhanced properties in real applications [1-2]. One of factors determining the performance of microbeam resonators is quality factor (*Q*-factor). A meaningful task for MEMS researchers and designers is to obtain a higher *Q*-factor of resonators. Thermoelastic damping (TED) is an intrinsic mechanism of energy lost in non-uniform microbeam resonators, which determines the upper limit of *Q*-factor. TED caused by an irreversible heat flux produced by temperature gradients cannot be completely avoided by the improved fabrication. Therefore, to accurately predict the TED value for the design and optimization of non-uniform microbeam resonators is of importance.

Zener [3-4], who firstly provided a 1D model of TED in the uniform cantilever, gave the TED formula of the beam with thickness $h$ which is expressed as follow

$$Q_{\text{Zener}}^{-1}=\frac{1}{2\pi}\frac{\Delta W}{W}=\frac{E\alpha^2T_0}{C_V}\frac{\omega\tau}{1+(\omega\tau)^2} \tag{1}$$

where $\Delta W$ represents the work lost per cycle, $W$ represents the maximum stored energy per cycle. The thermal relax time $\tau=h^2C_V/\pi^2\kappa$ , $E$ is the elastic modulus (Young's modulus), $T_0$ denotes the equilibrium temperature, $\kappa$ is the thermal conductivity, $\alpha$ denotes the coefficient of thermal expansion, $C_V$ denotes the specific heat, and $\omega$ is the vibrational frequency.

Zener's model and LR's model are popularly used to calculate TED for uniform microbeam resonators.

Devices with circular cross-section have been produced and studied, such as nanowires [5], microposts [6]. In 2010, Tunvir et al studied the TED on hollow microresonators with circular cross-section [7]. In 2016, Li et al investigated the TED in microrings with circular cross-section [8]. However, Zener's model and LR's model are failed to provide an appropriate estimation in some cases [9]. Li and Hu [10], and Zhou et al [11] all pointed that LR's model and Zener's model are not valid to evaluate TED in non-uniform microbeam resonators. In this study, as a typical microbeam with continuous variable cross-section, TED model for the cone microcantilever with circular cross-section is derived based on Zener's theory.

## 2. TED definition of cone microcantilever

Figure 1 shows a simple uniform cantilever and a cone cantilever with circular cross-section. The cantilever is continuous, homogeneous and isotropic. The $x$-axis is along the microcantilever length direction, and the $y$-axis is along the microcantilever transverse vibrational direction. The $r_0$ and $l$ represent right-end radius and length of microcantilever, respectively. The radius function $r(x)$ and area function $A(x)$ of cantilever shown in Figure 1.(b) are expressed as follows:

$$r(x)=r_0\frac{x}{l},\ A(x)=\pi r_0^2\left(\frac{x}{l}\right)^2 \tag{2}$$

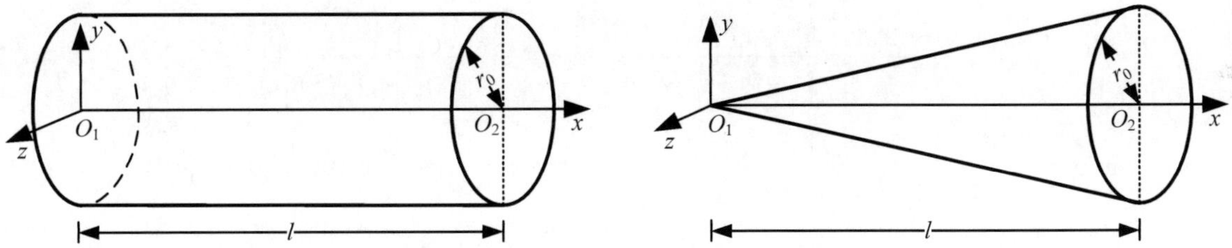


**Figure 1.** Schematic drawings of microcantilevers: (a) uniform microcantilever with circular cross-section, (b) cone microcantilever with circular cross-section.

Based on Euler-Bernoulli theory, the governing equation of free vibration in the $x-y$ plane of cone microcantilever can be expressed as

$$\frac{\partial^2}{\partial x^2}\left[M_M(x)+M_T(x)\right]+\rho A(x)\frac{\partial^2 y(x,t)}{\partial t^2}=0 \tag{3}$$

here $\rho$ denotes the mass density, $y(x,t)$ represents the transverse vibration displacement, $M_T(x)$ represents the thermoelastic moment , and $M_M(x)$ represents the mechanical moment.

The governing equation of 2D heat conduction is based on the Principle of Fourier heat conduction, which can be expressed in the cylindrical coordinate system

$$\frac{\partial\theta}{\partial t}=\chi\left(\frac{\partial^2\theta}{\partial r^2}+\frac{1}{r}\frac{\partial\theta}{\partial r}+\frac{\partial^2\theta}{r^2\partial\varphi^2}\right)+\frac{\Delta_E}{\alpha}\frac{\partial\left(y_{xx}(x,t)\right)}{\partial t}r\sin\varphi \tag{4}$$

Where $\Delta_E$ is called Zener's modulus and $\Delta_E=E\alpha^2T_0/C_V$. By substituting $\theta(r,\varphi,t)=\hat{\theta}(r,\varphi)e^{i\omega t}$ and $y_{xx}(x,t)=Y_{xx}(x)e^{i\omega t}$ into above equation, we can obtain

$$i\omega\hat{\theta}=\chi\left(\frac{\partial^2\hat{\theta}}{\partial r^2}+\frac{1}{r}\frac{\partial\hat{\theta}}{\partial r}+\frac{\partial^2\hat{\theta}}{r^2\partial\varphi^2}\right)+i\omega\cdot r\sin\varphi\cdot\frac{\Delta_E}{\alpha}Y_{xx} \tag{5}$$

In the paper [4], $\theta(x,y,t)$ is written as the total summation of thermal modes. It is worth noting that the first-order thermal mode affects thermoelastic energy dissipation seriously. Therefore, it is viable that $\theta(x,y,t)$ of cone microcantilever with circular cross-section can be written as the first-order thermal mode. Additionally, the thermal condition of the cone microcantilever surface is assumed to be adiabatic. At last, one can get the formula of $\theta_1(x,y,t)$

$$\hat{\theta}_1(x,r,\varphi)=\sin\varphi\frac{2r_0\Delta_E}{\alpha}Y_{xx}(x)\frac{1}{(a_1^2-1)J_1(a_1)}\frac{\mathrm{i}\omega\tau_1}{1+\mathrm{i}\omega\tau_1}J_1\left(\frac{a_1}{r_0}r\right) \tag{6}$$

here $\tau_1=\dfrac{r^2}{\chi a_1^2}$, and $a_1$=1.841 [8].

The total energy dissipation over one period due to TED can be calculated by [8]

$$\Delta W=\pi\iiint_V \hat{\sigma}\,\mathrm{Im}(\hat{\varepsilon}_{\text{thermal}})\mathrm{d}V=2\pi^2E\Delta_E\cdot\frac{[2J_1(a_1)-a_1J_0(a_1)]}{a_1^2(a_1^2-1)J_1(a_1)}\int_0^l\frac{\omega\tau_1}{1+(\omega\tau_1)^2}r^4(x)Y_{xx}^2(x)\mathrm{d}x \tag{7}$$

here the amplitude of thermal strain in cone microcantilever $\hat{\varepsilon}_{\text{thermal}}=\alpha\hat{\theta}$, the amplitude of mechanical stress $\hat{\sigma}=EY_{xx}(x)\cdot r\sin\varphi$, and Im represents the imaginary part.

The maximum energy stored $W$ in the microcantilever can be calculated by [8]

$$W=\frac{1}{2}\iiint_V \hat{\sigma}\hat{\varepsilon}_{\text{elastic}}\mathrm{d}V=\frac{E}{2}\int_0^l\int_0^{2\pi}\int_0^{r(x)}\sin^2\varphi\cdot Y_{xx}^2(x)\cdot r^3\mathrm{d}r\mathrm{d}\varphi\mathrm{d}x=\frac{\pi E}{8}\int_0^l r^4(x)Y_{xx}^2(x)\mathrm{d}x \tag{8}$$

According to the energy definition [12], TED model in the cone microcantilever with circular cross-section is given by

$$Q_{\text{TED}}^{-1}=\frac{1}{2\pi}\frac{\Delta W}{W}=\frac{8[2J_1(a_1)-a_1J_0(a_1)]}{a_1^2(a_1^2-1)J_1(a_1)}\cdot\Delta_E\cdot\frac{\int_0^l\frac{\omega\tau_1}{1+(\omega\tau_1)^2}r^4(x)Y_{xx}^2(x)\mathrm{d}x}{\int_0^l r^4(x)Y_{xx}^2(x)\mathrm{d}x} \tag{9}$$

According to $\dfrac{8[2J_1(a_1)-a_1J_0(a_1)]}{a_1^2(a_1^2-1)J_1(a_1)}=0.9877\approx 1$ with respect to $a_1$=1.841, the equation (9) reduces to

$$Q_{\text{TED}}^{-1}=\Delta_E\cdot\frac{\int_0^l\frac{\omega\tau_1}{1+(\omega\tau_1)^2}r^4(x)Y_{xx}^2(x)\mathrm{d}x}{\int_0^l r^4(x)Y_{xx}^2(x)\mathrm{d}x} \tag{10}$$

Then, the dimensionless form of equation (10) is written as follows

$$Q_{\text{TED}}^{-1}=\Delta_E\cdot\frac{\int_0^1\frac{\eta\cdot X^6}{1+\eta^2X^4}Y_{XX}^2(X)\mathrm{d}X}{\int_0^1 X^4Y_{XX}^2(X)\mathrm{d}X} \tag{11}$$

Where $\eta=\dfrac{\omega r_0^2}{\chi a_1^2}$. It is obvious that $Y(X)$ is an important step to calculate thermoelastic damping, and the formula of $Y(X)$ will be derived in Section 3.

## 3. Modal parameters

### *3.1. Modal shape function*

The non-dimensional radius function, non-dimensional area function, and non-dimensional inertia moment function are written as

$$r(X)=r_0X,\ A(X)=A_0X^2,\ I(X)=I_0X^4 \tag{12}$$

here $X=x/l$, $A_0=\pi r_0^2$, and $I_0=\pi r_0^4/4$. Because $M_T(x)$ is neglected compared to $M_M(x)$, substitude $y(x,t)=Y(x)\mathrm{e}^{\mathrm{i}\omega t}$ into equation (3), then one can obtain the non-dimensional form of equation (3):

$$\frac{\mathrm{d}^4Y(X)}{\mathrm{d}X^4}+\frac{8}{X}\frac{\mathrm{d}^3Y(X)}{\mathrm{d}X^3}+\frac{12}{X^2}\frac{\mathrm{d}^2Y(X)}{\mathrm{d}X^2}-\frac{\lambda^2}{X^2}Y(X)=0 \tag{13}$$

where $\lambda^2=\omega^2\rho A_0l^4/EI_0$ and $\omega$ is the natural angular frequency of the cantilever.

The solution of equation (13) is given by [13]

$$Y(X)=\frac{1}{X}\left[C_1J_2\left(2\sqrt{\lambda X}\right)-C_2I_2\left(2\sqrt{\lambda X}\right)\right] \tag{14}$$

here $J_2$ is the 2nd order Bessel function of the first kind with real argument, and $I_2$ is the 2nd order Bessel function of the first kind with imaginary argument. The $C_1$ and $C_2$ are undetermined coefficients.

### *3.2. Modal eigen frequency*

Considering boundary conditions of the cone microcantilever, the rotation and displacement of the right-end are zero, $Y(1)=Y'(1)=0$, then the following equations are

$$\begin{cases} J_2\left(2\sqrt{\lambda}\right)C_1-I_2\left(2\sqrt{\lambda}\right)C_2=0 \\ \left[\sqrt{\lambda}J_1\left(2\sqrt{\lambda}\right)-2J_2\left(2\sqrt{\lambda}\right)\right]C_1+\left[-\sqrt{\lambda}I_1\left(2\sqrt{\lambda}\right)+2I_2\left(2\sqrt{\lambda}\right)\right]C_2=0 \end{cases} \tag{15}$$

For meaningful solutions of $C_1$ and $C_2$, the eigen frequency equation is computed by

$$\begin{vmatrix} J_2\left(2\sqrt{\lambda}\right) & -I_2\left(2\sqrt{\lambda}\right) \\ \sqrt{\lambda}J_1\left(2\sqrt{\lambda}\right)-2J_2\left(2\sqrt{\lambda}\right) & -\sqrt{\lambda}I_1\left(2\sqrt{\lambda}\right)+2I_2\left(2\sqrt{\lambda}\right) \end{vmatrix}=0 \tag{16}$$

And equation (16) can be expanded to be

$$J_1\left(2\sqrt{\lambda}\right)I_2\left(2\sqrt{\lambda}\right)-J_2\left(2\sqrt{\lambda}\right)I_1\left(2\sqrt{\lambda}\right)=0 \tag{17}$$

Solve equation (17) in MATLAB and obtain the values of $\lambda$. By setting $C_2=1$ and $C_1=\gamma=I_2\left(2\sqrt{\lambda}\right)/J_2\left(2\sqrt{\lambda}\right)$, the final the square of $Y_{XX}(X)$ is obtained by

$$Y_{XX}^2(X)=\frac{1}{X^7}\left[\lambda X^{1.5}\left(\gamma J_0-I_0\right)+4\sqrt{\lambda}X\left(-\gamma J_1+I_1\right)-6\sqrt{X}\left(\gamma J_0+I_0\right)+\frac{6}{\sqrt{\lambda}}\left(\gamma J_1+I_1\right)\right]^2 \tag{18}$$

where $J_0$, $J_1$, $I_0$ and $I_1$ are all Bessel functions of the first kind with respect to $2\sqrt{\lambda X}$ .

By substituting equation (18) and structural geometry functions equation (12) into equation (8), then the correspongding value of TED can be computed by integration. Here we notice that

$$\omega=\lambda\sqrt{\frac{EI_0}{\rho A_0 l^4}} \tag{19}$$

and λ=8.7193, 21.1457, 38.4538,...

## 4. Validation and discussion

In this section, the results computed by the present model for cone microcantilevers with circular cross-section are compared with the FEM results obtained by ANSYS. Solid226, which is a 3D 20-Node element, is applied in coupled thermal-structural analysis. A harmonic force $p\sin(\omega t)$ excites the cone microcantilever. The single-crystal silicon is applied in simulation. And the material properties of silicon [14] are listed as follows: $E$=160GPa, $\rho$=2330Kg/m$^3$, $C_V$ =1.6×10$^6$J/m$^3$/K, $\kappa$=120W/m/K, $\alpha$=2.6×10$^{-6}$K$^{-1}$, $\chi$=7.4033×10$^{-5}$m$^2$/s, $\upsilon$=0.22, and $T_0$=300K. In real applications, only TED at fundamental frequencies is of interest [7, 11].

Firstly, according to equation (19), the first-order natural frequencies of cone microcantilever are calculated. Figures 2 and 3 show the fundamental frequency against the beam length with the constant $r_0$ is equal to 2.5μm and 5μm, respectively. It is worth noting that for the same $r_0$, the frequency of cone microcantilever is larger than that of uniform microcantilever. Therefore, when the resonators need to work at high frequencies, it is better to choose cone cantilever instead of uniform cantilever.

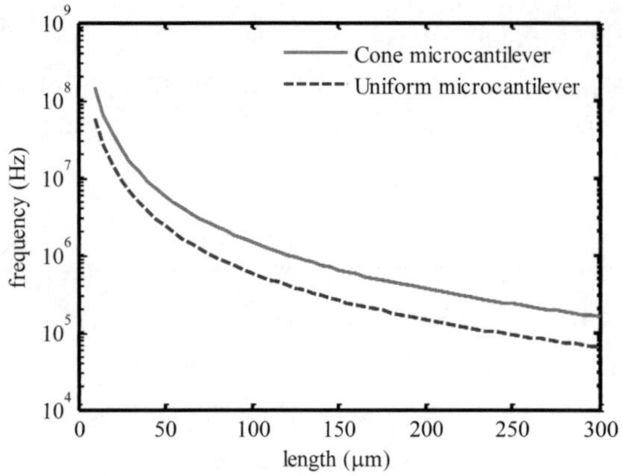


**Figure 2.** Comparison of first-order natural frequencies against the length. The radius $r_0$ of cone cantilever and uniform cantilever is 2.5μm.

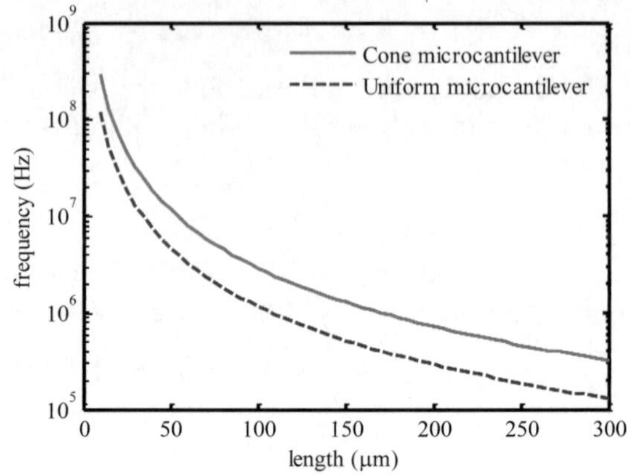


**Figure 3.** Comparison of first-order natural frequencies against the length. The radius $r_0$ of cone cantilever and uniform cantilever is 5μm.

Comparison of TED values obtained from the present model and FEM model are plotted in Figures 4 and 5 for the representative radius $r_0$. As expected, results based on the present model agree well with those based on the FEM model, and the relative errors in two cases are less than 3% over the wide range of beam length. As observed from Figure 5, TED values of cone microcantilever are larger than those of uniform microbeam with increasing aspect ratios ($l/r_0$). Figures 4 and 5 display the perfect agreement proving that the present model is correct and effective.

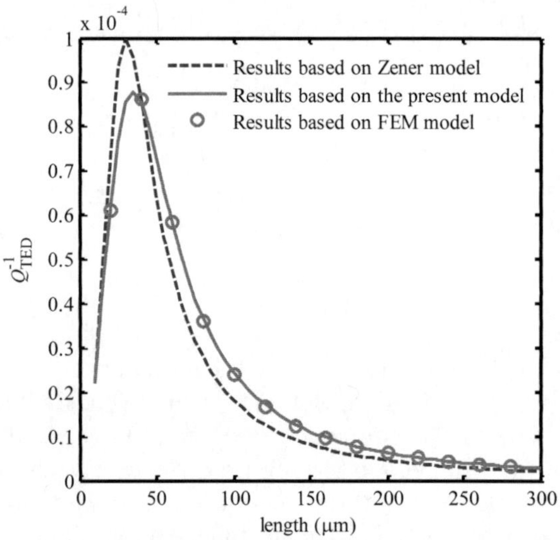


**Figure 4.** Comparison of TED values against the beam length. The radius $r_0$ of uniform beam and cone cantilever are both 2.5μm.

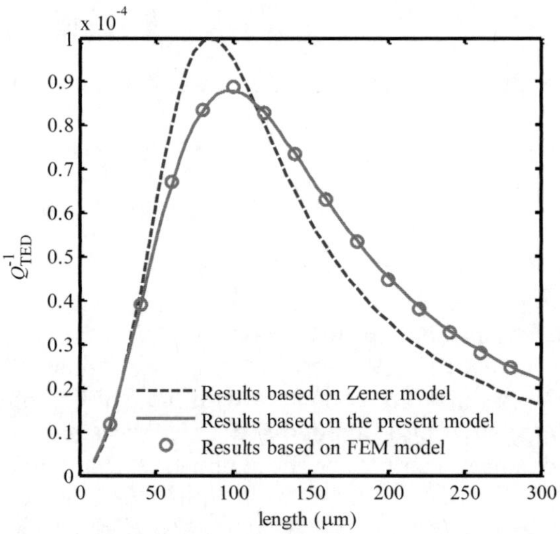


**Figure 5.** Comparison of TED values against the beam length. The radius $r_0$ of uniform beam and cone cantilever are both 5μm.

On the other hand, Figure 6 shows that the Debye peak value of cone microcantilever with circular cross-section is about $0.438\Delta_E$ at $\eta$=2, while that of Zener's model is $0.5\Delta_E$ at $\eta$=1. Moreover, the results suggest that for non-uniform microbeam resonators, TED is dependent of both the vibrational frequencies and the mode shape functions.

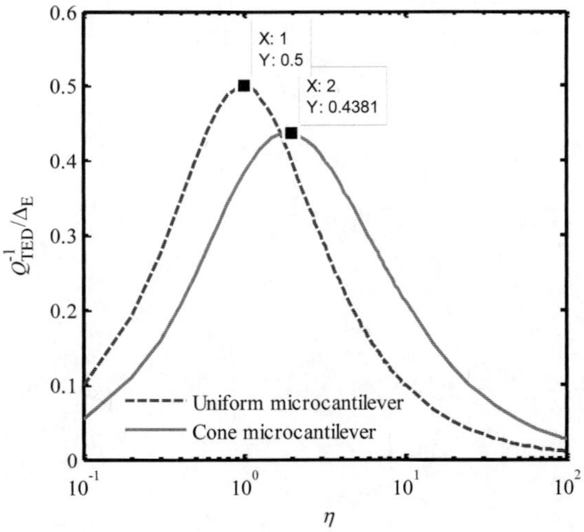


**Figure 6.** Variation of normalized TED curves for uniform microcantilever and cone microcantilever.

## 5. Conclusions

In this work, TED model for the cone microcantilever resonator with circular cross-section are presented and analyzed. Compared with numerical results obtained from the FEM model, and perfect agreements indicate that the proposed TED model is correct and reasonable. The current work could help MEMS designers to achieve better performance (higher $Q$-factor) by reducing the effect of thermoelastic damping.

**Acknowledgements**
We acknowledge the financial support from National Natural Science Foundation of China (Grant No. 51375091).

## References

[1] Mohanty P Harrington D A Ekinci K L et al 2002 Phys Rev B **66**(8) p 085416.
[2] Matova S P Renaud M Jambunathan M et al 2013 Smart Mater Struct **22**(7) p 075015.
[3] Zener C 1937 Phys Rev **52**(3) p 230-236.
[4] Zener C 1938 Phys Rev **53**(1) p 90-99.
[5] Hu J Q Bando Y Liu Z W et al 2003 J Am Chem Soc **125**(37) p 11306-11313.
[6] Khoo H S Liu K K and Tseng F G 2003 J Micromech Microeng **13**(6) p 822-831.
[7] Tunvir K Ru C Q and Mioduchowski A 2010 Physica E **42**(9) p 2341-2352.
[8] Li P Fang Y M and Zhang J R 2016 J Sound Vib **361** p 341-354.
[9] Candler R N Duwel A Varghese M et al 2006 J Microelectromech Syst **15**(4) p 927-934.
[10] Li P and Hu R F 2011 Jpn J Appl Phys **50**(7) p 077202.
[11] Zhou H Y Li P and Zuo W L 2016 *2016 IEEE International Conference on Mechatronics and Automation*, p 1590-1595.
[12] Nowick A S and Berry B S 1972 *Anelastic relaxation in crystalline solids* (New York: Academic Press).
[13] Kamke E 1977 *Ordinary differential equations manual* (Beijing: Science Press).
[14] Yoneoka S Roper C S Candler R N et al 2010 J Microelectromech Syst **19**(2) p 357-366.

IOP Conf. Series: Materials Science and Engineering **224** (2017) 012044 doi:10.1088/1757-899X/224/1/012044

# Small-time Scale Network Traffic Prediction Based on Complex-valued Neural Network

**Bin Yang**

School of Information Science and Engineering, Zaozhuang University, Zaozhuang, China

batsi@126.com

**Abstract.** Accurate models play an important role in capturing the significant characteristics of the network traffic, analyzing the network dynamic, and improving the forecasting accuracy for system dynamics. In this study, complex-valued neural network (CVNN) model is proposed to further improve the accuracy of small-time scale network traffic forecasting. Artificial bee colony (ABC) algorithm is proposed to optimize the complex-valued and real-valued parameters of CVNN model. Small-scale traffic measurements data namely the TCP traffic data is used to test the performance of CVNN model. Experimental results reveal that CVNN model forecasts the small-time scale network traffic measurement data very accurately

## 1. Introduction

Network traffic modeling and analysis play a major role in characterizing network performance. Accurate models have an important role in capturing the salient characteristics of the network traffic, analyzing for the network dynamic, and improving the forecasting ability for system dynamics. It has a fundamental meaning for many network designs and engineering problems, e.g., switcher designing, router, the management of devices and its software development [1].

The rapid development of the communication and network technologies results in the uncertain characteristics of the traffic network, especially the nonlinear time series. The nonlinear time series are more accurately predicted by using some models. The flexible neural tree (FNT) model was employed to predict the small-time scale traffic measurements data [2]. The FNT structure was developed by using the Genetic Programming, and the parameters are optimized by the Particle Swarm Optimization (PSO) algorithm. Ordinary differential equations (ODEs) were also presented to forecast and reflect the actual characterize of the real network traffic [3]. Jiang et al. proposes a new prediction algorithm to network traffic in the large-scale communication network in order to overcome the problem of network traffic prediction in the communication network [4].

Recently, complex-valued neural network (CVNN) has been proposed to predict the time series data. Compared with real-valued neural network, CVNN is more flexible and functional. In order to predict the small-time scale traffic measurements data, complex-valued neural network (CVNN) model is proposed to predict nonlinear small-time scale traffic measurements data. Artificial bee colony (ABC) algorithm is proposed to optimize the complex-valued and real-valued parameters of CVNN model. The TCP traffic data is used to test the performance of CVNN model.

The paper is organized as follows. In Section 2, the details of CVNN model are described. In Section 3, the examples are used to examine the effectiveness and veracity of the proposed method. Finally, Section 4 concludes this paper.

## 2. Method

### 2.1. Complex-valued neural network

In a complex-valued neural network, input layer, weights, threshold values and output layer are all complex numbers. A three-layer complex-valued neural network is described in Fig. 1, which has m input nodes, n hidden nodes and one output node. Suppose that input vector $[z_1, z_2, \cdots$ ]. The result of *i*-th hidden node is computed as followed.

$$h_i = f(W_i + W_{i1}z_1 + W_{i2}z_2 + \dots \qquad ). \tag{1}$$

Where $W_i$ is the threshold value of *i*-th hidden node and $W_{i1}, W_{i2}, \dots$ are the weights. $f$ is the activation function, which is

$$f(z) = \frac{z}{c + \frac{1}{r}|z|}. \tag{2}$$

Where $c$ and $r$ are the real variables.
The output layer is computed:

$$y = f(h_1V_1 + h_2V_2 + \dots \tag{3}$$

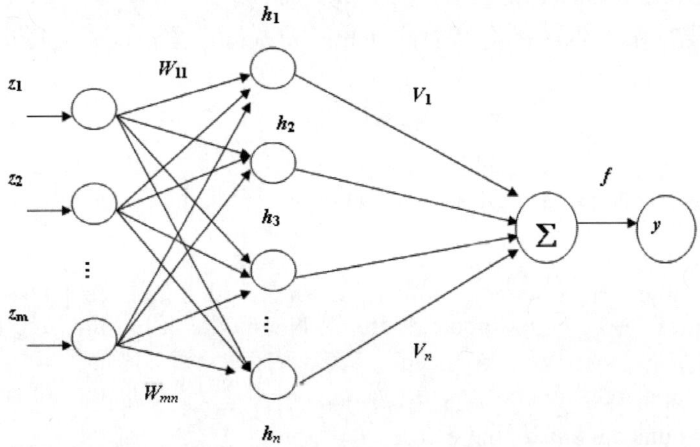


**Figure. 1.** A three-layer complex-valued neural network.

### 2.2. Optimization of parameters

In order to search the optimal parameters of CVNN model, there are many local and global evolutionary methods, such as PSO, genetic algorithm (GA) and simulated annealing (SA). In this paper, artificial bee colony (ABC) algorithm is selected because of its fast convergence, more accurate solution and stability [5-6].

ABC algorithm can be summarized as follows.

(1) Choose the initial value of algorithm parameter including maximum number of generation (*max_gen*), number of employed bees and onlooker bees (*num_employ* and *num_onlooker*), the maximum times that food sources do not change its location (*limit*).

(2) In a CVNN (Fig. 1), many parameters need to be optimized, containing weights ($W_{ij}$ and $V_i$), threshold value ($W_i$) and parameters in activation function ($c$ and $r$). Due to the fact that weights and threshold value are complex numbers, both real and imaginary parts need to be optimized. For a CVNN *m*-*n*-1 (*m* inputs, *n* hidden neurons and 1 output), the number of parameters is 2*nm*+6*n*+2 and parameter vector is

$[\mathrm{Re}(W_{11}), \mathrm{Im}(W_{11}), \ldots$ $\ldots$ $\ldots$
$r_n, \mathrm{Re}(V_1), \mathrm{Im}(V_1), \ldots$ $, \mathrm{Im}(V_n)]$.

Generation *num_employ* and *num_onlooker* solutions, calculate the fitness value, and select the best half of solutions as foods.

(3) In order to produce a candidate food position from the old one in memory, the ABC uses the following equation:

$$v_{ij} = x_{ij} + R_{ij}(x_{ij} - x_{kj}) \tag{4}$$

Where $v_{ij}$ is a new position of food, $R_{ij}$ is a random number in the range [-1,1], $k \in \{1,2,3,\mathrm{K}\ ,N\}$ and $k \neq i$. Calculate the fitness of $v_i$, and when the fitness of $v_i$ is better than $x_i$, replace $x_i$ with $v_i$. If $x_i$ is not changed, $failure_i = failure_i + 1$.

(4) The probabilities for onlookers are calculated as follows.

$$p_i = \frac{fit_i}{\sum_{n=1}^{N} fit_n} \tag{5}$$

Where $fit_i$ is the fitness value of i-th bee.

(5) According to $p_i$, onlooker bees search the area as employed bees.

(6) If $failure_i > limit$, then scout bee randomly generations a new solution according to the following equation:

$$x_{ij} = x_{min\ j} + rand(0,1)(x_{max\ j} - x_{min\ j}) \tag{6}$$

(7) Record the best solution at each generation.

(8) If *max_gen* is reached or a satisfactory solution is found, then stop; otherwise go to step (3).

### 2.3. Flowchart of method

(1) The traffic sample data is real number, so input and output data need be transformed to complex values before being inputted into CVNN model. The following transformation steps are used [7]:

- Suppose the input real number vector $[x_1, x_2, \ldots$ , and tally the maximum and minimum values of real numbers ( $max$ and $min$ ).
- *i*-th real number $x_i$ is transformed into complex number as followed.

$$\varphi_i = \frac{x_i - min}{max - min}(2\pi - \delta),$$
$$z_i = e^{i\varphi_i}. \tag{7}$$

Where $\delta$ is the shift angle and $i$ stands for the value of $\sqrt{-1}$.

(2) Train the complex-valued neural network according to the above complex numbers (input data) and cuckoo search, which is introduced in detail in Section 2.2. The output of CVNN is complex number, so the output value needs to be transformed into real number in order to evaluate model effectively. The inverse transformation shall be used:

$$\arg z = \varphi,$$
$$y = \frac{\varphi(max - min)}{2\pi - \delta} + min. \tag{8}$$

Where $\arg z$ is the argument of complex value $z$.

## 3. Experiments

Small-time scale traffic measurements data is from Meng [3]. The data are normalized in the interval [0, 1] with the following formula

$$x' = \frac{x - x_{\min}}{x_{\max} - x_{\min}} \tag{9}$$

The length of this traffic data set is 36000. The front 33000 data points are used as the training set and the last 3000 data points are used as the test set. In the paper, we use the front 10 variable to predict the current variable. The used instruction set for additive tree model

To access the performance of the different methods, the root mean squared error (RMSE), maximum absolute percentage error (MAP) and mean absolute percentage error (MAPE) are used to study the performance of the trained forecasting model for the test data, defined as:

$$RMSE = \sqrt{\frac{1}{N}\sum_{k=1}^{N}(y_{actual,k} - y_{predict,k})^2}$$

$$MAP = \max\left(\frac{|y_{actual,k} - y_{predict,k}|}{y_{predict,k}} \times 100\right)$$

$$MAPE = \sqrt{\frac{1}{N}\sum_{k=1}^{N}\left(\frac{|y_{actual,k} - y_{predict,k}|}{y_{predict,k}} \times 100\right)} \tag{10}$$

where $N$ is the total number of the data, $y_{actual,k}$ is the *k- th* actual data, and $y_{predic,k}$ is the *k-th* forecast value.

Table 1 summarizes the results achieved using three different methods, and a comparison of actual network traffic time series data and the predicted ones by CVNN model is shown in Fig. 2. From Table 1, it can be clearly seen that CVNN model can effectively improve the accuracy of small-time scale network traffic prediction.

The prediction error of CVNN model is shown in Fig. 3. From Fig. 3, it can see that CVNN could effectively predict the traffic data. The prediction error converges to the small value, even at zero

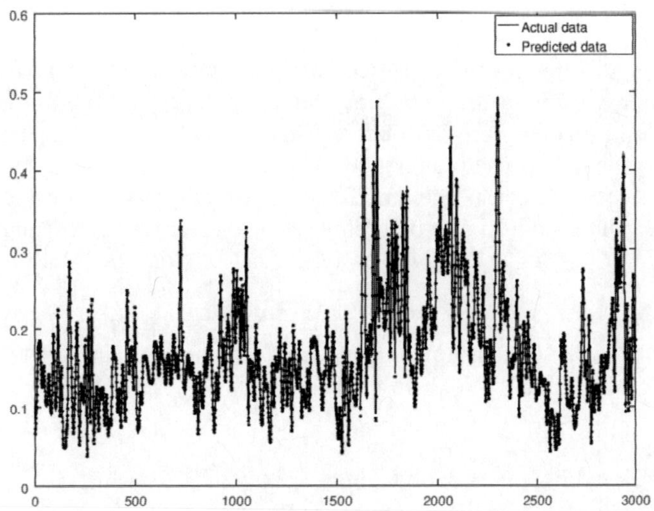


**Figure.2.** Comparison of actual data and the predicted ones by CVNN model.

**Table 1.** Statistical analysis of three learning methods (test data).

| Criterion | FNT [2] | ODE [3] | CVNN |
|---|---|---|---|
| MAP | 159.46 | 157.3 | 151.57 |
| MAPE | 5.01 | 4.84 | 4.34 |
| RMSE | 0.0129 | 0.0124 | 0.0113 |

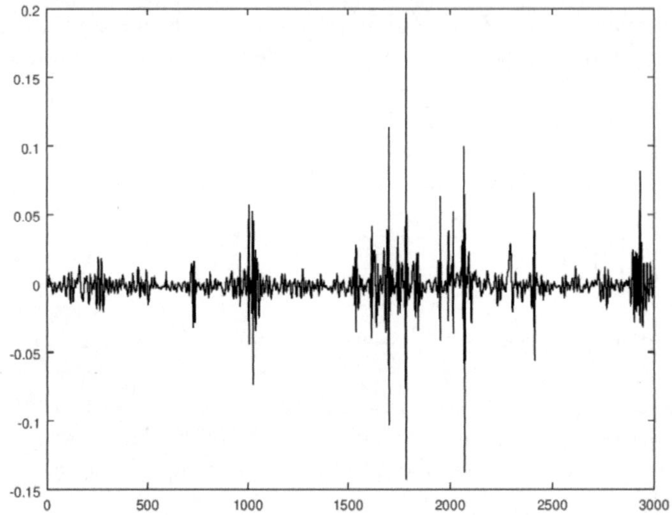


**Figure. 3.** The predicted errors.

## 4. Conclusions

In order to forecast small-time traffic data accurately, complex-valued neural network model is proposed to further improve the accuracy of time series forecasting. The complex-valued and real-valued parameters are optimized by artificial bee colony algorithm. Small-scale traffic measurements data namely the TCP traffic data is used to test the performance of CVNN model. Experimental results reveal that the CVNN model forecasts the small-time scale network traffic measurement data more accurately and FNT and ODE models. The prediction error converged to the small value, even at zero.

## Acknowledgments

This study was funded by Shandong Provincial Natural Science Foundation, China (No. ZR2015PF007), the PhD research startup foundation of Zaozhuang University (No.2014BS13), and foundation of Zaozhuang University (No.2015YY02).

## References

[1] Meng QF, Chen YH and Peng YH 2008 *Chinese Physics B* **18** 2196.
[2] Chen YH, Yang B and Meng QF 2012 *Applied Soft Computing* **12** 274.
[3] Chen YH, Yang B, Meng QF, Zhao YO and Abraham A 2010 *Information Sciences* **181** 106
[4] Jiang D, Xu Z and Xu H 2015 *Annals of Telecommunications* **70** 427.
[5] Cheng XY and Jiang MY 2012 *Lecture Notes in Computer Science* **7331** 326.
[6] Karaboga D, Gorkemli B, Ozturk C and Karaboga N 2014 *Artificial Intelligence Review* **42** 21.
[7] AizenbergI, Sheremetov L, Villa-Vargas L and Martinez-Muñoz J 2015 *Neurocomputing* **175** 980.

IOP Conf. Series: Materials Science and Engineering **224** (2017) 012045 doi:10.1088/1757-899X/224/1/012045

# Decoupling research on a flexible tactile sensor array with novel structure

**G Y Hu**[1,2], **Y Y Xu**[1,2] , **H Q Pan**[1*] **and F Shuang**[1,2,3]

[1] Institute of Intelligent Machines, Chinese Academy of Sciences, Hefei, China
[2] Department of Automation, University of Science and Technology of China, Hefei, China
[3] Department of Mechanical Engineering, Anhui Polytechnic University, Wuhu, China

*hqpan@iim.ac.cn

**Abstract.** This paper presents a flexible tactile sensor, which uses a novel structure and effective decoupling method to detect multi-point three-dimensional touch force. Firstly, the structure characteristics of the sensor are introduced. According to the theories of force-sensitive conductive rubber, the relationship between the resistance and the force exerted on the sensor is analyzed. Furthermore the mathematical model of the sensor based on finite element method is conducted. The simulation result is consistent well with the theory result. Then a decoupling method based on the RBF neural network is proposed in order to obtain the force from the resistance. The research indicates that the flexible tactile sensor with the novel structure and the decoupling method can accurately detect the three-dimensional force.

## 1. Introduction

Tactile as an important form of perception for the robots can directly perceive a variety of physical properties of environment and the target object. The flexible tactile sensor that can imitate human shin will play the key role in the robots[1,2], and be applied to other areas, such as surgery, rehabilitation, aerospace and so on. So the research of flexible tactile sensor is one of the hot spots in the field of robot sensors. The working principle of tactile sensor can generally be distinguished as pressure resistance, piezoelectric and capacitor, etc[3-6]. Koc. I. M et al.[7]use PVDF material to develop the piezoelectric tactile sensor with micro cylindrical structure, which can detect both force and force distribution at same time. Bao et al.[8]use PDMS material to create a highly sensitivity capacitive tactile sensor, which can clearly perceive the micro force produce by a fly or butterfly stay on the surface of sensor. F. Xu et al.[9] developed a three-dimensional force sensor based on conductive rubber, and the sensor with multi-layer array structure is similar to artificial skin tissue.

The complex coupling relationship generally exists in the multi-dimensional force measurement of the tactile sensor. It will affect the accuracy and real-time performance of the sensor. For the early tactile sensor array there exists the complex coupling relationship[3,10]. We redesigned a new microstructural sensitive unit for three-dimensional detection. Its mathematical model is simpler, and also easier to construct a larger scale of the sensor array. Moreover, a decoupling method based on RBF neural network with high dimensional nonlinear ability [11] is used in order that the sensor can detect the three-dimensional force information accurately.

IOP Conf. Series: Materials Science and Engineering **224** (2017) 012045 doi:10.1088/1757-899X/224/1/012045

## 2. Structural Design

### *2.1 Structure of Flexible Tactile Sensor Array*

In the model, the orderly microstructure unit array was adopted. It is the core component of the flexible tactile sensor. As shown in Figure 1, the microstructure unit is composed of three sections 1, 2 and 3, which denote the columns made of conductive rubbers, and 4, 5, 6 and 7 denote the electrode points on the surface of the columns.

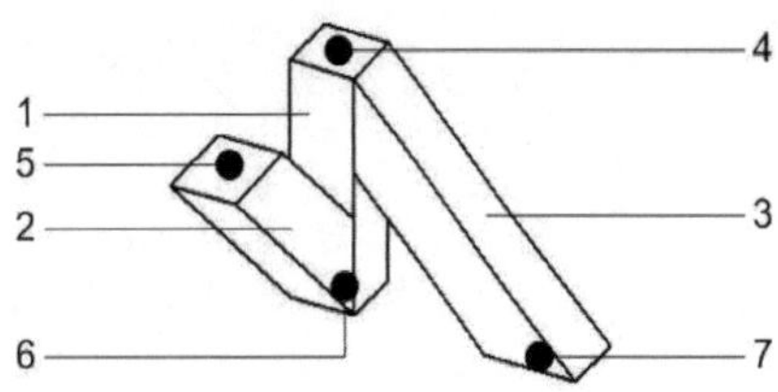


**Figure 1.** Microstructural unit

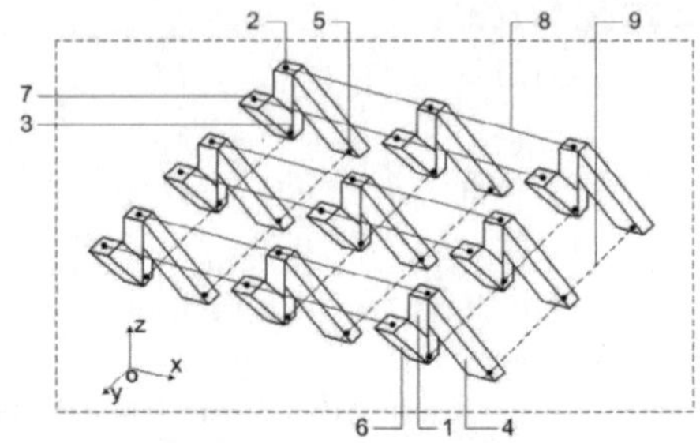


**Figure 2.** Structure of flexible tactile sensor array

Figure 2 shows the flexible tactile sensor array with $3\times3$ dimension of, in which, 1, 4 and 6 denote the columns made of conductive rubber; 2, 3, 5 and 7 denote the electrode points; 8 denotes the upper horizontal wire which connects the upper electrode points; 9 denotes the lower vertical wire which connects the lower electrode points. Insulating rubber is used to fill gap in the array.

When force is applied on the microstructure, the force can be decomposed into three different directions, $F_x$, $F_y$ and $F_z$. The resistance of each column is obtained by means of scanning, and then the resistance between the upper and the lower layer is detected one by one, so the resistance matrix $R_x$, $R_y$, $R_z$ are obtained, and corresponds to the resistance changes of three different directions, respectively. Finally, according to state equations $F_i = f(\Delta R_i)$, we can decouple the force of the sensitive unit.

### *2.2 Force Analysis of Sensor*

When the force exerts on the surface of flexible tactile sensor, the relevant columns in the microstructure are deformed, resulting in change in the resistance values. According to the properties of hyperelastic material, volume of the conductive rubber remains constant when it is deformed. Therefore, the resistance values of the three conductive column in each microstructure satisfy the following formula:

$$R_i = \rho\frac{l_i}{s} = \frac{\rho}{V_i} * l_i^2 \quad i = x, y, z \tag{1}$$

For each microstructure unit, there are three resistance outputs, $R_x$, $R_y$, $R_z$ respectively; They are consistent with three input of the three-dimensional force, $F_x$, $F_y$ and $F_z$ respectively. According to formula (1), there is a mapping equation:

$$\Delta l_i = l_i^{'} - l_i = (\sqrt{R_i^{'}} - \sqrt{R_i})\sqrt{V_i/\rho} \quad i = x, y, z \tag{2}$$

The conductive rubber can approximatively satisfy the condition: $F \propto \Delta l$ because of its piezoresistive properties, and then can be used to calculate the relationship between three-dimensional force and resistance of sensor.

$$F_i \propto (\sqrt{R'_i} - \sqrt{R_i})\sqrt{V_i / \rho} \quad i = x, y, z \tag{3}$$

Hence there is the following function relationship between the force exerted on the surface of tactile sensor and the resistance of conductive rubber.

$$\begin{aligned} F_x &= f_1(R_x) \\ F_y &= f_2(R_y) \\ F_z &= f_3(R_z) \end{aligned} \tag{4}$$

It can be seen that the structure of the flexible tactile sensor reduces coupling relation and degree of nonlinearity.

## 3. Finite Element Analysis

Using COMSOL software, the numerical model of flexible three-dimensional tactile sensor is constructed based on the coupling simulation method[12]. Generally, the lower surface is considered as the fixed constraint in the simulation. When the three-dimensional force is exerted to the microstructure unit in the second row and the second column of the 3*3 sensor array, distribution of force can be seen in the variation stress chart. Figure 4 shows the displacement diagram when *Px*=50MPa, *Py*=50MPa and *Pz*=50MPa. It illustrates that the deformation of microstructure unit in the force area is large, and the microstructure units of surrounding almost have no deformation.

In the experiment, the range of forces applied to the three directions is from 0 to 50MPa, with the increment of 4MPa. It will produce 1000 set of samples. Therefore, we get 1000 samples as training samples of RBF neural network, and 125 groups of samples are randomly generated as test samples of RBF neural network.

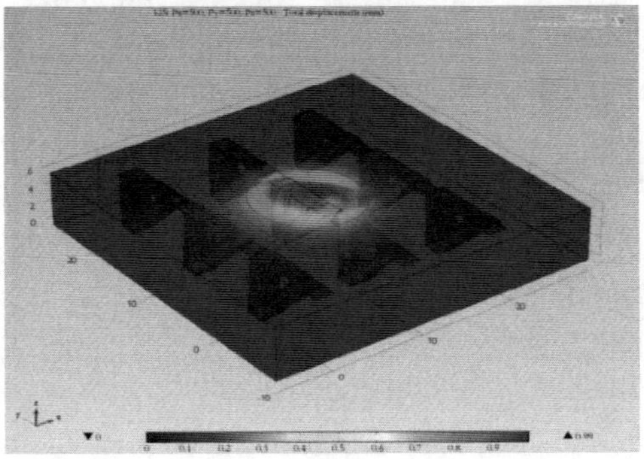

**Figure 3.** Deformation of the sensor

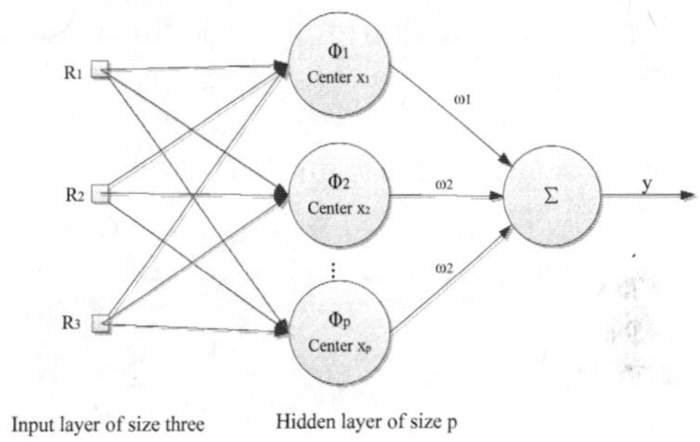


**Figure 4.** Structure of the RBF neural network

## 4. Decoupling of the flexible tactile sensor

### *4.1 Decoupling algorithm based on RBF neural network*

RBF neural network has 3 layers, which are input layer, hidden layer and output layer. The network has the characteristics of simple structure and strong nonlinear approximation ability. As shown in Figure 5, the input layer is composed of three neuron nodes and transformation function of hidden layer is radial basis function, and the transformation function of output layer is linear function. The RBF neural network is used to decouple the three-dimensional flexible tactile sensor, its essence is that the complex pattern classification problem nonlinearly is mapped to the high-dimensional space, and then is classified in the output layer.

There are kinds of radial basis functions in RBF neural network. In the paper, the radial basis function is mainly used in the form of Gaussian function as follows:

$$\varphi(x, x_j) = \varphi(\|x - x_j\|) = \exp(-\frac{1}{2\sigma_j^2}\|x - x_j\|^2) \tag{5}$$

Here, the vector x is the signal of the input layer; $X_j$ is Gaussian function center of j-th input sample; $\varphi(||x\text{-}x_j||)$ is a collection which contains many radial basis functions; $||x\text{-}x_j||$ is the center of the radial basis function; $\sigma_j$ represents the basis function extension of $X_j$ as a vector center. When the RBF neural network is used as a general function approximator, its output has following mathematical form:

$$F(x) = \sum_j \omega_j \varphi(x, x_j) = \sum_j \varphi(\|x - x_j\|) \tag{6}$$

Since the change in resistance resulted from pressure exerted on the tactile sensor is nonlinear, $\{x_i\}_{i=1}^{N}$ is a collection of N different points, it can be sure that ϕ is nonsingular matrix, which is reversible and has determination solution.

In the experiment, we construct 1000 sets of samples to train the RBF network, each sample contains two parts: The input vector is the 3-D matrix $R=[R_x, R_y, R_z]$, which is resistance matrix constituted by the three columns in the microstructure unit; The output vector represents the surface of the tactile sensor on which the three-dimensional force $F=[F_x, F_y, F_z]$ is exerted. Then use the trained RBF network model to decouple the three-dimensional force of the test samples, and solve the relative error. The result shows that the ability of the nonlinear mapping relationship based on the RBF neural network is very well, it can decouple the three-dimensional force information accurately.

*4.2 Analysis of the decoupling results*

In the RBF neural network, the most important factor affecting the performance of the network is the spread of the radial basis. The larger the spread is, the smoother the function fits, but the approximation error becomes large, the net requires more neurons, and it will leads to a large quantity of calculation. Conversely, the smaller the spread is, the more precise the decoupling results will be, but the approximation process is not smooth, the performance of network is poor, and there would produce over-fitting phenomenon, which is easy to produce larger training error. Therefore, we try to choose the different spread, as shown in Table 1, different mean square errors of the RBF neural network decoupling are given (the maximum number of iterations is 100).

**Table 1.** Mean squared error(MSE) of the RBF neural network decoupling

| Spread | MSE | Spread | MSE |
|---|---|---|---|
| **1.0** | 0.0144011 | 6.0 | $9.03\times10^{-7}$ |
| **2.0** | $3.21\times10^{-5}$ | 7.0 | $2.89\times10^{-6}$ |
| **3.0** | $9.79\times10^{-7}$ | 8.0 | $1.35\times10^{-6}$ |
| **4.0** | $3.60\times10^{-7}$ | 9.0 | $1.87\times10^{-6}$ |
| **5.0** | $3.74\times10^{-7}$ | 10.0 | $6.81\times10^{-6}$ |

In table 1, the MSE of the three-dimensional force trend to decrease as the increase of the spread of radial base expands. When spread = 4.0, the MSE is minimized. With the spread continues to increase, the neurons become less sensitive to the input, and the output signal of each neuron in the hidden layer tends to approximate value. even if increase the number of neurons, the error would not reduce significantly except increasing the amount of calculation. So we choose spread = 4.0, to construct the three layer RBF neural network.

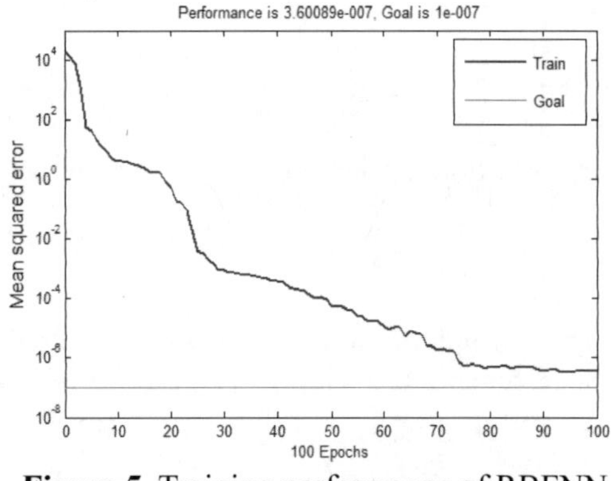


**Figure 5.** Training performance of RBFNN

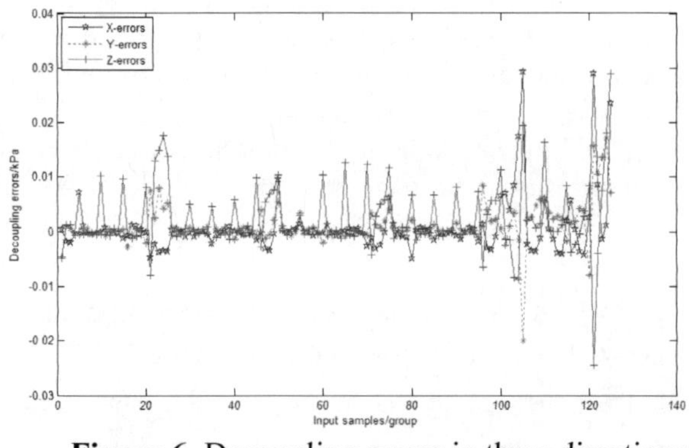


**Figure 6.** Decoupling errors in three directions

In the training process, we set the MSE of ideal target is $1\times10^{-7}$, Figure 5 shows the performance during the training process of the RBF neural network with spread = 4.0. As can be seen, when the training is finished, the final MSE is $3\times10^{-7}$, which satisfies the requirement of decoupling. Figure 6 shows the decoupling error of 125 sets of the samples with three different directions in the RBF neural network under the same conditions.

As shown in Figure 6, the decoupling effect of the flexible tactile sensor along the three directions is very good. The maximum absolute errors in directions of $F_x$, $F_y$ and $F_z$ are 0.029KPa, 0.0275KPa and

IOP Conf. Series: Materials Science and Engineering **224** (2017) 012045 doi:10.1088/1757-899X/224/1/012045

0.0377Kpa, respectively. Due to the maximum pressure we exerted on the sensor is 500KPa, it can be known that the relative error of the three-dimensional force satisfies the requirement of the precision of the actual needs. And we can see that the decoupling errors of the data points at right edge are significantly higher than the other part of data points. It is attributed to that the force is exerted at right edge, so the microstructure at the side would produce some interference signal during the process of the data acquisition.

## 5. Conclusions

Aiming at the complex nonlinear problem and difficult of decoupling the flexible tactile sensor has. In the paper, we designed the novel flexible tactile sensor, which can separate the three-dimensional force information from the structure, and reduce the difficult of decoupling method as far as possible. Combine with the RBF neural network, the three-dimensional force is decouple the optimized spread of the network and the faster speed of decoupling show that the RBF neural network can satisfy the need of real-time decoupling of the three-dimensional force.

## Acknowledgement

The article is supported by National Science Foundation of China (Nos. 61473272, 61201076, 61471155 and 61301060), and authors would like to thank the financial support of the leader talent plan of Universities in Anhui Province.

## References

[1] Dahiya R S, Valle M, Metta G, Lorenzelli L and Collini C *2007 proceedings of 2007 Ph.D Research in Microelectronics and Electronics conference*(Bordeaux)

[2] Eltaib M E and Hewit J R 2003 *Mechatronics*, vol. 13, no. 10, pp. 1163-1177

[3] Huang Y, Liu P, Zhang Y, Qiu H and Ge Y 2011 *Journal Wuhan University of Technology*, vol.26, no.3, pp. 443-448

[4] Kim H K, Lee S and Yun K S 2011 *Sensors and Actuators A: Physical*, vol.165, no.1 pp. 2-7.

[5] Sumer B, Aksak B, Sahin K, Chuengsatiansup K and Sitti M 2011 *Sensor Letters*, vol.9, no.2, pp. 457-463.

[6] Sekitani T, Noguchi Y, Hata K, Fukushima T, Aida T and Someya T 2008 *Science*, vol.321, no.5895, pp.1468-1472.

[7] Koc I M and Akca E 2013 *Tribology International*, vol.59, pp. 321-331.

[8] Lipomi D J, Vosgueritchian M, Tee B C, Hellstrom S L, Lee J A, Fox C H and Bao Z 2011 *Nature nanotechnology*, vol.6, no.12, (2011), pp. 788-792.

[9] Xu F, Ge Y, Shuang F, Ding J and Li S 2011 *Yi Qi Yi Biao Xue Bao/Chinese Journal of Scientific Instrument*, vol.32, no.6, pp. 1350-1356.

[10] Ding J, Ge Y, Li S, Xu F and Shuang F 2011 *Sensors and Transducers*, vol.124, no.1, pp. 37-47.

[11] Wang F, Sun X, Wang Y, Ding J, Pan H, Song Q, Yu Y and Shuang F 2013 *Sensors and Transducers*, vol.159, no.11, pp. 289-298.

[12] Behrouz T, “Thermomechanical transient analysis of a thick-hollow FGM cylinder”, 2016 *Engineering Solid Mechanics* 4 pp.25-32.

# Maintenance Optimization Model for One kind of Three-Dimensional Radar Antenna Array

**Min Wang, Jiang-ping Yang, Yong-pan Wang, Wei-jian Liu**

Air Force Early Warning Academy, Wuhan 430019, China

13995668024@163.com

**Abstract:** For three-dimensional radar, maintenance cost is high and maintenance time is difficult to be determined. What's more, model simulation computing is significantly complex. The subject of this paper is a new generation of meter wave active, phased array 3D radar. We put forth a concept that maintenance should be done after dividing into several regions to the asymmetrical distribution planar array antenna. First, a failure model of array elements is built to analyze the influence from the element to antenna. Second, the maintenance optimization model is established. Finally, computer simulations are conducted to verify the feasibility and effectiveness of the proposed model.

**Keywords:** three dimensional radar; antenna array; maintenance optimization

## 1. Introduction

The three-dimensional radar(3D) has the feature of small bulk, light weight and simple structure.It plays an important role in the modern high tech war. It is of great significance to maintenance the 3D radar. As the main subsystem of the 3D radar, the optimization of array antenna's maintenance gains more attention from military and scientific research units. Generally, the antenna has a large number of array elements, and is designed with a redundant structure, so it can be disposed as a $K/n$(G) system [1-2]. At present, there exists three main problems in the maintenance optimization of the antenna. The first one is the high maintenance costs and uncertain maintenance time; the second one is that the subject of this paper is a new generation of meter wave active, phased array 3D radar, which is referred to hereafter simply as 3D radar. Compared with conventional radar, it has the asymmetrical distribution of the antenna elements due to the use of amplitude weighting [3]; the last one is the difficult simulation and calculation of the optimal models because of the huge number of array elements. In recent years, there are few reports about how to solve these problems. In literature [4] a condition-based optimal maintenance model is proposed by taking the antenna as a uniform distribution array. This is why we take meter wave active, phased array 3D radar for example. This solves the first problem. However, the other two problems are not considered in literature [4]. To solve these problems, this paper proposes a maintenance optimization model. Firstly, the antenna will be divided into several regions according to the influence rule of the elements to the antenna performance, and then the proposed problems can be solved.

In order to study the relationship between elements' failure and antenna performance, some experts introduce the idea of density weighting matrix, and a series of research results are achieved [5-7]. These references consider that the effect from element in different positions on the antenna performance is the same. So they mainly study the effect of failure rate on antenna performance. However, the influence of element in different positions on antenna performance is not investigated. In fact, the influence of element in different position is different due to the amplitude weighting in array elements. Therefore, to divide the antenna, we should have to study the effect of different elements on antenna performance firstly, and then summarize the influence rule between array elements and antenna performance.

In this paper, the failure effect model of element in different locations is established, and the influence of array elements on antenna performance is studied with methods of theoretical derivation and simulation analysis. Then, maintenance optimization model is built. Finally, computer simulations and analysis are conducted to verify the feasibility and effectiveness of the proposed model.

## 2. Failure influence analysis of antenna element

### *2.1. Failure influence model of antenna element*

Assuming that the planar antenna is composed of $M\times N$ array elements and the antenna element is isotropic. All the array elements are arranged in a rectangular grid way, and the element distance is $dx=dy=d$. The sketch of one planar antenna is shown in Fig.1. Let $B(m,n)$ denote the element in the $n$th columns and the $m$th rows, where $m=1,2,\cdots,M$, $n=1,2,\cdots,N$.

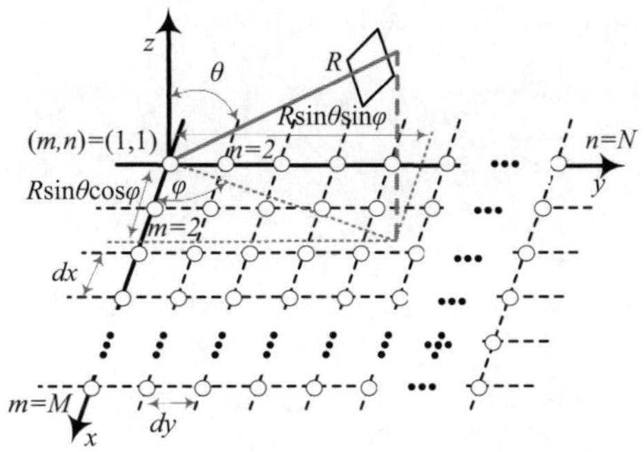


**Figure 1.** Sketch of planar antenna

Set the element factor $S_e(\theta)=1$, since the isotropic of the antenna array element. Therefore, the antenna pattern in the direction of $\theta$ can be expressed as [8]

$$S(\theta)=\sum_{m=1}^{M}\sum_{n=1}^{N} I_{mn}\exp(jknd\sin\theta) \tag{1}$$

where $k$ is the wave number, $k=2\pi/\lambda$. $\lambda$ is wavelength. $I_{mn}$ is the current amplitude value of $B(m,n)$. We uses the Taylor weighting in this paper.

Define the indicative functions $X(m)$ and $Y(n)$, where $m=1,2,\cdots,M$, $n=1,2,\cdots,N$. Set $X(m)=1$ and $Y(n)=1$ when element in $B(m,n)$ works. Set $X(m)=0$ and $Y(n)=0$ when element in $B(m,n)$ fails. Thus, the indicative matrix $H_{m\times n}$ reflecting the working state of the $M\times N$ elements is obtained.

$$H(m,n)=X(m)\cdot Y(n) \tag{2}$$

Further more, the failure influence model of element could be estabilished as

$$S(\theta)=\sum_{m=1}^{M}\sum_{n=1}^{N} H(m,n)I_{mn}\exp(jknd\sin\theta) \tag{3}$$

Considering the importance of the antenna gain and sidelobe level, influence of element in different locations on these two parameters is studied as follows.

### *2.2. Failure influence of element on antenna gain*

The power of a single array is proportional to the square of the amplitude of the array element. Set $Q$ be the proportional coefficient, then the power $P_E$ of the element $B(m,n)$ is

$$P_E(m,n)=Q\left|H(m,n)I_{mn}\right|^2 \tag{4}$$

Hence, the power $P$ of the whole planar antenna is [9]

IOP Conf. Series: Materials Science and Engineering **224** (2017) 012046 doi:10.1088/1757-899X/224/1/012046

$$P=\sum_{m=1}^{M}\sum_{n=1}^{N}P_E(m,n) \tag{5}$$

The gain of the element failure in the direction $\theta_0$ at the maximum lobe is

$$G(\theta_0)=\frac{4\pi\left|\sum_{m=1}^{M}\sum_{n=1}^{N}H(m,n)I_{mn}\right|^2}{Q\sum_{m=1}^{M}\sum_{n=1}^{N}\left|H(m,n)I_{mn}\right|^2} \tag{6}$$

When all the elements are in good condition, set $H(m,n)=1$. It follows that the maximum antenna gain $G_0(\theta)$ is

$$G_0(\theta_0)=\frac{4\pi\left|\sum_{m=1}^{M}\sum_{n=1}^{N}I_{mn}\right|^2}{Q\sum_{m=1}^{M}\sum_{n=1}^{N}\left|I_{mn}\right|^2} \tag{7}$$

The decrease rate α of antenna gain is defined as

$$\alpha=\frac{G_0(\theta_0)-G(\theta_0)}{G_0(\theta_0)}=1-\left|\frac{\sum_{m=1}^{M}\sum_{n=1}^{N}H(m,n)I_{mn}}{\sum_{m=1}^{M}\sum_{n=1}^{N}I_{mn}}\right|^2\cdot\frac{\sum_{m=1}^{M}\sum_{n=1}^{N}\left|I_{mn}\right|^2}{\sum_{m=1}^{M}\sum_{n=1}^{N}\left|H(m,n)I_{mn}\right|^2} \tag{8}$$

Assuming that the antenna array consists of $M\times N=100\times100=10000$ array elements, the influence of element failure on antenna gain is shown in Fig.2. The results indicate that the decrease rate of antenna gain is less than 0.2‰ when element in different locations fails. So the effect of a single element on antenna gain is little and can be ignored. When different numbers of array element break down, the decrease rate increases, which means that the antenna gain changes dramatically. When all the elements are invalid, the antenna gain becomes 0.

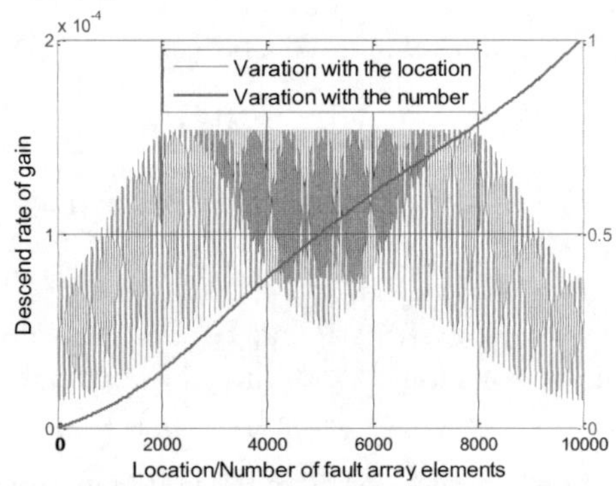


**Figure 2.** Influence of failure element on antenna gain

### *2.3. Failure influence of element on sidelobe level*

According to the failure influence model, the failure effect of element on sidelobe level can be studied by means of simulation. Assuming that there are $M\times N=100\times100=10000$ elements in the planar antenna, and the desired sidelobe level is -35dB, the element is subject to Taylor distribution. The relationship between element in different locations and peak values of the sidelobe is shown in Fig.3. It is shown that the failure influences of elements in symmetrical locations on sidelobe are almost the same. And in a certain range, the failure influences of different elements are also equal. Considering the large numbers of the array elements, the effect of single element on the side lobe level is little. But due to the high requirement of sidelobe, it is necessary to study the effects of element in different locations on sidelobe level.

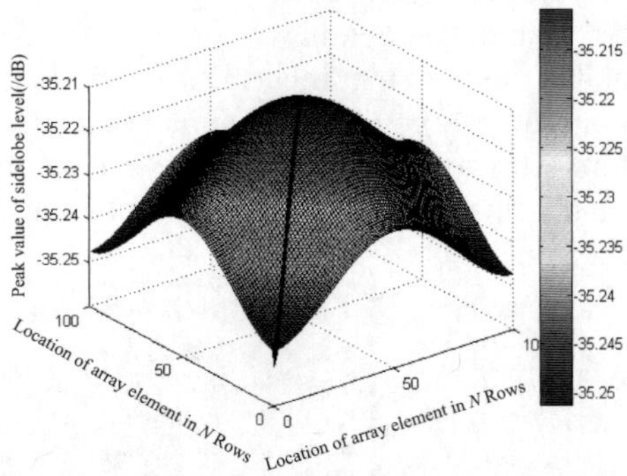


**Figure 3.** Influence of element in different location on sidelobe level

## 3. Maintenance Optimization model

Through the above analysis, we can know that the effect on antenna performance varies with the elements' position and number. But there exists a certain failure range of different elements (denoted as area R*i*), where the failure effect of element in different locations on antenna performance is basically the same. The influence rule can be extended to more antenna arrays with any array elements. Therefore, the antenna array can be divided into different R*i*， as shown in Fig.4.

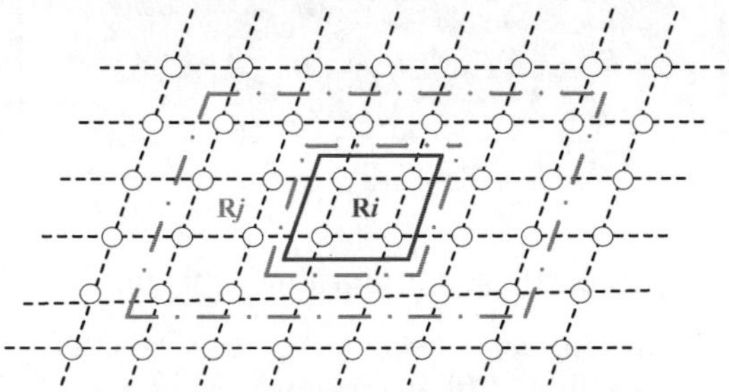


**Figure 4.** Sketch of divided antenna

After divided into several regions, the failure effect of element in different R*i* on sidelobe level is different. Yet, in the same region, the failure effect is regarded as the same. As the antenna is always regarded as a *K*/*N* system, the system can be decomposed into *r* different regions by determining the different R*i*（$i$=1,2,···,$r$）, and array elements is regarded as the same in the same area. This lays a foundation for solving the maintenance optimization problem of asymmetrical antenna. Then, the key problem is how to carry out the regional division work.

### *3.1. Regional division criteria*

The regional division criterion is on the base of the variation of the sidelobe level when one array element is invalid. The criteria has two purposes. One is that it can solve the maintenance optimization problem of the asymmetrical antenna. The other is that it can reduce the amount of model calculation. Combined with the front simulation and analysis, the quantitative criterion can be established as follows.

(1) In the same region R*i* , the variation range of the sidelobe level is not more than $\Delta B$ dB when one array element in different location is invalid.

(2) There is no intersection between R*i* and R*j*, and the total number of elements in all the region equals the number of elements in the antenna.

(3) The numbers of array elements in divided region R*i* can be controlled, and the calculation is in the operational range of person computer.

### *3.2. model steps*

According to the established regional division criterion, and combining with the failure influence rule of the different elements on sidelobe level, the maintenance optimization for the planar antenna

IOP Conf. Series: Materials Science and Engineering **224** (2017) 012046 doi:10.1088/1757-899X/224/1/012046

should be conducted. The detailed steps are given as follows

**Step1** Set the number of rows *M* and the columns *N*.

**Step2** Set regional *level* denotes the region number. Due to the large number of array elements, it is very difficult for a common computer to conduct the simulations and calculations, so it brings the problem of engineering application. To solve this problem, we can set the number of elements in each region to enhance the model's performance of engineering appliance. The method to confirm the regional division levels is given in Fig.5.

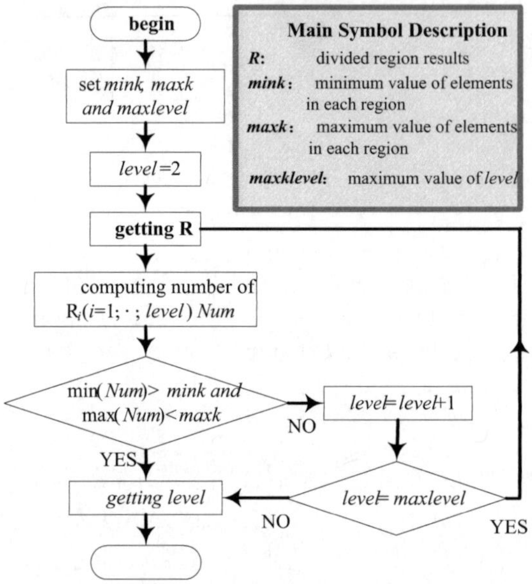


**Figure 5.** Method to confirm parameter level

**Step3** According to the failure influence model, the peak value of sidelobe level PSLL($m$,$n$) can be calculated when the element B($m$,$n$) in different location is invalid.

**Step4** Construct the peak value database PSLL of sidelobe level, which contains a number of $M \times N$ items.

**Step5** Calculate the maximum and minimum value of the peak value side lobe level, these values can be denoted as Max=$max$ {***PSLL***} and Min =$min$ {***PSLL***}, respectively.

**Step6** According to user needs, calculate the variation value $\Delta B$ of peak sidelobe level, where $\Delta B$=(Max-Min)/*level*.

**Step7** Achieve the final regional results R. Traversing the database of PSLL, searching and confirming the element B($m$,$n$) in the range of variation $\Delta B$. Finally, determining which region R the element B($m$,$n$) belongs to.

## 4. Simulations and analysis

As an example, we have known that an antenna array composed of $M \times N$= 100×100 =10000 elements, the desired sidelobe level is -35dB, and the feed amplitudes of the array elements follow a Taylor distribution. Set the parameter of *maxk*=2000 and *mink*=100. According to the proposed model, we can achieve the results easily by a simulation tool, such as MATLAB. Simulation results are shown in Fig.6. As a result, the planar antenna is divided into 9 regions, and there is no intersection between different regions.

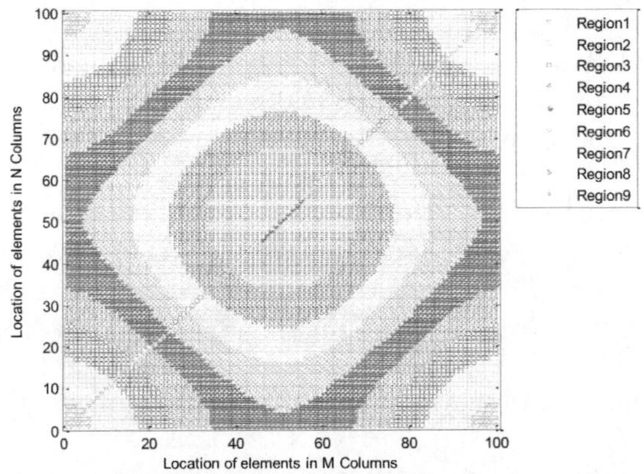


**Figure 6.** The final results

The detailed element number in each region is listed in Tab.1. From the table, we can see that the number of the array element is controllable, and the sum number of array elements in each region equals the total number of array elements.

**Table 1.** Numbers of array elements in each region

| Region1 | Region2 | Region3 | Region4 | Region5 |
|---|---|---|---|---|
| 142 | 710 | 900 | 1210 | 1994 |
| Region6 | Region7 | Region8 | Region9 | Total number |
| 1776 | 1298 | 1074 | 896 | 10000 |

Hence, when the large antenna is divided into several small regions, we can carry out the maintenance optimization research easily on this basis. Therefore, this paper has offered a scientific and reasonable theoretical basis for conducting the optimal maintenance work of the large planar antenna.

## 5. Conclusions

To solve the existing maintenance problems, at first , the failure influence model is established, and the influence rules are summarized. Then the maintenance optimization model is built. Finally, simulations and analysis are conducted to verify the presented model. Results show that the large planar can be divided into several small regions by using the proposed method.

However, there still exist some problems to be solved The main question is that the classification standard $\Delta B$ in regional division criterion is constant. To make the results more reasonable, we will design the parameter $\Delta B$ in the future research.

## Reference

[1] Karin S.de Smidt-Destombes, Matthieu C.van der Heijden and Aart van Harten. On the availability of a k-out-of-N system given limited spares and repair capacity under a condition based maintenance strategy[J]. Reliability Engineering and System Safety, 2004, 83: 287-300.

[2] Moghaddass Ramin, ZUO Ming-J., WANG Wen- bin. Availability of a general k-out-of-n:G system with non-identical components considering shut-off rules using quasi-birth- death process[J]. Reliability Engineering and System Safety, 2011: 96: 489-496.

[3] WANG Zhan. Design and implementation of man-pack UCB Frequency jammer[D]. Chengdu: University of Electronic Science and Technology of China, 2015..

[4] LU Lei, YANG Jiang-ping. Optimal analysis of condition-based maintenance model or large-scale phased array radar antenna[J]. Radar Science and Technology, 2014,12(3): 325-328.

[5] ZHANG Liang. Analysis on the reliability of low-sidelobe solid-state active phased array antenna[J]. Modern Radar,1992,(6):97-105,75.

[6] LI Jian-xin, GAO Tie. Statistical pattern characteristics of low sidelobe solid-state active phased array antennas [J].Chinese Journal of Radio Science, 1993,8(2): 64-69,77.
[7] GAO Tie, LI Jian-xin. Study of multi-step quantization and sidelobe characteristic for solid-state active phased arrays[J]. Acta Electronica Sinica, 1994, 22 (3):11-17.
[8] Hubregt J.Visser. Array and phased array antenna basics[M]. The Netherlands: John Wiley & Sons Inc, 2005.
[9] GUO Yan-chang, QIAN Ji-zeng, HUANG Fu- xiong, et al. Phased array and frequency scanning antenna[M]. Beijing:National Defence Industry Press, 1978:48-50.
[10] Yan Zhou-jie, Wang Hua-bing, Chen Yuan-zheng, Zhao Yan-li. Self-cross phase calibration method of oversized front distributed phased array radar[J]. Journal of Radars,2013,2(4):339-444.
[11] Yang Jian-yu. Development laws and macro trends analysis of radar technology[J]. Journal of Radars,2012,1(1):19-27.
[12] Yan zhou-jie, Wang Hua-bing et al. Self-cross phase calibration method of oversized front distributed phased array radar[J]. Journal of Radars,2013,2(4):439-444.
[13] Tang Yue, Mao Tian, Jiang Bing. Design and experiment of multi-resolution composite digital array antenna[J]. Journal of Radars,2016,5(3):265-270.
[14] Liu Xiao-meng, Gao Wen-jun, et al. Beam synthesis technique for array antenna based on projection matix algorithm[J]. Journal of Radars,2016,5(3):50-55.
[15] Li Shan-yan. Structrual design and analysis for one kind of 3-dimensional radar antenna array[D]. Chengdu: University of Electronic Science and Technology of China, 2015.

IOP Conf. Series: Materials Science and Engineering **224** (2017) 012047 doi:10.1088/1757-899X/224/1/012047

# The Ground Surface Displacement with a Circular Inclusion Buried in a Layered Half-space Impacted by SH Wave

**Y B Zhao, H Qi, X H Ding and D D Zhao**

College of Aerospace and Civil Engineering, Harbin Engineering University, Harbin, Heilongjiang, 150001, China

ZYB201507@126.com

**Abstract.** To analysis the ground surface displacement with buried circular inclusion under the action of SH wave, the large-arc assumption method is used to approximate the straight boundary by a circle with large radius. A numerical example demonstrates that the ground surface displacement amplification coefficient increases with the down of incident wave number $k_1$ and the hardness of the surface covered layer.

## 1. Introduction

With the development of modern society, a large number of population settle in cities and the cities scale are bigger and bigger. The harm of earthquake also cause everyone's attention. Elastic wave theory is an important theoretical method in seismic analysis and research. When the seismic wave propagated in soil meet with discontinuous place, scattering will occurs in the discontinuous place and the surface ground motion will be affected. Studies of elastic wave scattering has been a lot. In 1973, MOW, C. C. and PAO, Y introduced the theory of elastic wave scattering and dynamic stress concentration problem in their monograph[1].In 1982, the complex function method used for the solution of static stress concentration was generalized to the case of dynamic loading [2]. Based on this method, a lot of researches was given about the scattering of circular defects in the infinite elastic plane and elastic half space, as well as the quarter space and double-phase medium.Scientist Trifunac studied the surface motion of semi-cylindrical and semi-elliptical alluvial valley impacted incident plane SH waves[3, 4].In 1993, Lee and Karl given the analytic solution of P Wave and SH wave scattering about the single circular cavity in half-space based on the large arc hypothesis method[5].On this basis, the problem of circular scattering in layer half space incidented by SH wave was studied by Qi Hui[6, 7].The ground surface motion of a semi-elliptical hill and an alluvial valley in layered half-space for incident plane SH waves was researched by Liang[8, 9].This paper presents the influence of shallow buried circular inclusion to the ground surface displacement impacted by SH-wave based on the complex method and the great circle hypothesis method. The numerical result is analyzed by example.

## 2. Theoretical analysis

### *2.1. Problem model*

The simplified model of problem researched in this paper is shown in Figure 1. The problem could be described as a single circular inclusion located in the surface elastic covered layer and the SH wave incident from the elastic half-space. The circular inclusion is defined as the area I, and the overbarden is defined as the areaII, the area III represents the half-space. Respectively, their physical parameters are $G_1$, $k_1$, $\rho_1$, $G_2$, $k_2$, $\rho_2$, $G_3$, $k_3$, $\rho_3$, in which G means the shear modulus, k means the wave number of SH wave, ρ means the density of medium, and the subscript corresponds to the corresponding area. h represent thickness of the surface layer. r expresses the radius of the circular inclusion. TC stands for the edge of the circular inclusion, whose radius is r. Respectively, the upper and lower boundary are defined as TU and TD. $h_1$ and $h_2$ denote the distance from the central point of the circular inclusion to the upper and the lower boundary, respectively. To construct the scattering SH wave of the surface covered layer, the large arc assumption method is used to approximate $T_U$ and $T_D$, and the center of the great circle is defined as so the $T_U$ is approximately expressed as $TT_U$, and the $T_D$ is described as $TT_D$. The rectangular coordinate system $X_1O_1Y_1$ is established for the circular inclusion centered on the point $O_1$, which is the center of the circular inclusion. Correspondingly, the point $O_2$ is defined as the center of the great arc, which is the original point of the rectangular coordinate system $X_2O_2Y_2$. Corresponding to the rectangular coordinate system, the complex function $z_1=X_1+iY_1$ and $z_2=X_2+iY_2$ are set up to describe the complex planes $z_1$ and $z_2$. The schematic diagram is shown as Figure 1 and the relationship between various parameters could be described as :

$$h = h_1 + h_2,\quad R_U = h + R_D,\quad z_2 = z_1 + i(R_D + h_2) \tag{1}$$

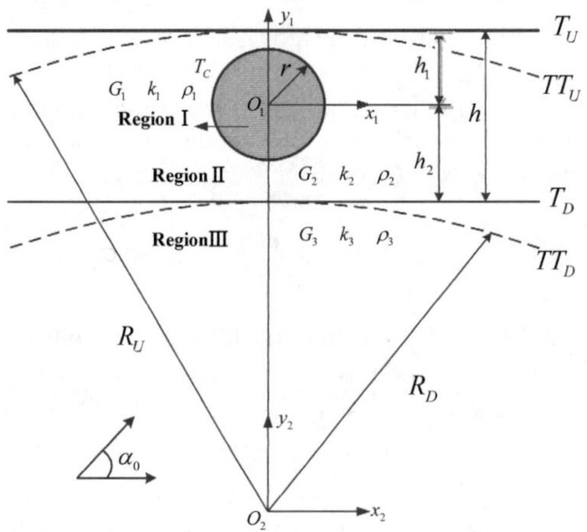


**Figure 1.** Schematic diagram of the model.

### *2.2. Control equations*

This paper studies the out-of-plane steady shear SH-Wave scattering. In the conventional rectangular coordinate system, the SH-Wave displacement field could be expressed as W(X, Y, t), whose movement is perpendicular to the X-Y plane and has nothing to do with the Z axis. The fields of displacement W(X, Y, t) must be satisfied with the Helmholtz equation in the rectangular coordinate system:

$$\partial^2 W/\partial X^2 + \partial^2 W/\partial Y^2 + k^2 W = 0 \tag{2}$$

where k=ω/c, $c^2=G/\rho$, ω=the circular frequency of displacement field W(X, Y, t), The correlation relationship between the displacement field W(X, Y, t) and the time is exp(-iωt), and the exp(-iωt) will not be considered in the following study because of the problem studied is steady state. In the complex plane, the formula (2) could be represented as:

$$4\partial^2 W/\partial z\partial\overline{z} + k^2 W = 0 \tag{3}$$

In the polar coordinates of complex plane, the stress-strain relationship could be described as:

$$\tau_{Zr} = G\left[(\partial W/\partial z)e^{i\theta} + (\partial W/\partial\overline{z})e^{-i\theta}\right],\quad \tau_{Z\theta} = iG\left[(\partial W/\partial z)e^{i\theta} - (\partial W/\partial\overline{z})e^{-i\theta}\right] \tag{4}$$

In which the θ means the angle from any point on the circumference of a circle in the polar coordinates.

### *2.3. Incident wave field and stress*

In the complex plane $z_2$, the incident wave displacement field and the stress could be expressed as:

$$W^{(I)}_{(z_2,\overline{z}_2)} = W_0 \exp\left[ik_3 \operatorname{Re}\left(z_2 e^{-i\alpha_0}\right)\right] \tag{5}$$

$$\tau^{(I)}_{Zr,(z_2,\bar{z}_2)} = ik_3G_3W_0 \exp\left[ik_3 \operatorname{Re}\left(z_2e^{-i\alpha_0}\right)\right]\operatorname{Re}\left(z_2e^{-i\alpha_0}/|z_2|\right) \tag{6}$$

$$\tau^{(I)}_{Z\theta,(z_2,\bar{z}_2)} = -ik_3G_3W_0 \exp\left[ik_3 \operatorname{Re}\left(z_2e^{-i\alpha_0}\right)\right]\operatorname{Im}\left(z_2e^{-i\alpha_0}/|z_2|\right) \tag{7}$$

Where the $W_0$ is the maximal displacement amplitude and the $\alpha_0$ is the incident angle.

### 2.4. Scattered wave field and stress

In the complex plane $z_2$, the displacement field and stress of scattered wave $W^{(S1)}$ caused by $TT_D$ in region III, the displacement field and stress of scattered wave $W^{(S2)}$ occurred in region II by $TT_D$, the displacement field and stress of scattered wave $W^{(S5)}$ occurred in regionII by $TT_D$ could be shown as formula (8)~(16). In the complex plane $z_1$, the displacement field and stress of scattered wave $W^{(S3)}$ occurred in regionII by $T_C$, the displacement field and stress of scattered wave $W^{(Z4)}$ occurred in region I by $T_C$ could be shown as formula (17)~(22):

$$W^{(S1)}_{(z_2,\bar{z}_2)} = \sum_{n=-\infty}^{n=+\infty} A_n H^{(2)}_n\left(k_3|z_2|\right)\left(z_2/|z_2|\right)^n \tag{8}$$

$$\tau^{(S1)}_{Zr,(z_2,\bar{z}_2)} = 0.5k_3G_3\sum_{n=-\infty}^{n=+\infty} A_n\left[H^{(2)}_{n-1}\left(k_3|z_2|\right) - H^{(2)}_{n+1}\left(k_3|z_2|\right)\right]\left(z_2/|z_2|\right)^n \tag{9}$$

$$\tau^{(S1)}_{Z\theta,(z_2,\bar{z}_2)} = i0.5k_3G_3\sum_{n=-\infty}^{n=+\infty} A_n\left[H^{(2)}_{n-1}\left(k_3|z_2|\right) + H^{(2)}_{n-1}\left(k_3|z_2|\right)\right]\left(z_2/|z_2|\right)^n \tag{10}$$

$$W^{(S2)}_{(z_2,\bar{z}_2)} = \sum_{n=-\infty}^{n=+\infty} B_n H^{(1)}_n\left(k_2|z_2|\right)\left(z_2/|z_2|\right)^n \tag{11}$$

$$\tau^{(S2)}_{Zr,(z_2,\bar{z}_2)} = 0.5k_2G_2\sum_{n=-\infty}^{n=+\infty} B_n\left[H^{(1)}_{n-1}\left(k_2|z_2|\right) - H^{(1)}_{n+1}\left(k_2|z_2|\right)\right]\left(z_2/|z_2|\right)^n \tag{12}$$

$$\tau^{(S2)}_{Z\theta,(z_2,\bar{z}_2)} = i0.5k_2G_2\sum_{n=-\infty}^{n=+\infty} B_n\left[H^{(1)}_{n-1}\left(k_2|z_2|\right) + H^{(1)}_{n+1}\left(k_2|z_2|\right)\right]\left(z_2/|z_2|\right)^n \tag{13}$$

$$W^{(S5)}_{(z_2,\bar{z}_2)} = \sum_{n=-\infty}^{n=+\infty} E_n H^{(2)}_n\left(k_2|z_2|\right)\left(z_2/|z_2|\right)^n \tag{14}$$

$$\tau^{(S5)}_{Zr,(z_2,\bar{z}_2)} = 0.5k_2G_2\sum_{n=-\infty}^{n=+\infty} E_n\left[H^{(2)}_{n-1}\left(k_2|z_2|\right) - H^{(2)}_{n+1}\left(k_2|z_2|\right)\right]\left(z_2/|z_2|\right)^n \tag{15}$$

$$\tau^{(S5)}_{Z\theta,(z_2,\bar{z}_2)} = i0.5k_2G_2\sum_{n=-\infty}^{n=+\infty} E_n\left[H^{(2)}_{n-1}\left(k_2|z_2|\right) + H^{(2)}_{n+1}\left(k_2|z_2|\right)\right]\left(z_2/|z_2|\right)^n \tag{16}$$

$$W^{(S3)}_{(z_1,\bar{z}_1)} = \sum_{n=-\infty}^{n=+\infty} C_n H^{(1)}_n\left(k_2|z_1|\right)\left(z_1/|z_1|\right)^n \tag{17}$$

$$\tau^{(S3)}_{Zr,(z_1,\bar{z}_1)} = 0.5k_2G_2\sum_{n=-\infty}^{n=+\infty} C_n\left[H^{(1)}_{n-1}\left(k_2|z_1|\right) - H^{(1)}_{n+1}\left(k_2|z_1|\right)\right]\left(z_1/|z_1|\right)^n \tag{18}$$

$$\tau^{(S3)}_{Z\theta,(z_1,\bar{z}_1)} = i0.5k_2G_2\sum_{n=-\infty}^{n=+\infty} C_n\left[H^{(1)}_{n-1}\left(k_2|z_1|\right) + H^{(1)}_{n+1}\left(k_2|z_1|\right)\right]\left(z_1/|z_1|\right)^n \tag{19}$$

$$W^{(Z4)}_{(z_1,\bar{z}_1)} = \sum_{n=-\infty}^{n=+\infty} D_n J_n\left(k_1|z_1|\right)\left(z_1/|z_1|\right)^n \tag{20}$$

$$\tau^{(Z4)}_{Zr,(z_1,\bar{z}_1)} = 0.5k_1G_1\sum_{n=-\infty}^{n=+\infty} D_n\left[J_{n-1}\left(k_1|z_1|\right) - J_{n+1}\left(k_1|z_1|\right)\right]\left(z_1/|z_1|\right)^n \tag{21}$$

$$\tau^{(Z4)}_{Z\theta,(z_1,\bar{z}_1)} = i0.5k_1G_1\sum_{n=-\infty}^{n=+\infty} D_n\left[J_{n-1}\left(k_1|z_1|\right) + J_{n+1}\left(k_1|z_1|\right)\right]\left(z_1/|z_1|\right)^n \tag{22}$$

In the complex plane $z_1$, the formulas (11)~(16) could be shown as:

$$W^{(S2)}_{(z_1,\bar{z}_1)} = \sum_{n=-\infty}^{n=+\infty} B_n H^{(1)}_n\left[k_2|z_1 + i(R_D + h_2)|\right]\left\{[z_1 + i(R_D + h_2)]/|z_1 + i(R_D + h_2)|\right\}^n \tag{23}$$

$$\tau^{(S2)}_{Zr,(z_1,\bar{z}_1)} = 0.5k_2G_2\sum_{n=-\infty}^{n=+\infty} B_n \left\{ \begin{array}{l} H^{(1)}_{n-1}\left[k_2\left|z_1+i(R_D+h_2)\right|\right]\left\{\left[z_1+i(R_D+h_2)\right]/\left|z_1+i(R_D+h_2)\right|\right\}^{n-1}e^{i\theta_1} \\ -H^{(1)}_{n+1}\left[k_2\left|z_1+i(R_D+h_2)\right|\right]\left\{\left[z_1+i(R_D+h_2)\right]/\left|z_1+i(R_D+h_2)\right|\right\}^{n+1}e^{-i\theta_1} \end{array} \right\} \quad (24)$$

$$\tau^{(S2)}_{Z\theta,(z_1,\bar{z}_1)} = i0.5k_2G_2\sum_{n=-\infty}^{n=+\infty} B_n \left\{ \begin{array}{l} H^{(1)}_{n-1}\left[k_2\left|z_1+i(R_D+h_2)\right|\right]\left\{\left[z_1+i(R_D+h_2)\right]/\left|z_1+i(R_D+h_2)\right|\right\}^{n-1}e^{i\theta_1} \\ +H^{(1)}_{n+1}\left[k_2\left|z_1+i(R_D+h_2)\right|\right]\left\{\left[z_1+i(R_D+h_2)\right]/\left|z_1+i(R_D+h_2)\right|\right\}^{n+1}e^{-i\theta_1} \end{array} \right\} \quad (25)$$

$$W^{(S5)}_{(z_1,\bar{z}_1)} = \sum_{n=-\infty}^{n=+\infty} E_n H^{(2)}_n\left[k_2\left|z_1+i(R_D+h_2)\right|\right]\left\{\left[z_1+i(R_D+h_2)\right]/\left|z_1+i(R_D+h_2)\right|\right\}^n \quad (26)$$

$$\tau^{(S5)}_{Zr,(z_1,\bar{z}_1)} = 0.5k_2G_2\sum_{n=-\infty}^{n=+\infty} E_n \left\{ \begin{array}{l} H^{(2)}_{n-1}\left[k_2\left|z_1+i(R_D+h_2)\right|\right]\left\{\left[z_1+i(R_D+h_2)\right]/\left|z_1+i(R_D+h_2)\right|\right\}^{n-1}e^{i\theta_1} \\ -H^{(2)}_{n+1}\left[k_2\left|z_1+i(R_D+h_2)\right|\right]\left\{\left[z_1+i(R_D+h_2)\right]/\left|z_1+i(R_D+h_2)\right|\right\}^{n+1}e^{-i\theta_1} \end{array} \right\} \quad (27)$$

$$\tau^{(S5)}_{Z\theta,(z_1,\bar{z}_1)} = i0.5k_2G_2\sum_{n=-\infty}^{n=+\infty} E_n \left\{ \begin{array}{l} H^{(2)}_{n-1}\left[k_2\left|z_1+i(R_D+h_2)\right|\right]\left\{\left[z_1+i(R_D+h_2)\right]/\left|z_1+i(R_D+h_2)\right|\right\}^{n-1}e^{i\theta_1} \\ +H^{(2)}_{n+1}\left[k_2\left|z_1+i(R_D+h_2)\right|\right]\left\{\left[z_1+i(R_D+h_2)\right]/\left|z_1+i(R_D+h_2)\right|\right\}^{n+1}e^{-i\theta_1} \end{array} \right\} \quad (28)$$

In the complex plane $z_2$, the formulas (17)~(19) could be shown as:

$$W^{(S3)}_{(z_2,\bar{z}_2)} = \sum_{n=-\infty}^{n=+\infty} C_n H^{(1)}_n\left[k_2\left|z_2-i(R_D+h_2)\right|\right]\left\{\left[z_2-i(R_D+h_2)\right]/\left|z_2-i(R_D+h_2)\right|\right\}^n \quad (29)$$

$$\tau^{(S3)}_{Zr,(z_2,\bar{z}_2)} = 0.5k_2G_2\sum_{n=-\infty}^{n=+\infty} C_n \left\{ \begin{array}{l} H^{(1)}_{n-1}\left[k_2\left|z_2-i(R_D+h_2)\right|\right]\left\{\left[z_2-i(R_D+h_2)\right]/\left|z_2-i(R_D+h_2)\right|\right\}^{n-1}e^{i\theta_2} \\ -H^{(1)}_{n+1}\left[k_2\left|z_2-i(R_D+h_2)\right|\right]\left\{\left[z_2-i(R_D+h_2)\right]/\left|z_2-i(R_D+h_2)\right|\right\}^{n+1}e^{-i\theta_2} \end{array} \right\} \quad (30)$$

$$\tau^{(S3)}_{Z\theta,(z_2,\bar{z}_2)} = i0.5k_2G_2\sum_{n=-\infty}^{n=+\infty} C_n \left\{ \begin{array}{l} H^{(1)}_{n-1}\left[k_2\left|z_2-i(R_D+h_2)\right|\right]\left\{\left[z_2-i(R_D+h_2)\right]/\left|z_2-i(R_D+h_2)\right|\right\}^{n-1}e^{i\theta_2} \\ +H^{(1)}_{n+1}\left[k_2\left|z_2-i(R_D+h_2)\right|\right]\left\{\left[z_2-i(R_D+h_2)\right]/\left|z_2-i(R_D+h_2)\right|\right\}^{n+1}e^{-i\theta_2} \end{array} \right\} \quad (31)$$

### *2.5. Equations set*

Because the displacement fields and stress fields expressions of incident and scattering waves have been constructed, according to the boundary conditions that the radial stress of $TT_U$ and $T_C$ is free and the continuous condition that the displacement and the radial stress of $TT_D$ are continuous, the set of equations could be expressed as equations set, multiply both sides by exp(-imφ1), integrate them in (-π, π) respective, So we can calculate the coefficient An~En, take them to the formulas and intercept limited items, all the unknown quantities will be found out.

### *2.6. The ground surface displacement amplification coefficient* ($W^*$)

Define W* as the displacement amplification coefficient:

$$W^* = \left|\left(W^{(S2)}_{(z_2,\bar{z}_2)} + W^{(S3)}_{(z_2,\bar{z}_2)} + W^{(S5)}_{(z_2,\bar{z}_2)}\right)/W_0\right|_{|z_2|=R_U} \quad (32)$$

## 3. Numerical Example

In this part, example of numerical computation was presented, the results are non-dimensionalized. It is supposed that r=1. Define the parameters combination $G^*=G_2/G_3$, $G^\#=G_1/G_3$, $k^*=k_2/k_3$, $k^\#=k_1/k_3$. If $k^*$ is less than 1， it means Region III is harder than region II, as well as $k^\#$. In this example, the incident angle is 90°, which means the incident wave vertical incidence. From figure 2 to figure 4 separately shows the ground surface displacement amplification coefficient $W^*$ with $x_2/r$ when the incident wave incoming from the hard area III, the circular inclusion is in the middle of the layer, whose thickness is 3 times of r, $\rho^\#=0.7$, $\rho^*=0.8$, the incident wave is in different frequency. Conversely, the figures 5, 6, 7 introduce the incident wave incoming from the soft area III.

It is observed that from figure 2 to figure 4, when the half space is harder than the layer area, the more soft of surface covered layer, the larger of the ground surface displacement amplification coefficient

$W^*$. When the incident wave number is 0.1, $W^*$ is largest and will reduce continuously with the increase of the incident wave number $k_3$. It can be perceived from figure 5 to figure 7, when the semi-space is softer than the surface elastic layer, the $W^*$ is less than the figure 2 to figure 4. The harder of the surface covered layer, the less of $W^*$. The W* will reduce with the increasement of incident wave number $k_3$.

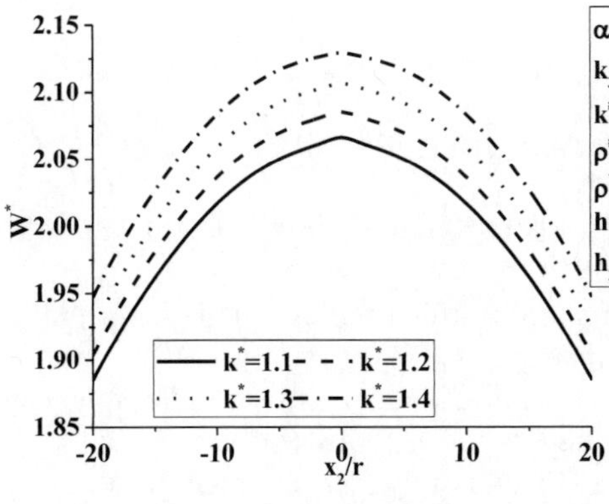


**Figure 2.** $W^*$ with $x_2/r(k_3=0.1, \rho^*=0.8)$.

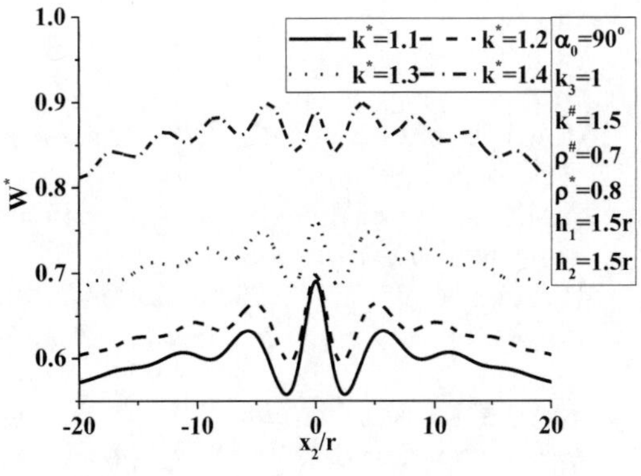


**Figure 3.** $W^*$ with $x_2/r(k_3=1, \rho^*=0.8)$.

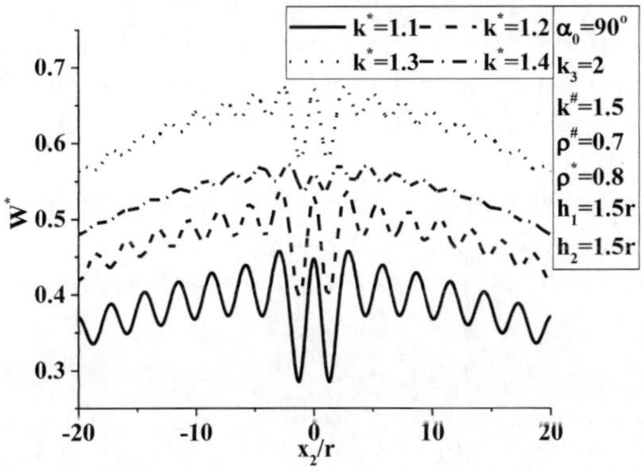


**Figure 4.** $W^*$ with $x_2/r(k_3=2, \rho^*=0.8)$.

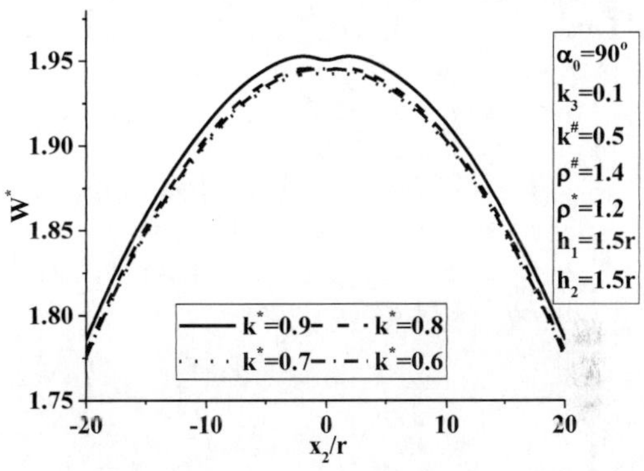


**Figure 5.** $W^*$ with $x_2/r(k_3=0.1, \rho^*=1.2)$.

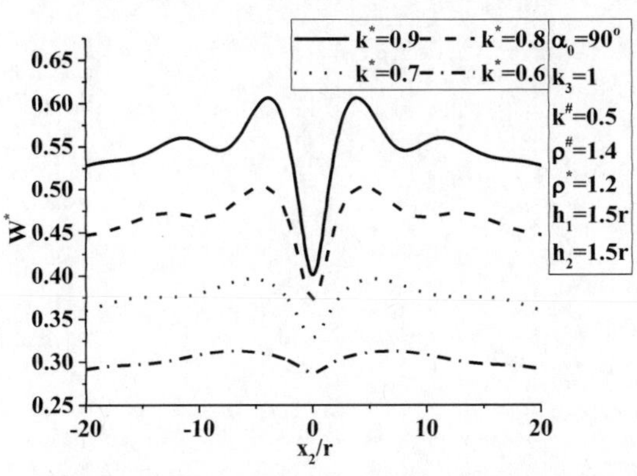


**Figure 6.** $W^*$ with $x_2/r(k_3=1, \rho^*=1.2)$.

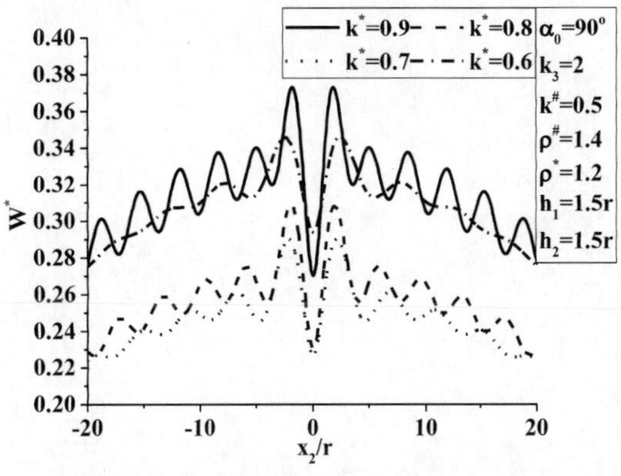


**Figure 7.** $W^*$ with $x_2/r(k_3=2, \rho^*=1.2)$.

IOP Conf. Series: Materials Science and Engineering **224** (2017) 012047 doi:10.1088/1757-899X/224/1/012047

## 4. Conclusion

Based on the large-arc assumption method, complex function and wave function expansion method, the ground surface displacement of shallow buried circular inclusion in the surface covered layer of half-space impacted by SH wave is studied. The numerical example indicates that the ground surface displacement amplification coefficient will reduce with the increasement of rigidity of the surface covered layer, and will be lower when the incident wave number increases from low frequency to high frequency.

## References

[1] Mow C C and Pao Y 1973 *Diffraction of elastic waves and dynamic stress concentrations*: Crane, Russak)

[2] Liu D, Gai B and Tao G 1982 Applications of the method of complex functions to dynamic stress concentrations *Wave Motion* **4** 293-304

[3] Trifunac M D 1971 Surface motion of a semi-cylindrical alluvial valley for incident plane SH waves. In: *Bulletin of the Seismological Society of America,* p 1770

[4] Wong H L and Trifunac M D 1974 Surface motion of a semi-elliptical alluvial valley for incident plane SH waves *Bulletin of the Seismological Society of America* **64** 1389-408

[5] Lee V and Karl J 1993 Diffraction of elastic plane P waves by circular, underground unlined tunnels *European Earthquake Engineering* **6** 29-36

[6] Chen D and Qi H 2014 Dynamic analysis for circular cavity and inclusion impacted by SH-wave in a layered half-space *Harbin Gongcheng Daxue Xuebao/Journal of Harbin Engineering University* **35** 171-6

[7] Chen D N, Qi H and Shi Y 2011 The effect on scattering of SH-wave in a layered half-space with the subsurface circular cavities. In: *2nd International Conference on Manufacturing Science and Engineering, ICMSE 2011, April 9, 2011 - April 11, 2011,* (Guilin, China: Trans Tech Publications) pp 1908-11

[8] Liang J and Ba Z 2007 Surface motion of an alluvial valley in layered half-space for incident plane SH waves *Journal of Earthquake Engineering & Engineering Vibration* **27** 1-9

[9] Liang J and Fu J 2011 Surface motion of a semi-elliptical hill for incident plane SH waves *Earthq Sci* **24** 447-62

IOP Conf. Series: Materials Science and Engineering **224** (2017) 012048 doi:10.1088/1757-899X/224/1/012048

# Dynamic Interaction of Interfacial Point Source Loading and Cylinder in an Elastic Quarter with Anti-plane Shear

**Gao Chun[1,2], Qi Hui[1], Pan Xiang Nan[1]and Zhao Yuan Bo[1]**

[1]College of Aerospace and Civil Engineering, Harbin Engineering University, Harbin 150001, China;

[2]Department of Civil Engineering, Harbin University, Harbin 150086, China

80680709@qq.com

**Abstract.** Theoretical steady state solution of a semi-circular cylinder impacted by an anti-plane point loading in a vertical bound of an elastic quarter is formulated in this paper through using image method and wave function expansion series. The elastic quarter is extended as a half space, and the semi-circular interfacial cylinder is extended as a circular cylinder. Displacement field is constructed as series of Fourier-Hankel and Fourier-Bessel wave functions. At last, circular boundary is expanded as Fourier series to determine coefficients of wave function. Numerical results show that material parameters have two widely divergent effects on the radial and circumferential dynamic stress distribution.

## 1. Introduction

As the simplest elastodynamics research object, anti-plane movement has an important theoretical significance. As the basic theory of anti-plane elastodynamics theory, SH wave propagation and scattering have a wide application in earthquake engineering, geotechnical engineering, material engineering and so on. In the early 1980s, Pao Y H, Mow C C[1] conducted systematic studied the SH wave scattering of circular cylinder. Since this century, Liu D K[2],Shi S X [3] Tian J Y[4,5] and Qi H[6-8] studied the scattering of plane SH wave and anti-plane line source load which influenced by interface circular cavity or cylinder and lining structure in full space. Qi H [9], Yang Z L[10] studied the interaction of convex structure with different foundation under the incidence of SH wave. In recent years, Qi H, Zh Y[11-13] ,Yang J[14-16] studied the scattering of SH wave and point source load which influenced by cylinder cavity or inclusion in elastic quarter space and bi-material. And they made certain achievements. But the research of anti-plane scattering of interface cylinder in the bi-material half space is still insufficient. In this paper, the steady state of interface cylinder which subjected by anti-plane point source load in elastic quarter space was studied. This is the basis to formulate Green function method for SH wave scattering of bi-material interface circular cylinder. Using the steady state solution of the point source load and superposition principle, the steady state response of an arbitrary distributed interface load to an elastic quarter space of interface semi-circular cylinder can be got. At the same time, through numerical calculating examples above, the conclusion can provide certain reference for theoretical research and engineering practice.

## 2. Descriptions and analysis

In this paper, we study anti-plane elastodynamics problems. Figure 1 shows an elastic quarter space. The horizontal boundary is plane $B_H$ , vertical boundary is plane $B_V$ . Embedded in a semi-circular cylinder, the center point $O$ is located in plane $B_V$ on the vertical boundary, and the direct distance between horizontal boundary plane $B_H$ is $h$. The shearing elasticity modulus of the quarter space is $\mu_1$, mass density is $\rho_1$. The shearing elasticity modulus of semi-circular cylinder is $\mu_2$, mass density is $\rho_2$. Build a plane rectangular coordinate system $(O,x,y)$, let the center point $O$ of semi-circular cylinder as the origin, inner normal orientation of vertical boundary plane $B_V$ as positive direction of axis $x$, and outer normal orientation of horizontal boundary plane $B_H$ as positive direction of axis y. Taking the origin point $O$ as the pole, and axis $x$ as the polar axis, the polar coordinate system $(O,r,\theta)$ ) is set up, according to the counterclockwise direction. Out-of-plane orientation ("◉"in figure1) as positive direction of anti-plane displacement, and in-plane orientation as negative direction.

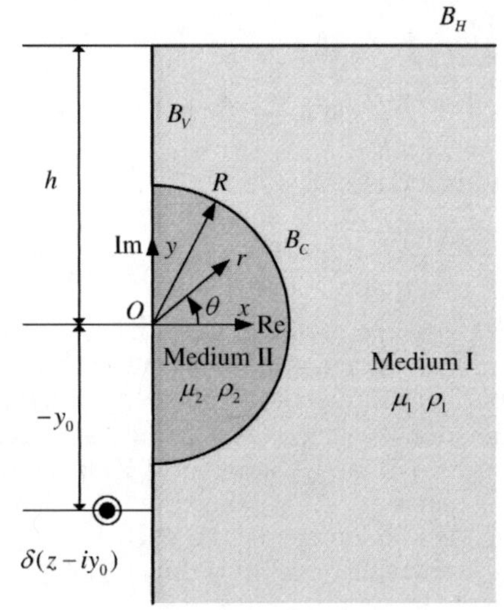


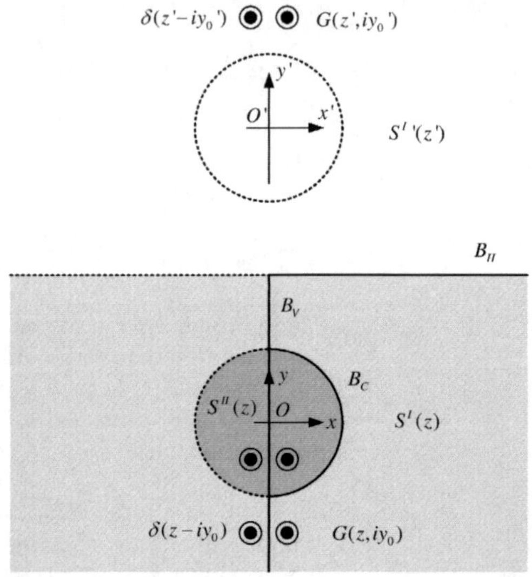


**Figure1.** Point source in the quarter space. **Figure2.** Application of mirror image method .

In this way, the mass point in the elastic quarter space and semi-circular cylinder satisfy the wave equation (1) and (2) respectively.

$$\mu_1 \Delta w + f = \rho_1 \frac{\partial^2 w}{\partial t^2} \tag{1}$$

$$\mu_2 \Delta w + f = \rho_2 \frac{\partial^2 w}{\partial t^2} \tag{2}$$

In the formula, $\Delta$ is a two-dimensional Laplace operator, $w$ is anti-plane displacement, $f$ is anti-plane force. Considering the steady state problem, separate the time variable $t$ and the space variable $(x,y),(r,\theta)$, obtain no homogeneous Helmholtz equation (3) and (4) of displacement.

$$\Delta w + k_1^2 w = -\frac{f^{\omega}}{\mu_1} \tag{3}$$

$$\Delta w + k_2^2 w = -\frac{f^{\omega}}{\mu_2} \tag{4}$$

$$k_1 = \frac{\omega}{c_1}, c_1 = \left(\frac{\mu_1}{\rho_1}\right)^{1/2} \tag{5}$$

$$k_2 = \frac{\omega}{c_2}, c_2 = \left(\frac{\mu_2}{\rho_2}\right)^{1/2} \tag{6}$$

In the formula, $k_1$ and $k_2$ are the numbers of shear waves in the quarter space and semi-circular cylinder respectively, $c_1$ and $c_2$ are the phase velocity of shear wave in the quarter space and semi-circular cylinder respectively, $f^{\omega}$ is out-of-plane loading which neglecting the time factor $\exp(-i\omega t)$. $\omega$ is angular frequency of steady state motion.

According to the description of functions (7), introducing complex variable $z$ and conjugate complex variable $\overline{z}$, create a complex plane $(z,\overline{z})$, among them, $\mathrm{i}=\sqrt{-1}$ is imaginary number.

$$\begin{cases} z = x + iy = re^{i\theta} \\ \overline{z} = x - iy = re^{-i\theta} \end{cases} \tag{7}$$

There is anti-plane steady point source load $\delta(z-iy_0)\exp(-i\omega t)$ on the Point $iy_0$ of the vertical boundary plane $B_V$, which is substituted into the no homogeneous Helmholtz equation (3) and (4). That is $f^{\omega}=\delta(z-iy_0)$, and the solutions are (8) and (9).

$$G(z,iy_0)=\frac{i}{4\mu_1}H_0^{(1)}\left(k_1\left|z-iy_0\right|\right) \tag{8}$$

$$G(z,iy_0)=\frac{i}{4\mu_2}H_0^{(1)}\left(k_2\left|z-iy_0\right|\right) \tag{9}$$

They are the basic solutions of point source functions in the whole space. Among them, $\delta(\bullet)$ is the Dirac function, and $H_0^{(1)}(\bullet)$is the Hankel function of the first kind of 0 order.

In this paper, the governing equations solution of the mathematical physics equations of the anti-plane elastodynamics are the governing equations (3) and (4), the solution conditions are on the horizontal, vertical and semi-cylinder boundaries. They are respectively homogeneous Neumann conditions (10) and (11) on the plane $B_H$ and plane $B_V$, and the homogeneous Dirichlet condition (12) and Neumann condition (13) on the circular cylinder $B_C$.

$$\tau_{yz}^{I}\Big|_{y=h}=\mu_1\frac{\partial w^{I}}{\partial y}\Big|_{y=h}=0 \tag{10}$$

$$\begin{cases} \tau_{xz}^{I}\Big|_{x=0}=\mu_1\dfrac{\partial w^{I}}{\partial x}\Big|_{x=0}=0 \\ \tau_{xz}^{II}\Big|_{x=0}=\mu_2\dfrac{\partial w^{II}}{\partial x}\Big|_{x=0}=0 \end{cases} \tag{11}$$

$$w^{I}\Big|_{r=R}=w^{II}\Big|_{r=R} \tag{12}$$

$$\tau_{rz}^{I}\Big|_{r=R}=\mu_1\frac{\partial w^{I}}{\partial r}\Big|_{r=R}=\tau_{rz}^{II}\Big|_{r=R}=\mu_2\frac{\partial w^{II}}{\partial r}\Big|_{r=R} \tag{13}$$

Here the components of stress and the superscript of displacements I and II are respectively represented in the quarter space Medium I and semi-circular cylinder Medium II.

The anti-plane displacement $w$ and the space variable $(x,y)$ in the plane of $(O,x,y)$ are decoupled. According to the method of mirror image, the definite solution can be constructed. As shown in Figure 2, taking the plane $B_V$ of the vertical boundary as the symmetry plane, extending the quarter space into half space, extending the semi-circular cylinder into cylinder, exerting symmetric anti plane steady point source loads $\delta(z-iy_0)\exp(-i\omega t)$ in the extended region. According to such extension processing, the exact solution of the problem is exactly equivalent, and satisfy the Neumann condition (11) on the plane $B_V$. This is steady state response of an elastic half space with a cylindrical inclusion on the surface of a symmetric plane. Here, the time harmonic function was omitted.

### 3. Point source load

The external force of the equivalent problem is the point source load $2\delta(z-iy_0)$. When it acts in the extended cylinder, that is $|y_0|<R$, the displacement field $G^{II}$ is in accordance with equation(13), Its radial stress component $\tau_{rz}$ is denoted as $G_{rz}^{II}$. According to equation(14), $H_{\#}^{(1)}(\bullet)$ is #order the first kind of Hankel functions.

When $y_0\le h$ and $|y_0|\ge R$, the influence of horizontal boundary should be considered. As shown in figure 2, built a plane rectangular coordinate system $(O',x',y')$, then the complex variable $z$ and the conjugate complex variable $\overline{z}$ can satisfy the equation (16). According to the method of mirror image, the displacement field $G^I$ of the point load $2\delta(z-iy_0)$ is the equation (17), radial stress component $\tau_{rz}$ is denoted as $G^I_{rz}$, that is the equation(18).Where $y_0'=-y_0$, $iy_0'$ is complex value of $iy_0$ that is the mirror point of the plane $B_H$ on the complex plane.

$$G^{II}(z,iy_0)=\frac{i}{2\mu_2}H_0^{(1)}\left(k_2\left|z-iy_0\right|\right) \tag{14}$$

$$G^{II}_{rz}(z,iy_0)=\frac{ik_2}{8}\left[H_{-1}^{(1)}\left(k_2\left|z-iy_0\right|\right)-H_1^{(1)}\left(k_2\left|z-iy_0\right|\right)\right]\left[\left(\frac{z-iy_0}{\left|z-iy_0\right|}\right)^{-1}\left(\frac{z}{\overline{z}}\right)^{1/2}+\left(\frac{z-iy_0}{\left|z-iy_0\right|}\right)\left(\frac{\overline{z}}{z}\right)^{1/2}\right] \tag{15}$$

$$\begin{cases} z'=x'+iy'=z-2ih \\ \overline{z}'=x'-iy'=\overline{z}+2ih \end{cases} \tag{16}$$

$$G^{I}(z,iy_0)=\frac{i}{2\mu_1}H_0^{(1)}\left(k_1\left|z-iy_0\right|\right)+\frac{i}{2\mu_1}H_0^{(1)}\left(k_1\left|z'-iy_0'\right|\right) \tag{17}$$

$$\begin{aligned}G^{I}_{rz}(z,iy_0)=&\frac{ik_1}{8}\left[H_{-1}^{(1)}\left(k_1\left|z-iy_0\right|\right)-H_1^{(1)}\left(k_1\left|z-iy_0\right|\right)\right]\left[\left(\frac{z-iy_0}{\left|z-iy_0\right|}\right)^{-1}\left(\frac{z}{\overline{z}}\right)^{1/2}+\left(\frac{z-iy_0}{\left|z-iy_0\right|}\right)\left(\frac{\overline{z}}{z}\right)^{1/2}\right]\\&+\frac{ik_1}{8}\left[H_{-1}^{(1)}\left(k_1\left|z'-iy_0'\right|\right)-H_1^{(1)}\left(k_1\left|z'-iy_0'\right|\right)\right]\left[\left(\frac{z'-iy_0'}{\left|z'-iy_0'\right|}\right)^{-1}\left(\frac{z}{\overline{z}}\right)^{1/2}+\left(\frac{z'-iy_0'}{\left|z'-iy_0'\right|}\right)\left(\frac{\overline{z}}{z}\right)^{1/2}\right]\end{aligned} \tag{18}$$

## 4. Displacement wave of cylinder

According to Fourier-Hankel wave function expansion, the extending of cylinder will engender the scattered wave in the half space. Its displacement field denoted as $S^I$ can be given in the equation (19). $A_n$ is the undetermined coefficient of wave function. Radial stress component $\tau_{rz}$ of scattering wave is denoted as $S^I_{rz}$, expressed as the series form in the equation (20), where $H_n^-(z)$ and $H_n^+(z)$ are the intermediate variable that can be obtained by differentiation, they can be defined as the equation (20).

$$S^I(z)=\sum_{n=-\infty}^{+\infty}A_nH_n^{(1)}\left(k_1|z|\right)\left(\frac{z}{|z|}\right)^n \tag{19}$$

$$S^I_{rz}(z)=\sum_{n=-\infty}^{+\infty}A_n\left[H_n^-(z)\left(\frac{z}{\overline{z}}\right)^{1/2}+H_n^+(z)\left(\frac{\overline{z}}{z}\right)^{1/2}\right] \tag{20}$$

$$\begin{cases} H_n^-(z)=\dfrac{k_1\mu_1}{2}H_{n-1}^{(1)}\left(k_1|z|\right)\left(\dfrac{\overline{z}}{z}\right)^{1/2}\left(\dfrac{z}{|z|}\right)^n \\ H_n^+(z)=-\dfrac{k_1\mu_1}{2}H_{n+1}^{(1)}\left(k_1|z|\right)\left(\dfrac{z}{\overline{z}}\right)^{1/2}\left(\dfrac{z}{|z|}\right)^n \end{cases} \tag{21}$$

By the method of mirror image, to built the mirror scattered wave that be engendered by the extending of cylinder which is the image of the plane $B_H$. The displacement field $S^I{}'$is expressed as the series form in the equation (22), radial stress component $\tau_{rz}$ is denoted as $S^I_{rz}{}'$, expressed as the series form in the equation (23), where $H_{-n}^-(z')$ and $H_{-n}^+(z')$ are the intermediate variable that can be obtained by differentiation, and they can be defined as the equation (24).

$$S^I{}'(z)=\sum_{n=-\infty}^{+\infty}A_nH_n^{(1)}\left(k_1|z'|\right)\left(\frac{z'}{|z'|}\right)^{-n} \tag{22}$$

$$S_{rz}^{I}{}'(z)=\sum_{n=-\infty}^{+\infty}A_n\left[H_{-n}^{-}(z')\left(\frac{z}{\overline{z}}\right)^{1/2}+H_{-n}^{+}(z')\left(\frac{\overline{z}}{z}\right)^{1/2}\right] \tag{23}$$

$$\begin{cases} H_{-n}^{-}(z')=-\dfrac{k_1\mu_1}{2}H_{n+1}^{(1)}\left(k_1\left|z'\right|\right)\left(\dfrac{\overline{z}'}{z'}\right)^{1/2}\left(\dfrac{z'}{\left|z'\right|}\right)^{-n} \\ H_{-n}^{+}(z')=\dfrac{k_1\mu_1}{2}H_{n-1}^{(1)}\left(k_1\left|z'\right|\right)\left(\dfrac{z'}{\overline{z}'}\right)^{1/2}\left(\dfrac{z'}{\left|z'\right|}\right)^{-n} \end{cases} \tag{24}$$

According to Fourier-Bessel wave function expansion, the extending of cylinder will engender the standing wave. Its displacement field denoted as $S^{II}$ can be given in the series form in the equation (25). Where, $B_n$ is the undetermined coefficient of wave function. $J_n(\bullet)$ is $n$ order Bessel functions. Radial stress component $\tau_{rz}$ of standing wave is denoted as $S_{rz}^{II}$, expressed as the series form in the equation (26), where $J_n^{-}(z)$ and $J_n^{+}(z)$ are the intermediate variable that can be obtained by differentiation, they can be defined as the equation (27).

$$S^{II}(z)=\sum_{n=-\infty}^{+\infty}B_nJ_n\left(k_2\left|z\right|\right)\left(\frac{z}{\left|z\right|}\right)^{n} \tag{25}$$

$$S_{rz}^{II}(z)=\sum_{n=-\infty}^{+\infty}B_n\left[J_n^{-}(z)\left(\frac{z}{\overline{z}}\right)^{1/2}+J_n^{+}(z)\left(\frac{\overline{z}}{z}\right)^{1/2}\right] \tag{26}$$

$$\begin{cases} J_n^{-}(z)=\dfrac{k_2\mu_2}{2}J_{n-1}\left(k_2\left|z\right|\right)\left(\dfrac{\overline{z}}{z}\right)^{1/2}\left(\dfrac{z}{\left|z\right|}\right)^{n} \\ J_n^{+}(z)=-\dfrac{k_2\mu_2}{2}J_{n+1}\left(k_2\left|z\right|\right)\left(\dfrac{z}{\overline{z}}\right)^{1/2}\left(\dfrac{z}{\left|z\right|}\right)^{n} \end{cases} \tag{27}$$

So, we can obtain the expressions of the equivalent displacement field $w$ and radial stress component $\tau_{rz}$ in the elastic half-space and the cylinder.

## 5. Conclusion

According to the mirror method, extend the elastic quarter space to half space, and then extend it to a circular cylinder. By the wave function expansion method, the steady state of the displacement wave is constructed. It is generated the equivalent problem. The Fourier series expansion is performed on the continuity condition of the extended cylindrical boundary, then the coefficient of wave function is determined, and the steady solution of the problem is obtained. The radial stress and circumferential stress components are introduced, and the dynamic stress of the extended cylinder is analyzed quantitatively. It describes the anti plane scattering of the steady state point source by an interface semi-circular cylinder.

Results show (1), the scattering of the point source load on the interface semi-circular cylinder is clearly divided into the projection side and the shadow side. There is a clear boundary between the projection side and the shadow side of the radial stress, and the maximum value of the projection side and the shadow side is obtained on the line of the center point and the point source. With the increase of the shear modulus of the semi-circular cylinder, the maximum value of the radial stress is almost several times that of the shield side.(2) The circumferential stress also achieves the maximum value at the projection side, but it is not in the way of point source load, but there is a deflection angle.(3) In general, with the increase of the elastic modulus, the radial stress will increase and the circumferential stress will increase.(4) Similar to the scattering of plane SH waves by a cylindrical surface, the relatively low frequency point source load can always cause the large relative dynamic stress on the cylinder surface.

**Reference**

[1] Pao Y H and Mow C C 1973 *Diffraction of Elastic Waves and Dynamic Stress Concentrations* (New York: Crane and Russak)

[2] Liu D K and Liu H W 1998 Scattering of SH wave and dynamic stress concentration near the interface *Journal of Mechanics* **30**(5) 597-604

[3] Liu D K and Shi S X 2002 Scattering of plane SH waves by circular lining structure *Journal of mechanics* **34**(5) 796-803

[4] Liu D K and Tian Jia Yong 1999 Scattering and dynamic stress concentration of SH wave in cylindrical elastic inclusion *Explosion and Shock Waves* **19**(2) 115-23

[5] Tian Jia Yong, LIU Dian kui and LIU Chun Xu 1999 Far field solution of SH-wave scattered by interface cylindrical rigid inclusion *Journal of Harbin Engineering University* **20**(4) 90-7

[6] Qi Hui, Liu D K and Shi S X 2001 The displacement field of elastic half space with half circular lining *Journal of Harbin Engineering University* **22**(2) 81-5

[7] Qi Hui, Wang Yan and Liu D K 2003 Scattering of SH waves by circular waves in a semi infinite space interface earthquake engineering and engineering vibration *Journal of Earthquake Engineering and Engineering Vibration* **23**(3) 41-6

[8] Zhao Jia Xi, Qi Hui and Su S W 2008 An approximate analysis of the scattering of SH waves by a cylindrical elastic inclusion with an interfacial debonding *Applied Mathematics and Mechanics* **29**(6) 705-12

[9] Qiu Fa Qiang, Wang Hui Wen and Wang Xue 2010 The interaction between non isosceles triangular structure and foundation under SH wave incidence *Journal of Harbin Institute of Technology* **42**(4) 634-8

[10] Yang Z L, Xv H N and Chen Z G 2011 Scattering of SH waves by an isosceles triangular bulge and a semi-circular concave topography *Journal of Harbin Institute of Technology* **S1** 6-11

[11] Shi Yong, Qi Hui and Yang Zai Lin 2008 Scattering and ground motion of circular holes in the right plane region by SH wave *Chinese Journal of Applied Mechanics* **25**(3) 392-8

[12] Qi H, Shi Y and Liu D K 2009 Interaction of a circular cavity and a beeline crack in right-angle plane impacted by SH-wave *Journal of Harbin Institute of Technology* **16**(4) 548-53

[13] Qi Hui, Shi Yong, Li Hong Liang and Zhao Jia Xi 2009 Dynamic analysis of the circular hole near the interface between two phases in a vertical half space with SH wave incidence *Explosion and Shock Waves* **29**(1) 73-9

[14] Qi H, Yang J and Shi Y 2011 Scattering of SH-wave by cylindrical inclusion near interface in bi-material half space *Journal of Mechanics* **27**(1) 37-45

[15] Qi H, Yang J, Shi Y and Tian J Y 2012 Dynamic analysis for circular inclusion near interfacial crack impacted by SH wave in half space *Journal of Mechanics* **28**(1) 143-51

[16] Qi H and Yang J 2012 Dynamic analysis for circular inclusions of arbitrary positions near interfacial crack impacted by SH-wave in half-space *European Journal of Mechanics* A/Solids **36** 18-24

IOP Conf. Series: Materials Science and Engineering **224** (2017) 012049 doi:10.1088/1757-899X/224/1/012049

# Study of Earthquake Disaster Prediction System of Langfang city Based on GIS

**HUANG Meng, ZHANG Dian, LI Pan, ZHANG YunHui, ZHANG RuoFei**

Institute of Disaster Prevention, HeBei SanHe, 065201

hm@cidp.edu.cn, 774847587@qq.com ,lipan@ cidp.edu.cn , 275885445@qq.com , 1246076188@qq.com

**Abstract.** In this paper, according to the status of China's need to improve the ability of earthquake disaster prevention, this paper puts forward the implementation plan of earthquake disaster prediction system of Langfang city based on GIS. Based on the GIS spatial database, coordinate transformation technology, GIS spatial analysis technology and PHP development technology, the seismic damage factor algorithm is used to predict the damage of the city under different intensity earthquake disaster conditions. The earthquake disaster prediction system of Langfang city is based on the B / S system architecture. Degree and spatial distribution and two-dimensional visualization display, comprehensive query analysis and efficient auxiliary decision-making function to determine the weak earthquake in the city and rapid warning. The system has realized the transformation of the city's earthquake disaster reduction work from static planning to dynamic management, and improved the city's earthquake and disaster prevention capability.

## 1. Introduction

With the rapid development of urban construction, there is an urgent need to assess and improve the city's earthquake and disaster prevention ability, strengthen the seismic key technology research and take effective measures to reduce the earthquake losses. Langfang City developed earthquake disaster reduction information system in 2012. However the earthquake disaster reduction professional model and IT technology and GIS technology is not tight enough in the system, and GIS spatial analysis function is weak, which often lead to earthquake disaster reduction information system lack of real-time characteristics. The earthquake disaster prediction system of Langfang city is the main task of the project ——"Seismic Activity Detection and Seismic Hazard Assessment of Hebei Province", organized by the Seismological Bureau of Hebei Province.

Based on the GIS spatial database, coordinate transformation technology, GIS spatial analysis technology and PHP development technology, the seismic damage factor algorithm is used to predict the damage of the city under different intensity earthquake disaster conditions. The earthquake disaster prediction system of Langfang city is based on the B / S system architecture. Degree and spatial distribution and two-dimensional visualization display, comprehensive query analysis and efficient auxiliary decision-making function to determine the weak earthquake in the city and rapid warning. This paper introduces the development process of this system in detail[1-3].

## 2. System overall design

The system structure of earthquake disaster prediction system of Langfang city is shown in Figure 1. The workflow is as follows: The spatial database of earthquakes prediction in Langfang City is established according to the distribution of the geologic disaster and the degree of danger, the basic situation of the building and the infrastructure (including name, spatial distribution, structure type, seismic fortification, current working status, etc.) in Langfang City. This system uses the MVC design pattern and GIS spatial analysis Model and the basic principles of earthquake engineering to predict the severity and spatial distribution of urban earthquake disaster and realize the change from static planning to dynamic management of urban earthquake disaster reduction.

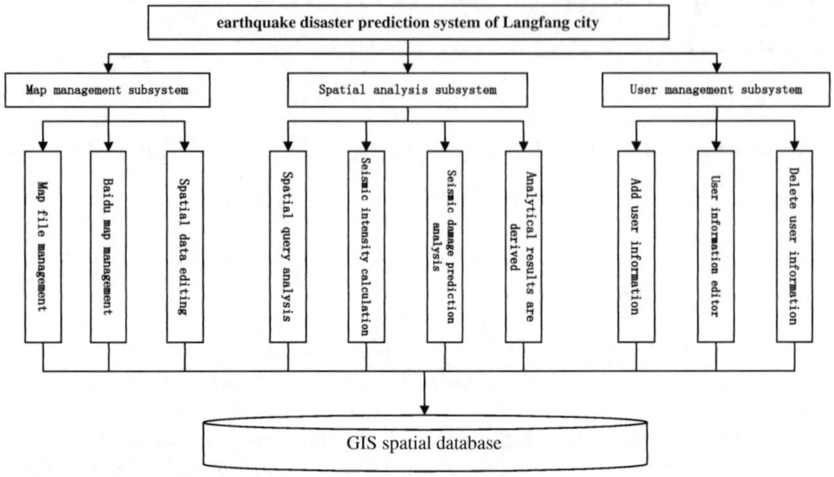


**Figure 1.** The system structure of earthquake disaster prediction system of Langfang city

## 3. System implementation

The earthquake disaster prediction system of Langfang city is based on PHP + MySQL database + Baidu map interface + ArcGIS shp layer (projection conversion lifeline data). The system fully utilize the GIS tool to deal with the spatial database,to improve the development efficiency of the application system. The application has a good appearance, more powerful database function, good reliability, and is easy to transplant and maintain.

### *3.1 Design of spatial database for the earthquake disaster prediction system of Langfang city*

Langfang earthquake damage forecasting space database, including the distribution of geological hazards and the degree of danger, the basic situation of buildings and infrastructure and lifeline and other information, the relevant content of the author in the previous article has been described in detail, will not repeat them here.

### *3.2 map projection conversion*

At present, the projection of data we collected is WGS84 coordinate system. However the system will eventually be displayed on Baidu map. Baidu map using Baidu coordinates (BD09). So we must use a coordinate system that has been used after at least two times of encryption. Coordinate conversion is very necessary. At present, Baidu provides a coordinate for the conversion of the API.Application of the API, this paper to achieve the WGS84 Baidu coordinates of the conversion.

### *3.3 seismic intensity attenuation*

China's vast territory, geological structure is complex, which, from the seismic intensity of the shape of the complex shape of the earthquake can also be reflected. In general, the seismic intensity of the isosceles line is oval, which is due to the impact of geological structure. along the direction of the earthquake direction of the intensity of attenuation is slow, and the vertical direction of the fault attenuation faster, decay of the direction The difference to the far field gradually disappear, isometric line and tend to round. Thus, the difference in the ratio of the major axis to the minor axis length of the isosceles of the seismic intensity in different regions can reflect the regional differences in seismic intensity attenuation, which is one of the bases of the seismic intensity attenuation relationship. In this

IOP Conf. Series: Materials Science and Engineering **224** (2017) 012049 doi:10.1088/1757-899X/224/1/012049

paper, we use the above two kinds of intensity calculation method to calculate, to provide technical support and assistance to earthquake experts.

*3.3.1 round model.* The seismic intensity attenuation includes two empirical formulas.
The first one: the intensity of meizoseismal area is equal to the magnitude plus 2 or 3, and the plain area (Langfang, for example) plus 2. For example, the system set the highest intensity 6 earthquake corresponding to 8 degrees, 7 corresponding to the earthquake intensity is 9 degrees, the highest intensity 8 earthquake corresponding to 10 degrees. The other one is an empirical formula. In general, the higher the intensity, the faster the decay. The distance that need when 10 degrees decay to 9 degrees is about 5km while 10km for 9 degrees to 8 degrees, 20km for 8 degrees to 7 degrees, 40km for 7 degrees to 6 degrees.

*3.3.2 linear focus model.* The relationship between the attenuation of the linear focus earthquake intensity in the North China area is as follows:

$$I_l = 2.033 + 2.010M - (0.923 + 0.151M)\ln(R' + 27.035)$$

$$I_s = 0.064 + 1.949M - (0.433 + 0.169M)\ln(R' + 13.073)$$

I is the intensity value corresponding to the focal depth, magnitude and epicentral distance and $R'$ is the distance from the source to the target area ( $R' = \sqrt{h^2 + D^2}$ ) , h is the focal depth, and D is the epicentral distance[4-5].
In addition to the three major factors of long axis distance, short axis distance and fault strike, it is necessary to determine the fault strike. We introduce the potential zoning map of the whole country into the ArcGIS software platform, and use the segment of the potential source area to simulate the earthquake fault, and finally get the fault to this element, and perfect the attenuation relationship under the line source. In the Langfang area simulation occurred a 7 earthquake, the effect shown in Figure 2.

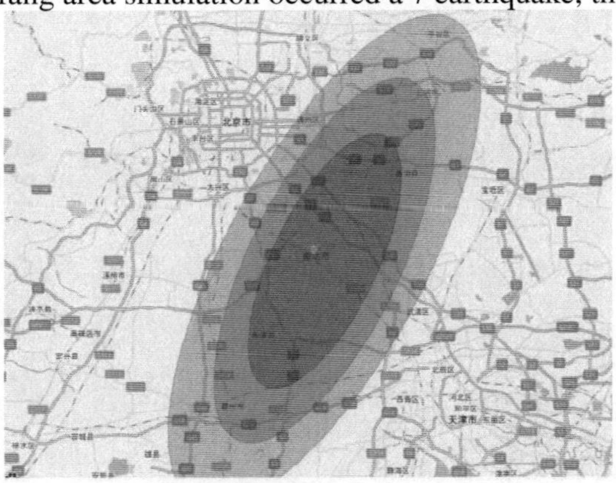

**Figure 2.** The linear focus Seismic Intensity Map

*3.4 Building damage prediction*
Langfang city has up to tens of thousands of buildings, one by one investigation is almost impossible. Therefore, according to the village distribution, street administrative area, scope of work and accuracy requirements, the Langfang City Planning Area is divided into 109 prediction unit grid.
The seismic damage prediction of the group building adopts the combination of the typical analysis and the group forecasting. Due to the large number of buildings in Langfang City, the group prediction method is adopted. And the damage factor analysis method is used to determine the damage state of the buildings under different earthquake intensities.
The basic idea of the damage factor method is to find out a number of factors that have great influence on the earthquake damage of the building on the basis of summing up the historical damage data and

IOP Conf. Series: Materials Science and Engineering **224** (2017) 012049 doi:10.1088/1757-899X/224/1/012049

to quantize it. We put these numerical factors as the earthquake damage factor, said d. After the damage factor is obtained, the factors are summarized as the earthquake damage factor table, and the damage index D of the building is obtained according to the formula (1). Through the statistical analysis of the previous earthquake damage, we can determine the range of the damage index D corresponding to the five damage levels which are basically intact, minor damage, moderate damage, serious damage and collapse. The calculated D value is compared with the range table to determine the damage of the building.

$$D = w\prod_{i=1}^{9}\prod_{k=1}^{2,3,4,5,6} d_{ik}^{m_{ik}} \tag{1}$$

W is the conversion factor of the peak acceleration of ground motion A, A＜0.05g，w=0；0.05≤A＜0.1g，w=0.4；0.1≤A＜0.2g，w=0.8；0.2≤A＜0.3g，w=1.6；0.3≤A＜0.4g，w=2.4；A≥0.4g，w=3.2； When the i item in the k class mik takes 1.when there is no k class mik takes 1.

There are two main advantages to the prediction of the earthquake damage of the group building by the seismic damage factor method. First, the damage factor method does not need very detailed investigation to the sampling data of buildings, to avoid the difficulties of collecting and processing a large number of building drawings and related information, which can improve the sampling rate and is also very easy to carry out practical work. Second, the precision of seismic damage prediction can be satisfied by the earthquake damage factor method. The system simulates the earthquake damage of buildings in Langfang area under the 7 earthquake, and the results are as follows:

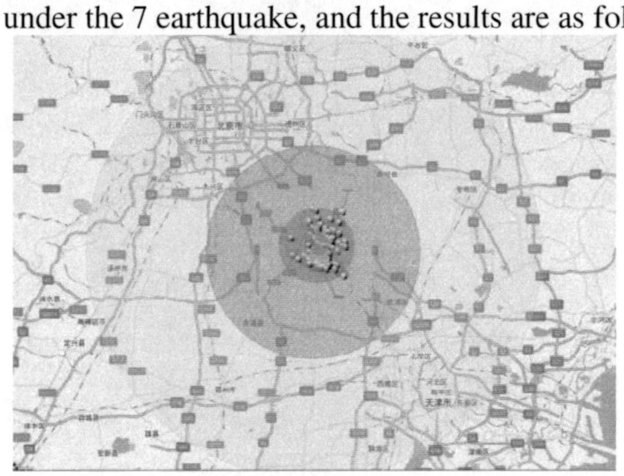

**Figure 3.** The earthquake prediction of buildings in Langfang City

*3.5 earthquake damage prediction analysis*

In order to improve the anti-earthquake and anti-disaster capability of Langfang, this system is based on the basic principle of earthquake prediction and spatial engineering of earthquake. The system is analyzed by accident tree model, AHP model and successive approximation model and GIS spatial analysis technique in the aspects of personnel casualty loss assessment, lifeline damage prediction, traffic system earthquake damage prediction and secondary disaster prediction. Through the analysis, we can find out the existing earthquake disaster prevention capacity in Langfang, find out the weak links and key links, and provide technical support for the decision-making departments to take the earthquake disaster prevention and control measures in advance.

*3.6 User Management*

This module mainly includes user management, that is, users add, edit and delete and other functions.

## 4. Conclusion

According to the current situation of improving earthquake disaster prevention capacity of the city to our country, we put forward the implement scheme of system of earthquake damage prediction of Langfang city based on GIS technology. We can predict the different intensity of earthquake disaster under the condition of the injury and city spatial distribution and show them on map to determine the city seismic weak link and quickly alert. The system has realized the transformation of the city's earthquake disaster reduction work from static planning to dynamic management, and improved the city's earthquake and disaster prevention capability.

**Acknowledgments**
This work was supported by Hebei Province Science and Technology Planning Project (15270132) and the Special Fund of Fundamental Scientific Research Business Expense for Higher School of Central Government (No. ZY20160106).

**Reference**

[1] Meng HUANG, Dian ZHANG, Maofa WANG,*A New Earthquake Analysis and Prediction System Based on OpenGIS*,IJSSST,Volume 17, Number 31 P35.1

[2] HUANG MENG, ZHAO LEI, LI PAN,*RESEARCH ON CITY GROUP BUILDINGS DAMAGE PREDICTION BASED ON 3D GIS,*Journal of the Balkan Tribological Association Vol. 22, No 2A-I,P1880-1889

[3] Q. Y. Liang, Y. Zhang, *Kunming Tourism Geographic Information System Based on Open Source Platform, Modern Computer*, vol. 12,pp. 76-79, 2015.

[4] S. Y. CHEN, N. CHEN, J. B. ZHENG, J.X. RUI, B. Q. SHI, *The design of Taining World Geopark tourism information platform based on open web GIS*, Journal of Shanghai Normal University(Natural Sciences), vol. 6, pp. 645-650, 2009

[5] Y. Ban, X. N. Wang, L. Ma, *Design and realization of earthquake emergency response system,* Science of Surveying and Mapping, vol.11, pp. 57-61, 2014.

IOP Conf. Series: Materials Science and Engineering **224** (2017) 012050 doi:10.1088/1757-899X/224/1/012050

# Design, Prototyping and Control of a Flexible Cystoscope for Biomedical Applications

**Canberk Sozer[1], Morteza Ghorbani[2,3], Gokhan Alcan[2,4], Huseyin Uvet[1], Mustafa Unel[2,4], Ali Kosar[2,3]**

[1]Mechatronics Engineering Department, Yildiz Technical University, Istanbul, Turkey
[2]Faculty of Engineering and Natural Science, Sabanci University, Istanbul, Turkey
[3]Sabanci University Nanotechnology Research and Application Center, Istanbul, Turkey
[4]Integrated Manufacturing Technologies Research and Application Center, Sabanci University, Istanbul, Turkey

canberksozer@gmail.com

**Abstract.** Kidney stone and prostate hyperplasia are very common urogenital diseases all over the world. To treat these diseases, one of the ESWL (Extracorporeal Shock Wave Lithotripsy), PCNL (Percutaneous Nephrolithotomy), cystoscopes or open surgery techniques can be used. Cystoscopes named devices are used for in-vivo intervention. A flexible or rigid cystoscope device is inserted into human body and operates on interested area. In this study, a flexible cystoscope prototype has been developed. The prototype is able to bend up to ±40° in X and Y axes, has a hydrodynamic cavitation probe for rounding sharp edges of kidney stone or resection of the filled prostate with hydrodynamic cavitation method and contains a waterproof medical camera to give visual feedback to the operator. The operator steers the flexible end-effector via joystick toward target region. This paper presents design, manufacturing, control and experimental setup of the tendon driven flexible cystoscope prototype. The prototype is 10 mm in outer diameter, 70 mm in flexible part only and 120 mm in total length with flexible part and rigid tube. The experimental results show that the prototype bending mechanism, control system, manufactured prototype parts and experimental setup function properly. A small piece of real kidney stone was broken in targeted area.

## 1. Introduction

Nowadays, robotics is widely used in production industry, defense industry and medical field. Many surgeries are performed by robots or robotic-assistant methods and open surgery is seldom in use. Devices used for surveillance and intervention in medical field are called as endoscope. If an endoscope is specialized for urogenital diseases it is called as cystoscope. The cystoscopes can be grouped as rigid and flexible cystoscopes. The rigid cystoscopes consist of just a rigid tube and the flexible cystoscopes include a flexible section in addition to the rigid tube.

The first primitive endoscopic device was created by Philipp Bozzini in 1806 [1], and the modern rigid cystoscope concept was presented by Maximilian Carl-Friedrich Nitze in 1879 [2]. Those devices have been evolving and as a result of continuous improvements, flexible cystoscope devices have been created [3]. The flexible section has significant advantages over the rigid cystoscope such as better postoperative hematuria, less pain and less need of medication, increasing patient and doctor comfort and compatibility [4-6].

Rigid cystoscopes include just a rigid tube and the operator has to reposition the whole device to reach the target. Flexible cystoscopes, on the other hand, have a flexible section in addition to the rigid tube. So, the operator just steers the bendable end-effector using joystick to reach the target.

The design and control of the flexible endoscopic robot arms have been widely studied in medical robotics field. Kato et al. (2015) developed forward kinematic mapping of a two segmented tendon-driven endoscopic robot arm which bends up to 180° [7]. Lei and Du (2010) presented bending mechanism of a tendon-driven flexible ureteroscope [8]. Seneci et al. (2014) described kinematic model and precise control strategy for a 2 DoF flexible robot arm for transluminal and endoluminal surgery [9]. Maeda et al. (1996) developed an active endoscope with 2 mm outer diameter that was actuated by a SMA coil spring [10]. De Sars et al. (2010) presented design and control of a 2 DoF active endoscope steered by two pairs of antagonist SMA wires [11].

In this study, a tendon wire driven cystoscope which has a flexible section has been developed. The aim of the cystoscope platform is to steer the hydrodynamic cavitation probe [12] to the target using a fully manual joystick. The prototype cystoscope has six channels; one for hydrodynamic cavitation tube, one for waterproof medical camera enclosed by fiber optic illumination system, and four 90° positioned channels for stainless steel tendon wires to transmit the deflection from Stewart platform to the end-effector. The probe is used for rounding the sharp edges of kidney stones and resectioning the prostate organ. The fiber optic illumination system is used for illuminating the dark space inside body and camera is used to provide visual feedback to the operator to show the position of the end-effector and the target. The flexible end-effector can bend up to ±40° in X and Y axes, and is controlled by fully manual joystick. The outer diameter of the platform is 10 mm and the length of the total prototype is 120 mm.

The rest of the paper is organized as follows: the prototype of the system is provided in Section 2, proposed control scheme is presented in Section 3, manufactured experimental setup is shown in Section 4 and finally the paper is concluded with some remarks and possible future directions in Section 5.

## 2. Design and Prototyping of a Flexible Cystoscope

The prototype aims steering the hydrodynamic cavitation probe to erode kidney stone surface and treat prostate diseases. To this end, design criteria have been defined by considering available components of the prototype; hydrodynamic cavitation probe with 1.58 mm outer diameter, a micro ScoutCam™ (Medigus Ltd., Israel) waterproof camera enclosed by an illumination system with 2 mm outer diameter, and four 90° circularly positioned stainless steel tendon wires with 1 mm outer diameter. To address all of the requirements, the prototype includes a 1.7 mm in diameter channel for hydrodynamic cavitation probe, a 2.1 mm in diameter channel for imaging and illuminating system and four 1.25 mm in diameter channels for stainless steel tendons. The whole prototype is 10 mm in outer diameter (Figure 1(a)).

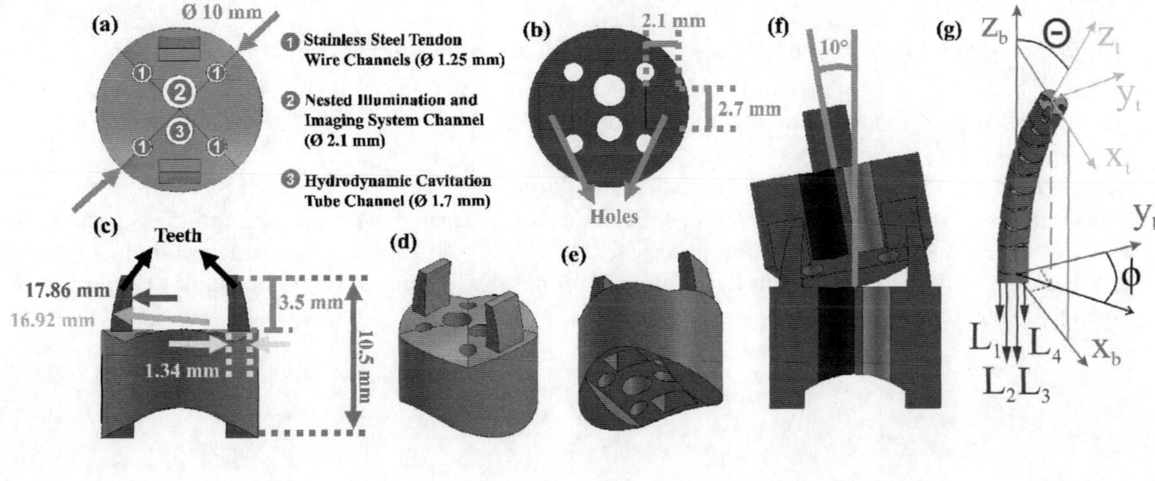

**Figure 1.** (a) The outer and channel diameters of bendable spacer in X axis; the placement of components, (b) dimensions of the holes, (c) length of the part, (d) and (e) perspective view of the part, (f) bending mechanism of a flexible pair, (g) the relationship between actuation space (L) and joint space (θ, ϕ)

The flexible part consists of 4 different micromanipulator pieces; base part, bendable spacer in X axis, bendable spacer in Y axis and distal end part. The base part has just teeth on the top surface, the X and Y bendable spacers have teeth on the top surface and holes on the bottom surface, the distal end part has just holes on the bottom surface. Thanks to the teeth-holes structure, the flexibility of the bending mechanism has been achieved. The holes are 2.1 mm in length, 2.7 mm in width and 2.72 mm in depth (Figure 1(b)). The teeth are 1.34 mm in length, 2.5 mm in width and 3.5 mm in depth with 16.92 mm in outer and 17.86 mm in inner radius of curvature. This curvy structure provides centering and bending. The base part is 12.5 mm in length with teeth, the spacers are 10.5 mm in length (Figure 1(c)) with teeth and the distal end part is 11 mm in length. When the flexible section is in 0° initial position, teeth get into the holes and the total length of the flexible section becomes 70 mm.

The flexible part of the prototype is able to bend up to ±40° in X and Y axes. The bending mechanism consists of 9 parts. These parts create 8 joints and the bending occurs at these points. Each joint can bend ±10° in a single axis (Figure 1(f)). This limitation is due to the physical restriction of the teeth-holes structure. Four of the eight joints provide bending in X axis and the remaining four joints provide bending in Y axis. When the parts are aligned in the proper order, the flexible end-effector can bend up to ±40° in X and Y axes. In 40° maximum bending angle, radius of curvature of the prototype is 80.3 mm.

To control position of the flexible section, HXP50 6-axis Hexapod (Newport, USA) was used as an actuator. The hexapod gets motion command from the operator via joystick and repositions the moving top plate. The motion of the top plate is transmitted by stainless steel tendon wires to the flexible part.

The relationship between actuation space ($L_1$ , $L_2$, $L_3$ and $L_4$) and joint space ($\theta$ *and* $\phi$) parameters is given in the following equations. These formulas [13] are used to determine bending angles in terms of linear displacement of the tendon wires (Figure 1(g)). When tendons are pulled 2.5 mm linearly, the flexible end-effector bends 40°.

$$L_1 = L_0 + 2N_f\left[bsin\left(\frac{\theta}{2}\right) - h_0 sin^2\left(\frac{\theta}{4}\right)\right] \tag{1}$$

$$L_2 = L_0 + 2N_f\left[asin\left(\frac{\theta}{2}\right) - h_0 sin^2\left(\frac{\theta}{4}\right)\right] \tag{2}$$

$$L_3 = L_0 - 2N_f\left[bsin\left(\frac{\theta}{2}\right) + h_0 sin^2\left(\frac{\theta}{4}\right)\right] \tag{3}$$

$$L_4 = L_0 - 2N_f\left[asin\left(\frac{\theta}{2}\right) + h_0 sin^2\left(\frac{\theta}{4}\right)\right] \tag{4}$$

where $a = \frac{d}{2}\sin(\phi)$ , $b = \frac{d}{2}\cos(\phi)$ are distances of the tendons to the bending neutral plane, $d$ is the distance between the tendons in a pair, $\phi$ is bending direction, $L_0$ is length of tendons at initial position, $N_f$ is free joint number, $\theta$ is total bending angle.

## 3. Control

The operator gets real time visual feedback from the camera inside the prototype and gives input to the joystick. The joystick commands are processed by MATLAB to produce hexapod commands which are then sent to the hexapod controller to reposition the flexible part through hexapod to approach the specimen (Figure 2).

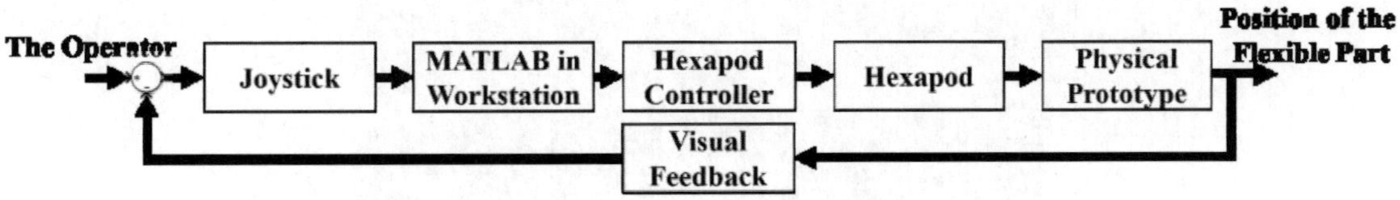


**Figure 2.** Control system for the prototype

As mentioned before, tendons must be pulled and released 2.5 mm to reach 40° maximum bending angle. When a tendon is pulled, cross tendon wire is released in the same amount. To achieve this, roll and pitch angles of hexapod were used. Both of those motions enable cross tendon's pull-release motion.

A perforated plexiglass plate is attached over the top platform of the hexapod and tendon wires were fixed (see Section 4). All of the selected holes on plexiglass plate are 35.86 mm away from the origin along both vertical and horizontal directions. To pull the tendon wire 2.5 mm, the hexapod must be bent approximately 4° in roll or pitch angles around X and Y axes, respectively.

## 4. Experimental Setup and Results

The parts of the flexible section have been designed in SolidWorks 3D CAD design software. The first prototype was manufactured with plastic based VisiJet M3 Proplast material by ProJet 3500 HDMax (3D System, USA) 3D printer and after its bending confirmation, the parts were produced with steel based MaragingSteel MS1 material using M 290 (EOS, Germany) DMLS 3D printer.

A plastic based piece was placed into a 10 mm in outer and 8 mm in inner diameter stainless steel tube to separate tendons, imaging and illuminating system and hydrodynamic cavitation tube. The manufactured parts were put at the end of the tube in the proper order and tips of 4 tendon wires were attached to the distal end part of the prototype.

**Figure 3.** Experimental setup

Aluminum profiles were attached each other to construct the base of the experimental setup. The hexapod, bowden cable holders for guiding tendon wires and a prototype holder were fixed on the setup. The perforated plexiglass plate was placed on the moving top plate of the hexapod to fix tendon wires. The bowden cables were inserted into the bowden cable holders. After the prototype inserted into the prototype holder, other tips of tendons were fixed to the plexiglass plate by passing through prototype holder and bowden cables, respectively (Figure 3).

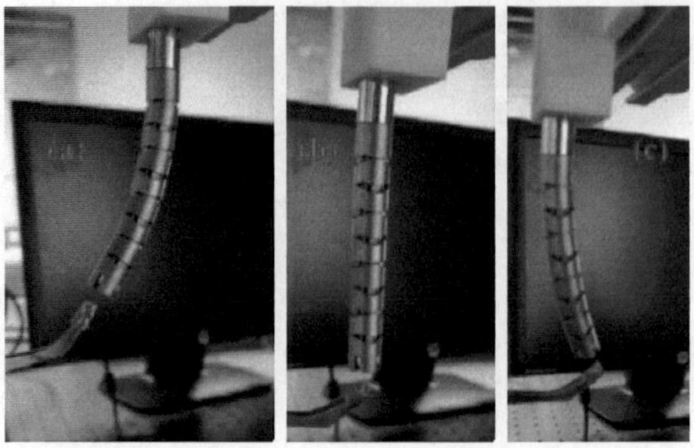

**Figure 4.** (a) 40° bending angle in +X axis (b) 0° initial position (c) 30° bending angle in -X axis

The prototype system was tested for in-vitro experiments. The real kidney stone was fixed at different points in the workspace of the prototype. The flexible end-effector was locked at different angles from 0° to 40° according to the position of the stone using joystick and cavitation was initiated (Figure 4). When the flexible end-effector was unlocked, it turned to its initial position. In addition, the flexible part was moved on the stone to reach different points on the stone surface. As a result, the prototype successfully broke the targeted point on the kidney stone using hydrodynamic cavitation.

## 5. Conclusion and Future Directions

We have now presented mechanical design, prototyping and control of a cystoscope for biomedical applications. The fully manual joystick controlled prototype steers the hydrodynamic cavitation probe toward the specimen (e.g. kidney stone, prostate). To provide bending up to ±40° in X and Y axis, a suitable teeth – holes structure has been designed. The relationships between tendon wires linear displacements and bending angles were utilized to determine the bending angles. Joystick commands are processed in MATLAB and hexapod commands are generated. The prototype was used for in-vitro experiments with real kidney stone and a small piece of kidney stone was successfully broken with hydrodynamic cavitation.

As future directions, semi-automatic and fully-automatic vision based control will be implemented in the proposed flexible cystoscope system, which will include real-time processing of the images acquired by the camera placed in the flexible part of the micromanipulator.

## Acknowledgment

This project was supported by TUBITAK (The Scientific and Technological Research Council of Turkey) under Grant No 113S092.

## References

[1] Doglietto F, Prevedello D M, Jane Jr J A, Han J, and Laws Jr E R 2005 A brief history of endoscopic transsphenoidal surgery—from Philipp Bozzini to the First World Congress of Endoscopic Skull Base Surgery *Neurosurgical focus* **19(6)** 1-6

[2] Herr H W 2006 Max Nitze, the cystoscope and urology *The Journal of urology* 1313-16

[3] Kavoussi L R and Clayman R V 1988 Office flexible cystoscopy *The Urologic clinics of North America* **15(4)** 601-608

[4] Erdem M R, Topaloğlu K, Yücebaş Ö E, Şengör F, Aslan A R, Tosun Ç, Ramazanoğlu M A and

Duvar S 2015 Better Comfort for Both Patient and Doctor: Prospective Randomized Study about Flexible and Rigid Cystoscopy *Journal of Urological Surgery* **2** 65-8

[5] Flannigan G M, Gelister J S K, Noble J G and Milroy E J G 1988 Rigid versus Flexible Cystoscopy A Controlled Trial of Patient Tolerance *British journal of urology* **62(6)** 537-40

[6] Denholm S W, Conn I G, Newsam J E and Chisholm G D 1990 Morbidity following cystoscopy: comparison of flexible and rigid techniques *BJU International* **66(2)** 152-4

IOP Conf. Series: Materials Science and Engineering **224** (2017) 012050 doi:10.1088/1757-899X/224/1/012050

[7] Kato T, Okumura I, Song S E, Golby A J, and Hata, N 2015 Tendon-driven continuum robot for endoscopic surgery: preclinical development and validation of a tension propagation model *IEEE/ASME Transactions on Mechatronics* **20(5)** 2252-63

[8] Lei M C and Du R 2010 *International Conference on Control Automation and Systems (ICCAS)* IEEE Gyeonggi-do pp 2019-2023

[9] Seneci C A, Shang J, Leibrandt K, Vitiello V, Patel N, Darzi A, Teare J and Yang G Z 2014 *IEEE/RSJ International Conference on Intelligent Robots and Systems (IROS)* IEEE Chicago pp 1314-21

[10] Maeda S, Abe K, Yamamoto K, Tohyama O and Ito H 1996 *Micro Electro Mechanical Systems, 1996, MEMS '96, Proceedings. An Investigation of Micro Structures, Sensors, Actuators, Machines and Systems. IEEE, The Ninth Annual International Workshop on* IEEE pp 290-5

[11] De Sars V, Haliyo S and Szewczyk J 2010 A practical approach to the design and control of active endoscopes *Mechatronics* **20(2)** 251-64

[12] Perk O Y, Şeşen M, Gozuacik D and Koşar A 2012 Kidney stone erosion by micro scale hydrodynamic cavitation and consequent kidney stone treatment *Annals of biomedical engineering* **40(9)** 1895-902

[13] Li Z, Feiling J, Ren H and Yu H 2015 A Novel Tele-Operated Flexible Robot Targeted for Minimally Invasive Robotic Surgery *Engineering* **1(1)** 73-78

IOP Conf. Series: Materials Science and Engineering **224** (2017) 012051 doi:10.1088/1757-899X/224/1/012051

# Study on the Automatic Detection Method and System of Multifunctional Hydrocephalus Shunt

**Xuan Sun, Guangzhen Wang, Quancheng Dong,Yuzhong Li**

School of Mechanical Engineering, University of Jinan, Jinan, China,250022

377094468@qq.com, 1907268702@qq.com

**Abstract.** Aiming to the difficulty of micro pressure detection and the difficulty of micro flow control in the testing process of hydrocephalus shunt, the principle of the shunt performance detection was analyzed.In this study, the author analyzed the principle of several items of shunt performance detection,and used advanced micro pressure sensor and micro flow peristaltic pump to overcome the micro pressure detection and micro flow control technology.At the same time,This study also puted many common experimental projects integrated, and successfully developed the automatic detection system for a shunt performance detection function, to achieve a test with high precision, high efficiency and automation.

## 1. Introduction

The utility model relates to a hydrocephalus shunt, which is a one-way hydraulic valve which can be implanted into a patient's intracranial for a long time,The main structure of the interior is adjustable spring and ball core valve,it can control the opening and closing pressure of the valve.Hydrocephalus shunt performance test is very strict.The inspection of the quality of the hydrocephalus shunt related works start late domestic, almost no reference can be consulted.Now,we still use laboratory common equipments for testing, which has low efficiency, poor accuracy and large workload.It is an urgent needing that developing a automatic, high efficiency and high precision multifunctional hydrocephalus shunt performance detection system.Through the analysis of the detection principle of physical characteristics of several common shunts, and integration completes the assignment.

## 2. System composition and principle

### *2.1.Structure and working principle of hydrocephalus shunt*

As shown in the Figure 1, hydrocephalus shunt was a spring type hydraulic one-way valve, special equipment pre adjustment the shunt of the opening and closing pressure after the end of the hose to the water hose parts of intracranial, terminal patients in abdominal cavity. In this way, when the patient's intracranial pressure exceeds the preset pressure value of the shunt, it can automatically open, and the hydrocephalus is drained into the patient's abdominal cavity through the shunt to achieve the purpose of intracranial decompression[1].

**Figure 1.** Shunt structure of hydrocephalus

IOP Conf. Series: Materials Science and Engineering **224** (2017) 012051 doi:10.1088/1757-899X/224/1/012051

*2.2.Artificial test system analysis hardware composition*
The physical characteristics of the diverter detected include: the detection pressure value of the opening and closing , the flow-pressure characteristic test, the test of the long-term stabilityand characteristics and the test of flow-pressure characteristics. As shown from figure 2 to Figure 4, the schematic diagram of the test system for the manual testing of the four test items.

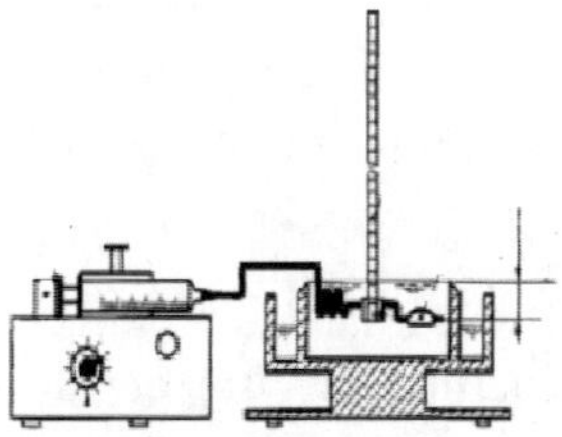

**Figure 2.** Open pressure, pressure flow characteristics test system

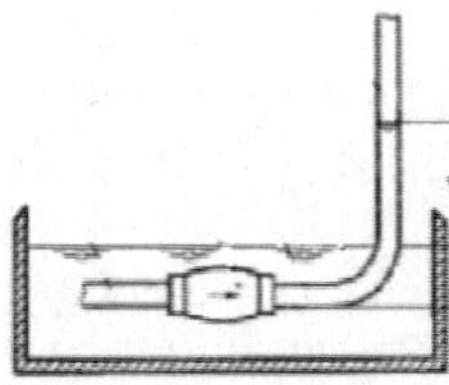

**Figure 3.** Back flow characteristic test system

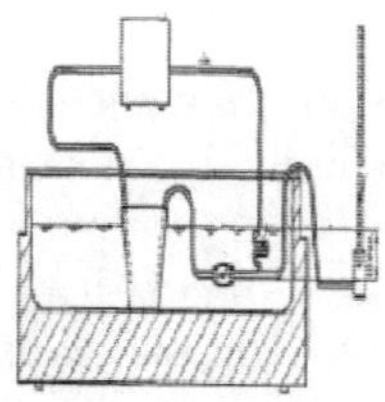

**Figure 4.** Long term stability test system

To analyze the constituent elements of the test system of four test items , it was not difficult to find that the following test conditions in common with test system for the four test items: (1) To test the flow in the 5-60mL/h adjustable flow range; (2) To detect read shunt entrance or exit test of hydraulic pressure. The accuracy of 0.5mm water column; (3) To maintain the temperature in constant temperature water bath 37 ±2°C.Therefore, in considering the detection precision, efficiency and automation and other factors, development of hydrocephalus shunts by precision pressure sensor, data acquisition, data processor, controllable precision test flow pump was a main component of the automatic detection system.

## 3.The hardware configuration of the automatic detection system for diverter

*3.1.Hardware composition*
The hardware structure of the automatic detection system for hydrocephalus shunts was shown as Figure 5, which consists of a pressure sensor, a controllable flow pump, a data acquisition card and a data processing computer. According to the function, it can be divided into two parts: data acquisition and data processing.

*3.1.1. Data acquisition section.*The micro pressure sensor for detecting bottom piezoelectric film, because the test fluid bring weak stress can cause slight changes in the film shape, using the piezoelectric effect of the films will convert the pressure into electric signals, through the transmission circuit, voltage signal was converted to 0-5v. The 6 sensor signal lines were connected to the 6 input channels of the A/D module of the data acquisition card. Acquired by industrial computer, the corresponding pressure. The data acquisition system was composed of pressure sensor and data acquisition card.

*3.1.2. Data processing.* As the core equipment of the IPC controlling input and output data processing , internal PCI interface data acquisition card will install the pressure data measured by sensors to CPU, by supporting the development of data acquisition and processing software test project four. There are parameter setting options on the operating interface, the user can adjust the parameters according to the different experiments and different experimental samples. The software also had the function of automatic analysis and calculation of test data, real-time display, production test curve, test report, print, historical data record and query.

*3.2. Design of the main hardware*

IOP Conf. Series: Materials Science and Engineering **224** (2017) 012051 doi:10.1088/1757-899X/224/1/012051

*3.2.1.Pressure senso. P*ressure detection using the Japanese MTO-39 micro pressure sensor [2-3], the ideal measurement accuracy can reach 0.25%.

*3.2.2.Data acquisition card.* Data acquisition using PCI1716L multi-function data acquisition card, installed in the host computer motherboard, exchange and IPC data through PCI interface, this card is a 16 bit high precision and multifunction analog input and output data acquisition card. It comes with a 250K/s 16 bit A/D converter that provides either a single ended analog input or a differential analog input of the 8 channels, or a combination of the inputs of the. It comes with 2 bit D/A output channels, a 16 bit digital input / output channel and a channel with 1 16 bit counters.

*3.2.3.Data processor.* Using Advantech IPC 610L, internal PCI and ISA interface for convenient expansion, and can satisfy the requirement that long-term stability of experimental project 28 days continuous operation without stopping .

*3.2.4.Controllable flow pump.* Flow control using Trevor peristaltic pump [4-5], the flow range of 1mL/h-100mL flow, the accuracy of 1%FS.

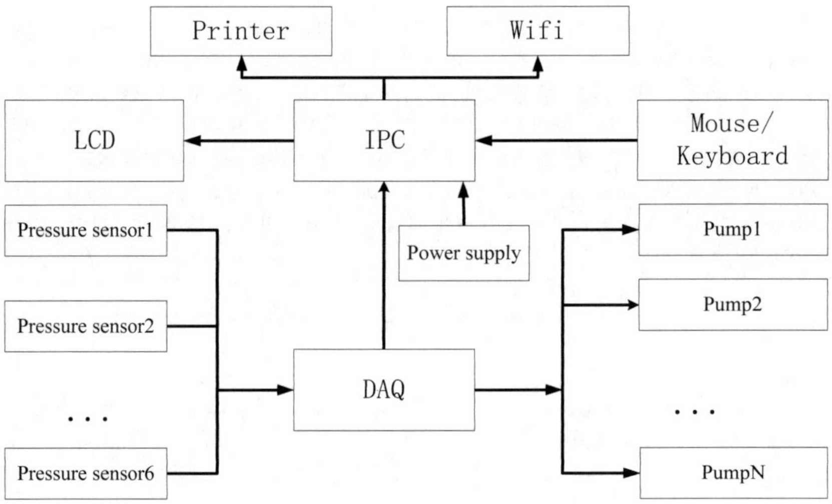


**Figure 5.** Hardware structure of detection system

## 4.Test of the system function and performance

### *4.1.Opening and closing pressure test*

As shown in Figure 6 and Figure 7, taked a splitter for the same time for 6 consecutive experimental data and analysis reports. From the table 1 data, the opening pressure value was stable at 45-50mm$H_2O$ pressure, the closing pressure was stable at 25-30 mm$H_2O$, slightly lower than the opening pressure, in line with the shunt of the opening and closing pressure characteristics. The system automatically calculated the average value, maximum value, minimum value and standard deviation according to the experimental data, was convenient for collecting data to determine the opening pressure and closing pressure of a specific design of the valve manufacturers; in addition, as a type of validation, it can verify the opening and closing pressure valve.

IOP Conf. Series: Materials Science and Engineering **224** (2017) 012051 doi:10.1088/1757-899X/224/1/012051

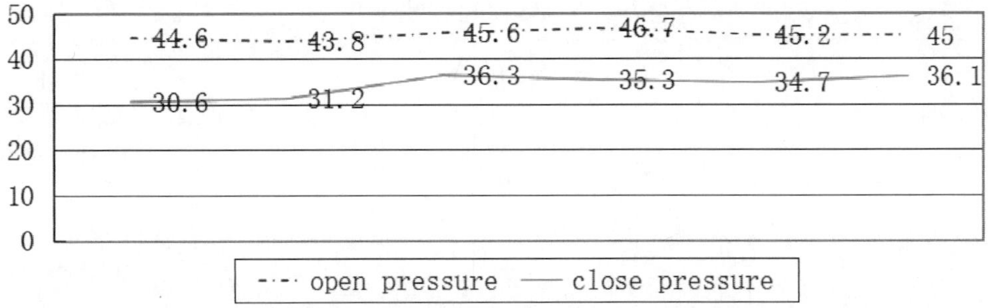


**Figure 6.** The real time data of opening and closing pressure test

Conclusion: The automatic detection system can satisfy the need of opening and closing pressure test, and reduce the workload.

*4.2.The characteristic test of the flow-Pressure*

The purpose of the flow-pressure characteristic experiment was to investigate the resistance of the diverter valve at different flow rates. By adjusting the flow pump, made the pressure transmitter to zero, and then, increased the flow of the pressure on the test piece by 50ml/h, 45ml/h, 40ml/h, 35ml/h and so on, until to 5ml/h decline. According to the experimental data, the pressure data obtained at different flow rates were consistent with the flow characteristics of the shunt. Peristaltic pump can meet the requirements of the minimum 5ml/h accuracy required by the experiment, and can easily control the flow rate adjustment from 5ml/h to 50ml/h.

Conclusion:The using of the pump to drive multiple pump head of the peristaltic pump for micro flow control can not only meet the flow - pressure test project accuracy requirements, and multi-channel at the same time to improve the detection efficiency.

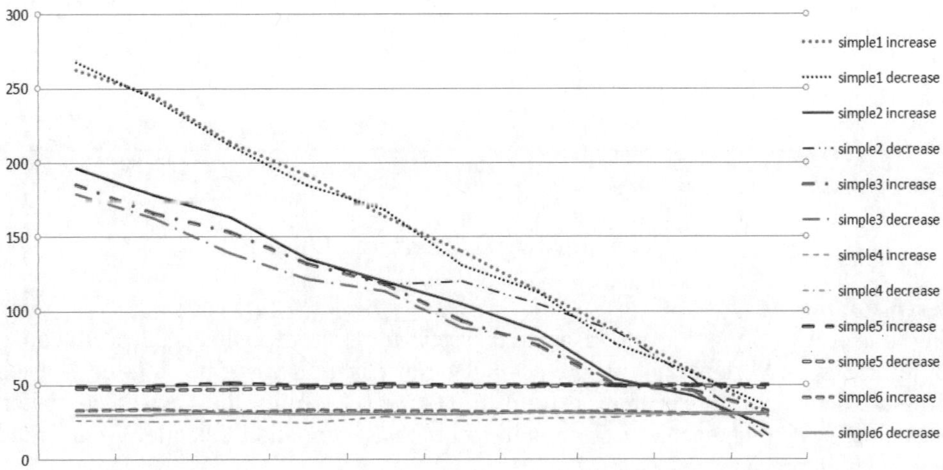


**Figure 7.** The characteristic test of the flow-Pressure

*4.3.Test of the back flow*

Using a pump,having three channels, taking 3 pieces of shunt, at the same time, does the experiment. As shown in the table 1, the back-flow pressure value was measured, found from the data obtained, the reverse pressure up to 500mm$H_2O$,back-flow numerical value only 0.02mL can be regarded as the reverse closure, all test pressure values were in line with the characteristics of reverse flow diverter, and detected the reverse flow of artificial observation was difficult to find the micro pressure detection system can detect 0.02mL, project flow characteristics can meet the requirement of automatic verification.

**Table 1.** Test of the back-flow

| Simple Number | Static Pressure | | | Unit |
|---|---|---|---|---|
| | 10 | 50 | 100 | Pressure: $mmH_2O$ |
| 1 | 0 | 0 | 0 | |
| 2 | 0 | 0.02 | 0.02 | Back -flow: mL |
| 3 | 0 | 0.01 | 0.01 | |

Conclusion: In the experiment, the micro pressure sensor not only meets the needs of the experiment, but also goes beyond the traditional detection accuracy.

*4.4.Long term stability test*

By 20mL/h flow,did a long term stability test.Record pressure versus time curve.As shown in the figure, within 30 days of the measured time, the pressure was stable, the work was normal, the actual test showed that the system can complete the long-term stability of the experimental task,and achieve automatic and efficient detection requirements.

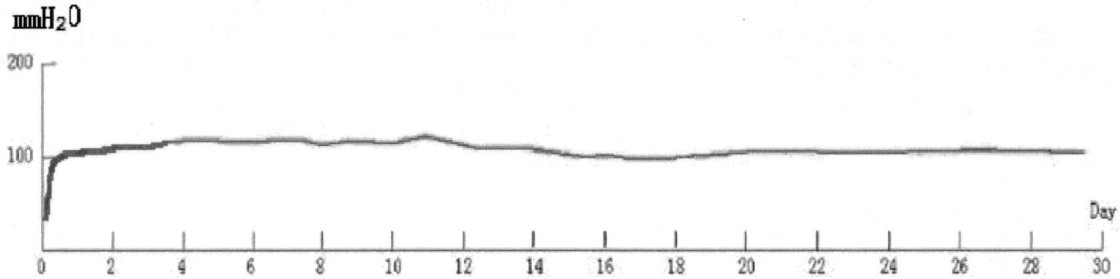


**Figure 8.** Long term stability test

## 5. Research summary

By testing, it is proved that automatic inspection system for hydrocephalus shunt well realizie four test tasks. In terms of micro flow control, the accuracy can reach at least 5mL/h. And in terms of the micro pressure detection ,the accuracy can up to $0.01mmH_2O$. With the help of computer detection technology , the system can achieve the requirements that automation detection, multi-function, high efficiency and high precision .

## Reference

[1] Zhenan Zou,Dengbin Zhang,Hongyuan Guan. Application of hydrocephalus shunt to lateral ventricle,Treatment of hydrocephalus with skin cavity shunt[J].Journal of Mudanjiang Medical College,2012,(02):41-42.
[2] Shengguo Wan. Study on the control system of high pressure micro flow constant current pump[D]. Nanjing University of Aeronautics and Astronautics.2012.
[3] Jiangtao Pang,Changqing Zhan,Zhimin Tan. Journal of functional materials and devices[J]. 2016, (02): 30-33.
[4] Qian ying Sun ,The application of peristaltic pump[J].Nursing Journal,2016,(08):35.
[5] Xianjun Liu,The principle and application of peristaltic pump[J], Fluid machinery,2015, (12):38- 40.

IOP Conf. Series: Materials Science and Engineering **224** (2017) 012052 doi:10.1088/1757-899X/224/1/012052

# Finite Element Analysis of Absorbable Sheath to Prevent Stress Shielding of Tibial Interlocking Intramedullary Nail

**Yansheng Dong[1], Yongqing Wang[2], Limin Dong[1], Peng Jia[3] and Fengcheng Lu[4]**

[1] Tianjin Key Laboratory of the Design and Intelligent Control of the Advanced Mechatronical System, School of Mechanical Engineering, Tianjin University of Technology, Tianjin 300384, China
[2] Dept of Orthopaedics, the Fourth Central Hospital of Tianjin, Tianjin 300140, China
[3] Tianjin University of Traditional Chinese Medicine, Tianjin 300314, China
[4] Dept of Orthopaedics, Tianjin Medical University, Tianjin 300314, China

13820789626@139.com

**Abstract.** The nail with absorbable sheath (AS nail) is designed to reduce the stress shielding effect of internal fixation with interlocking intramedullary nail. In order to verify its feasibility, two types of the finite element models of internal fixation of tibia with the AS nail and the common metal nail (CM nail) are established using the Softwares of Mimics, Geomagic, SolidWorks and ANSYS according to the CT scanning data of tibia. The result of the finite element analysis shows that the AS nail has great advantages compared with the CM nail in reducing the stress shielding effect in different periods of fracture healing. The conclusion is that the AS nail can realize the static fixation to the dynamic fixation from the early to the later automatically to shorten the time of fracture healing, which also provides a new technique to the interlocking intramedullary nail.

## 1. Introduction

Domestic and foreign scholars believe that the fixation of interlocking intramedullary nail belongs to rigid internal fixation and the disadvantage is the stress shielding effect can be generated [1]. At present, the main solution is to realize the static and dynamic transformation of the interlocking intramedullary nail. The commonly used method in clinic is the screws of one end of the intramedullary nail are removed in the metaphase of fracture healing to make fracture ends produce fretting and stimulate the growth of callus, but this requires patients to undergo surgery again [2].

In order to avoid the defects mentioned above, the nail with absorbable sheath (AS nail) is designed by combining the absorbable sheath with the common metal nail (CM nail) as shown in figure 1. The basic principle is sheathed the absorbable sheath on metal screws using the degradation of absorbable materials to reduce the stress shielding effect. This paper is to verify whether the AS nail can reduce the stress shielding effect of tibial locking nail using the method of finite element analysis (FEA).

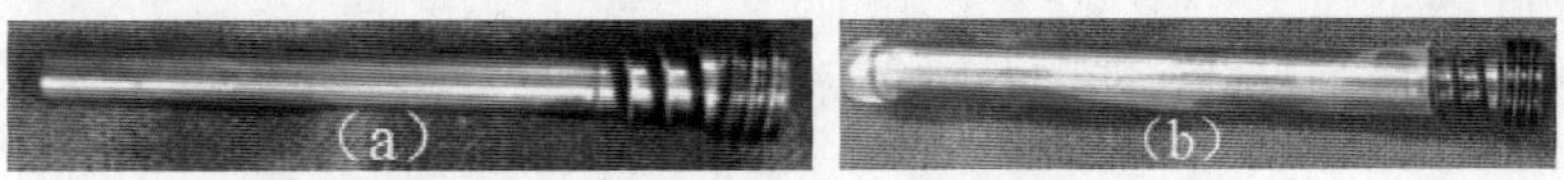


**Figure 1.** (a) the CM nail; (b) the AS nail.

## 2. Materials and methods

### *2.1. Data acquisition*

A healthy adult volunteer is selected, aged 25 years old, 60kg body weight. The LightSpeed VCT is used for continuously scanning the left lower limb from the ankle to the knee joint, and the scanning results are stored in the format of DICOM (Digital Imaging and Communications in Medicine).

### *2.2. Establishment of finite element model of tibial internal fixation*

*2.2.1. Establishment of the 3D model of middle tibial fracture with callus.* First of all, the CT scanning data of tibia is imported into the Mimics. The triangular surface of tibia is calculated after threshold setting (226-2311HU), region growing and holes filling. And the NURBS (Non-Uniform Rational B-Splines) surface is generated by surface thinning, contour editing and surface fitting in Geomagic. Then the 3D solid model is generated in SolidWorks. Finally, the transverse fracture of the 1.5mm [3] is removed by transverse stretching in the middle, and the bony callus of the same size is lofted in the gap. So far, the 3D model of middle tibial fracture with callus is established.

*2.2.2. Establishment and assembly of the model of internal fixation.* In this study, the diameter of CM nail is 4.0mm and 5.0mm; the inner and outer diameter of the absorbable sheath is 4.0mm and 5.0mm. The AS nail is got by combining the absorbing sheath and the CM nail with a diameter of 4.0mm. All the internal fixation devices are established in SolidWorks including the absorbable sheath, which is divided into A, B, C three parts because of the different absorption rate in bone and tissue fluid. The thicknesses of them are calculated by the absorption rate [4] in Table 1. Then the reamed tibial model with screw holes is established named no fixed group after the Boolean operation as shown in Figure 2a. Finally, models of internal fixation are assembled as shown in Figure 2b and Figure 2c.

**Table 1.** Absorption rate and thickness of the absorbable sheath.

| Grouping(months) | Partition | 1 | 3 | 6 | 9 | 12 |
|---|---|---|---|---|---|---|
| Absorbility | B | 0.51 | 2.02 | 4.79 | 12.64 | 26.49 |
| (%) | A, C | 0.53 | 2.68 | 6.57 | 16.58 | 30.66 |
| Thickness | B | 0.9949 | 0.9798 | 0.9521 | 0.8736 | 0.7351 |
| (mm) | A, C | 0.9947 | 0.9732 | 0.9343 | 0.8342 | 0.6934 |

*2.2.3. Parameter setting and meshing.* The models of tibial internal fixations (a total of 15) are imported into ANSYS respectively and given the following assumptions [5]: The components of the models are homogeneous, continuous and isotropic linear elastic materials, and tibia is totally composed of cortical bone. The material properties of callus and absorbable sheath in different periods of fracture healing [4, 6] are listed in Table 2. Among them, the material of absorbable sheath is poly l-lactic acid (PLLA); the interlocking nail and CM nail are Ti6Al4V. Then the 10 node tetrahedral element solid 92 [7] is selected to mesh the model which is shown in Figure 2d. The material properties [4, 8], the number of nodes and elements in the rest of the model are shown in Table 3.

**Table 2.** Material properties of callus and absorbable sheath.

| Grouping(months) | Material properties | 1 | 3 | 6 | 9 | 12 |
|---|---|---|---|---|---|---|
| Callus | Young's modulus(MPa) | 2 | 800 | 1000 | 6000 | 17000 |
| tissue | Poisson's ratio | 0.167 | 0.3 | 0.3 | 0.3 | 0.3 |
| Absorbable | Young's modulus(GPa) | 5.23 | 4.14 | 1.88 | 0.42 | 0 |
| sheath | Poisson's ratio | 0.3 | 0.3 | 0.3 | 0.3 | 0 |

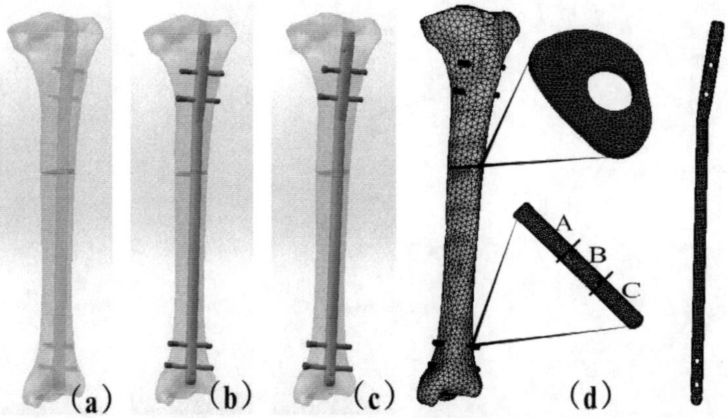


**Figure 2.** (a), (b), (c) are the models of tibial internal fixation with no fixed, AS nail and CM nail; (d) is AS nail group after meshing.

**Table 3.** Material properties, nodes and elements in the model.

| Material types | Young's modulus (MPa) | Poisson's ratio | Nodes | Elements |
|---|---|---|---|---|
| Cortical bone | 17000 | 0.3 | 624769 | 256993 |
| Callus | — | — | 12051 | 6699 |
| Interlocking nail | 112000 | 0.3 | 76356 | 50013 |
| AS nail | — | — | 53605 | 37379 |
| CM nail | 112000 | 0.3 | 40144 | 27464 |

### *2.3. Contact conditions*

The absorbable sheath is bonded with the CM nail by the glue command in ANSYS; the relationship of contact between absorbable sheath, interlocking intramedullary nail and tibial screw holes, the callus and the fracture sections are bonded in AS nail group. All types of contact are set to the bond in CM nail group. The friction between intramedullary nail and medullary cavity is ignored for them.

### *2.4. Loading and boundary conditions*

Human knee joint bears about 85.6% of body weight when human normal walks [9]. The weight of the volunteer is 60kg. So about 250N is loaded on the unilateral tibial plateau vertically in a way of uniformly distributed load. And the degree of freedom of all nodes in the ankle joint is set to zero as a boundary condition. Then the three groups are calculated after setting load step and other parameters.

## 3. Results

### *3.1. Average stress of callus in different periods of fracture healing*

The stress nephogram of callus is obtained according to the results of FEA as shown in Figure 3. The average stress of callus is calculated based on the stress values of all nodes in callus at 1, 3, 6, 9 and 12 months respectively, and the results are shown in Table 4, in which showed that the average stress of callus of the AS nail group is higher than that of the CM nail group in different periods of fracture healing. The results indicated that the degradation of absorbable sheath reduced the gap of fracture ends, which caused the callus to produce larger stress and reduced the stress shielding effect.

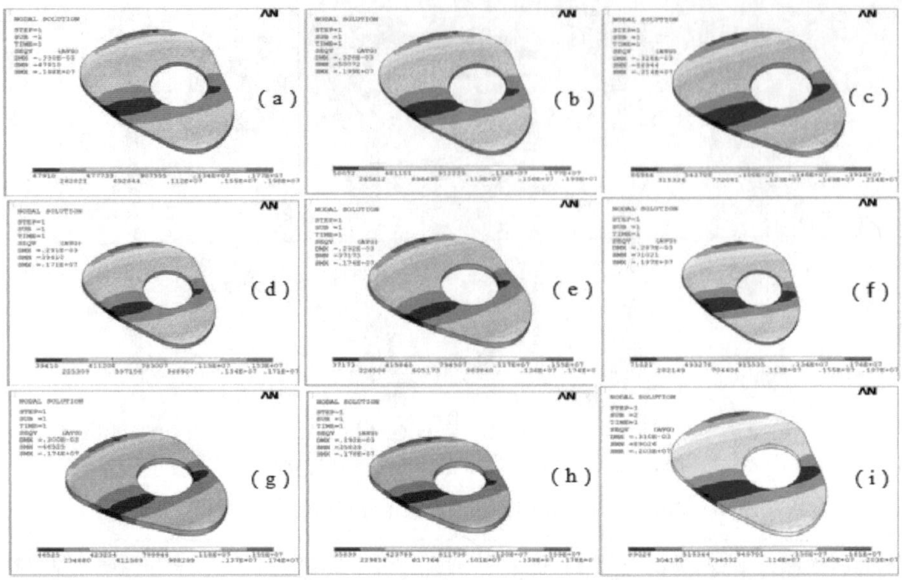


**Figure 3.** (a, b, c), (d, e, f), (g, h, i) are the no fixed group, the CM nail group and the AS nail group of the stress nephograms of callus at 3, 6, 9 month respectively.

**Table 4.** Average stress of callus in different periods of healing (Pa).

| Grouping(months) | 1 | 3 | 6 | 9 | 12 |
|---|---|---|---|---|---|
| No fixed group | 58613.3 | 799875.8 | 802293.4 | 903767.7 | 1285344.1 |
| CM nail group | 34050.7 | 613306.9 | 638626.0 | 690224.5 | 959393.0 |
| AS nail group | 40944.8 | 687347.7 | 719961.9 | 791171.0 | 1014994.6 |

*3.2. Stress shielding rate of callus caused by internal fixation*

The stress shielding rate is used for representing the stress shielding effect in orthopaedics, and the two groups are calculated by the equation (1) [10]. The results are shown in Figure 4.

$$\eta = (1 - \sigma_{\text{fixed}}/\sigma_{\text{no fixed}}) \times 100\% \quad (1)$$

where the $\eta$ represents the stress shielding rate, the $\sigma$ represents the average stress of callus.

The analysis showed that the stress shielding rate of AS nail group is significantly lower than that of CM nail group in different periods of fracture healing. And most obvious, the reduction percentage is about 50% compared with the CM nail group at 6 month. The simulation result shows that AS nail can reduce the stress shielding rate of the interlocking nail, and has more advantages than the CM nail.

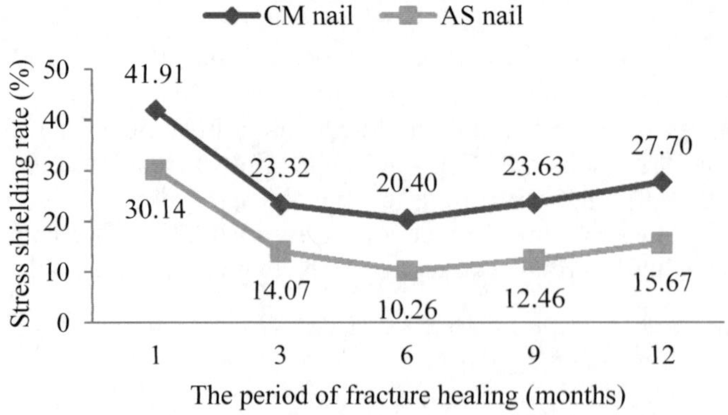


**Figure 4.** Stress shielding rate in different periods of healing.

## 4. Conclusion

The result shows that the average stress of callus of the AS nail group is always higher than that of the CM nail group without fretting effect in the different periods of fracture healing.

The average stress of callus in AS nail group is 20% and 12% higher than that of CM nail group in the first month and the third month, 12% higher in the sixth month, 14% and 6% higher in ninth month and the twelfth month. These results indicate that the interlocking intramedullary nail with AS nail has high stress, high strength and good healing quality. However, stress shielding rate instead, AS nail group is lower than that of CM nail group, in the early low 28% and 39%, in the medium low 49%, in the late low 47% and 43%. These results demonstrate that AS nail can reduce the stress shielding effect of tibial interlocking intramedullary nail on fracture ends and promote fracture healing.

These results also provide a theoretical basis for the subsequent animal experiments, and lay a solid foundation for the clinical application of AS nail as soon as possible.

## Acknowledgments

This study is supported by the Science and Technology Foundation of Tianjin Municipal Health Bureau (Nos.15KG122). And thanks for Tianjin Jinxingda Industrial Co., Ltd. provides related data.

## References

[1] Jr J C A, Lindsey R W, Hipp J A, et al 2008 *Clinical Biomechanics*. **23** 839.
[2] Samiezadeh S, Tavakkoli A P 2015 *Journal of Biomechanical Engineering*. **137** 307–315.
[3] Cheng B, Ding Z, Yao X, Chen B, et al 2016 *Chin J Bone Joint Injury* 839-842.
[4] Peng W, Zheng W, Shi K, et al 2015 *Biomedical Materials* **10** 065007.
[5] Yu H, Li S, Zhao C, Yan J 2013 *Chinese Journal of Tissue Engineering Research* 7571-7580.
[6] Byrne D P, Lacroix D, Prendergast P J 2011 *Journal of Orthopaedic Research* **29** 1496-1503.
[7] Chen L, Zhang J 2014 *Computer simulation* 280-283.
[8] Miyoshi S, Takahshi T, Ohtani M, et al 2002 *Clin Biomech* **17** 521-525.
[9] Trent PS, Wolf B.Ligament *Essential Basic Sciences for Orthipaedics* **14** 94-109.
[10] Liu X, Chen L 1995 *Chinese journal of traditional traumatology & orthopedics* 14-17.

IOP Conf. Series: Materials Science and Engineering **224** (2017) 012053 doi:10.1088/1757-899X/224/1/012053

# Calculations of the Acceleration of Centrifugal Loading on Adherent Cells

**Kang Chen, Yang Song, Qing Liu and Chunqiu Zhang**

Tianjin Key Laboratory of the Design, Intelligent Control of the Advanced Mechatronical System, School of Mechanical Engineering, Tianjin University of Technology, Tianjin 300384, China

13012251281@126.com

**Abstract.** Studies have shown that the morphology and function of living cells are greatly affected by the state of different high acceleration. Based on the centrifuge, we designed a centrifugal cell loading machine for the mechanical biology of cells under high acceleration loading. For the machine, the feasibility of the experiment was studied by means of constant acceleration or variable acceleration loading in the Petri dish fixture and/or culture flask. Here we analyzed the distribution of the acceleration of the cells with the change of position and size of the culturing device quantitatively. It is obtained that Petri dish fixture and/or culture flask can be used for constant acceleration loading by experiments; the centripetal acceleration of the adherent cells increases with the increase of the distance between the rotor center of the centrifuge and the fixture of the Petri dish and the size of the fixture. It achieves the idea that the general biology laboratory can conduct the study of mechanical biology at high acceleration. It also provides a basis for more accurate study of the law of high acceleration on mechanobiology of cells.

## 1. Introduction

With the development of aerospace industry, the performance of fighter aircraft has been greatly improved. As one of the most common and strongest stress factors during the process of fighter flight, the high positive acceleration (forward acceleration, + Gz) produced by the modern high performance fighter has the characteristics of rapid growth rate (6G/s), high +Gz value (+9Gz) and long acting time (15~45s), which has exceeded the physical limits of the human body [1]. Research shows that repeated high positive acceleration (+Gz) exposure of the brain can produce similar effects of ischemia and reperfusion, and thus cause brain injury in the continuous high acceleration condition [2]. Li Wenbing et al. [3] used adult male Wistar rats for testing, and the growth rate of +Gz was 1G/s, the peak values were +6Gz and +10Gz, the peak time was 3min, 30 min interval, and repeated 5 times. It was found that the liver cells were arranged in disorder, irregular in shape, and the cell's space was not clear, and the +10Gz group was more significant than the +6Gz group after loading. The effects of shear stress on the morphology of cells are widely founded in organisms. Liu et al. [4] applied shear strength of 1.2 Pa, 1.6 Pa and1.9 Pa to osteoblasts of adult rats, 1 hour later, they found the morphology of cells was spindle shaped in the two groups of 1.6 Pa and 1.9 Pa. Furthermore, the results of quantitative analysis of cell morphology index also showed that the overall morphology of the cells can be extended along its long axis under suitable shear stress, whose arrangement direction also occurred from the random positioning at the beginning to achieve unity in the end.

Over the last ten years, the biomechanical research has reached the level of cell and molecule, and gradually formed a new research field, "mechanobiology", which is the study of the effects of mechanical environment on the health, disease and injury of organisms. It studies the mechanism of mechanical signal perception and response to elucidate the relationship between the mechanical processes and biological processes such as growth, reconstruction, adaptive change and repair, then to develop a new therapeutic or diagnostic technology [5].

In this paper, we design and manufacture a kind of centrifugal loading machine, therefore, the current culture conditions and the commonly used culture cell containers in the general laboratory can be used to carry out centrifugal high acceleration loading on the cell. In addition, we can describe the change of the high acceleration with the change of cell's position by changing the size of the high acceleration loader quantitatively. The above study provides a new idea and theoretical basis for studying the mechanical and biological characteristics of cells.

## 2. Design and application of centrifugal loading machine

Mechanical factors exist in all the microenvironment of cells, when we construct an engineered tissue in vitro, the lack of mechanical stimulation can lead a result that cells of seed can't play its function fully, thus, the biological activity of the tissue is low. However, the bioreactor can simulate a variety of mechanical environments easily, so as to promote the function of seed cells in biological materials [6].

Centrifugal high acceleration loading machine, which is a common instrument used in the laboratory to study the high acceleration mechanical environment, and the device is placed indoor, for the device has the advantages of good working environment, small occupied space, accurate control of the dynamic experiment process, and its applicability is very strong.

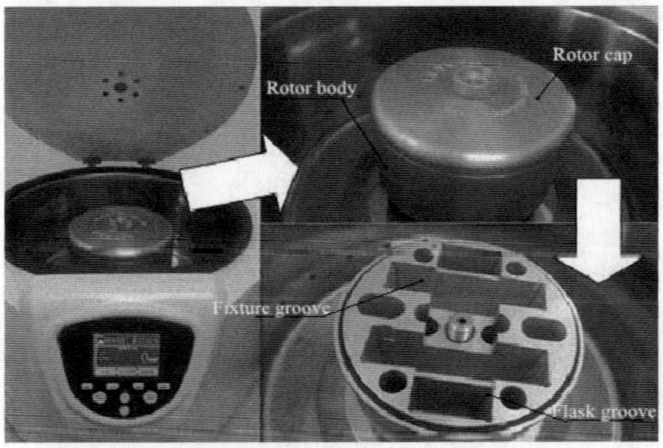


**Figure 1.** Centrifuge rotor.

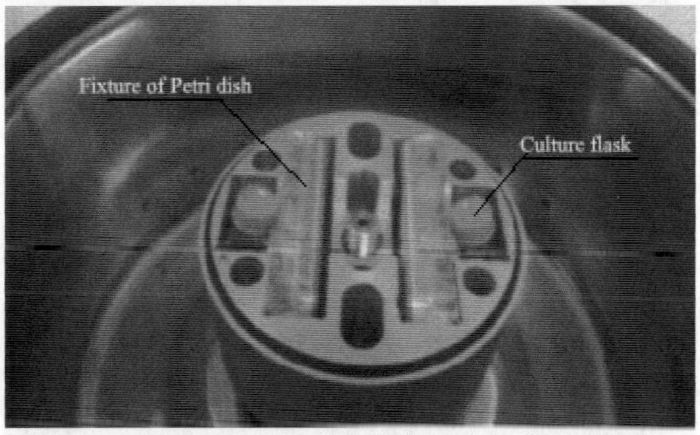


**Figure 2.** Fixture of Petri dish and Culture flask.

The centrifugal high acceleration loading machine is mainly composed of a rotary motor, a principal axis, a rotor, an inner shell of a centrifugal loading machine and a casing. The rotor is mainly composed of a rotor body and a rotor cap, which is made of aluminum alloy for aviation. The rotor body is provided with four rectangular grooves, which are mutually corresponding to each other. The grooves are used for placing the Petri dish fixture and the culture flask (Figure 2). The overall structure of the centrifuge rotor is shown in Figure 1. We have to balance the rotor after the processing is completed, which is the so-called centrifugal balancing, and centrifugal balancing refers to the whole rotor in geometry symmetry and mass balance, thus to ensure homogeneous distribution of the rotor load [7]. Then we carry out the loading experiment.

We used grey PVC to manufacture the Petri dish fixtures (Figure 3). There are six grooves in one fixture body (Figure 3A). Six Petri dishes whose specifications are 35mm can be fitted with it accordingly. We sealed the Petri dish with medical tape to prevent the leakage of the culture medium. The sealed Petri dish can be placed in the Petri dish fixture, and thus can ensure that the experiment can be carried out smoothly. Then we cover the lid of Petri dish fixture and put it into the groove of the rotor body.

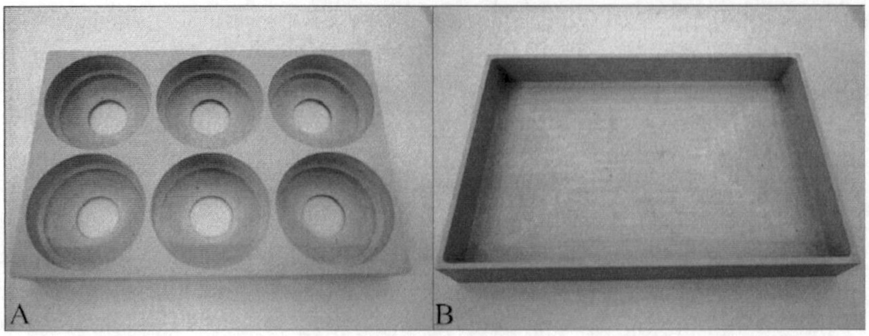


**Figure 3.** Fixture of Petri dish. (A):Fixture body (B):Petri dish fixture lid

In this paper, the Petri dish can be loaded on the Petri dish fixture directly; at the same time, it can also be used to test the acceleration of cells under different conditions when the cells were loaded with centrifuge, which makes the study of mechanical biology in the condition of high acceleration more convenient.

## 3. Study on the centripetal acceleration of cells in different positions

The Petri dish is in the same plane in the fixture, and the distribution of the adherent cells is different in different positions. If the cells are loaded with centrifugal high acceleration at the position of point A in Figure 6, the centripetal acceleration of the cell ($a_{n1}$) was perpendicular to the culture dish. If the cells are loaded at point B in Figure 6, the centripetal acceleration of the cell ($a_{n2}$) can be decomposed into the acceleration along the inner wall of the culture dish ($a_1$) and the acceleration perpendicular to the inner wall of the culture dish ($a_2$). We can change the size and position of the fixture to describe the change of the acceleration of the cells, which provides a reference for the future study of mechanical biology of cells.

### *3.1. Change of centripetal acceleration*

We can see that the centripetal acceleration of cells is different in different locations (Figure 4). With changing the distance between the center of the rotor and the fixture ($L_1$) or the distance between A and B ($L_3$), the centripetal acceleration of the cells in different positions can be calculated quantitatively.

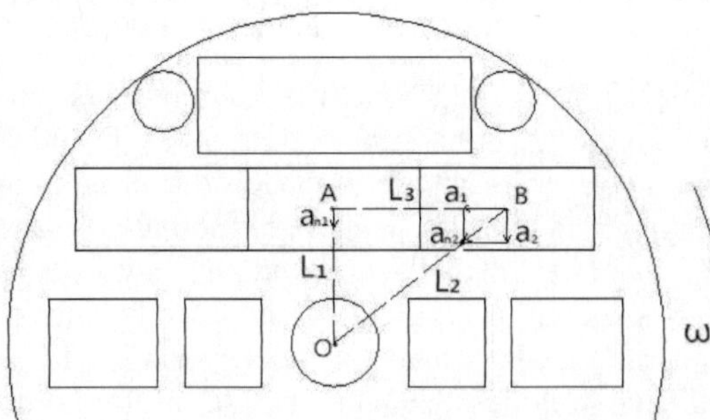


**Figure 4.** Sketch of centrifuge.

In a rotating state, the magnitude of centripetal acceleration of a point is

$$a_n = r\omega^2 \tag{1}$$

Among them, ω represents angular velocity, we take ω=80rad/s, $L_3$=41.5mm, and we know

$$L_2 = \left(L_1^2 + L_3^2\right)^{1/2} \tag{2}$$

We take the range of $L_1$ from 10mm to 50mm. The graph of $a_{n1}$, $a_{n2}$ curve with the change of $L_1$ is shown in Figure 5. We can know that $a_{n1}$ and $a_{n2}$ have corresponding changes with the increase of $L_1$, and $a_{n1}$ is linear increase, $a_{n2}$ is nonlinear increase. With the change of distance, the direction and the magnitude of the acceleration of the cells in different positions are changed, and with the increase of $L_1$, the gap between them has a narrowing trend.

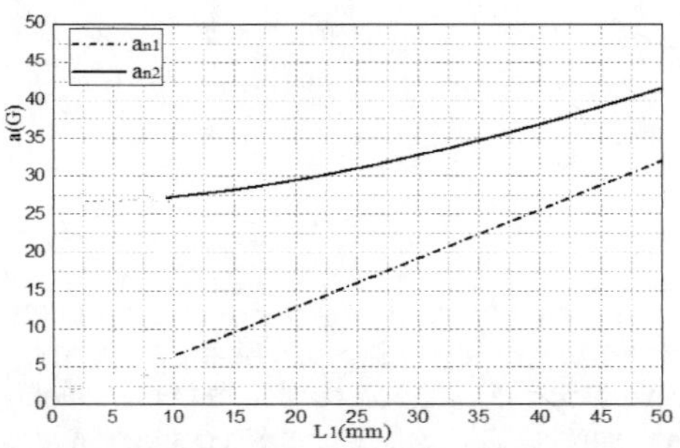


**Figure 5.** The curve of $a_{n1}$ and $a_{n2}$ with the change of $L_1$.

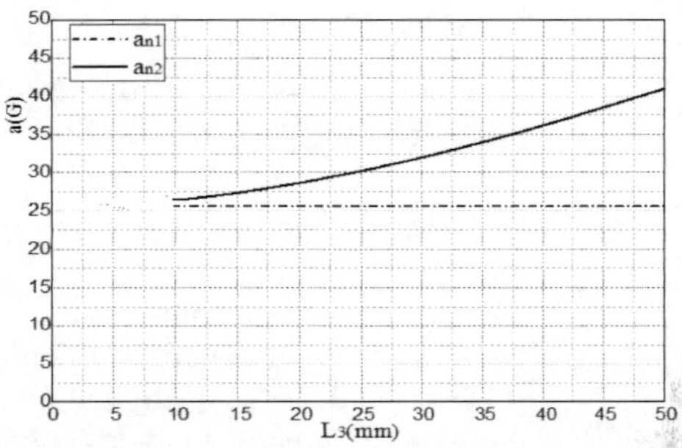


**Figure 6.** The curve of $a_{n1}$ and $a_{n2}$ with the change of $L_3$.

With the change of the fixture's size, the distance between the two points of A and B will also change. We take the range of $L_3$ from 10mm to 50mm, and $L_1$=40mm, then $a_{n1}$ will remain unchanged, $a_{n2}$ will change with the change of $L_3$, the changed curve is shown in Figure 6. With the increase of $L_3$, the gap between them has an increasing trend.

## 4. Conclusion

The dish fixture and/or the flask could be used for the constant acceleration ((0-40) ×g) or variable acceleration loading of the centrifugal loading machine. The Petri dish and/or the culture flask are in the same plane in the fixture, and the distribution of the adherent cells in different positions is different. The centripetal and tangential acceleration of the cell increases with the increase of the vertical distance between the fixture and the rotating center. In this paper, we describe the centripetal acceleration of the cells with the change of position quantitatively, and this provides the basis for a more accurate study of the mechanical and biological law of high acceleration.

## Acknowledgments

The project was partly supported by the National Natural Science Key Foundation of China (Nos.11432016) and the National Natural Science Foundation of China (Nos. 11672208, 11572222).

## References

[1] Zonglai Jiang 2017 *Advances In Mechanics*. **47** 1-24

[2] Peng Li, Weidong Li, Qingping Zhao and Shuhai Zhang 2011 *Space Medicine & Medical Engineering*. **24** 69-71

[3] Liping Jiang, Xiaoyu Wang, Yunyi Li, Wenge Tong, Ping Yao, Nan Shen and Moran Chen 2003 *Journal of Jilin University (Medicine Edition)*. **29** 528-529

[4] Wenbing Li, Yalin Kong, Xiaojun He, Gang Zhao, Huaicheng Zhu, Shen Hu, Lei Liu and Hongyi Zhang 2014 *Medical Journal of Chinese People's Liberation Army*. **39** 240-244

[5] Xiaoli Liu, Xu Zhang and Lee Imshik 2010 *Acta Biochim Biophys Sin*. **42** 195-201

[6] Yan Huang and Yubo Fan 2010 *Journal of Biomedical Engineering*. **27** 920-923

[7] Mo Zhou, Lu Yao, Jian Gao and Zeshi Cui *2015 China medical equipment*. **12** 20-24

IOP Conf. Series: Materials Science and Engineering **224** (2017) 012054 doi:10.1088/1757-899X/224/1/012054

# Implementation of Fixed-point Neuron Models with Threshold, Ramp and Sigmoid Activation Functions

**Lei Zhang**

School of Engineering and Applied Science, University of Regina, Regina, Saskatchewan, Canada, S4S0A2

lei.zhang@uregina.ca

**Abstract.** This paper presents the hardware implementation of single-neuron models with three types of activation functions using fixed-point data format on Field Programmable Gate Arrays (FPGA). Activation function defines the transfer behavior of a neuron model and consequently the Artificial Neural Network (ANN) constructed using it. This paper compared single neuron models designed with bipolar ramp, threshold and sigmoid activation functions. It is also demonstrated that the FPGA hardware implementation performance can be significantly improved by using 16-bit fixed-point data format instead of 32-bit floating-point data format for the neuron model with sigmoid activation function.

## 1. Introduction

Although the calculation speed of computer hardware has been growing exponentially for decades, human brain is still far more efficient than any sequential computer in terms of power consumption per calculation. A human brain has 100 billions of neurons and over 100 trillions of connection, forming the largest and most complex parallel architecture, even a tenth of which is hard to simulate. Recent research shows the brain simulation of 20,000 neurons and 51 billions synapses on an ARM processor at speeds close to biological real-time[1]. This is an infinitesimal fraction of the human brain. Parallel computing devices such as FPGA can better represent the parallel architecture of the neural network and provide hardware acceleration in order to simulate a larger brain neural network. There has been a number of hardware-based neuron models published in recent research. An FPGA implementation of a neuron model with 64-bit double precision floating point data format is presented in [2], and a second-order piecewise nonlinear approximation function was used to simplify the calculation. An FPGA implementation of a single artificial neuron using various activation functions and XOR Gate is presented in [3]. A high accuracy adaptive exponential integrated neuron model is presented in [4]. An efficient neural architecture for FPGA-based spiking neural networks is presented in [5]. An FPGA implementation of Hodgkin-Huxley Neuron model is presented in [6]. Sigmoid function is the most important and commonly used activation functions in an artificial neural network (ANN) architecture [7]. There are several approximation methods used for simplify the implementation of sigmoid function, such as piece-wise linear (PWL), look-up table and CORDIC. The FPGA implementation of bipolar sigmoid activation function is presented in [8]. The previous research commonly represents neuron models with floating-point data format to achieve high precision, which is however inefficient in FPGA hardware implementation. In this paper, 16-bit fixed-point data format is used for the design of a single neuron model with three different types of activation functions: threshold, ramp and sigmoid.

## 2. General Neuron Model for ANN

An ANN is a network of interconnected neurons arranged in multiple layers, including one input layer, one output layer and a number of hidden layers. A neuron is the basic element in the construction of an ANN. The model of a single neuron is illustrated in Fig.1. A neuron has multiple synaptic inputs with individual weight associated with each input. A simple 3-layer ANN architecture is illustrated by Fig.2, with 2 inputs, 4 neurons in the hidden layer and 1 output.

The calculation of a neuron output can be summarized into 3 sequential steps. First is to multiply the inputs with synaptic weights to calculate the products; Second is to add the products to calculate the weight sum; last is to apply a selected activation function to the summed weights and generate the neuron output. A general mathematic representation of an individual neuron within an ANN architecture is shown by equation (1).

$$a_j^l = \sum_{i=1}^{N_l} \omega_{j,i}^l x_i + b_{j,0}^l, \quad j = 1,2,\dots,N_{l+1}, \quad y_j^l = f_l(a_j^l) \tag{1}$$

where $N_l$ is the number of neurons at $l$-layer. $j$ is the index of the neurons at $l$-layer. Each hidden neuron j receives the output of each input neuron $i$ from the input layer multiplied with a weight of $\omega_{j,i}^l$. The sum of all weighted inputs is used by an activation function $f_l$ to produce the output of the hidden layer neuron and feed it forward to the output layer. A similar weighted sum is generated for each output neuron. $b_{j,0}^l$ is the bias of the $j$th neuron at the $l$th layer, which are added as noise to randomize the initial condition in order to get better chance to converge when training ANN with back propagation algorithm. FPGA is particularly suitable for hardware implementation of the successive multiply-accumulation (MAC) operation of multiple neurons in an ANN.

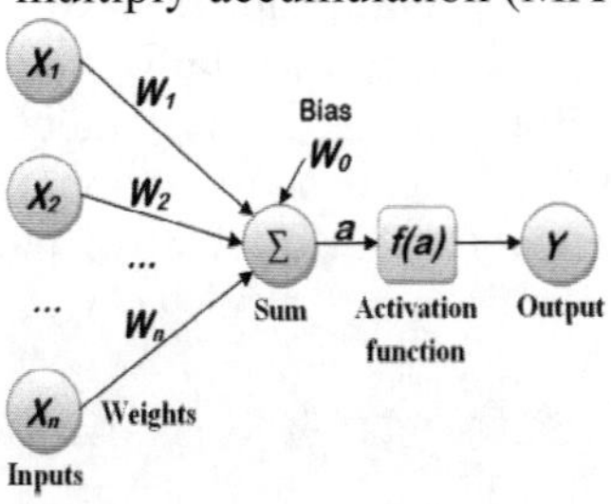


**Figure 1.** Single Neuron

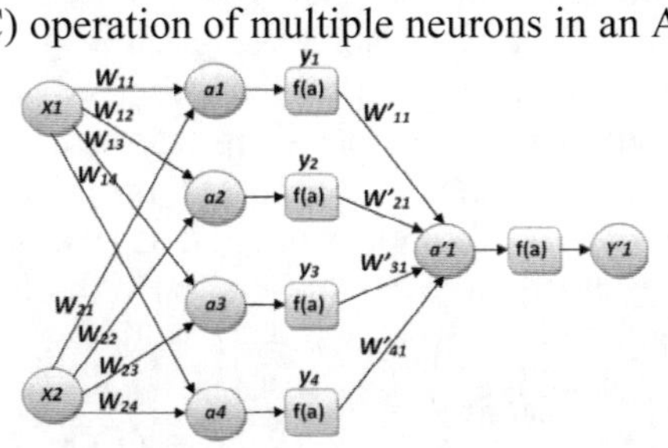


**Figure 2.** A 3-layer ANN

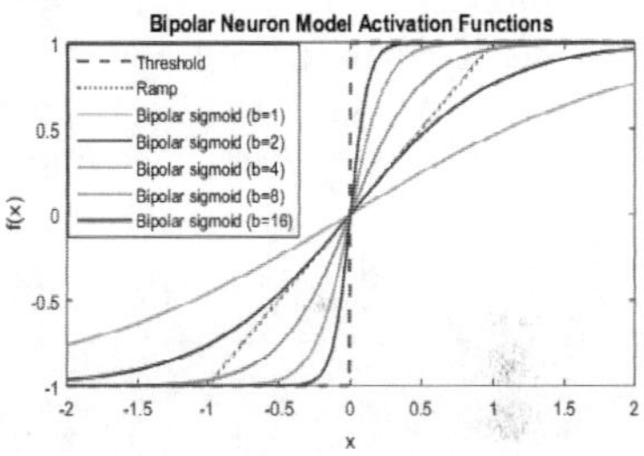


**Figure 3.** Bipolar Activation

## 3. Activation Functions

The accumulated synaptic inputs of a neuron, represented by the summed weights, is transfered to output via an activation function. Common activation functions include threshold, ramp and sigmoid function. These functions can be used to represent either unipolar or bipolar neuron model. Unipolar model has a single polarity, normally positive, while bipolar model has two polarities, both positive and negative. The output of the unipolar activation function ranges from 0 to +1 and is symmetrical to the (0,0.5) point. The corresponding bipolar activation functions are illustrated in Fig.3. The output of the bipolar activation function ranges from -1 to +1 and is symmetrical to the origin.

### *3.1. Threshold Activation Function*

The ramp function is a piece-wise linear function saturated at upper and lower limits. The unipolar and bipolar ramp activation functions are shown in equation (2). The ramp function is used to design ANN for fitting applications. The summed weights in the ANN architecture need to be normalized within the linear range of the activation function in order to represent linear patterns.

$$f_u(x) = \begin{cases} 1.0, & x \geq x_{th} \\ 0.0, & x < x_{th} \end{cases}; \quad f_b(x) = \begin{cases} 1.0, & x \geq x_{th} \\ -1.0, & x < x_{th} \end{cases} \tag{2}$$

IOP Conf. Series: Materials Science and Engineering **224** (2017) 012054 doi:10.1088/1757-899X/224/1/012054

### *3.2. Ramp Activation Function*

The ramp function is a piece-wise linear function saturated at upper and lower limits. The unipolar and bipolar ramp activation functions are shown in equation (3). The ramp function is used to design ANN for fitting applications. The summed weights in the ANN architecture need to be normalized within the linear range of the activation function in order to represent linear patterns.

$$f_u(x) = \begin{cases} 1, & x > 1 \\ x, & 0 \le x \le 1; \\ 0, & x < 0 \end{cases} \qquad f_b(x) = \begin{cases} 1, & x > 1 \\ x, & -1 \le x \le 1 \\ -1, & x < -1 \end{cases} \tag{3}$$

### *3.3. Sigmoid Activation Function*

The unipolar and bipolar sigmoid activation functions are represented by equation (4). The bipolar sigmoid function is also called a hyperbolic tangent sigmoid function or a logistic sigmoid function.

$$f_u(x) = \frac{1}{1+e^{-\beta x}}; \qquad f_b(x) = \frac{1-e^{-\beta x}}{1+e^{-\beta x}} = \frac{2}{1+e^{-\beta x}} - 1 \tag{4}$$

where $\beta$ represents the slope of the sigmoid function. The outputs of sigmoid functions (unipolar and bipolar) approach to the threshold function when the value of $\beta$ increases to the positive infinity ($\beta \to +\infty$). When $\beta$ =0, the outputs is a horizontal line (x axis). Sigmoid function is a continuous non-linear function, which maps an infinite input domain ($-\infty, \infty$) to a finite output domain: (0,1) for unipolar and (-1,1) for bipolar. This property of sigmoid function provides generalization in ANN models to approximate any function. Sigmoid function is biologically plausible and commonly used for ANN design with back propagation training. Back propagation algorithm is based on gradient decent which requires calculating the differential output of the activation function. The sigmoid function is easy to differentiate and its differential function is also continuous. The derivative of sigmoid function is shown in equation (5), which can be simply implemented on an FPGA using a multiplier and a subtracter (2's complement adder).

$$f'(x) = \frac{df(x)}{dx} = \left(\frac{1}{1+e^{-x}}\right)' = f(x)(1-f(x)) \tag{5}$$

## 4. Hardware Models and Implementation

A model-based hardware design approach is employed for the neuron model FPGA implementation. A hardware model is created using Simulink and Xilinx System Generator for threshold, ramp and sigmoid activation function respectively. These models are firstly simulated in Matlab/Simulink software environment. The neuron models are designed with two inputs, which can be easily extended by adding inputs and weights. The multiplier block `Mult' is used for calculating the product of input and weight. The adder block `Add' is used to calculate the summed weights. The input $x_{in}$ is a series of linear increment values range between (-2, 2) for evaluating the outputs of the activation function. The 16-bit fixed-point bipolar threshold, ramp and sigmoid models are shown in Figure 4(a), (b) and (c) respectively. After the design function is verified by the behavior simulation, a design file (netlist) is then exported to generate a VHDL-based Vivado project for FPGA implementation. The neuron models are implemented on a Zeboard with a Zed7020 FPGA device. Xilinx Vivado 2015.4 was used to implement the model-based designs. The FPGA implementation results are shown in Table.1.

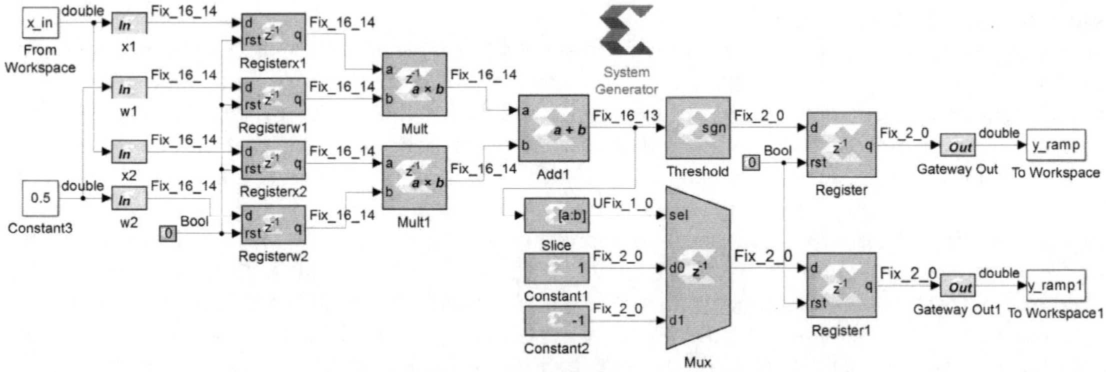


(a) Bipolar Threshold Activation Functions

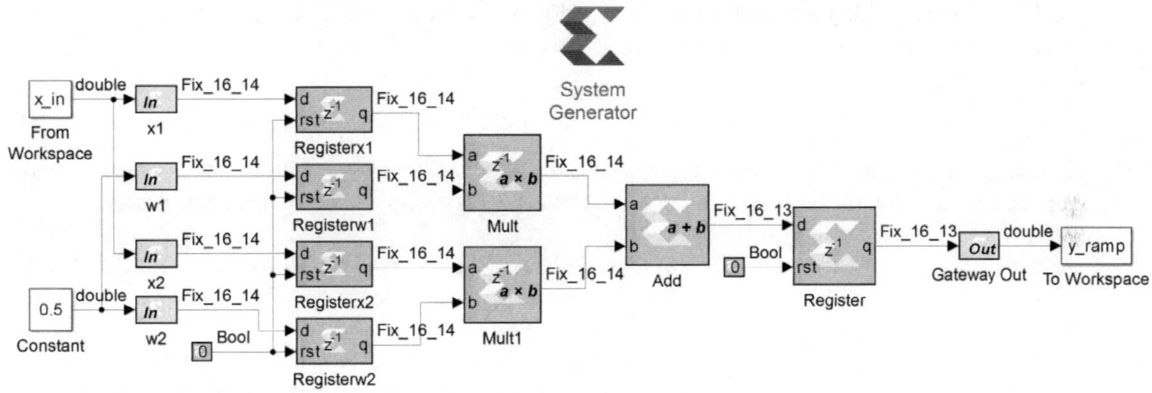


(b) Bipolar Ramp Activation Functions

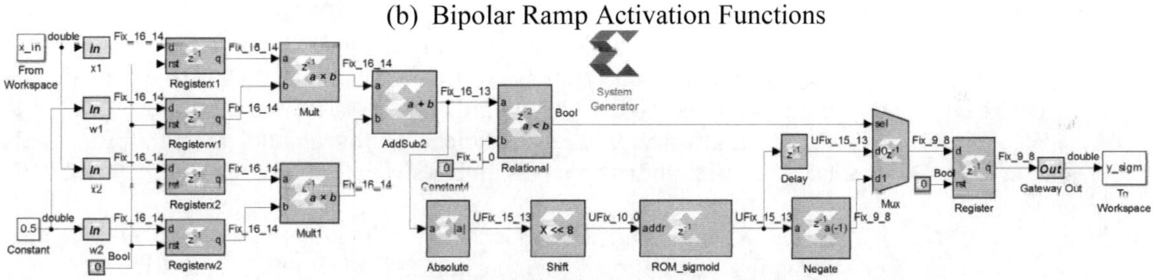


(c) Bipolar Sigmoid Activation Functions

**Figure 4.** 16-bit Fixed-point System Generator Neuron Models

**Table 1.** FPGA Implementation Results.

| Activation Function | Threshold | | | Ramp | | Sigmoid | | | |
|---|---|---|---|---|---|---|---|---|---|
| Data Format | **16-bit Fixed-point** | | | **16-bit Fixed-point** | | **32-bit Floating-point** | | **16-bit Fixed-point** | |
| Option | **Syn** | **Mux** | **Mux(DSP)** | **Mult(logic)** | **Mult(DSP)** | **Mult(logic)** | **Mult(DSP)** | **D-RAM** | **BRAM** |
| Worst Negative Slack (ns) | (-0.269) | 0.034 | 1.195 | (-0.388) | 1.28 | -44.537 | -43.312 | 0.703 | 1.165 |
| Maximum Frequency (MHz) | (97.38) | 100.34 | 113.57 | (96.26) | 114.68 | (18.34) | (18.76) | 107.56 | 113.19 |
| Look-up Table (LUT) | 590 | 590 | 24 | 598 | 32 | 3133 | 2018 | 272 | 97 |
| Registers/Flip-flop (FF) | 98 | 100 | 100 | 112 | 112 | 356 | 356 | 147 | 135 |
| Slices | 192 | 193 | 37 | 196 | 40 | 1001 | 712 | 112 | 52 |
| DSP Elements | 0 | 0 | 2 | 0 | 2 | 15 | 21 | 2 | 2 |
| **Block RAM** | 0 | 0 | 0 | 0 | 0 | 1 | 1 | 0 | 0.5 |
| **Total On-chip Power (W)** | 0.132 | 0.132 | 0.126 | 0.148 | 0.141 | 0.232 | 0.218 | 0.146 | 0.137 |

## 5. Conclusion and Future Research

In conclusion, this paper presented the 16-bit fixed-point FPGA hardware implementation of a single neuron model with three different activation functions: threshold, ramp and sigmoid function. Xilinx System Generator is used to create hardware-based models for the hardware implementation. The performance of the sigmoid neuron model is significantly improved by using 16-bit fixed-point data format and memory based design compared to 32-bit floating point model. The neuron models are further optimized by using FPGA dedicated hardware resource DSP48E1 and block RAM. The maximum operating frequency is increased by more than six times from 18 MHz to 113 MHz. The FPGA logic resource utilization is reduced by 93\% from 712 to 52. The optimized performance achieved by the 16-bit fixed-point models also reflexes the characteristics of biological neurons, which need to maintain simplification in order to deliver efficiency. The simplified neuron model will increase the processing speed for an ANN. The bipolar activation function models can be easily modified for unipolar activation functions. These models will be use in future research for the design and FPGA hardware implementation of ANN topologies in brain simulation and machine learning applications. The fixed-point sigmoid hardware implementation will also be used for the design of large ANN topology.

## References

[1] J. C. Knight, P. J. Tully, B. A. Kaplan, A. Lansner, and S. B. Furber, "Large-scale simulations of plastic neural networks on neuromorphic hardware," Frontiers in Neuroanatomy, vol. 10, p. 37, 2016.

[2] S. Kawamura, M. Saito, and H. Yoshida, FPGA Implementation of Neuron Model Using Piecewise Nonlinear Function on Double-Precision Floating-Point Format. Springer International Publishing, 2016, pp. 620–629.

[3] S. Abrol and R. Mahajan, "Implementation of single artificial neuron using various activation functions and xor gate on fpga chip," in 2015 Second International Conference on Advances in Computing and Communication Engineering, May 2015, pp. 118–123.

[4] A. Makhlooghpour, H. Soleimani, A. Ahmadi, M. Zwolinski, and M. Saif, "High accuracy implementation of adaptive exponential integrated and fire neuron model," in 2016 International Joint Conference on Neural Networks (IJCNN), July 2016, pp. 192–197.

[5] L. Wan, Y. Luo, S. Song, J. Harkin, and J. Liu, "Efficient neuron architecture for fpga-based spiking neural networks," in 2016 27th Irish Signals and Systems Conference (ISSC), June 2016, pp. 1–6.

[6] M. Lu, J.-L. Wang, J. Wen, and X.-W. Dong, "Implementation of hodgkin-huxley neuron model in fpgas," in 2016 Asia-Pacific International Symposium on Electromagnetic Compatibility (APEMC), vol. 01, May 2016, pp. 1115–1117.

[7] A. N. Ide and J. H. Saito, FPGA Implementations of Neural Network. Springer, 2006, ch. 7. FPGA Implimentations of Neocognitrons, pp. 197–224.

[8] I. Sahin and I. Koyuncu, "Design and implementation of neural networks neurons with radbas, logsig, and tansig activation functions on fpga," Elektronika ir elektrotechnika, vol. 120, no. 4, pp. 51–54, Apr. 2012.

IOP Conf. Series: Materials Science and Engineering **224** (2017) 012055 doi:10.1088/1757-899X/224/1/012055

# The Design of Case Products' Shape Form Information Database Based on NURBS Surface

**LIU Xing [2], LIU Guo-zhong [3], XU Nuo-qi [4]and ZHANG Wei-she [1,5]**

[1] School of Engineering Machinery, Chang' an University, Xi'an 710064, Shan Xi, China.

[2] 843492563@qq.com, [3]1582355099@qq.com, [4] 496882236@qq.com,
[5] zwshe@chd.edu.cn

**Abstract.** In order to improve the computer design of product shape design,applying the Non-uniform Rational B-splines(NURBS) of curves and surfaces surface to the representation of the product shape helps designers to design the product effectively.On the basis of the typical product image contour extraction and using Pro/Engineer(Pro/E) to extract the geometric feature of scanning mold,in order to structure the information data base system of value point,control point and node vector parameter information,this paper put forward a unified expression method of using NURBS curves and surfaces to describe products' geometric shape and using matrix laboratory(MATLAB) to simulate when products have the same or similar function.A case study of electric vehicle's front cover illustrates the access process of geometric shape information of case product in this paper.This method can not only greatly reduce the capacity of information debate,but also improve the effectiveness of computer aided geometric innovation modeling.

## 1. Introduction

Since NURBS was proposed,its unique characteristics have been widely recognized and NURBS become the only standard for the Initial Graphics Exchange Specification(IGES) gradually,and so there are more and more correlation studies about NURBS curves and surfaces[1].The application of NURBS surface reconstruction is extensive and it has a very wide range of application in the appearance of the products' three-dimensional model[2] and reverse design[3].At present,there are a number of independent research and development of NURBS surface modeling of the geometric characteristics of the relevant computer aided system[4],which can provide a theoretical reference for the development of complex surface modeling by using programming language.

This paper,based on the analysis of the geometric feature information of the appearance of the product,proposes the unified expression of NURBS curves and surfaces and its simulation presentation methods for the products with the same or similar functions by using the typical image contour extraction software,Pro/E,MATLAB software.And According to the digital characteristics of NURBS curves and surfaces, the information database system is constructed.

## 2. Geometric shape feature extraction

### *2.1. Product morphological features and elements*

An industrial product is a three-dimensional object with a certain geometry and product appearance modeling is based on certain technology and art to achieve.The shape of the product is composed of simple geometric shapes.Although the composition,structure and function of each product are

IOP Conf. Series: Materials Science and Engineering **224** (2017) 012055 doi:10.1088/1757-899X/224/1/012055

different,the basic form of the appearance is the same.The morphological categories of industrial products has figurative form, analog form, symbolic form and abstract form.The geometric shape of the product is formed by different sets of points, lines and surfaces as a result of being decomposed.The line is not only the characteristic element, but also can influence the surface and body feature,so the line is the smallest factor that affects the geometric shape of the product.

### *2.2. Geometric feature line extraction of product modeling*

The geometric shape of the appearance of the product is formed by connecting a plurality of shapes,so the feature line is the key to the shape and style of the product. The contour boundary line is the first characteristic line of the product. The following is an example of the front cover of electric vehicles, which shows that the process of product feature extraction. The geometric shape of the product is formed by different sets of points, lines and surfaces as a result of being decomposed. The line is not only the characteristic element, but also can influence the surface and body feature, so the line is the smallest factor that affects the geometric shape of the product.

By using image contour extraction software to extract feature line of electric vehicle front cover,figure 1 shows the characteristic line of the prototype which is vague but has less error.

The point cloud model of the front shell of the electric vehicle is obtained through three-dimensional scanning,and the grid surface is shown in Figure 2.Then we import it in Pro/E by IGES data format.As shown in Figure 3,referring to figure 1 (b) to construct the cut plane,we extract feature lines in cut plane and reconstruct of spline curves based on extracted feature lines.

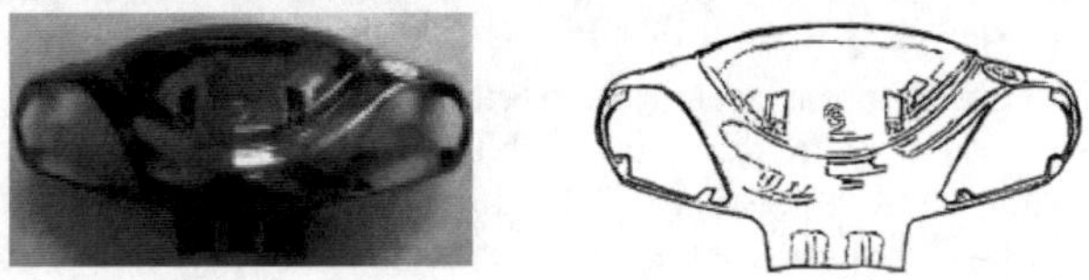

(a) (b)

**Figure 1.** Prototype and characteristic line

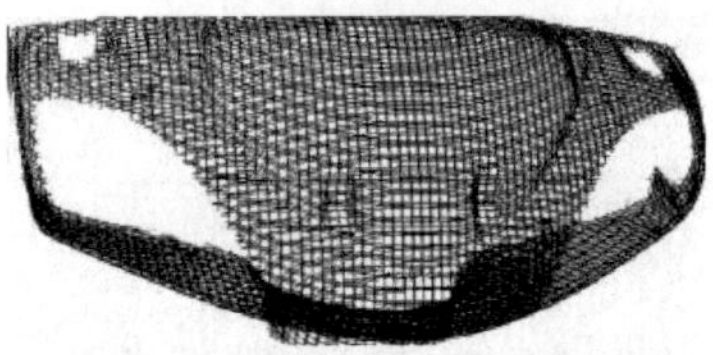

**Figure 2.** Extracting feature lines

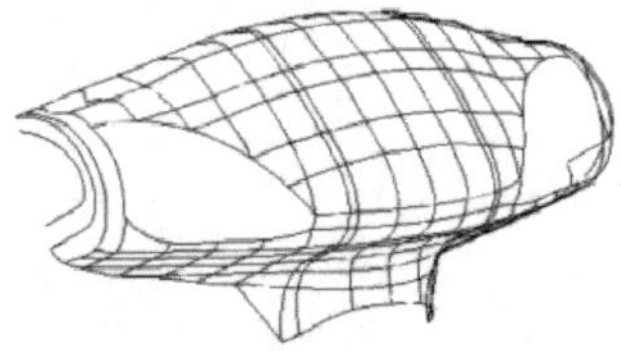

**Figure 3.** Reconstruction spline

## 3. Unified expression of NURBS surface

The appearance of the product can reflect the internal structure to a certain extent.When the products have the same type of internal structure,they also have a similar appearance profile.In order to describe the contour of the product better,it's necessary to summarize the product modeling,and describe the similar geometry in a unified way.

### *3.1 The inverse of the three NURBS interpolation curve*

By using the type value points on the curve,the control process is:①NURBS curve value point $q_i$;②using parametric method to calculate node $u_i$;③compute node vector $\boldsymbol{U}$;④compute its basis function $N_{i,3}(u)$;⑤compute linear equations of control point matrix;⑥get control points $d_i$.

For the three NURBS interpolation curve,we use the method of accumulating chord length to obtain the corresponding parameter of value points.

$$\begin{cases} u_0 = u_1 = u_2 = u_3 = 0 \\ u_{i+3} = u_{i+2} + |p_i - p_{i-1}| \Big/ \sum_{i=1} \sqrt{|p_i - p_{i-1}|} \quad (i=1,2,\ldots,n\text{-}1) \\ u_{n+3} = u_{n+4} = u_{n+5} = u_{n+6} = 1 \end{cases} \tag{1}$$

For the three NURBS open curve,the node is generally 4,and the starting and ending points are between 0 and 1.The spline basis is：

$$N_{i,3}(u) = \frac{u_{i+4}-u}{u_{i+4}-u_{i+1}} \frac{u_{i+4}-u}{u_{i+4}-u_{i+2}} \frac{u_{i+4}-u}{u_{i+4}-u_{i+3}},\ N_{i+1,3}(u) = \frac{u-u_{i+1}}{u_{i+4}-u_{i+1}} \frac{u_{i+4}-u}{u_{i+4}-u_{i+2}} \frac{u_{i+4}-u}{u_{i+4}-u_{i+3}}$$
$$+ \frac{u_{i+5}-u_i}{u_{i+5}-u_{i+2}} \frac{u-u_{i+2}}{u_{i+4}-u_{i+2}} \frac{u_{i+4}-u}{u_{i+4}-u_{i+3}} + \frac{u_{i+5}-u}{u_{i+5}-u_{i+2}} \frac{u_{i+5}-u}{u_{i+5}-u_{i+3}} \frac{u-u_{i+3}}{u_{i+4}-u_{i+3}},$$
$$N_{i+2,3}(u) = \frac{u-u_{i+2}}{u_{i+5}-u_{i+2}} \frac{u-u_{i+2}}{u_{i+4}-u_{i+2}} \frac{u_{i+4}-u}{u_{i+4}-u_{i+3}} + \frac{u-u_{i+2}}{u_{i+5}-u_{i+2}} \frac{u_{i+5}-u}{u_{i+5}-u_{i+3}} + \frac{u_{i+6}-u}{u_{i+6}-u_{i+3}}$$
$$\times \frac{u-u_{i+3}}{u_{i+5}-u_{i+3}} \frac{u-u_{i+3}}{u_{i+4}-u_{i+3}},\ N_{i+3,3}(u) = \frac{u-u_{i+3}}{u_{i+6}-u_{i+3}} \frac{u-u_{i+3}}{u_{i+5}-u_{i+3}} \frac{u-u_{i+3}}{u_{i+4}-u_{i+3}} \tag{2}$$

The solution of the closed curve control points is:

$$a_i = \frac{(u_{i+3}-u_{i+2})^2}{u_{i+3}-u_i},\ b_i = \frac{(u_{i+3}-u_{i+2})(u_{i+2}-u_i)}{u_{i+3}-u_i} + \frac{(u_{i+2}-u_{i+1})(u_{i+4}-u_{i+2})}{u_{i+4}-u_{i+1}},$$
$$c_i = \frac{(u_{i+2}-u_{i+1})^2}{u_{i+4}-u_{i+1}}, e_i = (u_{i+3}-u_{i+1}) q_{i-1} \tag{3}$$

The solution of open curve control points is:

$$\begin{cases} a_1 = 1,\ b_1 = c_1 = 0,\ \vec{e}_1 = \vec{q}_0 + \dfrac{\Delta_3}{3} \vec{q}_0 \\ a_{n-1} = b_{n-1} = 0,\ c_{n-1} = 1, \vec{e}_{n-1} = \vec{q}_n + \dfrac{\Delta_n}{3} \vec{q}_n \end{cases} \tag{4}$$

*3.2. NURBS surface inverse calculation*

NURBS Surface expression is:

$$p(u,v) = \frac{\sum_{i=0}^{n}\sum_{j=0}^{m} \omega_{i,j} d_{i,j} N_{i,k}(u) N_{j,l}(v)}{\sum_{i=0}^{n}\sum_{j=0}^{m} \omega_{i,j} N_{i,k}(u) N_{j,l}(v)} \tag{5}$$

The inverse calculation of NURBS surface control points is the extension of the inverse of the curve control points, which are calculated in two directions, $u$ and $v$.Firstly, the control vertex of the NURBS surface and its power factor[5],the power of parameter $u$ and $v$,and the node vector $\boldsymbol{U}$ are determined.Then calculating $u$ direction in the method of the inverse of the curve.Then,the control points of the curves are given as known quantities and we take the same sequence as the $v$ value of the point of the curve sequence and simulate curve inversely.Finally,the control points of the surface are obtained.

*3.3.The control points inverse calculation of modeling NURBS curve control points inverse calculation*

As shown in Figure 4,we extract the feature points of the front shell of the electric vehicle by using Pro/E.The extracted feature points are regarded as the value points.

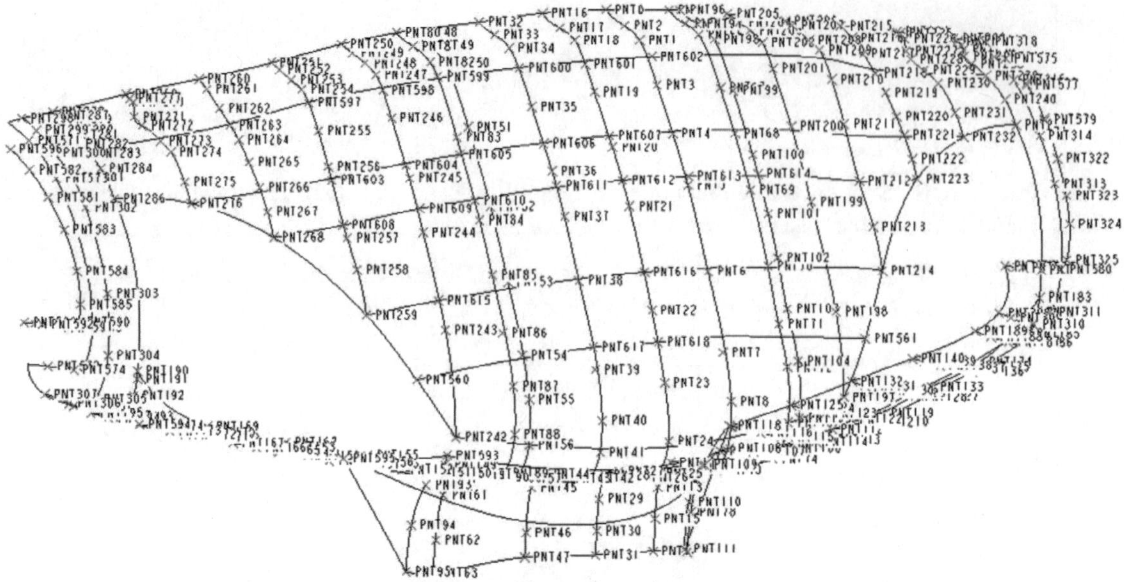

**Figure 4.** Extracting feature points on the feature line

The algorithm of the three NURBS curve control points is illustrated by an example of characteristic line in the Figure 4.The coordinates of the feature points on the feature line are (7.5606,11.4546),(8.2126,11.5759),(8.8613,11.6227),(9.9596,11.1586),(11.0924,10.4693),(12.04, 9.7446),(12.8427,8.9772),(13.2756,8.4897),(13.448,7.8535),(13.117,7.139),(12.4096,6.7871),(11.5 628,6.4084),(10.7712,5.9399),(10.4282,5.5006),(10.3743,4.7424),(10.4042,4.2318).The node vector by using the method of accumulating chord length is U=[0,0,0,0,0522 ,0.1034,0.1973, 0.3017,0.3956,0.4711,0.5344,0.5863,0.6483,0.7105,0.7836,0.8560,0.8999,0.9597,1,1,1,1].And then the control points $d_i$ of the feature line are (7.5606,11.4546),(7.5606,11.4546),(8.2997,11. 5987),(9.0039,11.6927),(10.0282,11.1263),(11.0610,10.5181),(12.0417,9.7828),(12.7255,9.0811), (13.3243,8.523),(13.5395,7.8112),(13.1853,7.0429),(12.3506,6.7889),(11.5676,6.4070),(10.8410,6. 0718),(10.3203,5.4422),(10.3656,4.9381),(10.4042,4.2318),(10.4042,4.2318).The coordinates of control points on other feature lines are obtained in like manner.

## 4. Structure the information database of geometric shape

Because the appearance of the product has a variety of geometric shapes for many cases,it is necessary to construct the relevant geometry information database,which can be used to store the information of the type value points,control points,node vectors and so on.We use Microsoft Office Access,MATLAB and other related software to construct the information database.The flow is:①extracting feature points in Pro/E;②recording feature point coordinates(*PX*, *PY*, *PZ*);③inverse control point coordinates (*DX*, *DY*, *DZ*);④recording control point coordinates (*DX*, *DY*, *DZ*);⑤simulating information base management system;⑥using MATLAB software to simulate.

### *4.1. Designing the information bank for modeling parts' feature lines*

The geometry of the front cover of the electric vehicle is divided into 4 parts,each of which has the same number of points.The same table structure can be designed.Other parts of tables can be constructed and the like.

The numerical calculation of the characteristic line is carried out on the MATLAB software.We need to connect the database and MATLAB,and read the relevant information in the MALAB to complete the calculation and store the results in the information bank.Table 1 shows the result.

### *4.2. Surface simulation based on information database*

IOP Conf. Series: Materials Science and Engineering **224** (2017) 012055 doi:10.1088/1757-899X/224/1/012055

As shown in Table 1,the geometry information of the product modeling is stored in the repository.And then according to the MATLAB software,the surface simulation is carried out,and the surface or curve shape of the model is presented in the form of the effect diagram.As shown in Figure 5 is the front cover of the electric vehicle shell surface in the Pro/E software interception of a section of the characteristic line.By comparing the value of the type and the shape of the feature line,we can get the spline curve which can be used to accurately represent the original feature line of the model.

The characteristic lines of the front cover of the whole electric vehicle are displayed in the same window,and the result is shown in Figure 6.Compared with figure 3,the spline curves constructed by NURBS curve have high similarity.

As shown in Figure 7,MATLAB software reproduces the shape of the surface by reading the corresponding data in the database.Compared with the original surface shown in Figure 8,it has high similarity.

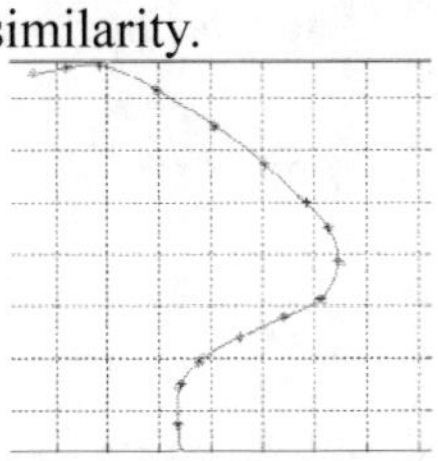

**Figure 5.**Characteristic lines and characteristic points

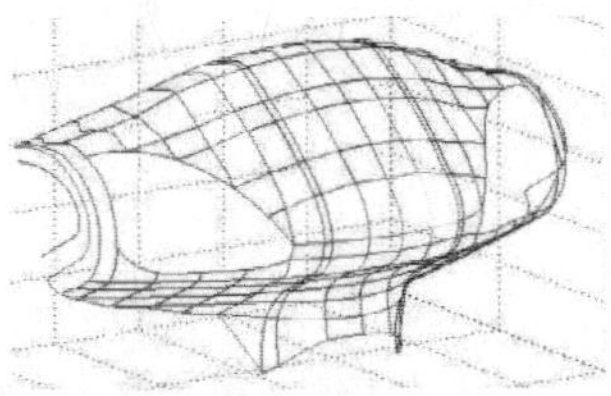

**Figure 6.** *u* and *v* direction interpolation spline curve

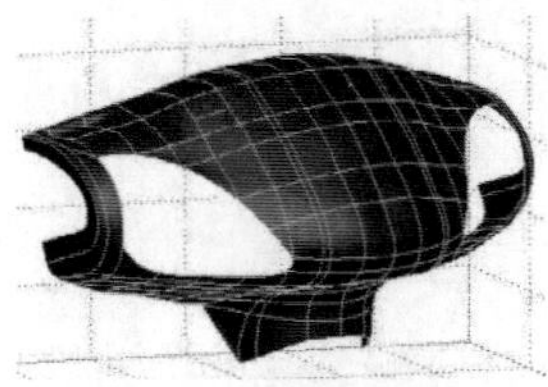

**Figure 7.** Simulation surface

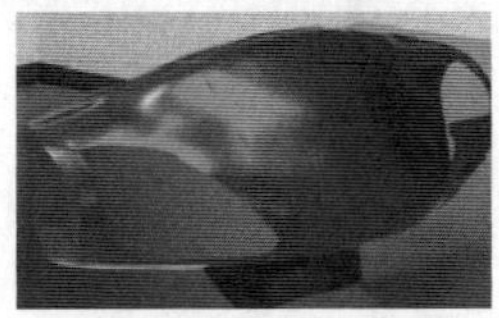

**Figure 8.** Primitive surface

**Table 1**. The value points of electric vehicle front shell type , control points and spline matrix

| *Ta* | *PX* | *PY* | *PZ* | *DX* | *DY* | *DZ* | $N_1$ | $N_2$ | $N_4$ | $N_5$ | $N_6$ | $N_7$ | $N_8$ | $N_9$ | $N_{10}$ | $N_{11}$ |
|---|---|---|---|---|---|---|---|---|---|---|---|---|---|---|---|---|
| 1 | 6.17 | 9.87 | 2.42 | 6.17 | 9.87 | 2.42 | 1 | 0 | 0 | 0 | 0 | 0 | 0 | 0 | 0 | 0 |
| 2 | 7.05 | 10.01 | 2.42 | 6.17 | 9.879 | 2.421 | 0 | 0 | 0 | 0 | 0 | 0 | 0 | 0 | 0 | 0 |
| 3 | 8.01 | 10.0 | 2.42 | 7.215 | 10.07 | 2.421 | 0 | 0.075 | 0 | 0 | 0 | 0 | 0 | 0 | 0 | 0.075 |
| 4 | 8.968 | 9.713 | 2.421 | 7.979 | 10.06 | 2.421 | 0 | 0.174 | 0.167 | 0 | 0 | 0 | 0 | 0 | 0 | 0.174 |
| 5 | 9.768 | 9.118 | 2.421 | 9.003 | 9.758 | 2.421 | 0 | 0 | 0.647 | 0.186 | 0 | 0 | 0 | 0 | 0 | 0 |
| 6 | 10.06 | 8.501 | 2.421 | 9.741 | 9.249 | 2.421 | 0 | 0 | 0.103 | 0.643 | 0.257 | 0 | 0 | 0 | 0 | 0 |
| 7 | 9.855 | 7.869 | 2.421 | 10.15 | 8.525 | 2.421 | 0 | 0 | 0 | 0.139 | 0.690 | 0.170 | 0 | 0 | 0 | 0 |
| 8 | 9.271 | 7.058 | 2.421 | 9.926 | 7.789 | 2.421 | 0 | 0 | 0 | 0 | 0.172 | 0.667 | 0.160 | 0 | 0 | 0 |
| 9 | 8.651 | 7.229 | 2.421 | 9.243 | 7.498 | 2.421 | 0 | 0 | 0 | 0 | 0 | 0.166 | 0.683 | 0.150 | 0 | 0 |
| 10 | 7.808 | 6.835 | 2.421 | 8.671 | 7.241 | 2.421 | 0 | 0 | 0 | 0 | 0 | 0 | 0.238 | 0.587 | 0.178 | 0 |
| 11 | 0 | 0 | 0 | 7.808 | 6.835 | 2.421 | 0 | 0 | 0 | 0 | 0 | 0 | 0 | 0 | 1 | 0 |
| 12 | 0 | 0 | 0 | 7.808 | 6.835 | 2.421 | 0 | 0 | 0 | 0 | 0 | 0 | 0 | 0 | -1 | 0 |

## 5. Conclusion

The appearance of the products which have same function is similar,and it can be applied to NURBS curve and surface.A unified data table structure can be constructed according to the digital characteristics of NURBS curves and surfaces.Based on this, the information base system can be built up.The integration of MATLAB software, Access database, Pro/E software and image contour extraction software can effectively realize the geometric feature line and feature points extraction,storage and surface reconstruction.The information system which has a large number of cases of product features of the geometric shape can effectively help designers innovate for product modeling.

**Reference**

[1] ChenQingxue, HanQingyao 2009 *Researchof NURBS-face Modeling Based on MATLAB.Electric Power Science and Engineerin.*.**(11)**70-72.

[2] ChengZhenBo, KeShanJun. 2016 *Application of geometric relationship of NURBS curve control points in product 3D modeling. JOURNAL OF MACHINE DESIGN.***(7)**105-108.

[3] HuoQingLi, XieJinSong,LiGen, FengLiang 2015 *An Introduction of Surface Reconstruction of Reve-rse Engineering.The Journal of New Industrialization.*.**(6)**47-52.

[4] WuGuangLing, ZhangQiuJu 2011 *Study on Computer Aided NURBS Surface Modeling Technology and Its Implementatio. Journal of Beijing University of Civil Engineering and Architecture.***(4)**158-161.

[5] LiDeXin, XieJing, JiaJie 2004 *Research on NURBS Curved Surface Weighted Factor and Its Applicati -on.Journal of Xi'an University of Technology.***(2)**154-158.

IOP Conf. Series: Materials Science and Engineering **224** (2017) 012056 doi:10.1088/1757-899X/224/1/012056

# Design and Implementation of Campus Application APP Based on Android

**Zhu dongxu, liu yabin, PI xian lei, Zhou weixiang and Huang meng**

Institute of Disaster Prevention, HeBei SanHei, 065201

zdx_y@icloud.com,amyabin@163.com,1934075569@qq.com,1126563580@qq.com, hm@cidp.edu.cn

**Abstract.** In this paper, "Internet + campus" as the entrance of the Android technology based on the application of campus design and implementation of Application program. Based on GIS(Geographic Information System) spatial database, GIS spatial analysis technology, Java development technology and Android development technology, this system server adopts the Model View Controller architectue to realize the efficient use of campus information and provide real-time information of all kinds of learning and life for campus student at the same time. "Fingertips on the Institute of Disaster Prevention Science and Technology" release for the campus students of all grades of life, learning, entertainment provides a convenient.

## 1. Introduction

With the popularity of the network and mobile network speed, making people tend to use the way in the mobile client to easily access information. Campus is the place to learn every student life, rich and diverse campus information is bound to provide more convenient for our study and life, and the current market, the main function of Application can provide college students with more learning and entertainment, so we develop the establishment of Android-based campus learning life Application ---- "fingertips on Institute of Disaster Prevention Science and Technology " , Specifically for the campus students to provide all kinds of learning and living real-time information. This paper introduces the development process of this system in detail[1-2].

## 2. System analysis

The system from the students themselves, with the campus panorama, free classroom, campus lectures, knowledge world, express speed, campus recruitment, campus weather and knowledge forums and other basic functions.

"fingertips on Institute of Disaster Prevention Science and Technology " is through the integration of GIS to provide the query, analysis and other map functions, to provide students with campus building, cafeteria, dormitory, road search function, path analysis function, peripheral query function and thematic map, heat Figure display function. The spatial analysis, buffer analysis and optimal path analysis of GIS are used to provide information such as building, canteens, dormitories, roads and the optimal path and surrounding food in the form of map display. Compared with the traditional way of obtaining information, the campus information obtained through the mobile client not only has the characteristics of fast speed, full content and new information, but also can be used as information publishers, at any time you see and hear the information published for other Students to facilitate. Specifically shown in Figure 1:

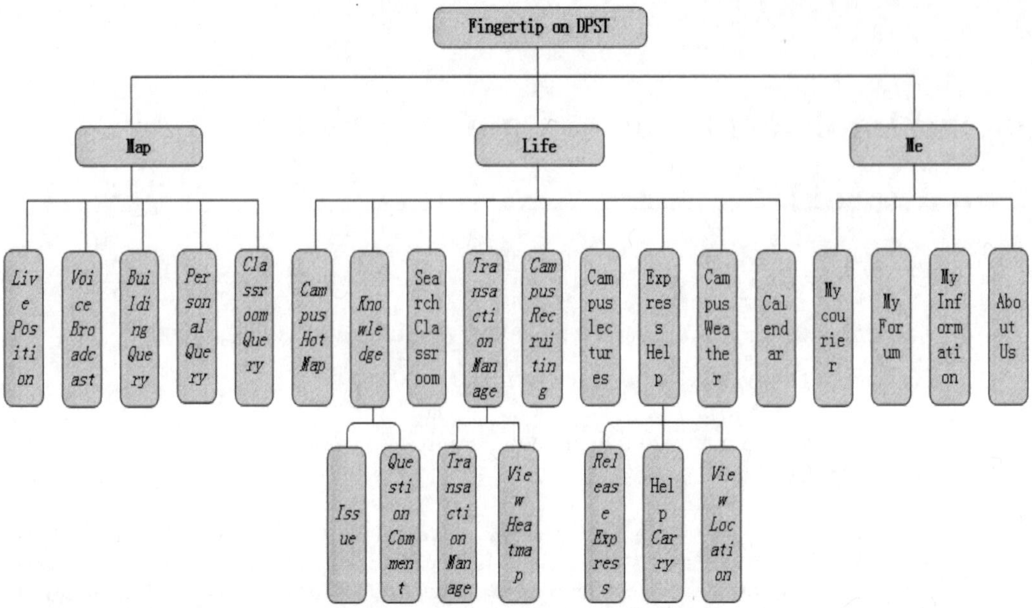


**Figure 1**. Functional structure diagram

## 3. System design

### *3.1.Data design*

A well-designed and powerful system requires full data as a basic support, so we collected data on campus and surrounding , for example teaching buildings, cafeterias, dormitories, attractions, hotels, food, roads and so on , obtaining the relevant data by Google map and Tencent map. Then we create a new data source, import the campus map data to register the map as a digital basis. Meanwhile transforming the coordinate system that will be transformed to Xi'an 80 coordinate system. Draw the base map, roads, trees and other simple data sets by a registered map. According to various data resources to build eight entity attributes different form, including things for information, classroom information, express information, knowledge forum reviews, campus lectures, campus building information, personal location information and recruitment information.

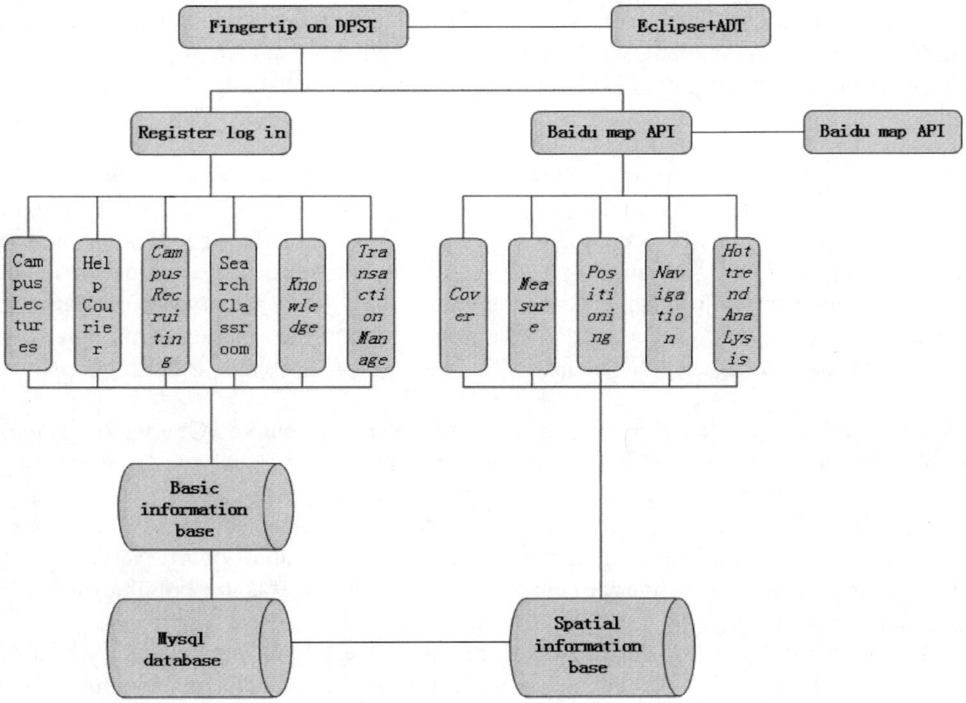


**Figure 2**. Data design diagram

### *3.2. Structure design*

This system adopts the Model View Controller design pattern, which is implemented through the servlets + JavaBean + Dao pattern, and organizes the idea of "model establishment, algorithm research-system implementation"[3]. The specific development of the Android platform based on the Java language native Application, it provides developers with notify the manager and event manager, window manager, such as top ten components, as a foundation for the development of this software application. Specific project implementation design basic data for the research through the north and the south campus, unified coordinate system, the design based on MySQL 5.6 version of the GIS spatial information database, data encapsulation in the javabeans, and convenient query results between the Servlet and Android page, use the Dao to connect to the database and database query, delete, change operations, such as through the Servlet validation data, instantiation JavaBean, called Dao connect to the database and business logic and so on. Make campus maps and add campus space and attribute data; Using GIS spatial analysis function, combining teaching buildings, dining halls, dormitories, roads, etc. In order to increase the interaction with the user, the system also designs the functions such as personal user management, knowledge and delivery[4-5]. The optimal path analysis for campus roads is designed and implemented; According to the hot map, you can analyze the flow of people. The design implements the personal location real-time location query, the path navigation voice broadcast, the study room real-time query, the building location query; The design realizes the campus life function of the delivery of the express and the location of real-time location.

## 4. System function implementation

Based on the system function design, this system uses a SuperMap iDesktop 8 c software and SuperMap iServer 8 c software has carried on the original map matching, vector map, the layer style Settings, thematic map Settings and release, etc; Use SuperMap iMobile 8C SP2 for Android and baidu map Android SDK(Software Development Kit) v3.7.0 for secondary development; Call the SuperMap iClient for JavaScript interface and the baidu map HeatMap interface real-time hot map query. Using MVC(Model View Controller) architecture, Android, Servlet, Json, Php, Java, Xml, etc. As the client, server and data interactive development technology, using MyEclipse10 main development environment as the background, use Wamp auxiliary development boom to function integrated development environment, the background for Tomcat7 main server, server for Apache2.4 heat to generate real-time request. Use MySQL 5.6 as the system database.

### *4.1.Quick delivery*

In view of such situation and we develop "fingertips disaster prevention", we develop the function of "speed take express", aimed at the north and the south campus students in disaster prevention institute of science and technology to express the great convenience. Students only need to use the spare time to release their own express information at any time. Each student can also be used as a help take delivery of the receiver, if you happen to be on the way to take delivery, you can order at any time, if the other party issued a reward, the two sides after completion of the deal approval, you also can get the rewards. This saves everyone's time and labor costs, but also creates a harmonious and loving campus atmosphere.

Click on the module to refresh the real-time SQL(Structured Query Language) query asynchronously for a few hours of delivery information. The user can click on the lower right corner of the red "+" to help to take delivery, by students at this time when help take express students click on the order, the Android client onCreate () to create events, onResume () for events, click on the focus and sends a request to the courierServlet server response, find to request courierServlet through Xml configuration files, and send the delivery to the client through Json status symbol of a variable.

### *4.2.Campus hot map*

According to the hot map, you can see where the students in the school usually have more times in the number of times, and the number of students in the school can be changed by the number of students. Infer the students' daily life activities.

We developed the campus thermal map based on the Baidu map Android SDK. Because the collected GIS spatial data projection is the WGS84 coordinate system, we need to transform the projection coordinates into the Baidu coordinates (BD09), which is Baidu coordinates Baidu in the coordinates of Mars (domestic publishing of various map systems (including electronic form), must be used at least GCJ-02 coordinate system on the geographical location of the non-linear addition of the first encryption) based on the secondary encryption used after the coordinate system.

### *4.3.Transaction processing*

Transaction management and students are closely related to daily life, through the practical application of Application, to provide users with the most efficient itinerary, a great convenience of the user's efficiency to save the user's work time. Through the BaiduMapNavigation class under the openBaiduMapWalkNavi (para, this) method to walk navigation Baidu map, according to the user specified two coordinates, passed DistanceUtil. GetDistance (p1, p2) method, the actual calculation of the two points Geographic distance, calculate an optimal route, display the route on the map, and provide you can use voice navigation.

## 5. Conclusion and prospect

The system uses Android technology to achieve the campus thermal map, self-study room inquiries, campus lectures, knowledge world, express speed, campus recruitment, campus weather and knowledge forums and other basic functions. In addition, the system interface is not particularly beautiful, in the future research and development will be further improved and optimized.

## Acknowledgments

This work was supported by the Special Fund of Fundamental Scientific Research Business Expense for Higher School of Central Government (No. ZY20160106) and 2016 Innovative Training Program For College Students(12140801/033).

## Reference

[1] Liu Zeheng . *Research and Implementation of Security Detection System Based on Android Smartphone* [D]. Harbin Institute of Technology .2011

[2] Huang Xing .*Android platform remote terminal control system research and design [D]. Beijing University of Posts and Telecommunications* .2013

[3] Q. Y. Liang, Y. Zhang, *Kunming Tourism Geographic Information System Based on Open Source Platform*, Modern Computer, vol. 12,pp. 76-79, 2015.

[4] ZHANG Peng-peng, MAO Shan-jun,*Design and Practices on Coal Reserves Management and Information System Based on GIS*, Coal Science and Technology， 2009.7

[5] Huang Meng，Zhang Zhen，*The Research on Earthquake Emergency Response System Based on GIS for the North China Plain* [J]， LNEE， 2012.1

IOP Conf. Series: Materials Science and Engineering **224** (2017) 012057 doi:10.1088/1757-899X/224/1/012057

# Development of Self-made LSM Software using in Neuroscience

**Maxim Doronin[1] and Sergey Makovkin[1] and Alexander Popov[1,2,a]**

[1] Institute of Neuroscience, Lobachevsky State University of Nizhny Novgorod, 603950, Nizhny Novgorod, Russia
[2] Department of computer control systems and telecommunications, Volga State University of Water Transport, 603950, Nizhny Novgorod, Russia

[a] popov@neuro.nnov.ru

**Abstract.** One of the main and modern visualization method in neuroscience is two-photon microscopy. However, scientists need to upgrade their microscopy system so regular because they are interested to get more specific data. Self-developed microscopy system allows to modify the construction of microscope in not-complicated manner depending on specialized experimental models and scientific tasks. Earlier we reported about building of self-made laser scanning microscope (LSM) using in neuroscience both for *in vivo* and *in vitro* experiments. Here we will report how to create software AMAScan for LSM controlling in MATLAB. The work was performed with financial support of the government represented by the Ministry of Education and Science of the Russian Federation, the unique identifier of the project is RFMEFI58115X0016, the agreement on granting a subsidy №14.581.21.0016 dated 14.10.2015.

## 1. Introduction

Application of modern techniques offer high resolution properties for living systems research is significant key for neurobiology progress. there are certain peculiarities arising from the specifics of experiments using this technique. In one case, for example, the analysis of the morphological cell data of hippocampal slices of the brain of rodents, other experiments *in vivo* depending on the design using the full-sphere of the virtual reality or, for example, in the form of a cylindrical hover platform. Working with selected methods require both of high-technology equipment availability and knowledge and skills to applying non-trivial mathematical algorithms for control two-photon microscopy system [1] and for well-formed data analysis [2,3].

To improve the quality of the resulting images it is necessary to use not only advantages of the technical (physical) characteristics of the microscope [4–7], but also good program support. Control of system is realized either exactly by using digital-to-analog (DAC) and analog-to-digital (ADC) converter(s) (e.g. National Instruments boards and others) or by using control element of microscope components (e.g. USB, RS232 connection). However, development of cross functional and convenient software to control the system is very important object for researchers. For achieving these goals developers could use commercially available software (ScanImage, Micro-Manager, YouScope or others) or create their own software using Matlab, LabView, Python or others. In case of software development researcher could create few sub-programs for each of elements controlled by software (e.g. manipulators, focus, lasers, scanners, acousto-optic modulator (AOM), detectors and others). After this step sub-programs could be combine to one main program which will control all of them and synchronize their integrated work. Having practical experience researchers could develop design of their software more friendly for end user than commercially available software because of their sector-specific goals. We are presenting all the microscope components interaction at the Figure 1.

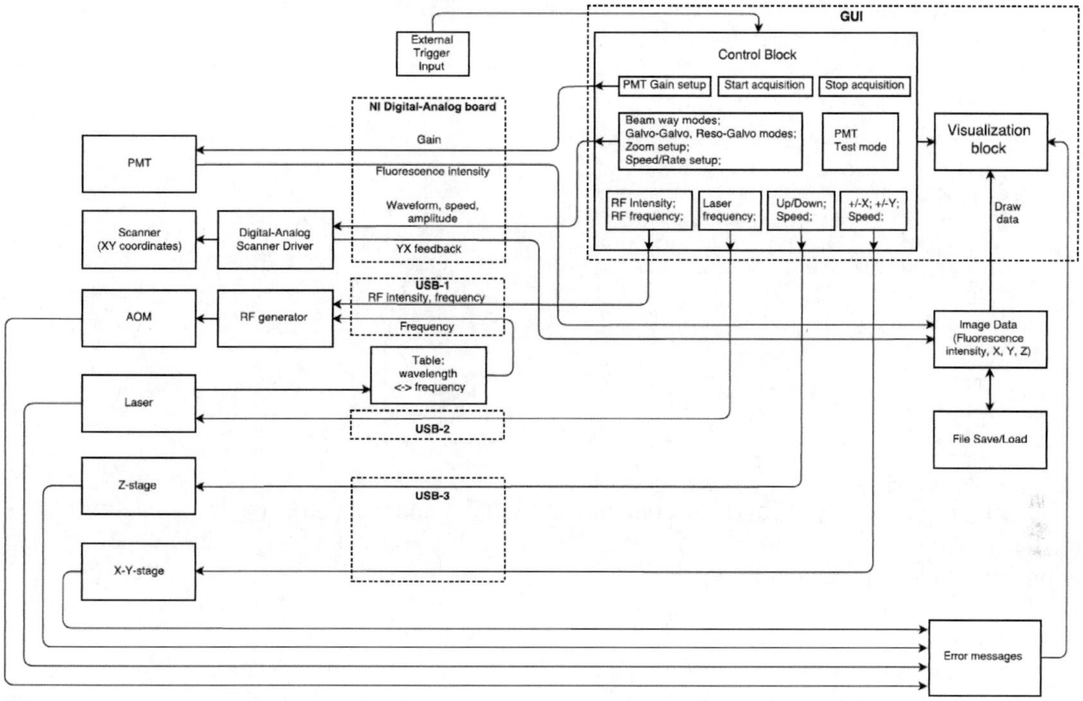


**Figure 1.** Block-scheme of microscope components interaction.

Software architecture is divided by program logical and graphical parts. Each part is working separately. This modularity allows to unify system and create it more flexible for operators. Moreover, software (full edition or program logics part without graphical part) could be started in command line. This article may be regarded as a quick reference to laboratory staff who are wishing to control their own microscopy system and modernization of this software for achieving different scientific goals.

## 2. Focus and region of interest

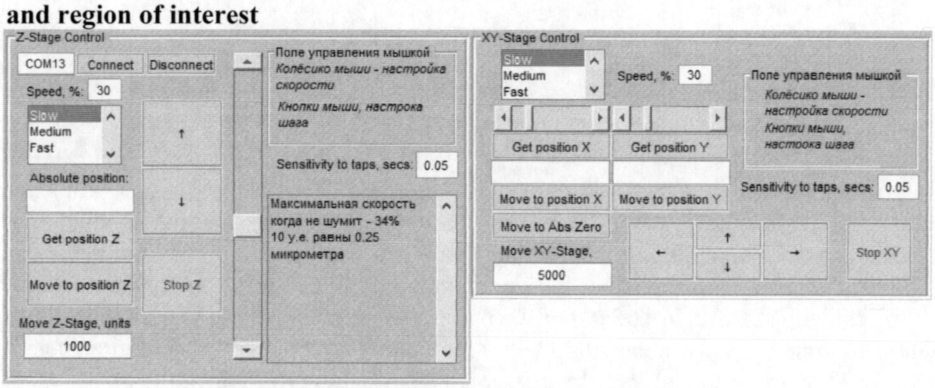


**Figure 2.** Z and X-Y motion control windows.

Motion control windows are identical for motion of system parts for Z and X-Y directions (Figure 2). Z translation motorized stage is using for vertical movement of platform with fixed objective and photomultiplier tubes (PMTs). Using this stage researcher can focus on object using differential interference contrast (DIC). X-Y translation motorized stage is using for finding region of interest (ROI) in axial plane. Motion control windows for Z and X-Y directions movement are separate in main software window, however they are using same principles of controlling by researcher. Researcher can set movement speed manually or using pre-defined modes ("Slow", "Medium",

"Fast"). Also researcher can change movement speed and step size in special area using PC mouse buttons. Clicking to mouse left button, researcher double increase step size, clicking to mouse right button, researcher double decrease step size, scrolling of mouse wheel, researcher change movement speed. Speed value and absolute position of stages could be also found in motion control window.

Modular-programmable product is working using RS-232 signal channel and can get three-dimensional position value (Z and X-Y). This channel could be used for setting high-precision step, movement speed, movement coordinate values independently for each from three translation motorizes stages.

## 3. Scan system

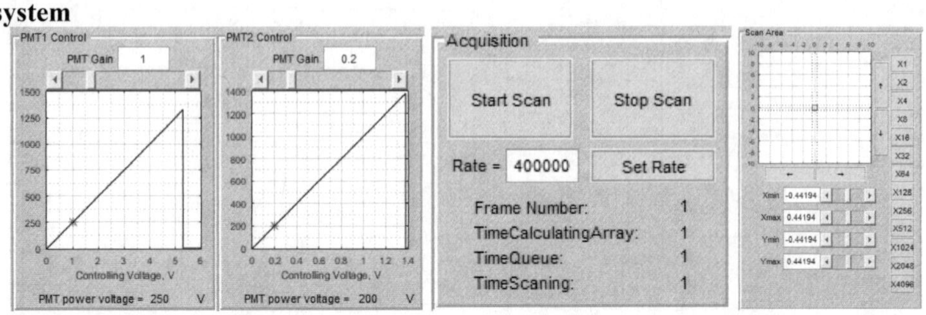


**Figure 3.** PMTs control (left), Acquisition (middle) and Scan Area (right) windows.

After focusing using DIC mode and finding ROI researcher should change microscope settings to multiphoton mode. In this mode researcher is using central control system (PC software). At first researcher is choosing scan mode, size of ROI (μm$^2$), resolution (pixel$^2$), scanning speed (μs/pixel or/and frame/sec).

After that researcher is setting supply voltage of PMTs which is defined gain of PMTs. However, researcher should remember that gain increasing is inducing noise/signal ratio. PMTs control windows are identical for both PMTs (Figure 3, left "PMT Control"). As this software is module based, researcher can increase number of channels (PMTs) up to unlimited value (usually 2–4 PMTs). Researcher also can choose colour scheme (hue-saturation-value (HSV), jet, hot, colour shadows (gray, red, blue, green etc)) for drawing results of scanning.

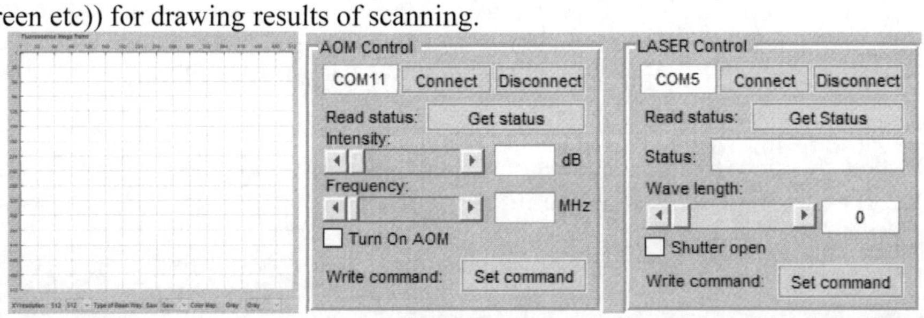


**Figure 4.** Fluorescence Image Frame (left), Laser and AOM control (right) windows.

Using buttons "Start Scan" and "Stop Scan" in Acquisition window researcher can start and stop scanning process. Also researcher can set scanning rate (pixels/sec) (Figure 3, middle). Scanning rate is defined by DAC (in our case National Instruments board) and in our case could be in the range of 100 000 to 5 000 000 samples (pixels) per second. Also in this window scanning frame number (Frame Number), time for movement of X and Y galvano mirrors (TimeCalculatingArray), time for sending scanning trajectory to National Instruments PCI-6110 board (TimeQueue), scanning time for one frame (TimeScanning) are shown.

Scale ratio of ROI could be set at Scan Area window (Figure 3, right). There researcher could change size of ROI (blue square) by changing zoom value (buttons X1…X4096), its position using arrow buttons or absolute position buttons.

Fluorescent image is presenting on a Fluorescence Image Frame window as a two-dimensional array of fluorescent values in each point of scanning image received from PMTs (Figure 4, left). Each point is represented as PMT anode output current (in mA). Researcher can change scanning type, scanning

IOP Conf. Series: Materials Science and Engineering **224** (2017) 012057 doi:10.1088/1757-899X/224/1/012057

rate, colour scheme, scanning image resolution during scanning process. Each frame is saving in defined folder automatically (researcher should choose this folder in Save/Load window) as a 14- or 16-bit *.TIF image depending on using digitizer.

## 4. Laser settings

Laser beam is characterized by Chameleon laser settings. Software AMAScan allow for set and change wavelength of interest. Acousto-optic modulator (AOM) allow for change laser beam intensity before hitting biological object (e.g. brain slice). Laser and AOM control windows are shown at the Figure 4, right. There researcher can connect and disconnect to laser or/and AOM, get status (connected or disconnected, current parameters, e.g. wavelength, intensity and frequency of AOM sound wave), turn on or turn of AOM, open and close laser shutter. Software is using RS-232 signal channel for setting up these parameters.

## 5. PMTs, filters and mirrors

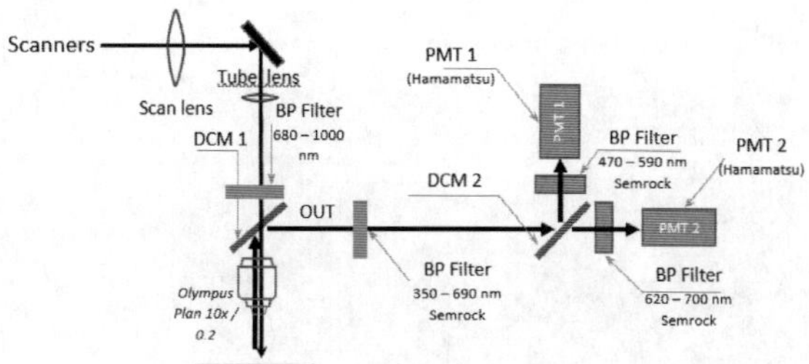


**Figure 5.** Detectors optical scheme.

Non-descanned detectors (NDDs) are using for detecting of fluorescence. Each NDD based on pair of bandpass filter to extract dye absorption wavelength and photomultiplier to detect fluorescent signal. Photodetectors are installed in close proximity to the aperture of lenses, besides, the fluorescence beam doesn't pass the way back through the scanning module, but directly gets onto the detector. The use of several dyes simultaneously is often the necessary requirement in multiphoton microscopy. Therefore the design of the microscope detector allows to make fast change of emission filters and dichroic mirror (Figure 5). The standard cube for filters mounted on the slider like "swalow-tail" that is easily inserted and taken out from the detector is used for this purpose and significantly facilitates the work of the researcher. The multiphoton detector must also be equipped with means of electronic protection and a shutter preventing the damage of the photo multiplier and increase in its performance term. Modern microscopy systems allow to mount up to 4–6 detectors depending on researcher goals.
PMTs efficiency depends on gain value which is represented as supply voltage. PMTs are operated at high voltage potential (usually up to 1 200 V). Further, the metal housing of the PMT is connected to the photocathode (potential) so that it becomes a high voltage potential when the product is operated at a negative high voltage (anode grounded). High voltage power supply is controlled by input voltage (usually 5 V). Regulation of this input voltage is possible using analog output of National Instruments board.

## 6. Service

For aligning laser beam and adjusting optics for experiments, researcher need to do all the calibration procedures using scanners coordinates as (0; 0). In this position laser beam should pass through centers of all optical elements. After that researcher should scan circular area and measure any aberrations under the objective at the focal plane. There is necessary to detect any defects of optical way during setting system up. This manipulations could be started at Beam Calibration Setup service form.

## 7. Conclusion

As a result of development of the own two-photon system taking into account all special characteristics allows to lower the expenses and to increase the quality of methods helping to improve

the quality of the obtained data, enables to conduct researches on live samples with the minimum damage of fabric, makes possible to carry out researches of the data of morphological slices of mice or rats hippocampus, as well as researches on live samples, that is carrying out experiments *in vivo*.
Control of system is realized either exactly by using DACs and ADCs or by using control element of microscope components (e.g. USB, COM connection). However, development of cross functional and convenient software to control the system is very important object for researchers. For achieving these goals developers could use commercially available software or create their own software using Matlab, LabView, Python or others. In case of software development researcher could create few sub-programs for each of elements controlled by software. After this step sub-programs could be combine to one main program which will control all of them and synchronize their integrated work. Having practical experience researchers could develop design of their software more friendly for end user than commercially available software because of their sector-specific goals.
This article may be regarded as a quick reference to laboratory staff who are wishing to develop their own microscopy system and create own software for LSM controlling. Such example is shown at the Figure 6 (AMAScan software).

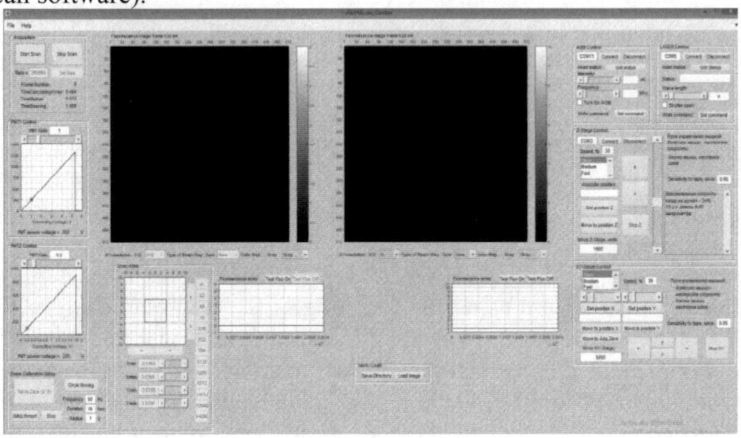

**Figure 6.** AMAScan software window.

## Acknowledgments
The work was performed with financial support of the government represented by the Ministry of Education and Science of the Russian Federation, the unique identifier of the project is RFMEFI58115X0016, the agreement on granting a subsidy №14.581.21.0016 dated 14.10.2015.

## References
[1] Svoboda K., Yasuda R. Principles of two-photon excitation microscopy and its applications to neuroscience. // Neuron. 2006. Vol. 50, № 6. P. 823 - 839.
[2] Danielyan A., Katkovnik V., Egiazarian K. BM3D frames and variational image deblurring // IEEE Trans. Image Process. 2012. Vol. 21, № 4. P. 1715 - 1728.
[3] Danielyan A. et al. Cross-color BM3D filtering of noisy raw data // 2009 Int. Work. Local Non-Local Approx. Image Process. LNLA 2009. 2009. № 118312. P. 125 - 129.
[4] Doronin M., Popov A. Design and building of modern two-photon laser scanning microscope for using in neuroscience // MATEC Web Conf. 2016. Vol. 77. P. 11002.
[5] Popov A., Doronin M. Design, upgrading and creation of the state-of-the-art two-photon laser microscope view of features used in neuroscience // 2016 6th Int. Work. Comput. Sci. Eng. WCSE 2016. 2016. P. 673–677.
[6] Doronin M., Popov A. Development and design of advanced two-photon microscope used in neuroscience // J. Phys. Conf. Ser. 2016. Vol. 741, № 1. P. 12070.
[7] Doronin M., Popov A. Development and design of up-to-date laser scanning two-photon microscope using in neuroscience // Proc. SPIE - Int. Soc. Opt. Eng. 2017. Vol. 10250. P. 102501M.

IOP Conf. Series: Materials Science and Engineering **224** (2017) 012058 doi:10.1088/1757-899X/224/1/012058

# Study on Roof Safe Thickness of Rock Foundation with Karst Cave and Karst Cave- surrounding Rock Stability

**Dongqin Chen, Cao Yi**

HuBei Urban Construction Voctional And Technological College

yyyiicao3@163.com

**Abstract.** The deformation of karst cave- surrounding rock and its roof thickness are the key indicators to evaluate the stability of karst cave foundation. According to engineering characteristics of pile foundation in karst area, Using mechanical analytical method, cave-surrounding rock elasticity theory and Griffith criterion to calculate the safe thickness of Karst cave roof. According to karst cave- surrounding rock unstable and failure conditions, such as groundwater, different pile diameter, different thickness of Karst cave roof.Through comparing and analyzing engineering cases, the results show that the mechanical analytical method to determine the safety thickness of Karst cave roof can not meet the engineering requirements , determine the safe thickness of Karst cave roof Should be based on stability of cave-surrounding rock.

## 1. Introduction

Karst landform is widespread in China. due attention has been paid to the influence of karst collapse on the foundation stability.

The evaluation method of karst foundation stability can be classified into qualitative analysis and quantitative analysis. Qualitative analysis is the basis for quantitative analysis.. The decisive method is quantitative analysis[1].

## 2. Quantitative analysis on the stability of karst cave foundation roof

Methods of calculating the safe thickness of the roof are as follows:

### *2.1.The mechanical analytic method*

Calculated as cantilever beam when there is fracture in the roof span and rocks at the standoffs of both sides of the roof are firm and complete:[2]

$$M = \frac{1}{2} pl^2 \tag{1}$$

Calculated as simply supported beam when there is fracture in the Standoffs and the roof is complete:

$$M = \frac{1}{8} pl^2 \tag{2}$$

Calculated as fixed beam when strata of the Standoffs and of the roof are complete:

$$M = \frac{1}{12} pl^2 \tag{3}$$

Checking computation of shear strength:

$$H \geq \sqrt{\frac{4Q}{\tau}} \tag{4}$$

In the formula(1) (2) (3) (4), M stands for bending moment, KN. m; p for total load of the roof, kN/m; l for cave span, m; σ for the computed bending resistance of the rock mass and estimated as 1/8 of the allowable compressive strength, kPa; b for the width of the beam and slab and estimated as 1.0m; H for thickness of roof strata, m; Q for shearing force of the standoffs, Q=1/2pL; $\tau$ for computed shearing strength of the strata (in the case of limestone, estimated as 1/12 of the allowable compression strength), kPa.

### *2.2. stability quantitative analysis of the karst cave- surrounding rock*

*2.2.1.Stress redistribution of circular caves elastic surrounding rock.* Suppose under external force the surrounding rocks of underground caves deforms elastically and can be regarded as continuous elastic body of same nature and uniformity[3]. As shown by figure 1, there is a horizontal cave with a radius of $R_0$ buried in elastic rock. Suppose in this plane strain problem both the horizontal and vertical stress is main stress and put to the cave, the stress distribution formula is shown in (5):

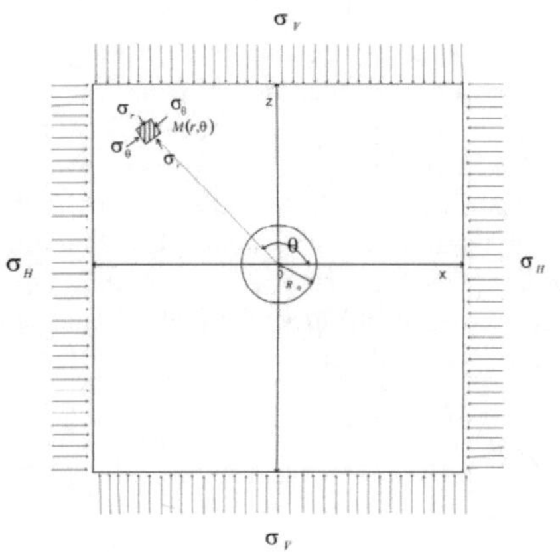


**Figture 1.** simulated diagram of stress analysis of circular caves surrounding rock

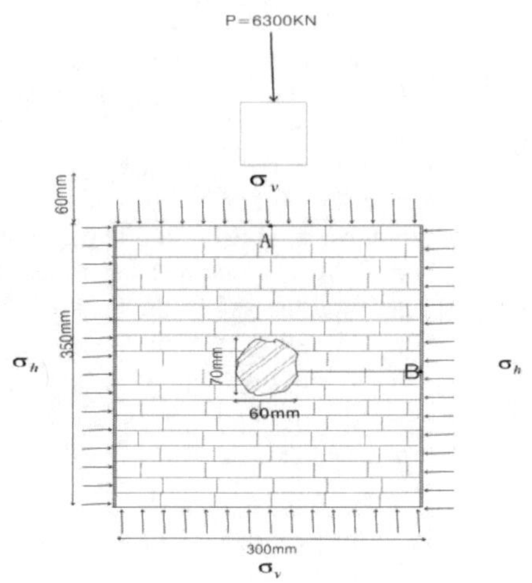


**Figture 2**. pile foundation of the cave

$$\begin{cases} \sigma_r = \dfrac{\sigma_h + \sigma_v}{2}\left[\left(1 - \dfrac{R_0^2}{r^2}\right) + \left(1 + \dfrac{3R_0^4}{r^4} - \dfrac{4R_0^2}{r^2}\right)\cos 2\theta\right] \\ \sigma_\theta = \dfrac{\sigma_h + \sigma_v}{2}\left[\left(1 + \dfrac{R_0^2}{r^2}\right) - \left(1 + \dfrac{3R_0^4}{r^4}\right)\cos 2\theta\right] \\ \tau_{r\theta} = -\dfrac{\sigma_h - \sigma_v}{2}\left[\left(1 - \dfrac{3R_0^4}{r^4} + \dfrac{2R_0^2}{r^2}\right)\sin 2\theta\right] \end{cases} \tag{5}$$

In the formula（5）, $\sigma_r$, $\sigma_\theta$, $\tau_{r\theta}$ stands for radial stress, hoop stress and shear stress on the point M respectively with compressive stress as positive and tensile stress as negative; $\sigma_v$ for vertical stress and $\sigma_h$ for horizontal stress; θ for the polar angle at M with counter-clockwise direction as positive; r for vector radius and $R_0$ for karst cave radius. The formula of the karst cave surrounding stress ($r=R_0$) and the stress on the wall of the circular karst cave are as formula（6）:

$$\begin{cases} \sigma_r = 0 \\ \sigma_\theta = \sigma_h(1 - 2\cos 2\theta) + \sigma_v(1 + 2\cos 2\theta) \\ \tau_{r\theta} = 0 \end{cases} \tag{6}$$

Theoretically, the stress redistribution is considered within the radius range six times the karst cave radius ($r=6R_0$) and surrounding rocks outside the range is considered under little influence of the stress distribution. This paper analyzes the stress $5R_0$ away from the cave center. The vertical and horizontal stresses of the karst cave ($r=5R_0$) are calculated according to the following formulas(7):

IOP Conf. Series: Materials Science and Engineering **224** (2017) 012058 doi:10.1088/1757-899X/224/1/012058

$$\begin{cases} \sigma_v = a_A P_0 + \sigma_A \\ \sigma_h = \beta(a_B P_0 + \sigma_B) \end{cases} \tag{7}$$

In the formula(7), $\alpha_A$ and $\alpha_B$ for the additional stress coefficient at A and B stand respectively; $p_0$ for the additional stress of the substrate; $\sigma_A$ and $\sigma_B$ for the self-weight stress from the soils and rocks at A and B respectively; β—lateral pressure coefficient of the rock mass, β=μ/1-μ (μlimestone regarded as 0.2)[4].

*2.2.2. Stress redistribution of other shapes caves* . The underground caves of karst terrain are usually not of regular circular structure. Suppose the karst cave is of oval structure, the horizontal and vertical diameters are b and a respectively, the horizontal and vertical stresses of the karst cave surrounding rocks are $\sigma_h$ and $\sigma_v$ respectively, the stress component of the karst cave chambers can be calculated by the following formula(8):

$$\left.\begin{aligned} \sigma_\theta &= \frac{\sigma_v[m(m+2)\cos^2\theta - \sin^2\theta] + \sigma_h[(2m+1)\sin^2\theta - m^2\cos^2\theta]}{\sin^2\theta + m^2\cos^2\theta} \\ \sigma_r &= \tau_{r\theta} = 0 \end{aligned}\right\} \tag{8}$$

In the formula(8), m for the ratio of the ellipse axes, m=a/b; θ for the eccentric stands angel of the calculation position around the karst cave.

*2.2.3. Judge the karst cave surrounding rocks stability.* When the the karst cave surrounding rocks at the position $r=5R_0$ do not wreak damage under the redistribution of stress, the karst cave is stable and the expression of judging its stability in accordance with the Griffith Criterion is the following formulas (9):

$$\begin{cases} \dfrac{(\sigma_1 - \sigma_3)^2}{\sigma_1 + \sigma_3} = 8\sigma_t & \sigma_1 + 3\sigma_3 \geq 0 \\ \sigma_3 = -\sigma_t & \sigma_1 + 3\sigma_3 < 0 \end{cases} \tag{9}$$

In the formula(9), $\sigma_1$ and $\sigma_3$ for the maximal and minimal stress respectively with the compression as positive; $\sigma_t$ for uniaxial tensile strength of the rock with a negative sign . In terms of the stress around the cave, the tangential stress $\sigma_\theta$ and the radial stress $\sigma_r$ are the maximal and the minimal stress respectively, shearing stress $\tau_{r\theta}=0$. Hence,$\sigma_\theta=\sigma_1$ and are Principal stress, namely $\sigma_r=\sigma_3=0$, which is substituted in the criterion (8). Judge the stability with the first formula if $\sigma_1+\sigma_3\geq0$ and with the second formula if $\sigma_1+\sigma_3<0$.

## 3. Engineering examples

### *3.1. Project Overview*

There is a building for both commercial and residential purposes in the area located near the middle and lower reaches of Yangtze River, which has karst development. After survey, it is found that there is a cave of irregular round form under the 3# foundation. The institute of design suggests that 1500mm pile bored piles be used, the design value of single pile be 6300kN, the macro and minor axis be 0.7m (the height of the cave) and 0.6m (the cave span) respectively, the pile foundation be 0.5m under the fossilized limestone surface, as shown in picture1. Take 31000kpa as the mono-axial compression strength value and 1950kpa as the mono-axial tensile strength value.

IOP Conf. Series: Materials Science and Engineering **224** (2017) 012058 doi:10.1088/1757-899X/224/1/012058

**Table 1.** Parameter of the strata of and above the cave roof

| The number | Soil lithology | Thickness (m) | Volume weight (KN/m³) |
|---|---|---|---|
| 1 | Miscellaneous fill | 4 | 18 |
| 2 | Silty clay | 3 | 17.64 |
| 3 | Silt | 1.5 | 17.2 |
| 4 | Silty clay | 1.5 | 17.64 |
| 5 | Slightly weathered carbonate | 2.5 | 26 |
| | Total thickness of the strata | 12.5 | |
| | Weighted average volume weight | | 20 |

*3.2. checking computation of cave roof's thickness*

Given the survey that the karst cave roof is complete, but there is weathering fracture development, according to the above geological condition, calculate the safe thickness of the karst cave roof:

1) considered as simple-supported beam, the safe thickness of the karst cave roof is:

$$H \geq \sqrt{\frac{6\times\frac{1}{8}(26\times 2+10500)\times 0.6^2}{1\times\frac{1}{8}\times 31000}} \geq 0.86m$$

2) considered as fixed beam, the safe thickness of the cave roof is:

$$H \geq \sqrt{\frac{6\times\left[\frac{1}{8}\times 6300\times 0.6+\frac{1}{12}(26\times 2)\times 0.6^2\right]}{1\times 0.1\times 31000}} \geq 0.96m$$

3) checked as shear strength, the thickness should meet the following calculations:

$$H \geq \sqrt{\frac{4Q}{\tau}} = \sqrt{\frac{4\times\frac{1}{2}(10500+26\times 2)\times 0.6}{\frac{1}{12}\times 31000}} \geq 2.21m$$

Take1.2 as safety coefficient, the safety thickness should be 2.65m.According to the preliminary design, the pile foundation is 0.5m embedded in the limestone surface, and the thickness of the thinnest part is 2.0m, which doesn't meet the safety requirement. Thus, improvement is needed in the pile design. Through analyzing related factors influencing the cave stability.

*3.3. Karst cave surrounding rock stress calculation and stability judgment*

*3.3.1. Calculate Karst cave surrounding rock stress.* (1) Calculate $\sigma_\theta$ of surrounding rocks ($r=5R_0$)
Calculate the vertical and horizontal stress according to formula(7), the results are that $p_0$=3354KN/m²;$\alpha_A$=0.756, $\alpha_B$=0.065;$\sigma_A$=229kPa;$\sigma_B$=275kPa; take 0.25 as β value, the results are as followed:

$$\sigma_v = 2765\,kpa \qquad \sigma_h = 123\,k\,p\,.$$

Calculate the hoop stress $\sigma_\theta$ at A point as in Fig.2, namely the maximal stress of the karst cave roof (θ=90°or 270°), Apply formula(8): $\sigma_{\theta A}$= -2355kpa
Likewise, calculate the hoop stress $\sigma_{\theta B}$ at B point, namely the maximal stress on both sides of the karst cave(θ=0°or 180°), $\sigma_{\theta B}$=7395kpa

*3.3.2. Judge stability of the karst cave surrounding rock.* Use formula(9) to judge the stress value at A and B
at A point (θ=90°or 270°), $\sigma_{\theta A}$=-2355kpa, $\sigma_{rA}$=0kpa; on the roof appears large tensile stress.

$|\sigma\theta A|$=2355kpa>1950kpa (unstable)

at B point (θ=0°or 180°), $\sigma_{\theta B}$=7395kpa, $\sigma_{rB}$=0kpa;

$$\left|\frac{-(\sigma_1-\sigma_3)^2}{8(\sigma_1+\sigma_3)}\right| = 924kpa < 1950kpa \qquad \text{(stable)}$$

The results show that tensile failure appears on the karst cave roof.

*3.3.3. Influence of Groundwater to the stability of the karst cave.* The influence is insignificant for the karst cave if the water level rises. If the water level declines below the cave, the cave would be under inhaling vacuum corrosion and negative pressure would appear with a value of $10^5$pa, which equals large-area additional load of 100kpa on the ground surface.

$\sigma_v$=2865kpa, $\sigma_h$=148kpa;

Apply formula(10) to calculate $\sigma_{\theta A}$ at A point and $\sigma_{\theta B}$ at B point as shown in picture.2 as followed:

$\sigma_{\theta A}$= -2372kpa; $\sigma_{\theta B}$=7642kpa;

The results show that $\sigma_\theta$ at the karst cave floor and at both sides increase, the tensile stress at A point changes from 2355kpa to 2372kpa,$\sigma_{\theta B}$ at B point changes from 7395kpa to 7642kpa. The degree of change is slight and the inhaling corrosion of vacuum produced by the dynamic change of the groundwater level has little influence on the stability of the karst cave surrounding rocks.

*3.4 .Improved methods of pile foundation*

*3.4.1. Influence of pile diameter to the surrounding rock.* When the load doesn't change while the pile diameter changes, the corresponding change of $\alpha_A$,$\alpha_B$,$\sigma_v$,$\sigma_h$,$\sigma_{\theta A}$ and $\sigma_{\theta B}$ are shown in Table 2 (the influence of groundwater considered):

In table 2. the results show that after the change of pile diameter, the tensile stress of the roof and floor

**Table 2.** Various pile diameters and the corresponding calculation results

| **Pile diameter (m)** | $\alpha_A$ | $\alpha_B$ | $\sigma_v$ **(kpa)** | $\sigma_h$ **(kpa)** | $\sigma_{\theta A}$ **(kpa)** | $\sigma_{\theta B}$ **(kpa)** |
|---|---|---|---|---|---|---|
| 1.3 | 0.729 | 0.043 | 2774 | 130 | -2341 | 7412 |
| 1.2 | 0.701 | 0.037 | 2680 | 125 | -2264 | 7162 |
| 1.0 | 0.646 | 0.028 | 2496 | 117 | -2006 | 6397 |

of the surrounding rocks declines to different degrees, yet the cave is still unstable. When the pile diameter is 1.0m, the tensile stress of the roof is 2006kpa >1950kpa, which means that the size of pile bottom has great influence on the cave stability. Pile diameter out of a certain range would surpass the standard bearing capacity value of the limestone pile tip.

*3.4.2. Influence of roof thickness to the karst cave surrounding rock.* When the load and the pile diameter do not change while thickness of the roof changes, the corresponding change of $\alpha_A$,$\alpha_B$,$\sigma_v$,$\sigma_h$,$\sigma_{\theta A}$ and $\sigma_{\theta B}$ are shown in Table 2. (the influence of groundwater considered):

**Table 3.** Various thicknesses of the roof and the corresponding calculation results

| **Roof thickness (m)** | $\alpha_A$ | $\alpha_B$ | $\sigma_v$ **(kpa)** | $\sigma_h$ **(kpa)** | $\sigma_{\theta A}$ **(kpa)** | $\sigma_{\theta B}$ **(kpa)** |
|---|---|---|---|---|---|---|
| 2.3 | 0.646 | 0.060 | 2503 | 146 | -2017 | 6659 |
| 2.4 | 0.55 | 0.058 | 2184 | 145 | -1701 | 5793 |
| 2.5 | 0.43 | 0.055 | 1784 | 143 | -1308 | 4707 |

The results show that when roof thickness is 2.3, there is a large tensile stress, $|\sigma_{\theta A}|$=2017kpa>1950kpa, the cave is unstable. Apply formula (9) to judge the stability of the karst cave surrounding rocks when the roof thickness is 2.40m:

① at A ($\theta$=90°or 270°), $\sigma_{\theta A}$=-1701kpa; $\sigma_{rA}$=0kpa. $|\sigma_{\theta A}|$=1701kpa<1950kpa (stable)

② at B ($\theta$=0°or 180°), $\sigma_{\theta B}$=5793kpa, $\sigma_{rB}$=0kpa.

$$\left|\frac{-(\sigma_1-\sigma_3)^2}{8(\sigma_1+\sigma_3)}\right| = 724kpa < 1950kpa \qquad \text{(stable)}$$

The results show that after the change of roof thickness, the surrounding rocks are in a state of stability. When the roof thickness is 2.4m, the top and floor and both sides of the karst cave are not damaged. When the safety factor is 1.2, the roof thickness needed is 2.9m. Thus, the safety roof thickness of 2.65m calculated by mechanical analysis cannot guarantee cave stability and the foundation must be improved.

IOP Conf. Series: Materials Science and Engineering **224** (2017) 012058 doi:10.1088/1757-899X/224/1/012058

## 4. Conclusions

(1) Apply elastic theory of surrounding rocks to judge the foundation stability when the supporting layer of the foundation is strata of karst cave. The safety thickness of the karst cave roof chosen in this way can guarantee both the roof safety and cave stability.

(2) The case study shows that within the standard value of the foundation supporting layer, the smaller the foundation bottom, the more stable the cave is. Also, an effective method of ensuring the foundation stability is to change the safety thickness of roof.

(3) A conclusion drawn from the calculations above is that the inhaling corrosion of vacuum caused by groundwater is not a key factor[5].

(4) According to Regulation 6.5.3 in GB50007-2002 Code for Design of Building Foundation, the influence of vertical grike with a width less than 1m and of the section near sinkholes on the foundation stability can be ignored. This paper suggests foundation stability be re-judged in those situations.

## References

[1] Basso, A; Bruno, E; Parise, M,etc, Morphometric analysis of sinkholes in a karst coastal area of southern Apulia (Italy). Environmental Earth Sciences, 2013,vol.70(6),pp: 2545-2559

[2] Chong Jiang, Ming-hua Zhao, "Stability analysis of subgrade cave roofs in karst region," J. Cent. South Univ. Technol, vol. 15,pp. 038−044, 2008.

[3] Mohamed I.El-Sayed.Karastic features associated with unconformity surfaces,a case study from the United Arab Emirates.Journal of Arid Environments,2000,Vol.46(3)P:295~312

[4] B.J.Barfielda,GK.Feltonb, E.W.Stevensc et al.A simple model of karst spring flow using modified NRCS procedures.Journal of Hydrology,2004,Vol.287,P:34-48

[5] Mngtang Lei，Xiaozhen Jiang，LiYu New advances in karst collapse research in China. Environmental Geology, Angust2002, Vol. 42, Numbers, pp:462−468

# Analysis of the Key and Difficult Points in the Engineering Construction Technology of the Steel Structures of a Super High-rise Building

**Lijun Dou** [1], **Hengxu Li** [1]

[1] School of Civil Engineering, Changchun Institute of Technology, Changchun, China

Ljdou63@163.com, 634692900@qq.com

**Abstract:** In recent years, with the rapid development of the economy of China, the traditional building structure has not been able to meet the current people's demands and the super high-rise building has become a symbol of a city. In the current period, the research on the super high-rise building in the architectural industry of China is late and the technical blanks exist in some construction difficulties in the super high-rise steel structures. Based on the above, a brief analysis and discussion on the difficult construction technology in the steel structures of a super high-rise building and some measures are presented for reference of the relevant personnel in this paper.

## 1. Factors to be considered in the deepening of the design

For considering the machining and making process of the steel structure, before the deepening of the design, the design deepening personnel and technicians should be familiar with the drawing of the structure, collect the information in the drawing, analyze the making process of the key parts, determine some points such as special plates, plate width requirements and test requirements, etc. and present the relevant suggestions.

### *1.1. Disposal of the transitional node structure of the variable section*

In order to assure the quality of the members, the transitional node of the variable section should be made by the machining factor. In order to assure the steady force transmission of the steel column, not less than 1:6 transitional section should be established between the round tubes with the different diameters, shown in Figure1. [1]

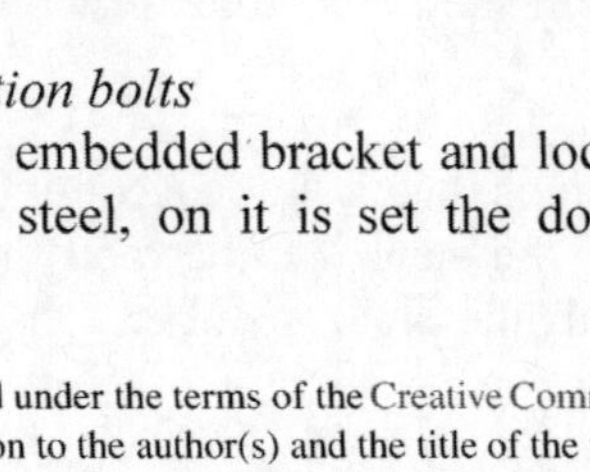

**Figure 1**. Disposal of the Transitional Node Structure of the Variable Section

### *1.2. Reinforcing design of the foundation bolts*

In the installation of sill anchors, the embedded bracket and locating plate should be set, the locating bracket column is made of angular steel, on it is set the double-layer locating ring plate. In the deepening, the checking calculation of the force stability should be made first to assure the rational structure, with the detailed form shown in Figure 2[2].

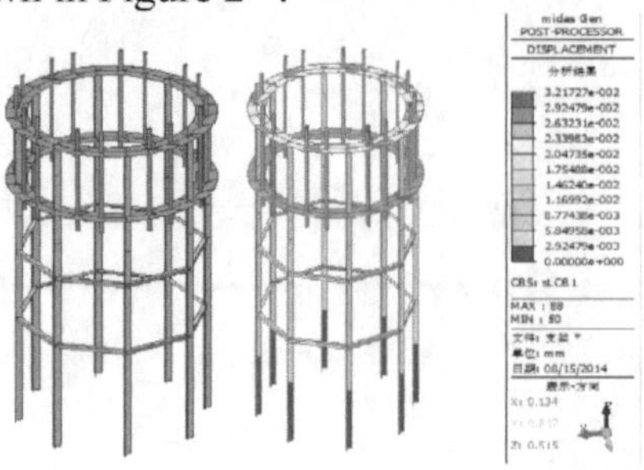

**Figure 2**. Design and Checking Calculation of the Sill Anchor Locating Bracket

*1.3. Design of the temporary support for installation of the truss on the site*

An engineering truss has the high weight and big span, so it should be hoisted in the sections. The lattice column is used as the temporary bracket for closure. In order to assure the stability of its installation, the form, dimensions and quantity of the lattice bracket should be taken into account in the deepening and the checking calculation of its forces should be carried out. The form of the bracket is shown in Figure 3[3].

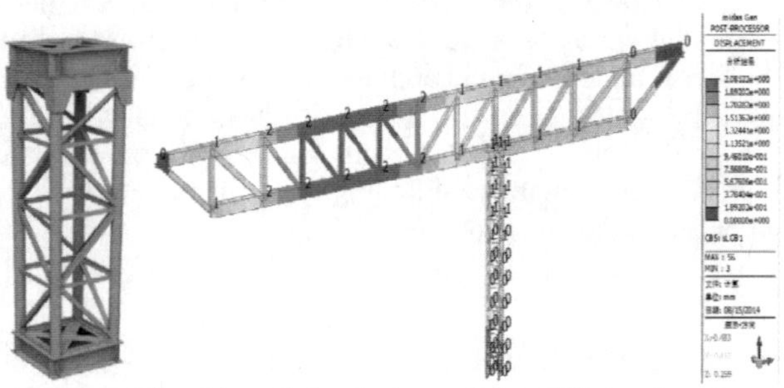

**Figure 3.** Design and Checking Calculation of the Sill Anchor Locating Bracket

*1.4. Cooperation with the electromechanical specialty*

The holes for the electromechanical equipment, screen wall and decoration should be reserved and the relevant reinforcing measures should be taken. All the specialties should be cooperated to set the connecting parts and temporary fixing parts, etc. to the steel member, shown in Figure 4[4].

**Figure 4**. Setting of the Reserved Holes

**2. Assembly of the steel member**

The purpose of pre-assembly is that the member may not be smoothly connected on the construction site due to the accumulated errors on the premise of meeting the standard because the errors are inevitable in the machining. In order to assure the installation accuracy on the site, for the part with more connecting rods, the member should be pre-assembled before it leaves the factory, the machining accuracy should be tested and production should be guided according to the pre-assembly result[5].

**3. Transportation of the member**

The general thought of transportation is taken into account from the perspective of safety and rapidness according to the features of the steel member and the transport experience of the similar members. It is suitable to transport all the steel members on the land. The special transport team should be organized to be responsible for loading and unloading of the steel member and the transport company with the rich transport experience and complete big-part transport vehicles for the long-term cooperation should be chosen[2].

Typical member transport method: in order to assure the transport safety and prevent the steel member from being damaged, in addition to the strict implementation of the loading, reinforcing and binding scheme, the special personnel should be designated to escort all the transport vehicles. For example, the packaging forms of some typical members of a project are as follows: the steel columns and small steel beam bearing plates are bundled and packaged, shown in Figure 5. The truss should not be packaged in the transport, shown in Figure 6. Some parts such as high-strength bolt anchor of the connecting plate, etc. should be packaged in the packing box, shown in Figure 7.

**Figure 5**. Bundling Packaging of Steel Columns and Small Steel Beam Bearing Plates

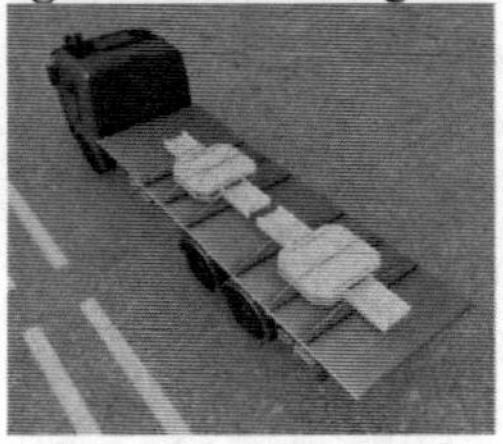

**Figure 6**. Unpackaging of the Truss

**Figure 7**. Packing Boxes of High-strength Bolt Anchors and other Parts

## 4. Analysis of the key and difficult points in the installation and construction of the steel structure

### *4.1. Deepened design of the steel structure*
For the complex structures of the ring truss nodes of a project, a lot of screen wall steel structure coordinates and more difficulties in the disposal of the intersection nodes, the relevant resolution is to be carefully familiar with the structure of each node, build the model according to 1: 1, determine the making, welding and installation processes, take full into account each welding seam of the nodes, assure the worker's operation space, consider the making, transport and installation aspects in the deepening and achieve the deepening of the guidance of the production.

### *4.2. Making of the truss nodes*
For more intersections of the truss node members of a project, which brings about more difficulties in the assembly and welding, the relevant resolution is to build the 1: 1 model before the members are machined, use the models as the reference to make the detailed and rational assembly and welding sequence, take into full account the welding operability; set some deformation resistance measures such as welding bonding plate and welding bonding support to assure the dimensional accuracies of the members.

### *4.3. Pre-assembly of the steel structure*
For the complexity of the steel structure of a project, it is required that the steel structure should be pre-assembled before the steel structure with the complex construction process leaves the factory so as to assure that the machining accuracy and interface should meet the installation requirements on the site.

### *4.4. Truss installation*
Because the ring truss installation of the project has the big workload and more difficulties in the construction, the annex roofing truss structure span is installed at the elevation of 22m and the construction and safety protection are difficult, the relevant resolution is to carry out the pre-assembly of the entity after the truss machining is completed, assure the completion of the installation on the site once, set the special piling and assembly site, complete the assembly of the truss on the ground, reduce the overhead workload, connect the steel beams between the trusses after the installation of the roofing truss is completed and carry out welding and fixing after the settlement is stable[6].

### *4.5. High assembly and installation accuracy of the screen wall steel structure*
The screen wall steel structure of the project is the 3D structure, with a lot of coordinates and complex form, high making accuracy requirements and installation accuracy, the simulation analysis of the deformation of the structure as a whole is made via the infinite element before the installation, the reversible deformation is made to offset the down deflection value; use the 3D laser scanner for pursuit in the whole process of installation and compare the scanning result with the model and achieve the adjustment of the deviation at any time.

### *4.6. Measuring accuracy control*
In the installation of the steel structure, the structure is affected by not only the wind load but also some climatic changes such as sunshine and temperature, etc., making the structure in the state of dynamic change all the time and presenting the high requirements for the measuring control method and measuring accuracy. The relevant resolution is to make use of the laser collimator to pursue and test the verticality of the steel column in 24 hours; use GPS for rechecking in the different weather conditions at the different time, test the influence of the temperature difference on the building and take the corrective methods to eliminate the structural deformation caused by the environment.

### *4.7. Construction safety protection and management of the super high-rise steel structure*
In the construction of the steel structure, the pertinent laying of safety protection measures and the safety management of operators are the key points of the construction; in the construction of the super high-rise steel structure and the high altitude dropping and welding fire are the key points of management. The “three-level education” of personal access and safety publicity and inspection should be strictly implemented; the horizontal channel, external protection and steel column hoisting protection measures, etc. should be set on the site; the fire protection measures should be taken in the welding of the steel structure and some measures such as application of fireproof asbestos cloth and fire basin, etc. should be taken[2].

## 5. Conclusion
On the whole, the occurrence of the super high-rise buildings has resolved the sharp contradiction between the rapid increase of the population in the world and intense supply of land resources and continuous increase of the prices. The development of the super high-rise buildings will make people develop toward the high altitude. Therefore, it will promote the economic development and greatly improve urbanization.

## Acknowledgment
This work was financially supported by the Jilin Science and Technology Developme -nt Planned Topics 《Tectonic Connection Technology and Economical Efficiency Study of Outer Enclosure System of Light Steel Residence in Jilin Province》 [20140203014SF] and Jilin Education Development Planned Topics《Research on the Tectonic Connection Technology of the Envelope Systems with the Steel Frame in Cold Region》 [2015296]

**References**

[1] Ministry of Housing and Urban-Rural Development of the People's Republic of China. GB50755-2012 *Code for construction of the steel structures* [S]. Beijing: China Building Industry Press, 2012.

[2] Jianjiang Du. Research on the Key Technology of Construction of the Steel Structures of the Super High-rise Buildings [J]. *Structural Construction*, 2015, 37:61.

[3] Wei Wei. The key technology of high-rise building steel structure construction analysis [J]. *Science and technology innovation herald*, 2016, (13) : 98-99.

[4] Lijuan He. Key techniques and measures for construction of steel structures in super-tall buildings [J]. *Jiangxi building materials*, 2016, (13) : 96.

[5] Shoaling Yang. Key techniques and measures for the construction of steel structures in super-tall buildings [J]. *Building materials and decoration*, 2016, (26) : 72-73.

[6] Gajian Li. Research on construction technology of super-high steel structure [J]. *Chinese and foreign architecture*, 2016, (06) : 164-165.

IOP Conf. Series: Materials Science and Engineering **224** (2017) 012060 doi:10.1088/1757-899X/224/1/012060

# Determination of Copper in Metal Processing Wastewater by Stripping Voltammetry

**Baoyou Liu**

Key Laboratory of pollution prevention and control in Hebei Province, College of Environmental Science and Engineering, Hebei University of Science and Technology, Shijiazhuang Hebei 050018, China

liuzhang544@163.com

**Abstract.** In this paper, the trace copper in metal processing wastewater was determined by stripping voltammetry, and the influence of instrument condition on the analysis was discussed. The standard working curve was established and the wastewater samples were determined. The results showed that the optimized conditions of instrument determination were as follows: electrolysis initial voltage was -1.3V, electrolysis time was1 minute, balance time was 10 S, the rate of stirring was high speed and scanning speed was 0.08 V/S. Under the above experimental conditions, the peak shape was good, the peak current was large and the peak potential was stable at -0.1V. The linear equation of standard working curve was $y = 0.1683X + 0.0121$, the standard deviation was $R^2 = 0.9974$, the detection limit was 0.2μg / mL, the recoveries of the wastewater samples were 96.6% -100.5%. The method has the advantages of good accuracy, high precision and high sensitivity, and is suitable for the determination of copper in metal processing wastewater.

## 1. Introduction

### *1.1. The origin and harm of the copper in the environment*

With the progress of science and technology and people's living standards, environmental awareness and self-care awareness, people's understanding of copper gradually deepened, its impact on human health and the environment has been gradually recognized. Copper is a natural element of the normal composition, but the copper concentration is too high will be harmful to humans, animals and plants. Copper mining, smelting and rock weathering can result in increased local copper concentrations in the local environment. Copper pollution in the atmosphere is mainly caused by non-ferrous metal smelting. Copper pollution is caused by smelting, metal processing, machine building, organic synthesis and other industrial copper-containing wastewater discharge. The copper content in the waste water by smelting and electroplating is the highest, ranging from dozens to hundreds of ppm. The main source of soil copper pollution is copper mine and smelter discharge wastewater, in addition to industrial dust, urban sewage and copper pesticides can cause soil copper pollution. Copper is one of the trace elements necessary for life, and plays an important role in human hematopoiesis, cell growth, activity of certain enzymes and endocrine gland function. Copper is not easy to dissolve, less toxic than copper salts, especially in the soluble salt in copper salt such as acetic acid copper and copper sulfate is toxic. Human copper poisoning is due to excessive intake of copper due to the performance of abdominal pain, rash, diarrhea, vomiting and other symptoms, a large number of long-term ingestion can lead to hepatitis and hemolytic anemia, and Parkinson's disease . The potential toxicity of copper has attracted great concern, and the regulatory framework for controlling its concentration levels to a greater extent has been enforced in most developed countries. Therefore, in the water quality, environment and food testing, the content of copper has always been an important indicator of environmental pollution and food quality, so sensitive, accurate and rapid detection of copper content is particularly important.

### *1.2. Existing methods of measurement*

Trace copper determination methods are atomic absorption spectrometry, visible spectrophotometry, flow injection chemiluminescence, fluorescence, differential potential dissolution and so on. Atomic absorption spectrometry (AAS) method has the advantages of low detection limit, high accuracy, good selectivity and fast analysis speed. AAS method is mainly applied to the analysis of trace and trace components in samples, and it has become the most widely used analytical method for the quantitative analysis of inorganic elements. Among the analytical methods for copper, the most commonly used is the spectrophotometric method. It has many advantages such as wide application, high sensitivity, good selectivity, high accuracy, wide application range, low analysis cost and easy operation. The spectrophotometry is based on the selective absorption of electromagnetic radiation to establish a qualitative and quantitative analysis method by different molecular structures., this method is simple, easy to promote . Differential potentiometric stripping method of copper, less sampling, high sensitivity, good reproducibility, the reagent used in low cost, fast and simple, easy to popularize [1-3]. For example, Fang Bin [1] by differential potentiometric stripping method on the determination of copper in human hair,accuracy and precision are good, high sensitivity. With a mercury plating method can be used to overcome a lot of secondary pollution problem caused by mercury liquid, at the same time also can reduce the operation process, improve the efficiency of measurement [2]. In this paper, the copper content in metal processing wastewater was determined by stripping voltammetry, so as to realize the rapid determination of the sample in the scene[4].

## 2. Experimental

### *2.1. Instruments and reagents*

Lk98A microcomputer electrochemical analysis system (Tianjin Lanlike company); Three electrode system (saturated calomel electrode as the reference electrode, silver electrode as the working electrode, platinum electrode as the electrode); The exact analytical balance (accuracy 0.0001g); SYC-15 super thermostatic water-bath boiler, 100mL electrolytic cup, pipette 25mL-1mL; Volumetric flask; Copper sulfate, NaAc, HAc, HgSO4 are analytical reagent and highly pure water for our laboratory.

*2.1.1.* Reference method [3] preparation. 3.9000 g of copper sulfate volume to 100mL volumetric flask, remove 1mL into 1L volumetric flask, set volume again, get 15.6μg • mL-1 $Cu2^+$ ion standard solution

*2.1.2.* Configure the HAc-NaAc buffer solution (pH≈5.6). 905mL2mol•L-1NaAc solution was mixed with 95mL 2mol•L-1HAc solution to obtain HAc-NaAc buffer solution (pH≈5.6).

*2.1.3.* 2 × 10-2mol • L-1 HgSO4 solution preparation.

*2.1.4.* Standard sample configuration. Respectively, to take a certain amount of standard solution, diluted with high purity water to 0.5,1,1.5,2,4 μg • mL-1.

*2.1.5.* Simulate the sample to be measured (about 2μg • mL-1Cu2 + ions), digested with reference to the literature, and then diluted to the corresponding concentration.

### *2.2. Experimental procedures*

*2.2.1. Connect the instrument and enter the following parameters on the keyboard*:

Baseline deduction(off) IR Compensation (off)
The initial voltage (V) -1.0000 Electrodeposition voltage (V) - 1.0000
Scanning speed (V/s) 0.08 The termination voltage (V) was 0.2000
Electrodeposition time (s) 60 Equilibrium time (s) 10

*2.2.2.* Add 25 mL of water sample and 1 mL of HAc-NaAc solution (six drops of Hgcl) into the electrolytic cup, and place the reference electrode and platinum electrode into the solution (try not to touch the stir bar).

*2.2.3.* As far as possible without changing the electrode position in the electrolytic cup by adding a certain amount of Cu2 + ion standard solution, press the "start" button, record several stripping voltammetric curve to obtain a stable peak current. Press the Pause key.

*2.2.4.* Repeat the above steps to draw the standard working curve at different concentrations, the quantitative determination of unknown samples. Determination of the process can adopt quantitative determination by standard addition method. Standard addition method is calculated as follows:

$$c_x = \frac{c_s V_s h_x}{H(V_x + V_s) - h_x V_x}$$

Cs, Vs are the concentration and volume of the standard solution, respectively; H is the peak height of the dissolution peak after adding the standard solution, cx, vx, hx are the concentration, volume and peak height of the dissolution peak, respectively; .

## 3. Results and discussion

### *3.1. Instrument condition setting*

*3.1.1. Effect of the concentration potential* It can be seen from the figure 1, CU ion peak position in the vicinity of -0.1 volts. As the potential of the enrichment changes from -0.8 to -1.3 volts, the peak current of the CU ions increases, and the front has a tendency to shift to the left, but the peak shape becomes wider and the symmetry of the peak becomes worse. It can be seen that when the concentration of the potential is set at -1.3 volts, a large current peak of Zn can be seen in the vicinity of -1.1 V in the figure.

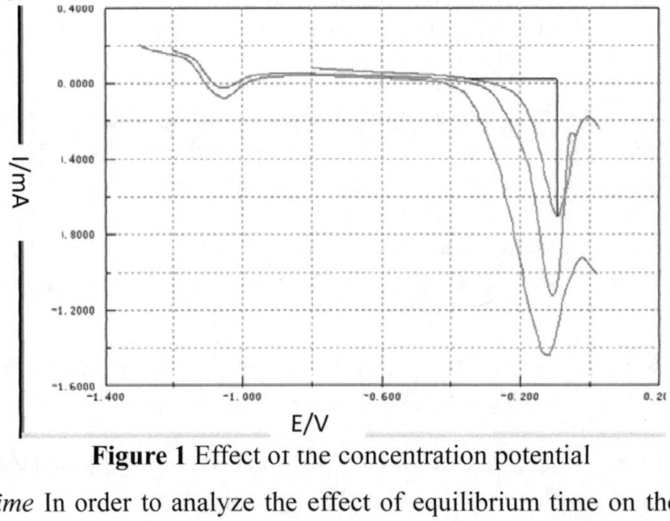


**Figure 1** Effect of the concentration potential

*3.1.2. Enrichment time* In order to analyze the effect of equilibrium time on the two metal ions, the enrichment potential of the equilibrium time experiment was also chosen to be -1.3 volts. The enrichment time was 1 minute, 1.5 minutes and 2 minutes. It was found that the peak height increased with the enrichment time, and the position of the front shifted to the right, but the symmetry of peak shape became worse in figure 2. Considered, set in 1 minute is more appropriate.

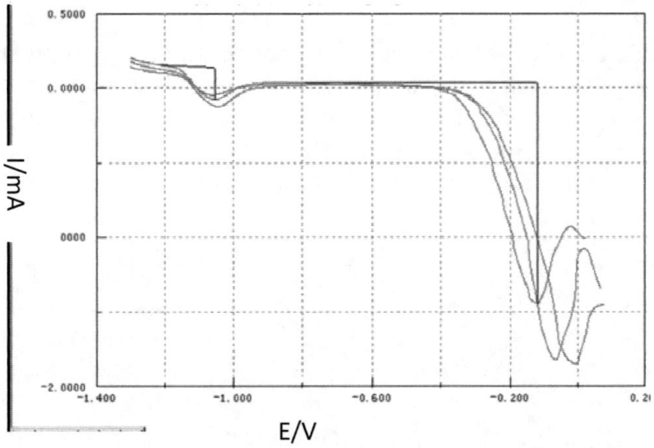


**Figure 2** Enrichment time

*3.1.3. Balance time* The enrichment time of the experiment was chosen to be -1.3 volts and the enrichment time was 1 minute. The effect of the static equilibrium time was 10S, 20S, 30S respectively. It can be seen that the longer the stationary time, deformation width, peak symmetry worse in figure 3. Hence, balance time of 10S was chosen in the following experiments.

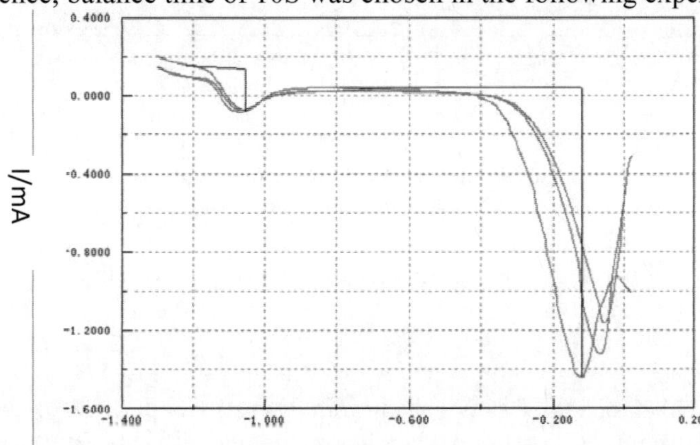


**Figure 3** Balance time

*3.1.4. Stirring speed* The results show that the stirring speed is high and the peak current is big, so it is convenient to analyze and measure. So the high speed stirring is chosen in this experiment.

*3.1.5 Scan speed* Respectively, the scanning speed of 0.06,0.08,0.1V / S three cases. It can be seen that as the scanning speed increases, the current peak position is shifted to the left and the peak current value increases, but the increase of 0.08 and 0.1V / S is not obvious, and the scanning speed is 0.08V / S More complete, therefore, this paper that the scanning speed of 0.08V / S is more appropriate.

*3.2. Determination of wastewater samples*

*3.2.1. Standard working curve and linear range* A series of copper standard solutions with different concentrations were prepared. Under the optimized experimental conditions, the linear relationship between the peak current peak current and the concentration at 0.5, 1, 1.5, 2, 4, ug / mL concentration was studied. The standard working curve is obtained: y = 0.1683X + 0.0121, R2 = 0.9974. The linear range was 1 order of magnitude and the detection limit was 0.2 μg / mL.

*3.2.2. Spike recovery* The results showed that the method has high sensitivity and good selectivity, and it can be applied to the determination of the actual samples. The method is simple, accurate and reproducible.

## 4. Conclusion

HAc-NaAc was used as the buffer solution, and trace copper in the metal processing wastewater was determined by stripping voltammetry[5]. The presence of Zn ions during the assay did not affect the assay. The instrument conditions of the experiment were optimized, and the peak current shape was obtained. The linear regression equation y = 0.1683X + 0.0121, r2 = 0.9974 in figure 4. The detection limit was 0.2 μg / mL. The recoveries of copper in water samples ranged from 96.6% to 100.5% with RSD <3.6%.

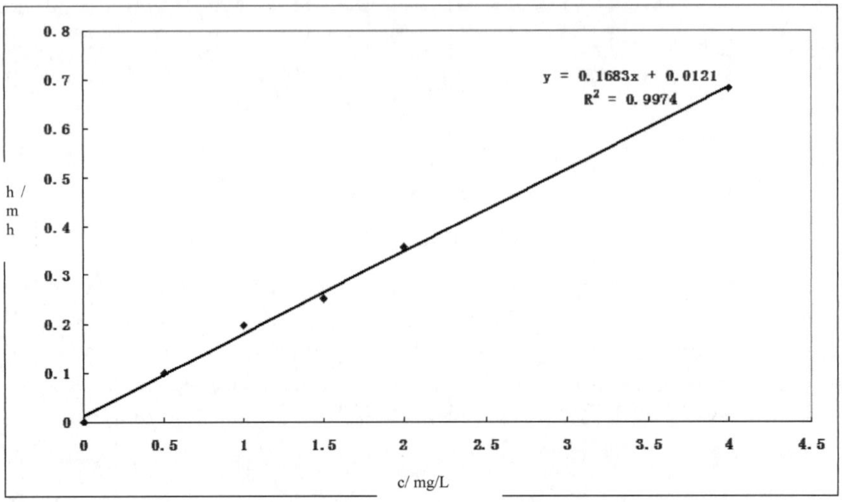


**Figure 4** The linear regression equation

**Acknowledgments**

The authors gratefully acknowledge environment engineering key subject construction of Hebei Province for their financial supports.

**References**

[1] B Fang; J Y Jin; Y Z Zhang; J L Wu; Determination of Copper, Lead, Cadmium and Zinc in Human Hair by Differential Potentiometric Stripping Analysis Laboratory,

[2] H J Zhu, D D Pan, Y Y Gu, etc. Anodic Stripping Voltammetry with Hg Mercury for Determination of Cadmium, Lead and Copper in Milk, Food Science, 2014, **35**, 121-124

[3] Y Z Zhou, Determination of Copper, Cadmium, Lead, Zinc and Environmental Engineering in Tap Water by Anodic Stripping Voltammetry, 1997, **15**, 55-58

[4] J Han, F X Wei, L L Hao, etc. Electrochemical Behavior of Hydrolyzate Catalyzed by Acetylcholinesterase, Analytical Laboratories, 2005, **24**, 62-64

[5] Y Q Hao , *Practical Operations Research*, (Beijing:) Machinery Industry Press, 2016.9

**Institute of Physics**
Dirac House, Temple Back
Bristol BS1 6BE UK

ISSN: 1757-8981
ISBN 978-1-5108-4646-3

9 781510 846463

CURRAN ASSOCIATES INC.
proceedings.com